Advances in Remote Sensing for Forest Monitoring

Advances in Remote Sensing for Forest Monitoring

Edited by Prem C. Pandey and Paul Arellano

WILEY

This edition first published 2023

Registered Office(s)
John Wiley & Sons, Inc., 111 River Street, Hoboken, NJ 07030, USA
John Wiley & Sons Ltd, The Atrium, Southern Gate, Chichester, West Sussex, PO19 8SQ, UK

Editorial Office
Boschstr. 12, 69469 Weinheim, Germany

For details of our global editorial offices, customer services, and more information about Wiley products visit us at www.wiley.com.

Wiley also publishes its books in a variety of electronic formats and by print-on-demand. Some content that appears in standard print versions of this book may not be available in other formats.

Library of Congress Cataloging-in-Publication Data

Names: Pandey, Prem C., 1985- editor. | Arellano, Paul, 1967– editor.
Title: Advances in remote sensing for forest monitoring / edited by Prem C. Pandey & Paul Arellano.
Description: Hoboken, NJ : Wiley, 2023. | Includes bibliographical references and index.
Identifiers: LCCN 2022040573 (print) | LCCN 2022040574 (ebook) | ISBN 9781119788126 (cloth) | ISBN 9781119788133 (adobe pdf) | ISBN 9781119788140 (epub)
Subjects: LCSH: Forests and forestry–Remote sensing. | Forest monitoring.
Classification: LCC SD387.R4 A36 2023 (print) | LCC SD387.R4 (ebook) | DDC 634.9–dc23/eng/20220829
LC record available at https://lccn.loc.gov/2022040573
LC ebook record available at https://lccn.loc.gov/2022040574

Cover Design: Wiley
Cover Image: © Vonkaral/Getty Images

Set in 9.5/12.5pt STIXTwoText by Straive, Pondicherry, India

Dedicated to the loving memory of my father, 1941–2019, and mother, 1946–2013.who inspired me to follow the dreams for success.

— Dr. Prem C. Pandey

This book is dedicated to my six-year-old beloved son Paul Nicolás, the sweetest little sunshine of my life.

— Dr. Paul Arellano

Contents

List of Contributors

Zulyan Afif
Regional Hazard Mitigation Agency (BPBD), Kotawaringin Timur, Sampit, Indonesia.

Paul Arellano
College of Forest Resources and Environmental Sciences, Michigan Technological University, Houghton, Michigan, USA.

Jagannath Aryal
Department of Infrastructure Engineering, The University of Melbourne; Centre for Spatial Data Infrastructures and Land Administration (CSDILA), The University of Melbourne; and Centre for Disaster Management and Public Safety (CDMPS), The University of Melbourne, Melbourne, Parkville, Victoria, Australia.

Surbhi Barnwal
Centre for Oceans, Rivers, Atmosphere, and Land Sciences (CORAL), Indian Institute of Technology Kharagpur, Kharagpur, India.

Mukunda D. Behera
Centre for Oceans, Rivers, Atmosphere, and Land Sciences (CORAL), Indian Institute of Technology Kharagpur, Kharagpur, India.

Soumit K. Behera
Plant Ecology and Climate Change Science Division, CSIR–National Botanical Research Institute, Lucknow, India.

Bimal K. Bhattyacharya
Space Applications Centre, ISRO, Ahmedabad, India.

Aashri Chauhan
Center for Environmental Sciences & Engineering, School of Natural Sciences, Shiv Nadar Institution of Eminence (Deemed to be University), Greater Noida, Uttar Pradesh, India.

Anup Kumar Das
Space Application Center, ISRO, Ahmedabad, Gujarat, India.

Tanmoy Das
Department of Geography, Faculty of Natural Sciences, Jamia Millia Islamia, New Delhi, India.

Soumesh K. Dash
Department of Civil and Environmental Engineering, Birla Institute of Technology, Mesra, Ranchi, Jharkhand, India.

Pawan Ekka
Department of Environmental Sciences, Central University of Jharkhand, Brambe, Ranchi, Jharkhand, India.

Sophie Fabre
Office National d'Études et de Recherches Aérospatiales (ONERA), Toulouse, France.

Lachezar Filchev
Space Research and Technology Institute, Bulgarian Academy of Sciences, Bulgaria.

Ayushi Gupta
Institute of Environment and Sustainable Development, Banaras Hindu University, Varanasi, Uttar Pradesh, India.

Yephi Haptadi
Regional Hazard Mitigation Agency (BPBD), Kotawaringin Timur, Sampit, Indonesia.

Bodi S.P.C. Kishore
Department of Geoinformatics, Central University of Jharkhand, Brambe, Ranchi, Jharkhand, India.

Amit Kumar
Department of Geoinformatics, Central University of Jharkhand, Brambe, Ranchi, Jharkhand, India.

Gajendra Kumar
Department of Geoinformatics, Central University of Jharkhand, Brambe, Ranchi, Jharkhand, India.

Rahul Kumar
ICFRE-Institute of Forest Productivity, Lalgutwa, Ranchi, Jharkhand, India.

Neeta Kumari
Department of Civil and Environmental Engineering, Birla Institute of Technology, Mesra, Ranchi, Jharkhand, India.

Rina Kumari
School of Environment Science and Sustainable Development, Central University of Gujarat, Gujarat, India.

Guillaume Lassalle
Office National d'Études et de Recherches Aérospatiales (ONERA), Toulouse, France.

Salim Lamine
Faculty of Biological Sciences, University of Sciences and Technology Houari Boumediene (USTHB), Algiers, Algeria; and Department of Geography and Earth Sciences, University of Aberystwyth, Ceredigion, Wales, UK.

Nikhil Lele
Space Applications Centre, ISRO, Ahmedabad Gujarat, India.

Neeraj K. Maurya
Center for Environmental Sciences & Engineering, School of Natural Sciences, Shiv Nadar Institution of Eminence (Deemed to be University), Greater Noida, Uttar Pradesh, India.

Parul Maurya
School of Environment Science and Sustainable Development, Central University of Gujarat, Gujarat, India.

Mthembeni Mngadi
School of Agricultural, Earth, and Environmental Sciences, Discipline of Geography, University of KwaZulu-Natal, Pietermaritzburg, South Africa.

Pegah Mohammadpour
University of Coimbra, ADAI, Department of Mechanical Engineering, Pólo II, Coimbra, Portugal; and Universidad de Alcalá, Department of Geology, Geography and Environment Science, Alcala de Henares, Spain.

Onisimo Mutanga
School of Agricultural, Earth, and Environmental Sciences, Discipline of Geography, University of KwaZulu-Natal, Pietermaritzburg, South Africa.

Mohd Waseem Naikoo
Department of Geography, Faculty of Natural Sciences, Jamia Millia Islamia, New Delhi, India.

Shah Al Nawajish
Department of Civil and Environmental Engineering, Birla Institute of Technology, Mesra, Ranchi, Jharkhand, India.

John Odindi
School of Agricultural, Earth, and Environmental Sciences, Discipline of Geography, University of KwaZulu-Natal, Pietermaritzburg, South Africa.

Prem C. Pandey
Center for Environmental Sciences & Engineering, School of Natural Sciences, Shiv Nadar Institution of Eminence (Deemed to be University), Greater Noida, Uttar Pradesh, India.

Soumya Pandey
Department of Civil and Environmental Engineering, Birla Institute of Technology, Mesra, Ranchi, Jharkhand, India.

Dyah R. Panuju
Geospatial Information and Technologies for the Integrative and Intelligent Agriculture (GITIIA), Center for Regional Systems Analysis, Planning and Development (CrestPent), Bogor Agricultural University, Bogor, Indonesia; and Department of Soil Science and Land Resources, Bogor Agricultural University, Dramaga, Bogor, Indonesia.

Somnath Paramanik
Centre for Ocean, River, Atmosphere and Land Sciences (CORAL), Indian Institute of Technology Kharagpur, Kharagpur, India.

Subhashree Patra
Department of Environmental Sciences, Central University of Jharkhand, Brambe, Ranchi, Jharkhand, India.

Atiqur Rahman
Department of Geography, Faculty of Natural Sciences, Jamia Millia Islamia, New Delhi, India.

Abbas Rajabifard
Department of Infrastructure Engineering, The University of Melbourne; Centre for Spatial Data Infrastructures and Land Administration (CSDILA), The University of Melbourne; and Centre for Disaster Management and Public Safety (CDMPS), The University of Melbourne, Melbourne, Parkville, Victoria, Australia.

Mohd Rihan
Department of Geography, Faculty of Natural Sciences, Jamia Millia Islamia, New Delhi, India.

Purabi Saikia
Department of Environmental Sciences, Central University of Jharkhand, Brambe, Ranchi, Jharkhand, India.

K.V. Satish
Institute of Environment and Sustainable Development, Banaras Hindu University, Varanasi, Uttar Pradesh, India.

Abdul R.M. Shariff
Faculty of Engineering, University Putra Malaysia, UPM, Serdang, Selangor Darul Ehsan, Malaysia; Smart Farming Technology Research Centre, Faculty of Engineering, Universiti Putra Malaysia, Serdang, Malaysia; and Institute of Plantation Studies (IKP), Universiti Putra Malaysia, Serdang, Malaysia.

Jigarkumar B. Solanki
School of Environment Science and Sustainable Development, Central University of Gujarat, Gujarat, India.

Prashant K. Srivastava
Institute of Environment and Sustainable Development, Banaras Hindu University, Varanasi, Uttar Pradesh, India.

Shahfahad
Department of Geography, Faculty of Natural Sciences, Jamia Millia Islamia, New Delhi, India.

Saroj K. Sharma
Department of Infrastructure Engineering, The University of Melbourne; Centre for Spatial Data Infrastructures and Land Administration (CSDILA), The University of Melbourne; and Centre for Disaster Management and Public Safety (CDMPS), The University of Melbourne, Melbourne, Parkville, Victoria, Australia.

Swapan Talukdar
Department of Geography, Faculty of Natural Sciences, Jamia Millia Islamia, New Delhi, India.

Amit K. Tripathi
Center for Environmental Sciences & Engineering, School of Natural Sciences, Shiv Nadar Institution of Eminence (Deemed to be University), Greater Noida, Uttar Pradesh, India.

Bambang H. Trisasongko
Geospatial Information and Technologies for the Integrative and Intelligent Agriculture (GITIIA), Center for Regional Systems Analysis, Planning and Development (CrestPent), Bogor Agricultural University, Jalan Pajajaran, Bogor, Indonesia; and Department of Soil Science and Land Resources, Bogor Agricultural University, Dramaga, Bogor, Indonesia.

Roma Varghese
Centre for Oceans, Rivers, Atmosphere, and Land Sciences (CORAL), Indian Institute of Technology Kharagpur, Kharagpur, India.

Carlos Viegas
University of Coimbra, ADAI, Department of Mechanical Engineering, Pólo II, Coimbra, Portugal.

Foreword

Forest ecosystems are a vital part of our biosphere, and forest ecosystem services and resources are the foundations of our civilized societies. The need to monitor and manage global forest resources is becoming increasingly important and urgent as the deleterious impacts of human activities on our planet's ecosystems are becoming widespread at an alarming rate.

Traditionally, forest resource monitoring and management are accomplished using conventional methods; however, advanced remote sensing technologies have emerged as indispensable tools for forest resource monitoring and management over the past several decades. More recently, the advancement and availability of remote sensing data obtained from various sensors including air-borne, space-borne, and terrestrial-handheld instruments, along with a wide range and improved spectral, spatial, and temporal resolutions have significantly added to our capabilities for monitoring and managing forest resources.

The editors and authors of this book have strived to provide an extensive discussion of the use of various advanced techniques in remote sensing that are relevant for forest resource management. In addition they have expertly synthesized various case studies to provide insights into biophysical and biochemical characteristics of forest ecosystems and their utilization in a sustainable manner.

I commend the editors Dr. Prem C. Pandey from the Center for Environmental Sciences & Engineering, Shiv Nadar Institution of Eminence (Deemed to be University), India, and Dr. Paul Arellano at Michigan Technological University, in the US for this outstanding publication. The book contains 14 thoughtfully organized chapters that cover various aspects of forest monitoring and their sustainable management utilizing remote sensing data and techniques to illustrate the relevant advancements made in recent years.

The editors are also to be commended for including an impressive international list of contributors from several countries: Algeria, Australia, Bulgaria, Ecuador, France, Greece, India, Indonesia, Malaysia, Portugal, Thailand, South Africa, and the United Kingdom.

I consider this comprehensive publication on remote sensing technology for forest ecosystems a very valuable contribution to both the remote sensing and forest management literature.

I am confident that this book will be an excellent resource for the students and researchers alike.

I congratulate Dr. Pandey and Dr. Arellano for undertaking an important task of editing this book, which will serve a wide range of research and professional communities.

Kamlesh Lulla, Ph.D.
NASA Medalist
Dr. Kamlesh Lulla serves as Director, Research Collaboration, and Partnership at NASA Johnson Space Centre, Houston Texas. Dr. Lulla served as Chief Scientist for Earth Observation at NASA Johnson for over 25 years.
March 2022

Preface

After the success of the first edited book on "Advances in Remote Sensing for Natural Resource Monitoring" in 2021, we were encouraged to continue the Series for other individual topics. In order to highlight the importance of remote sensing and its advancement in the different research themes, we have selected individual topics in upcoming editions. This is the second volume dealing with the advancement in remote sensing for forest monitoring. A careful attempt is taken in this volume to ensure the coverage of all topics related to forest monitoring. To achieve sustainable forest management across the globe, information of all parameters to be observed from remotely sensed images, and an urgent review of the present and future scenario is required.

Recent developments in Sustainable Development Goals (SDGs) have gathered people collectively to manage forest resources, by implementing long-term plans so that forests may fulfill their expected functions through forest management based on the inherent resilience of a healthy forest ecosystem. The forests function in several ways, a few major ways which are directly linked to goals of SDGs are water resource quantification, climate actions, forest carbon stock assessment, biodiversity conservation, and forest management for timber production at a large scale. SDGs No. 6 (Clean Water and Sanitation), 12 (Responsible Consumption and Production), 13 (Climate Actions), and 14 (Life on Land) will be directly and indirectly linked to the forest functions, and thus monitoring will help to achieve these goals through forest monitoring through Earth Observation (EO) datasets. Thus, the curiosity and contribution toward the environment may be achieved by anyone, and will help in achieving SDGs for future generations. This is supported by the Earth observation technology at local, regional, and global scales with high spatial/spectral and temporal datasets.

Therefore, an attempt has been made to deliver both the basic and advanced methods that are very much awaited from end-users to understand technology supporting SDGs. Moreover, how the EO technologies are employed in forest resource monitoring and management, and how they play an effective role in policy implementation at different scales; root level to regional to global scale. EO provides a clear concept in different fields, such as vegetation, water, soil, and disasters related to them, with advanced techniques implemented and their outcome will help to understand the feasibility of remote sensing in the future in terms of reliability, accuracy, and cost-effective applications. The Editors believe that this effort will help readers in understanding the advancement of remote sensing and offer practical guidance toward their research. The Editors hope that the present book will be a valuable asset for researchers working toward the sustainable and judicious use of forest resources using RS technology. Moreover, taking advantage of the methods and technologies, researchers will be able to scientifically address the issues involved with forest research.

The updated knowledge from countries including Algeria, Australia, Bulgaria, Ecuador, France, Greece, India, Indonesia, Malaysia, Portugal, The Netherlands, Thailand, South Africa, and The United Kingdom is demonstrated in this edited book through research and selected case studies for monitoring forests, to elaborating the scientific understanding of advancement in remote sensing and forest parameters assessment. This book is primarily focused on the *Advances in Remote Sensing for Forest Monitoring*; it provides a detailed overview of the potential applications of advanced satellite data including spaceborne, airborne, and handheld instruments employed in the assessment of forest parameters for their monitoring. Further, this book determines how environmental–ecological knowledge and satellite-based–drone-based information could be effectively combined to address a wide array of current management needs and requirements. Each chapter covers different aspects of remote sensing approaches to monitor the forest parameters effectively, to provide a platform for conservation and planning.

This book is structured as a set of 14 contributed chapters addressing the advancement of remote sensing to monitor and manage forest related research themes. This book is divided into five major sections; each section has specific chapters to present the knowledge and concepts of forest monitoring, and delivers the findings in a vivid way that is useful for a wide range of readers including undergraduate, graduate, and researchers. While a wide range of applications and the latest innovative technologies are constantly emerging, this book provides a crisp summary of key case studies to provide the most recent and widely used methods and information.

Section I, *General-Editorial Message – Introduction to Forest Monitoring*, includes an Editorial message which provides an insight to forest monitoring; importance, needs, and requirements, along with the use of advanced remote sensing technology for assessment of forest parameters discussed in chapters contributed by several authors and scientists. This Section also discusses sustainable forest management to enhance Ecosystem Services and livelihood security. There is a brief discussion about the ecological disturbances of forests, Forest ecosystem services (FESs), as well as detailed information about the Landscape-based Approach (LbA) and Ecosystem-based Approach (EbA) of Sustainable Forests Management (SFM). This section provides traditional knowledge in forest management, along with how forest resources help in the livelihood of the people.

Section II: Forest Parameters – Biochemical and Biophysical Parameters, contains three chapters that acclaim the virtue of Machine Learning (ML) methods and algorithms to assess biochemical and biophysical parameters of forests. This section employed different datasets; Multispectral, Synthetic Aperture Radar, MODIS, Global orbiting carbon observatory-2 SIF (GOSIF), and handheld instruments such as spectroradiometers, and SPAD for retrieval of parameters and their assessment. Section II also presents Mapping Oil Palm Plantations in naturally protected areas, to illustrate the importance of monitoring vegetation even in the protected areas. This section presents several machine learning methods to distinguish carotene and xanthophyll contents in the leaves of riparian forest, review on modeling of abiotic stress of conifers, Retrieval of Mangrove Forest Properties, and photosynthetic variables estimation in a mangrove forest using advanced remote sensing datasets. This section also discusses strategies for evaluating and quantifying carbon stock allocation and variability across different species, structural components, and age groups of reforested trees.

Section III: Remote Sensing Technology for Forest Fire Monitoring, illustrates forest fire susceptibility mapping and Landscape-Level Bushfire Severity assessment using remote sensing data. ML methods such as Random Forest (RF) and Support Vector Machines (SVMs) were incorporated for fire susceptibility mapping and validation. This section presents cloud-based computing of remotely sensed datasets for Landscape-Level Bushfire Severity assessment. It also incorporates the outcomes on temporal assessment of vegetation recovery in Australian bushfires. This section demonstrates the response of multi-sensors data and effectiveness in estimating the forest fires and bushfire recovery/post fire damage assessment.

Section IV: Advancement in RS – Drones and Multi-Sensors–Multi-Source for Forest Monitoring exemplifies recent advances in remote sensing technology for forest parameters estimation. This section incorporates the widely used drones/UAVs for this purpose. There are opportunities in drone-based studies, which can be mounted with RGB camera, multispectral sensors (MICA sense), thermal sensors, and LiDAR sensors for easy access to data and temporal studies as compared to spaceborne datasets, in term of spatial resolution up to 10 cm, and temporal resolution as and when required. This section exemplifies recent advances in drones and data fusion approaches for forest monitoring utilizing earth observation datasets.

Section V: Opportunities, Challenges and Future Aspects in Forest Monitoring provides a discussion on the current issues and reviews challenges in sustainability of forest and vegetation management due to natural and anthropogenic activities. This section deals with the forest components in brief, and presents literature on the challenges faced during ground/field surveys, and satellite use for forest monitoring. The main components of forests, which are experiencing challenges with monitoring are species distribution modeling and mapping, Land Use and Land Cover (LULC) and changes, climate monitoring, wildlife monitoring, biodiversity, livelihood forest production monitoring and others. This section provides a discussion on the current status, future trends, and prospects of remote sensing methods in forest monitoring, and underlines the scientific challenges that need to be addressed. It also discusses the future possibilities in forest monitoring with high-resolution satellites and cameras, drones and aircraft, LiDAR, and carbon credits while attracting funds for nations. There are scope of development and additions of new technologies, algorithms, and methods for future advancement of remote sensing for forest monitoring. Finally, it discusses the importance of RS technology, how it has evolved with time and spread its wings in the research domains, and is still evolving and emerging to its global height.

The book attempts to match user needs with the level of technology required for forest monitoring, management, and planning. We believe that this endeavor shall provide a valuable scientific basis to students and researchers to address future challenges in forestry research. We further hope this book will be a valuable reference and provide practical guidance for all who work toward the themes mentioned.

We thank all the authors for their enthusiastic efforts in completing the book with the quality of their chapters. We are deeply indebted and thankful to the reviewers who took pains to review the chapter manuscripts, and for their voluntary support. On behalf of the team of authors, we express our gratitude to the entire crew of Wiley (Andrew, Frank, Merryl, and Athira) for all sorts of assistance to make this a successful endeavor. We are thankful to Shiv Nadar Institution of Eminence (Deemed to be University), and our colleagues for their support and help throughout the progress of the work. Last, but not the least, the editors are heartily thankful to our publisher, Wiley, for providing an opportunity to gather the thoughts of several contributors into a book. Dr. Pandey is grateful to his beloved wife and daughter Adele for all their support and emotions during the development of this book at the final stage. We are grateful to all persons and individuals who overcame the great challenges faced during the world wide pandemic of COVID-19 for more than two years, among those are a few authors who lost their lives, and could not contribute in this volume/series.

We hope this Preface has successfully provided some insight into the breadth of the advancement of remote sensing applications and related topics covered in this book. Users of this book are encouraged to adapt to it and use it in the way it best fits their own needs to help them in understanding the capabilities and potentials of natural resources monitoring and its applications, of which this book is concerned. Users of this book can inform the editor of any errors, suggestions, or comments at prem26bit@gmail.com or prem.pandey@snu.edu.in and paularellano@hotmail.com.

Editors
Greater Noida, India
Quito, Ecuador
March 2022

Prem C. Pandey
Paul Arellano

List of Abbreviations

Acronym	Description
2D	Two-Dimensional
3D	Three-Dimensional
AATSR	Advanced Along-Track Scanning Radiometer
ADCM	Ancillary Data Classification Model
AE	Assimilation Efficiency
AECL	Atomic Energy Canada Limited
AF	Absorbed Part of Internal Irradiation
AGB	Above Ground Biomass
AI	Artificial Intelligence
AIRSAR	Airborne synthetic aperture radar
AISA	Airborne Imaging Spectrometer for Applications
ALOS-1/2	Advanced Land Observing Satellite-1/2
ALS	Airborne Laser Scanning
ANN	Artificial Neural Network
ANOVA	Statistical Analysis of Variance
AO	*Avicennia officinalis*
APAR	Absorbed Photosynthetically Active Radiation
AR–CDM	Afforestation–Reforestation Clean Development Mechanism
Arc-GIS	Aeronautical Reconnaissance Coverage Geographic Information System
ARI2	Anthocyanin Reflectance Index2
ARIES	Australian Resource Information and Environment Satellite
ARVI	Atmospherically Resistant Vegetation Index
ASAR	Advanced Synthetic Aperture Radar
ASI	Italian Space Agency
ASM	Angular Second Moment
ASTER	Advanced Spaceborne Thermal Emission and Reflection Radiometer
ATSR	Along Track Scanning Radiometer
AUC	Area Under Curve

AVHRR	Advanced Very High-Resolution Radiometer
AVIRIS	Airborne Visible/Infrared Imaging Spectrometer
AWiFS	Advanced Wide Field Sensor
BCC	Biochemical Content Classifier
BWG	Biota Working Group
BWS	Bhitarkanika Wildlife Sanctuary
C	Carbon
C&Is	Criteria and Indicators
CASI	Canadian Aeronautics and Space Institute (https://casi.ca/)
CASI	The Compact Airborne Spectrographic Imager
CBI	Composite Burn Index
C-cycle	Carbon Cycle
CCC	Canopy Chlorophyll Content
CCM	Chlorophyll Content Meter
CDM	Clean Development Mechanism
CER's	Certified Emission Reductions
CFCs	Chlorofluorocarbons
Chl-a	Chlorophyll a
Chl-b	Chlorophyll b
CHM	Canopy Height Model
CHRIS	Compact High Resolution Imaging Spectrometer
CNES	Centre National d'études Spatiales - French Space Agency
CNN	Convolutional Neural Network
CO_2	Carbon dioxide
COIS	Coastal Ocean Imaging Spectroradiometer
CR	Continuum Removal
CRI2	Carotenoid Reflectance Index 2
CRR	Continuum-Removed Reflectance
CWC	Canopy Water Content
DART	Discrete anisotropic radiative transfer
dB	Decibel
DBH	Diameter at Breast Height
DELWP	Department of Environment, Land, Water, and Planning
DEM	Digital Elevation Model
DHP	Digital Hemispherical Photography
dNBR	Difference Normalized Burn Ratio
dNDVI	Differenced Normalized Difference Vegetation Index
DNN	Deep Neural Networks
DOS	Dark Object Subtraction
DRAP	Durban Research Action Partnership
DSM	Digital Surface Model
DT	Decision Tree
DTM	Digital Terrain Model
EA	*Excoecaria agallocha*
EbA	Ecosystem-based Approach

EBV	Essential Biodiversity Variable
EC	European Commission
ECVs	Essential Climate Variables
EIA	Environmental Impact Assessment
ELVIS	Elevation and Depth – Foundation Spatial Data
EMR	ElectroMagnetic Radiation
EMRAS	Environmental Modeling for Radiation Safety
ENET	Elastic net
EO	Earth Observation
EOS	End of the Season
ERICA	Environmental Risk from Ionising Contaminants – Assessment and Management
ERTS	Earth Resources Technology Satellite
ESA	European Space Agency
ESD	Ecosystem Syndrome Distress
ESRI	Environmental Systems Research Institute
ESs	Ecosystem Services
ESU	Elementary Sampling Unit
ET	Evapotranspiration
ETM	Enchanced Thematic Mapper
ETM+	Enhanced Thematic Mapper Plus
EUFORGEN	European Forest Genetic Resource Program
EVI	Enhanced Vegetation Index
EWT	Equivalent Water Thickness
FAO	Food and Agriculture Organization
FAPAR	Fraction of Photosynthetically Active Radiation
*f*APAR	Fraction of Absorbed Photosynthetically Active Radiation
FAR	False Alarm Rate
FBD	Fine beam dual polarization
FBP	Canadian Fire Behavior Prediction
FBS	Fine beam single polarization
*f*COVER	Fraction of Vegetation Cover
FESs	Forest Ecosystem Services
FnF	Forest/non-Forest
FORTRAN	FORmula TRANslation
*f*PAR	Fraction of Absorbed Photosynthetically Active Radiation
FSDAF	Flexible Spatiotemporal Data-Fusion
FSI	Forest Survey of India
FTHSI	Fourier Transform Hyperspectral Imager
GBH	Girth at Breast Height
GCOM	Global Change Observation Mission
GCOS	Global Climate Observing System
GCP	Ground Control Points
GDP	Gross Domestic Product
GEE	Google Earth Engine

GEOV1	Geoland2/Bio Par version 1
GHG/s	Greenhouse Gas/es
GLCF	Global Land Cover Facility
GLCM	Grey Level Co-occurrence Matrix
GOSAT	Greenhouse gases Observing SATellite
GOSIF	Global Orbiting Carbon Observatory-2 SIF
GPP	Gross Primary Productivity
GPM (10)	Global Precipitation Measurement
GPS	Global Positioning Systems
GRD	Ground Range Detected
GSD	Ground Sampling Distance
HERO	Hyperspectral Environment and Resource Observer
HF	*Heritiera fomes*
HH	Horizontal transmit, Horizontal receive
HIRIS	High Resolution Imaging Spectrometer
HKH	Hindu Kush Himalaya
HPLC	High-Pressure Liquid Chromatography
HRG	High Resolution Geometric
HRV	High Resolution Visible
HSI	Hyperspectral Imager
HSI transformation	Hue–Saturation–Intensity transform
HSR	High Spatial Resolution
HV	Horizontal transmit, Vertical receive
IAEA	International Atomic Energy Agency
ICP "Forest"	International Co-operative Programme on Assessment and Monitoring of Air Pollution Effects on Forests
IHR	Indian Himalayan Region
IMD	India Meteorological Department
INS	Inertial Navigation System
InSAR	Interferometric Synthetic Aperture Radar
IPCC	Intergovernmental Panel on Climate Change
IPS	Invasive Plant Species
IR	Infra-Red
IRS	Indian Remote Sensing
IRS-1B	Indian Remote Sensing satellite-1B
IRS-1C	Indian Remote Sensing satellite-1C
IRS-1D	Indian Remote Sensing Satellite-1D
IRS-P6	Indian Remote-Sensing Satellite-P6
ISFR	India State of Forest Report
ISRIC	International Soil Reference and Information Centre
IUCN	The International Union for Conservation of Nature
IW	Interferometric wide swath
JAXA	Japan Aerospace Exploration Agency
JERS-1	Japanese earth resources satellite-1
JF	January and February

JFM	Joint Forest Management (India)
K	Kappa Coefficient
L1TP	Level-1 Precision Terrain
L8SR	Landsat 8 Surface Reflectance
LAI	Leaf Area Index
Landsat	Land Remote-Sensing Satellite
Landsat TM	Land remote-sensing satellite Thematic mapper
Landsat ETM+	Landsat 7 Enhanced Thematic Mapper-Plus
LaSRC	Landsat Surface Reflectance Code
LbA	Landscape based Approach
LCC	Leaf Chlorophyll Content
LiDAR	Light Detection and Ranging data
LISS	Linear Imaging Self-Scanning Sensor
LISS III/IV	Linear Imaging Self-Scanning Sensor-III
LN	Linear Kernel
LPM	Laser Penetration Matrices
LR	Logistic regression
LSD	Landlord Simile Dynamic link
LST	Land Surface Temperature
LSU	Linear Spectral Unmixing
LUE	Light Use Efficiency
LULC	Land Use and Land Cover
LVIS	Land, Vegetation, and Ice Sensor
MAM	March, April, and May
MARS	Multivariate Adaptive Regression Splines
MCARI	Modified Chlorophyll Absorption Ratio Index
MCC	Means Clustering Classifier
MERIS	Medium Resolution Imaging Spectrometer
MFPs	Minor Forest Products
Mha	Million hectares
MISR	Multi-angle Imaging SpectroRadiometer
ML	Machine Learning
MLA	Machine learning algorithm
MLC	Maximum Likelihood Classifier
MLP	Multi-Layer Perceptron
MNDWI	Modified Normalized Difference Water Index
MODFLOW	MODular finite-difference FLOW model
MODIS	Moderate Resolution Imaging Spectroradiometer
MOS	Modular Optoelectronic Scanner
MSAVI	Modified Soil Adjusted Vegetation Index
MSI	Multispectral Image
MSS	Multispectral Scanner
MSS	Multispectral Scanner System
mtry	number of explanatory variables
MTVI	Modified Triangular Vegetation Index

MVC	Mapping Vegetation Community
NASA	National Aeronautics and Space Administration
NBR	Normalized Burn Ratio
NDCI	Normalized Difference Chlorophyll Index
NDREI	Normalized Difference Red-Edge Index
NDVI	Normalized Difference Vegetation Index
NDWI	Normalized Difference Water Index
NEMO	Naval Earth Map Observer
NetCDF	Network Common Data Form
NFFL	Northern Forest Fire Laboratory
NFI	National Forest Inventory
NIR	Near-Infra Red
NISAR	NASA-ISRO Synthetic aperture radar
NMP	New Millennium Program
NN	Neural Network
NOAA	National Oceanic and Atmospheric Administration
NPP	Nuclear Power Plant
NPV	Non-Photosynthetic Vegetation
NRMSE	Normalized Root Mean Square Error
NTFPs	Non-Timber Forest Products
ntree	number of trees
OA	Overall Accuracy
OCO-2	Orbiting Carbon Observatory-2
OLI	Operational Land Imager
OND	October, November, and December
OSAVI	Optimized-Soil-Adjusted Vegetation Index
PA	Producer Accuracy
PAI	Plant Area Index
PALSAR	Phased Arraytype L-band Synthetic Aperture Radar
PAN	peroxyacetyl nitrate
PAR	Photosynthetically Active Radiation
PB	propyl-peroxybutylenetrate
PCA	Principal Component Analysis
PEM	Production Efficiency Models
PL	Polynomial kernel
PLSR	Partial Least Square Regression
PPP	peroxypropionyltrate
PRI	Photochemical Reflectance Index
PROBA	Project for On-Board Autonomy
PROSAIL	PROSPECT and SAIL radiative transfer models
PRVI	Polarimetric radar vegetation index
PSRI	Plant Sense Reflectance Index
PVI	Perpendicular Vegetation Index
QC	Quality Control
QGIS	Quantum Geographic Information System

R	Coefficient of correlation
R^2	Coefficient of Determination
RADAR	Radio Detection and Ranging
RBF	Radial Basis Function
RdNBR	Relative Difference Normalized Burn Ratio
RDVI	Renormalized Difference Vegetation Index
REDD	Reducing Emissions from Deforestation and forest Degradation
REDD+	Reduce Emissions from Deforestation and Forest Degradation +
RF	Random Forest
RFR	Random Forest Regression
RGB	Red-Green-Blue
RMSE	Root Mean Square Error
ROC	Receiver operating characteristic
RPD	Residual Predictive Deviation
RS	Remote Sensing
RRF	Regularized Random Forest
RSADU	Remote Sensing Applications Development Unit
RSDP	Remote Sensing Data Policy
RST	Remote Sensing Technology
RT	Regression Tree
RTMs	Radiative Transfer Models
RUSLE	Revised Universal Soil Loss Equation
RVI	Ratio Vegetation Index
SAC	Spectral Angle Classifiers
SAIL	Scattering by Arbitrary Inclined Leaves
SAM	Spectral Angle Mapper
SAR	Synthetic Aperture Radar
SAVI	Soil Adjusted Vegetation Index
ScanSAR	Scanning synthetic aperture radar
SCOPE	Soil Canopy Observation, Photochemistry and Energy fluxes
SDGs	Sustainable Development Goals
SDM	Species Distribution Model
SDS	Scientific Data Sets
SENTINEL	Copernicus Programme satellite constellation conducted by the European Space Agency
SFM	Sustainable Forest Management
S*f*M	Structure from Motion
SGLI	Second Generation Global Imager
SI	Separability Index
SIF	Solar-Induced (chlorophyll) Fluorescence
SIG	Sigmoid kernel
SIP	Subpixel Inundation Percentage
SIPI	Structure Insensitive Pigment Index

SIR-B/C	Shuttle imaging radar-B/C
SLAR	Side Looking Airborne Radar
SLC	Scan Line Corrector
SLC	Single Look Complex
SLS	Spaceborne Laser Scanning
SNAP	Sentinel application platform
SNC	Sebangau National Park
SOS	Start of the Season
SPAD	Soil Plant Analysis Development
SPOT	Satellite for observation of Earth
SRTM	Shuttle radar topography mission
SVM	Support Vector Machines
SVR	Support Vector Regression
SWIR	Short Wave InfraRed
SWR	Stepwise Regression
TCI	Temperature Condition Index
TIS	Thermal Infrared Sensor
TLS	Terrestrial laser scanning
TM	Thematic Mapper
UA	User Accuracy
UAS	Unmanned Aircraft Systems
UAV	Unmanned Aerial vehicles
UAV/UAS	Unmanned Aerial Vehicle/System
UHF	Ultra-High Frequency
UK	United Kingdom
UML	Unified Modeling Language
UN	The United Nations
UNEP	United Nations Environment Programme
UNEP-WCMC	Environment Programme World Conservation Monitoring Centre
UNFCCC	United Nations Framework Convention on Climate Change
UNSDGs	United Nations Sustainable Development Group
USA	United States of America
USDOE	U.S. Department of Energy
USGS	United States Geological Survey
USSR	Union of Soviet Socialist Republics
VH	Vertical transmit, Horizontal receive
VH	Vertical transmit, Horizontal receive (Cross-polarization-Horizontal)
VI/s	Vegetation Index/Indices
VIF	Variance Inflation Factor
VIP	Variable Importance Projection
VIS	Visible
VNIR	Visible and Near-Infrared
VNIS	Visible and Near-infrared Imaging Spectrometer

VPD	Vapor Pressure Deficit
VTOL	Vertical Take-Off and Landing
VV	Vertical transmit, Vertical receive
VV	Vertical transmit, Vertical receive (Co-polarization-Vertical)
WAVI	Water-Adjusted Vegetation Index
WBI	Water Band Index
WRS-2	Wordwide Reference System-2
WUI	Wildland–Urban Interfaces
WV	WorldView
XGB	Extreme Gradient Boosting

Editors

Dr. Prem C. Pandey received his PhD from CLCR, University of Leicester, UK, and completed his Postdoctoral from Tel Aviv University, Israel. Currently he is working as Assistant Professor at the Center for Environmental Sciences & Engineering, School of Natural Sciences, Shiv Nadar Institution of Eminence (Deemed to be University), formerly known as Shiv Nadar University, Greater Noida, India. Previously, he had been associated with IESD, Banaras Hindu University India. He received his M.Sc in Environmental Sciences and M.Tech in Remote Sensing. He worked on remote sensing applications as Professional Research funded by NRSC Government of India. He has been a recipient of several awards including The Commonwealth award UK, INSPIRE fellowship, MHRD UGC fellowships, and SERB-NPDF from GoI India. He has published more than 45 peer reviewed journal papers, has edited five books, a few book chapters, and presented his work in several conferences. Additionally, he is a life member of the Indian Society of Geomatics, and the Indian Society of Remote Sensing, as well as a IUCN CEM Member (2017–2021; 2022–2025), and the Society of Wetland Scientists (2021–2022). He is serving as an associate editor for the *Journal-Geocarto International,* Taylor & Francis, and associated with *Remote Sensing MDPI* as guest Editor. Dr. Pandey is working in three projects related to Monitoring of wetlands/chilika lake mainly focusing on Ramsar sites along with other natural resources based research work funded by NGP and SERB Government of India. Dr. Pandey is also working with science-collaborators for real time Disaster monitoring at Himalayan regions. He has expertise in remote sensing of environment; his research interests include natural resources monitoring such as forestry, agriculture, urban studies, and atmospheric pollutant monitoring, mapping and modeling.

Dr. Paul Arellano received his Ph.D. from the Department of Geography, Geology, and Environment of the University of Leicester, UK. Currently, he is a Post-doctoral Researcher at Michigan Technological University in the US, where he is focused on developing new remote sensing algorithms to detect, map, and monitor peatlands and greenhouse gases (GHG) in the Andean and Amazon regions of South America. He was the Dean of the School of the Earth Sciences, Energy, and Environment from 2015 until 2019, and a member of the Board of Trustees of Yachay Tech University in Ecuador. He was an Honorary Research Visitor from the University of Leicester-UK and an Adjunct Professor at Arizona State University-US. He has been a recipient of several scientific awards including the NUFFIC Scholarship from the Netherlands Government, the National Recognition for a scientific contribution in Ecuador, and best papers awards in several scientific conferences. Dr. Arellano focuses his research on remote sensing for vegetation monitoring, natural disasters, climate change, greenhouse gases monitoring, pollution detection, and fire monitoring.

Section I

General Introduction to Forest Monitoring

1

Introduction to Forest Monitoring Using Advanced Remote Sensing Technology – An Editorial Message

Prem C. Pandey,[1] and Paul Arellano*[2]

[1] *Center for Environmental Sciences & Engineering, School of Natural Sciences, Shiv Nadar Institution of Eminence (Deemed to be University), Greater Noida, Uttar Pradesh, 201314, India*
[2] *College of Forest Resources and Environmental Sciences, Michigan Technological University, Houghton, Michigan, USA*

1.1 Introduction

Remote sensing has a wide range of applications in forest monitoring at local, regional as well as global scale in all ecosystems on Earth. Remotely sensed images acquired through optical (multispectral–hyperspectral), radar, thermal, microwave, and LiDAR (Light Detection And Ranging) sensors are being employed to study and assess natural resources for their monitoring and management across the globe (Pandey and Sharma 2021a, Pandey et al. 2020a). Pandey and Sharma (2021b) have already extolled remote sensing technology and its advancement in natural resource monitoring. In this edited volume, editors have focused on forest monitoring using remote sensing technology as well as traditional methods. This volume discusses several machine learning (ML) methods, spectral indices, and spaceborne–airborne–(unmanned aerial vehicle) UAV data that support research, algorithm development, monitoring, and management of forestry resources. Of interest here is the fact that several remotely sensed data sets acquired from different platforms for forest regions provide sufficient and reasonable cues in the form of representative databases. This will allow the support of monitoring, surveillance, and management at local, regional, and global levels in the context of an ecosystem approach for conservation and protection. Further, advancement in upcoming satellite missions will add to the advantages of remote sensing for forest monitoring. Detailed information about the future mission are available from Sharma et al. (2021) and Pandey et al. (2020b).

1.2 Forest Monitoring: Importance and Trends

Over the past decade, tremendous progress has been made in demonstrating the potentials and limitations of the applications of remote sensing in forestry. In the case of forest research, this is mainly focused on the estimation of its biophysical and biochemical parameters through

---

*Corresponding author's email: prem26bit@gmail.com

0000-0002-0049-1415 (Prem C. Pandey) and 0000-0003-2548-4550 (Paul Arellano)

Advances in Remote Sensing for Forest Monitoring, First Edition. Edited by Prem C. Pandey and Paul Arellano.

several remotely sensed images, and validated through field measurements. Remote sensing has been widely used for mapping the distribution of forest types, landscape, and global changes in plant productivity for seasonal, annual, and 3D forest structures, such as canopy height and canopy cover. The range and diversity of remote sensing systems, as well as the variety of applications, have evolved greatly over the last few decades.

Forest monitoring is the foremost requirement for the assessment of forest parameters and related research work. Forest monitoring includes traditional field survey as well as advanced earth observation with space-borne, airborne, drone-based, or terrestrial mounted sensors or scanners. Forest monitoring includes the capturing, measurement, and reporting of several forest parameters using field measurements and employing these measurements for verification and validation for models and algorithms developed using remotely sensed datasets. Monitoring also includes other estimation, assessment, and analysis such as forest disturbance arising from logging, burning, disease, or insect infestations, which can be monitored by remote sensing approaches. The concept behind forest monitoring is to identify the conservation priorities in global hotspots, rate of deforestation, and quantification of overall forest regions for its biophysical and biochemical parameters. Forest monitoring such as degradation monitoring is possible due to canopy gap analysis with temporal data, and clearing dense forest that changes the status from dense to open types of canopy cover. This is seen from datasets acquired from spaceborne, airborne, and UAVs, which illustrate the changes caused by natural and anthropogenic factors such as high-intensity logging causing fragmentation of canopy cover in dense forest. Forest monitoring involves other applications such as wildlife management, species mapping, and distribution at multiple scales; local, regional, and global. Recently, data acquisition by users are facilitated by the Remote Sensing Data Policy (RSDP) 2022 and Drone (Amendment) Rules, 2022 in India. The government of India has relaxed policy over RS data for the civilians and researchers, so that they may acquire and distribute the RS data from India and foreign satellites easily, and experience a smoother procurement process. Similarly, other countries are also planning for easy data delivery to users.

These forest monitoring systems with advanced earth observation help to generate high quality maps, and reliable digital data to be shared with or transferred to the end users and agencies for work plan and future actions. These measurements include different forest types, species mapping, estimating pigments, tree heights, crown density, basal cover, degradation area, biomass, and carbon mapping of the forest. Moreover, these parameters are assessed and estimated with different algorithms and models, with the use of single data or multiple data, such as multispectral data, hyperspectral data, microwave data, or a combination of other earth observation datasets. Remotely sensed digital outputs are accurate and highly reliable data on forests, which are very important in forest monitoring because of the advantages over traditional field methods. Remotely sensed data are acquired for a large area, hilly, or inaccessible regions with synoptic view, which makes this technology preferable over traditional methods. Forest monitoring is required for the point of view of conservation and protection as well as these derived outputs and information, which are imperative to assist policies, make decisions, and provide sustainable management plans for different forest types. Based on the results and outcome, management may plan policies and actions accordingly to each type of forests, and positive policies will have many benefits, such as non-carbon benefits, food security, forest health, and biodiversity richness and regularity, poverty alleviation, and improved restoration and land use land cover practices for protection of ecosystems and their services. We are dependent on the forest for several activities and resources; timber, fruits, medicinal plants, and Non-Timber Forest Products (NTFPs), as well as providing clean air and water, acting as climate moderator, rich biodiversity, home to several endemic flora and fauna, esthetic, and recreational values. Advanced levels of forest monitoring using earth observation data and sustainable forest management contributes to human well-being by meeting people's socioeconomic, ecological, cultural, and spiritual needs at the local, national, and global levels. Integrated sustainable forest management (SFM) enhances forests' carbon value, and maintains or improves other locally or globally important ecosystem services to support local livelihoods. As different forest types such as tropical, riparian, mangrove, temperate, and deciduous/evergreen are providing different opportunities to the researchers to assess and estimate their parameters, and in most cases, it is difficult to measure field data due to dense nature, inaccessibility, hilly terrain, or protected status of the forests by government.

1.3 Advances in Remote Sensing Technology for Forest Monitoring

This section deals with the advancement in RS technology for forest monitoring established through original research work and case studies. RS technology can identify, detect, discriminate, classify, evaluate, and measure several forest parameters in two ways; qualitative analysis and quantitative analysis. (i) Qualitative analysis generally involves the mapping and classification of forest covers types; riparian, tropical, or mangrove forest, while (ii) Qualitative analysis measures the statistical or estimated forest parameters involving biophysical and biochemical parameters; pigments, tree height, DBH (diameter at breast height), biomass, carbon, basal area, tree density, and so on. Estimation of forest parameters with RS technology generally employs correlation with spectral indices and modeling. Sometimes, high spatial resolution is required for accurate mapping and classification, while other times high spectral resolution is required for discrimination of species-based analysis and species distribution mapping. For the previously mentioned reasons, there are demands in the upgradation of sensors for specific analysis and day by day new algorithms are established by experts.

These book chapters consolidate multi-source, multi-sensor, and multi-temporal data that offers contemporary understanding to describe the way of forest monitoring. The following chapters establish the ongoing advancements in the field of earth observation for forest monitoring, and provide a comprehensive look into diverse forest regions such as riparian, mangrove, and conifer forests as well as oil palm plantations, etc., where these studies are being conducted by experts. These include handheld instruments, airborne data, spaceborne data, several machine learning algorithms, and modeling, as well as providing a comprehensive look into diverse forest parameter assessment, such as leaf area index (LAI), (DBH), biomass, pigments, carbon, canopy cover, abiotic stress, solar-induced chlorophyll fluorescence (SIF), gross primary productivity (GPP) among others.

Chapter 2 by Kumar et al. (2022) focuses on sustainable forest management using geospatial perspectives for ecosystem services as well as major ecological disturbances such as grazing, fuelwood collection, forage removal, forest fire, invasive plant species, and climate change impact. Kumar et al. also stress the sustainable use of forests for livelihood security of the surrounding area. The authors discuss the Landscape based Approach (LbA) and

Ecosystem-based Approach (EbA) of Sustainable Forests Management (SFM), which is the present focus for forest management as well as the need to maintain the integrity of ecosystems for the continuous supply of ecological services with sustainable development. Finally, the authors demonstrate how forest conservation is a current need that can be met by EbA, sustainable forest management, and through comprehensive forest resource management that addresses ecological, economic, and socio-cultural functions. Overall, ecological and sustainable use of forest resources along with satellite data used in research for monitoring forest ecosystem services and their mapping is presented.

Chapter 3 by Lassalle and Fabre (2022) demonstrates the efficiency of machine learning (ML) algorithms and their performance in distinguishing carotene and xanthophylls in the leaves of riparian forest species. They show the power of imaging spectroscopy in their study to distinguish the carotene and xanthophylls, with continuum removal and ML algorithms. The authors extoll the technology (handheld and remotely sensed) for their capability to distinguish different pigments. Lassalle and Fabre (2022) employ ML techniques to distinguish carotene and xanthophyll contents in the leaves of riparian forest species. These pigments are one of the requirements to understand and monitor plants' stress, and remain a challenge in remote sensing communities. A novel ML approach is presented with continuum removed field reflectance data in comparison with several other algorithms. The authors successfully illustrated the proposed ML algorithms such as partial least square regression (PLSR), support vector regression (SVR), random forest (RF), elastic net (ENET), and neural network (NN) along with their performance in distinguishing carotene and xanthophyll contents in the leaves of riparian forest species. Authors demonstrated how continuum removal (CRR) and ML approaches could be employed for monitoring of riparian forests for estimating the pigments, and this may be further utilized to study forest parameters.

Chapter 4 by Lachezar (2022) discusses the usefulness of remote sensing techniques with different modeling methods of abiotic stress assessment of coniferous forests, and demonstrates the use of different modeling approaches to assess the abiotic stress. Detailed literature review on the abiotic stress, loads, and response to these are well demonstrated in the chapter. Filchev discusses the natural and anthropogenic factors responsible for the abiotic stress, and deals with mainly natural components such as soil, atmosphere, and hydrosphere. The author pays special attention to multi-band, hyperspectral, fluorescent, and thermal infra-red satellite images for earth observation data, which can be readily available for conifer monitoring purposes, especially forest stress. The chapter discusses the abiotic factors in two major groups; natural and anthropogenic factors, and also devotes a section to provide insight to the critical loads and thresholds of various chemical constituents in soils and for conifers in particular, providing methods to deal with remote sensing methods and data used to discriminate them from space.

Chapter 5 by Solanki et al. (2022) demonstrates the retrieval of mangrove forest properties using synthetic aperture (SAR) data, focused mainly on the areal extent, tree density, basal area, tree height, DBH, LAI, and biomass. Along with these, authors discuss the role and importance of mangrove forests and the role of SAR data in monitoring mangrove forests across the globe for estimating their parameters. Authors mention the previous literature conducted for mangrove forest monitoring using SAR datasets, while presenting case studies using L-band advanced land observing satellite (ALOS-2) and C-band Sentinel-1 ground range detected (GRD) images for identification and monitoring of degraded regions in the mangrove forest.

Chapter 6 by Paramanik et al. (2022) analyzes the temporal pattern of the mangrove primary productivity in the Bhitarkanika Wildlife Sanctuary, a Ramsar Site (declared on 19 August 2022 as site no. 1205). The authors employed remotely sensed driven biophysical variables such as LAI, SIF, and GPP for the assessment, and demonstrate that the seasonal pattern could be well deciphered from long term analysis of moderate resolution imaging spectroradiometer (MODIS) derived and SIF derived products.

Chapter 7 by Mngadi et al. (2022) employs remote sensing datasets to quantify the carbons stock variability among species on reforested landscape. The authors adopted grey level co-occurrence matrix (GLCM) techniques to derive the texture measures along with RF machine learning to perform carbon stock variability assessment using sentinel-2 imagery in their studies. Results demonstrate that the carbons stock variability ranges from 3.97 to 6.96 t/ha in different species such as *Acacia robusta*, *Brideliar micarantha*, *Acacia caffra*, *Erythrinacaffra*, and *Syzygium cordatum*. These outcomes are required for the understanding of carbon stock in different species within a landscape and for species-wide monitoring in landscapes, thereby providing evidence-based species prioritization for reforestation. This is ultimately invaluable for promoting the value of urban ecosystem goods and services, carbon sink capacity, and climate resilient cities, particularly in the developing world.

Chapter 8 by Trisasongko et al. (2022) stresses the importance of closely monitoring the forest area, oil palm plantations, and protected areas in Indonesia. This is feasible with synoptic regional coverage by remotely sensed datasets and radar remote sensing which can penetrate clouds, providing an opportunity to gather time-critical information. Authors mapped the protected area using dual polarimetric, single look complex (SLC) Phased Array-type L-band SAR (PALSAR-2) data, and investigated five common land cover types in the study site. Also suggested is the power of incorporation of machine learning methods such as RF, Support Vector Machine (SVM) with horizontal transmit, horizontal receive (HH polarization) and horizontal transmit, vertical receive (HV polarization) backscatter coefficients, and their performance over other algorithms.

In Chapter 9 by Shahfahad et al. (2022), explains how field measured data and model outputs were combined, integrated, processed, and analyzed against ML algorithms in order to assess the susceptibility to forest fire and to generate risk maps using a machine learning approach integrated with RS datasets.

Chapter 10 by Sharma et al. (2022) assesses the bushfire severity and postfire vegetation recovery using advantages of Google Earth Engine and Landsat images. Authors incorporated spectral indices derived from Landsat images such as relative difference normalized burn ratio (RdNBR), and difference normalized burn ratio (dNBR) along with machine learning RF models, and classified four different burn severity classes in a heterogeneous landscape. Sharma et al. show the influence of climatic and topographic variables on the bushfire activities, and utilized temporal NDVI for the duration 2008–2020 to assess the trajectory of post fire vegetation recovery. Authors include anomalies of climatic variables (precipitation, temperature, solar radiation, and soil water volume) and topographic variables (elevation, aspect, and slope) in their study to highlight the assessment of bushfire severity, to conclude the strong correlation between climatic and topographic variables and the recovery process, with the highest influence by temperature, followed by topography, solar radiation, soil water volume, and precipitation. This study can assess the impacts of the fires to forest ecosystems and on biodiversity spatially and temporally.

Chapter 11 by Maurya et al. (2022) reviews the role of drones in forest monitoring, incorporating the traditional methods with their merits and demerits. This chapter extolls the emerging role of drones, which is revolutionizing the research in the remote sensing communities.

Chapter 12 by Mohammadpour and Viegas (2022) highlights the role and applications of remote sensing for forest management purposes. Authors emphasize the recent breakthroughs on multi-source and multi-sensor data fusion approaches available in the literature of forest ecosystems, such as for forest species mapping.

Chapter 13 by Pandey et al. (2022) emphasizes sustainable forest monitoring through geospatial applications. The authors have done their best to review the sustainable management strategies of vegetation cover and the challenges faced in this process. They also reviewed the techniques regarding how remote sensing and geographic information systems (GIS) application help in overcoming the challenges of sustainable monitoring and management of vegetation and forest resources.

Finally, Chapter 14 by Gupta et al. (2022) provides opportunities, challenges, and future aspects of remotely sensed data for forest monitoring and brief on the challenges in forest monitoring and management. How the different aspects may affect the accurate deliverables and retrieval of forest parameters or discriminating features depends upon the spatial, spectral, and temporal in special cases.

1.4 Summary

Clearly visible with all above research and case studies, forest monitoring requires field measured data to validate the remotely sensed analyzed data. Currently, RS technology has the capability to provide the possible near real time data from space for forest monitoring such as burn, fire, and changes in forest cover or density. This can help ecologists, planners, decision makers, and conservationists create rapid responses to threats as mentioned above. Even illegal deforestation, clearing of forest, logging, and mining ongoing inside the dense forests may be easily revealed due to high spatial resolution. RS data may reveal any changes and alterations in the composition of the forest vegetation easily and rapidly. Therefore, easy access to RS data for end users and researchers could improve the forest parameters retrieval and data dissemination which is already contributing to scientists of different disciplines using updated information about ecosystem health, ecosystem services, and potential impacts of climate change on the bio-regions at local, regional and global scales. The accurate deliverables estimated through RS technologies may be employed for forest resource monitoring for decision making, and tracking changes through space and time for evaluating the effectiveness of management in future.

References

Gupta, A., Srivastava, P.K., Satish, K.V. et al. (2022). *Challenges and Future Possibilities Towards Himalayan Forest Monitoring*. In *Advances in Remote Sensing for Forest Monitoring*. 329–350. Wiley.

Kumar, A., Ekka, P., Patra, S. et al. (2022). *Geospatial Perspectives of Sustainable Forests Management to Enhance Ecosystem Services and Livelihood Security*. Wiley.

Lachezar, F. (2022). *Modelling of Abiotic Stress of Conifers with Remote Sensing Data*. In *Advances in Remote Sensing for Forest Monitoring*. 78–106. Wiley.Wiley.

Lassalle, G. and Fabre, S. (2022). *Distinguishing Carotene and Xanthophyll Contents in the Leaves of Riparian Forest Species by Applying Machine Learning Algorithms to Field Reflectance Data*. In *Advances in Remote Sensing for Forest Monitoring*. 45–77. Wiley.

Maurya, N.K., Tripathi, A.K., Chauhan, A., and Pandey, P.C. (2022). *Recent Advancement and Role of Drones in Forest Monitoring: Research and Practices*. In *Advances in Remote Sensing for Forest Monitoring*. 223–254. Wiley.

Mngadi, M., Odindi, J., and Mutanga, O. (2022). *Quantifying Carbon Stock Variability of Species Within a Reforested Urban Landscape using Texture Measures Derived from Remotely Sensed Imagery*. In *Advances in Remote Sensing for Forest Monitoring*. 150–166. Wiley.

Mohammadpour, P. and Viegas, C. (2022). *Applications of Multi-Source and Multi-Sensor Data Fusion of Remote Sensing for Forest Species Mapping*. In *Advances in Remote Sensing for Forest Monitoring*. 255–287. Wiley.

Pandey, P.C. and Sharma, L.K. (ed.) (2021a). *Advances in Remote Sensing for Natural Resource Monitoring*. Wiley.

Pandey, P.C. and Sharma, L.K. (2021b). Introduction to natural resource monitoring using remote sensing technology. *Advances in Remote Sensing for Natural Resource Monitoring* 1–8.

Pandey, P.C., Srivastava, P.K., Balzter, H. et al. (ed.) (2020a). *Hyperspectral Remote Sensing: Theory and Applications*. Elsevier.

Pandey, P.C., Balzter, H., Srivastava, P.K. et al. (2020b). Future perspectives and challenges in hyperspectral remote sensing. In: *Hyperspectral Remote Sensing*, 429–439. Elsevier.

Pandey, S., Kumari, N., Dash, S.K., and Nawajish, S.A. (2022). *Challenges and Monitoring Methods of Forest Management Through Geospatial Application: A Review*. In *Advances in Remote Sensing for Forest Monitoring*. 291–328. Wiley.

Paramanik, S., Varghese, R., Behera, M.D. et al. (2022). *Photosynthetic Variables Estimation in a Mangrove Forest*. In *Advances in Remote Sensing for Forest Monitoring*. 126–149. Wiley.

Shahfahad, T., Das, T., Naikoo, M.W. et al. (2022). *Forest Fire Susceptibility Mapping by Integrating Remote Sensing and Machine Learning Algorithms*. In *Advances in Remote Sensing for Forest Monitoring*. 181–195. Wiley.

Sharma, L.K., Gupta, R., and Pandey, P.C. (2021). Future aspects and potential of the remote sensing technology to meet the natural resource needs. *Advances in Remote Sensing for Natural Resource Monitoring* 445–464.

Sharma, S.K., Aryal, J., and Rajabifard, A. (2022). *Leveraging Google Earth Engine (GEE) and Landsat Images to Assess Bushfire Severity and Postfire Short-term Vegetation Recovery: A Case Study of Victoria, Australia*. In *Advances in Remote Sensing for Forest Monitoring*. 196–220. Wiley.

Solanki, J.B., Kumari, R., Das, A.K. et al. (2022). *Retrieval of Mangrove Forest Properties Using Synthetic Aperture Radar*. In *Advances in Remote Sensing for Forest Monitoring*. 107–125. Wiley.

Trisasongko, B.H., Panuju, D.R., Afif, Z. et al. (2022). *Mapping Oil Palm Plantations in the Fringe of Sebangau National Park, Central Kalimantan*. Indonesia: In *Advances in Remote Sensing for Forest Monitoring*. 167–178. Wiley.

2

Geospatial Perspectives of Sustainable Forest Management to Enhance Ecosystem Services and Livelihood Security

Amit Kumar[1], *Pawan Ekka*[2], *Subhashree Patra*[2], *Gajendra Kumar*[1], *Bodi S.P.C. Kishore*[1], *Rahul Kumar*[3], *and Purabi Saikia*[*,2]

[1] *Department of Geoinformatics, Central University of Jharkhand, 835205, Brambe, Ranchi, Jharkhand, India*
[2] *Department of Environmental Sciences, Central University of Jharkhand, 835205, Brambe, Ranchi, Jharkhand, India*
[3] *ICFRE-Institute of Forest Productivity, 835303, Lalgutwa, Ranchi, Jharkhand, India*

2.1 Introduction and Background

Sustainable Forest Management (SFM) is a key governing principle in forest management (Lattimore et al. 2009) to better fulfill today's demands without jeopardizing the future needs (UNCED 1992). The policies and laws that assist people are instrumental for boosting SFM along with inventories, monitoring, certification, stakeholders' participation, and forest management plans. Various authorities, including governments, non-governmental organizations, research groups, and private corporations have collaborated to develop SFM criteria and indicators (C&Is) using scientific, economic, and social values to elucidate SFM toward its implementation (Linser et al. 2018). SFM practices may be made in a more cost-effective way by supportive national legal, legislative, and institutional frameworks that stimulate the activities needed for proper implementation of SFM (Lovric et al. 2010). Independent third-party verification of conformance to a defined set of management requirements is provided by forest management certification to promote SFM (MacDicken et al. 2015). The three-dimensional problem of sustainable development is environmental deterioration, food security, and deforestation (Gil et al. 2019). Sustainability of forest landscapes includes social, ecological, and biodiversity demands in addition to commercial profits as a permanent supply of timber (de Groot et al. 2010). Conviction of forest sustainability has changed with the shift of social understanding of forests (Harrison 2002).

Natural catastrophes, human activities, and a combination of both have altered over 80% of terrestrial ecosystems (Zhu and Liu 2004). Recurrent disturbances shrink the growth and development of biomass and existing stands in natural forests (Hartmann 2011). There

* Corresponding author's email: purabi.saikia83@gmail.com, purabi.saikia@cuj.ac.in

0000-0002-4582-5677 (Amit Kumar), 0000-0002-6113-9626 (Pawan Ekka), 0000-0002-4624-6968 (Subhashree Patra), 0000-0003-4708-3918 (Gajendra Kumar), 0000-0003-1835-3595 (Bodi S.P.C. Kishore), 0000-0003-3259-9591 (Rahul Kumar), and 0000-0001-5481-282X (Purabi Saikia)

Advances in Remote Sensing for Forest Monitoring, First Edition. Edited by Prem C. Pandey and Paul Arellano.

is a rising concern about the long-term effects of extensive timber and fuelwood harvesting on-site productivity (Jang et al. 2015). The core idea behind disturbance management is that forest activities consistent with natural biological processes and within the local disturbance regimes, are more likely to preserve the ecosystem development processes and their functioning (Dove and Keeton 2015). Forest operations and anthropogenic dependencies must be confined within the natural disturbance regimes, to mitigate a few negative environmental effects of forest management practices (North and Keeton 2008). The frequency of disturbances, the degree of disturbances, its intensity and geographical dimensions, and the quantity and type of biological legacies that endure post-disturbance conditions are the major factors to consider to determine the variations between natural disturbance regimes and forest harvesting implications (Seymour et al. 2002). Understanding of post-disturbance ecosystem recovery and long-term process of stand growth is a critical piece of knowledge for disturbance management in natural forests (Franklin et al. 2002). Traditional wisdom and knowledge from around the world are working to ensure the long-term viability of natural forests (Parrotta et al. 2009). Forests can contribute significantly to the economy by contingency on a wide range of products and services that help people to derive livelihoods, while simultaneously safeguarding the environment (MacDicken et al. 2015). The natural balance of sustainability might be improved through better management of the agriculture and forestry sectors, which would assist to improve ecosystem services and livelihood security (Munang et al. 2011). For providing the most effective mitigation measures against ongoing natural and anthropogenic disturbances, forest managers must aim to establish techniques that limit the likelihood of long-term production losses.

Remote sensing images with periodic and seasonal time scales, unique spectral agility, and refined spatial resolution in combination with improved geospatial methodologies and innovations are greatly supported in estimation of forest resources, including its changes over the spatio-temporal patterns, that contributes effectively in improved forest management plans and its implementation (Roy et al. 2015). New pathways for accurate evaluation of forest yield, health, mapping, classification, yield management, planning (Laamanen and Kangas 2011), and emissions monitoring have been performed due to recent advancements in Geographic Information Systems (GISs), Global Positioning Systems (GPSs), and remote sensing. The geospatial techniques provide a systematic, consistent picture of earth landmass at regular intervals, which is very pertinent for detecting land cover changes and estimating biodiversity dimensions (Kumar 2011). High-resolution satellites offer multi-temporal and multispectral data gathered with the help of passive and active sensors (Solano-Correa et al. 2018). A plethora of high spatial and spectral resolution aerial and space-borne sensors have become accessible to investigate forest alterations, including optical techniques based on Moderate Resolution Imaging Spectroradiometer (MODIS) imagery, moderate-resolution LANDSAT imagery, high-resolution Satellite for observation of Earth (SPOT) data, Linear Imaging Self-Scanning Sensor-IV (LISS-IV), very high-resolution Synthetic Aperture Radar (SAR), WorldView-II (WV-II), and Light detection and ranging data (LiDAR) based perspectives (Kumar et al. 2012; Andersen et al. 2014). With the availability of high-resolution satellite data, remote sensing has aided in the detection, management, and recovery of forest disturbances (Hansen et al. 2013). Identification and concurrent monitoring of forest canopy density, vegetation pattern at landscape level, forest cover change (Lal et al. 2021a), vegetation indices, primary productivity, and forest disturbances (Justice et al. 2002) has been analyzed using time series remote sensing images (MODIS, MERIS, etc.). Radio detection and ranging (RADAR) and LiDAR data have been effectively utilized to assess vegetation biomass and canopy height (Kumar et al. 2019; Santoro et al. 2015). With the coarser spatial resolution (~250 m) datasets, NASA's Aqua and Terra satellites' MODIS sensor produces a global image of percent forest cover, vegetation fires, and land cover change (Savtchenko et al. 2004). On the other hand, Advanced Spaceborne Thermal Emission and Reflection Radiometer (ASTER) is a multispectral imager with 15–90 m spatial resolution and stereos, with a long track used for mapping land surface temperature, reflectance, and elevation; while with a spatial resolution of 5–20 m, the high spatial and spectral resolution European sensor-based SPOT may be employed to map ground vegetation changes (King 2002) (Table 2.1). While the hyperspectral remote sensing images from space (Hyperion) and Airborne Visible/Infrared Imaging Spectrometer (AVIRIS) have been successfully used for vegetation type mapping and species identification (Kishore et al. 2020) combining various techniques and machine learning algorithms.

2.2 Major Ecological Disturbances of Forests

2.2.1 Livelihood Dependencies

Indigenous populations living in and around forests are intricately reliant on the forests for food, fodder, agriculture, housing, and several marketable minor forest products (MFPs) that can potentially degrade forests due to their unsustainable harvesting (Banerjee and Madhurima 2013). MFPs include a collection of edible fruits, flowers, tubers, roots, and leaves for food and medicine; firewood for cooking, heating, and selling in the local market; agricultural implements; house construction; fencing; and livestock fodder (TRIFED 2021). As the human and animal populations grow, so does the need for forest goods and resources by the indigenous communities due to the severe decline in per capita land availability (Kumar and Saikia 2020) that degrades the forest. Plant community structure in the forests is greatly influenced by ecological disturbances of both natural and anthropogenic origin (Yadav and Gupta 2006) as it creates canopy gaps and lowers leaf fall, which has a negative impact on nitrogen return to the soil and increases forest dryness (Singh et al. 1984), resulting in poor forest tree regeneration. Major ecological disturbances of forests due to livelihood dependencies include livestock grazing, fuelwood, fodder, and MFPs collection.

2.2.1.1 Livestock Grazing

Livestock grazing is considered as an act of feeding domesticated animals freely upon grasses, saplings, and other parts of plants growing in a particular locality (Filazzola et al. 2020). It is ubiquitously unregulated (Diaz et al. 2007) land use activity by humans, accounting for more than 25% of earth's land surface being grazed by livestock (Schieltz and Rubenstein 2016). Selective browsing, soil erosion, nutrient depletion, soil acidity, removal of plant biomass, compaction of topsoil, creation of hydromorphic humus, and damage to tree roots are all consequences of uncontrolled cattle grazing and overstocking of feed (Glatzel 1999). Overgrazing in forests results in slow regeneration of a majority of plants species and poor forest health, causing biodiversity loss, decrease in plant biomass, alteration in ecological succession, nutrient imbalance, and landscape heterogeneity (Huntly 1991). By directly destroying reproductive structures, grazing animals

Table 2.1 Remote sensing sensors and its applications in sustainable forest management.

		Resolution				
Type	**Sensor**	**Spatial**	**Spectral**	**Temporal**	**Application**	**References**
Multispectral remote sensing	MODIS	1 km, 500 m, 250 m	36	Daily	AGB, BP, LULC, FF, FH, FP, Ph	Justice et al. 2002
	SPOT (1, 2, 3, 4, 5)	20 m, 10 m; 10 m, 5 m	5	26 d	LULC, FD, FF, FR, FT	Chevrel et al. 1981
	MERIS	300 m	15	3 d	LULC, FC, FF, FT	Verstraete et al. 1999
	ASTER	90 m, 30 m, 15 m	14	4–16 d	AGB, FC, FF, IPS	Muukkonen and Heiskanen 2005
	Landsat (MSS; TM; ETM+; OLI)	80 m; 30 m; 15–60 m; 30 m	4; 7; 8; 11	16 d	AGB, BP, LULC, FC, FD, FH, FP, FT, Ph, GM	Banskota et al. 2014
	Sentinel-2	10–60 m	13	5 d	AGB, BP, LULC, FC, FD, FH, FP, FSM, FT, IPSM, Ph, GM	Phiri et al. 2020; Misra et al. 2020
	IKONOS	4 m	4	3–5 days	LULC, FD, FS	Kayitakire et al. 2006
	QuickBird	2.44–2.88 m	4	1–3.5 d	LULC, FC, FSM, FT	Abdollahnejad et al. 2017
Hyperspectral remote sensing	Hyperion	30 m	220	16 days	AGB, BP, BC, LULC, FFT, FH, FS, FT, FSM, IPSM	Ghiyamat and Shafri 2010
	ER-AVIRIS; AVIRIS-NG	20 m; 4–9 m	224; 430	Flight per request	AGB, BP, BC, LULC, FFT, FH, FS, FT, FSM, IPSM, GM	Smith et al. 2021; Kishore et al. 2020; Martin and Aber 1997
	HyMap	2–10 m	126	flight per request	BP, FSM, FT	Lucas et al. 2008a; Schlerf et al. 2005
	CASI	1–2 m	288	Flight per request	AGB, BP, FS	Lucas et al. 2008b; Sampson et al. 2003
Synthetic aperture radar	ERS-SAR	30 m	L&C	3, 35, 336 d	AGB, FF, SM	Zink et al. 1996

(Continued)

Table 2.1 (Continued)

Type	Sensor	Resolution			Application	References
		Spatial	Spectral	Temporal		
	ALOS PALSAR 1; 2	10–100 m; 3–100 m	L	46 d; 14 d	AGB, LULC, FC	Zhang et al. 2019; Shimada et al. 2014; Lal et al. 2021b
	Sentinel-1	10–20 m	C	12 d	AGB, FD, Ph, GM	Reiche et al. 2021; Forkuor et al. 2020; Frison et al. 2018; Kumar et al. 2019

AGB: Above ground biomass; BC: Biochemical parameters; BP: Biophysical parameters; FC: Forest cover; FD: Forest disturbances; FH: Forest health; FF: Forest fire; FFT: Forest fuel type; FP: Forest productivity; FR: Forest regeneration; FS: Forest structure; FSt: Forest stress; FSM: Forest species mapping; FT: Forest type; IPSM: Invasive plant species mapping; LULC: Land use land cover; Ph: Phenology; SM: Soil moisture; GM: Grassland monitoring.

Figure 2.1 Cattle are on the way to the nearby forest for grazing at the same time locals are carrying fuelwood back to home in Bero block of Ranchi, Jharkhand.

(Figure 2.1.) can reduce flower and seed output, which reduces energy available for developing seeds, eventually inducing stress on plant growth and regeneration (Amadi and Aleru 2021). Grazing can increase the populations of invasive species by dispersing and transporting seed in animals' furs, fleeces, hairs, feet, or digestive tracts. Long-term continuous grazing reduces vegetative dynamics and degrades topsoil, eventually resulting in increased wind and water erosion (Belnap and Gillette 1998). Continued overgrazing of selective plant species harms the structure of plant communities by altering plant height, canopy cover, biomass, and fuel load (Huntly 1991) that may lead to frequent return of forest fire-cycles, increased intensity, and rapid spreading patterns (U.S. Fish and Wildlife Service 2021). Although grazing induces negative effects on the forest composition and structure, livestock rearing can support the livelihood and food security of forest dwelling peoples with provisions of milk, meat, cheese, wool, leather, manure, and monetary support (Langat et al. 2016). Therefore, proper management strategies and collaborations should be developed between forest management authorities and livestock holders for minimizing the negative effects of grazing (Varga et al. 2020). Integration of scientific approach and traditional knowledge in sustainable and managed grazing and wise use of forest resources could be an integral part of sustainable forest management to fulfill ecological services of natural forest and livelihood security of forest dependent people (Öllerer et al. 2019) (Figure 2.2). Although measuring livestock grazing is very challenging through remote sensing images, yet some indirect measurement has been used to monitor grazing that are primarily related with variety of vegetation biomass metrics to deduce stocking rate at the pasture scale (40–250 ha; a management indicator)

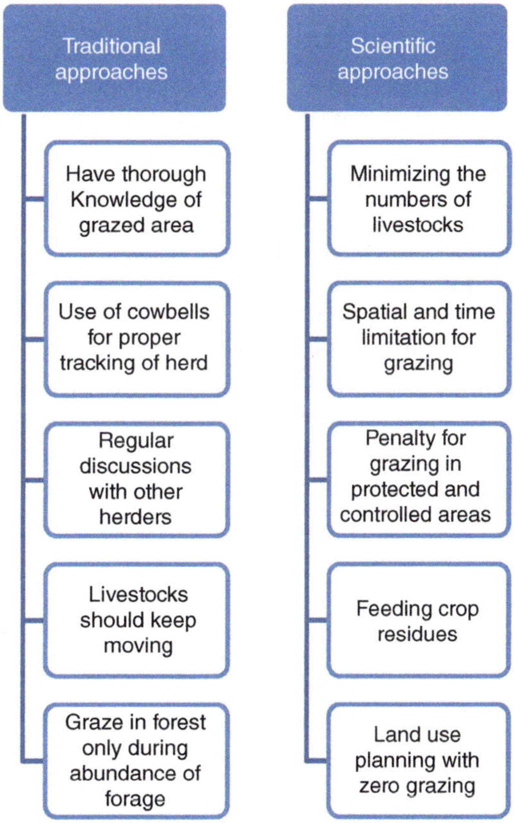

Figure 2.2 Traditional and scientific approaches that should be practiced by herders for preventing overgrazing. Source: Adapted from Varga et al. 2020; Pietikäinen 2006.

and field-based utilization estimates at the plot scale (25–50 m; a grazing indicator) (Jansen et al. 2021). While GPS and ground-based sensor technologies are used to manage grassland and livestock with incredible precision (Ali et al. 2016).

2.2.1.2 Fuelwood Collection

Fuelwood is the most prevalent and primary source of energy for rural inhabitants in developing countries (Allen et al. 1988), used for cooking, heating rooms and water, or power generation during the winter season, and ~54% of the total global harvest of forests is for fuelwood purposes (Nautiyal and Kaechele 2008). Dried leaves are collected from the forest floor and used as fuel or converted into briquettes (artificial fuelwood) in most tribal dominating Indian states such as Jharkhand, Odisha, Chhattisgarh, and Madhya Pradesh (Ahuja and Bhaskar 2021). More than 1.7 billion people around the world rely on forest resources for their livelihoods, either directly or indirectly, of which fuelwood collection is one of the major economic activities that support sustenance of forest dwelling populations (Yadav et al. 2019) (Figure 2.1). The consequences of intensive fuelwood collection has led to forest degradation, fuelwood crisis, loss of biodiversity, deterioration of watershed, carbon emission, and soil erosion (Heltberg et al. 2000). The majority of fuelwood used in traditional stoves for cooking in developing countries have low efficiencies and incomplete combustion leading to the emission of greenhouse gases (GHGs) including CO, CH_4, NO_2, and particulate matters (Lam et al. 2012). Since the 1990s, exploitation of forest resources for fuelwood and housing construction has resulted in more than 30% deforestation (Kumar and Saikia 2020). The villagers of Garhwal, Uttarakhand extract ~91% of the total fuelwood from forests for household consumption throughout the year in the temperate zone of Himalaya (Sharma et al. 2009). Fuelwood collection and consumption can be an approach of natural resource and fire management by using crop residues, animal dung, biogas, solar energy, and wind energy as alternatives for rural domestic energy demand (Heltberg et al. 2000). In addition, government policies and schemes (e.g. Pradhan Mantri Ujjwala Yojana by the Govt. of India launched in 2016, providing cooking fuel (liquefied petroleum gas, LPG) for rural people), a total of 89 553 012 connections have been released up to 24 December 2021 (MPNG, GoI 2021). These approaches would reduce pressure on natural forests and decrease the fuel demand in rural areas, which may be considered as a way toward sustainable forest management. The spatial distributions of supply and demand of these bioresources (fuel wood, charcoal, and other biofuels) are greatly aided by GIS and remote sensing due to high spatial-spectral agility and temporal availability, which is useful in inventory and mapping of natural resources (Ramachandra 2010).

2.2.1.3 Forage Removal

Forage is edible plants or plant parts (Figure 2.3) of which there are many types with different vegetative terms such as browse (leaves and tender twigs of shrubs, xerophytes, trees, etc.), herbage (edible roots and tubers), green chops, and fodder, and used by herbivore animals as their food (Oregon State University 2021). Particularly in rural areas, livestock farming is an important source of employment and income (Chandramolly and Islam 2015) and is closely connected to the forests, as forests and grasslands together account for ~30–50% of total animal feed (Balachander and Ganesan 1993). Throughout the world, rural farmers are dependent on forests for fulfilling their forage needs, especially during

(a) (b)

Figure 2.3 Freshly collected mango and jackfruit rachis from the local forest are used as forage for feeding goats in Itki, Ranchi.

the dry winter seasons, causing forest degradation due to massive stockpiles during the off seasons (Peprah et al. 2021). Forage removal can sometimes improve the plant quality, when old fodder is removed from the forests (Georgiadis et al. 1989). The impact of unregulated forage removal from natural forests can be alleviated by improving grazing land and pastures with quality forage species to produce fodder. Forage removal is one of the major economic activities in the forests of Madhya Pradesh (22.53%) and Jharkhand (47.83%) (Kumari et al. 2019). Incorporation of quality forage crops into farming systems can lessen the burden on forest resources for livestock production (Crotty et al. 2014). Alternative forage species with high nutritional content, have been added to agropastoral systems in Mexico to feed animals during all cultivation cycles (Sánchez et al. 2015).

2.2.1.4 Minor Forest Products (MFPs) Collection

Minor Forest Products (MFPs) are goods other than timber of biological origin with social, cultural, or religious significance; extracted from natural, modified, or managed forests (Pandey et al. 2016); and utilized to fulfill household needs or sell in the nearby markets for economic sustainability (Marshall et al. 2003). They supply items such as leaves, flowers, twigs, medicinal plants, resins, spice and condiments, fruits, nuts, edible fungi, wild edible vegetables, fish, bamboos, rattans, palms, honey, lac, and grasses (Kumar and Saikia 2020) for food, fibers, housing materials, medicines, energy, cultural artifacts, and other NTFPs (Dao and Hölscher 2018; Figure 2.4). Collection of raw materials for various industries including bamboos for paper and pulp, Kendu leaves for bidi making industries, Sal leaves

Figure 2.4 Minor forest products like Sal leaves, twig brush (dattun), fuelwoods, and bamboo baskets are prepared and sold at local markets as NTFPs in Itki, Ranchi.

for plate manufacturing, and herbal medicines are some of the important MFPs collected by the underprivileged families, forest dwellers, and farmers as a vital source of their income (Kumar and Saikia 2020). Approximately 37 lakh people are employed in the trade of Kendu leaves (Kerketta et al. 2018), while silk culture processing and trading employs about 7.85 million people in 51 000 villages throughout India (Central Silk Board 2017). Overexploitation, decreases in local abundance, and extinction of some essential species that offer highly valued goods can all result from unsustainable MFP collecting for commercial purposes (Vasquez and Gentry 1989). Intensive harvesting of seeds, fruits, and flowers may result in a reduction in the diversity of species in natural forests (Witkowski et al. 1994). Unsustainable harvesting of MFPs from the Indian Himalayan region has led to habitat loss, land use change, forest degradation, and deforestation (Uprety et al. 2010). Over the last few years, honey, soapnut, and Indian gooseberry extraction patterns have shifted from subsistence to large-scale commercial extractions. Strengthening of long-term national programs to increase regional and international cooperation in research, education, and training on the use of forest resources can reduce overexploitation of useful plants that produce various MFPs (FAO 2014). Knowledge of significant plant population status, habitat modification and uniqueness, a limited distribution range, and the impacts of overgrazing may help in protecting plants from further exploitation (Kumar and Saikia 2020). Existing populations of valuable plants should be conserved to ensure future generations' livelihood security and ecological stability to achieve environmental health, economic profitability, and social equity (Brodt et al. 2011).

2.3 Forest Fires

Forest fires are complicated phenomena that originate from a combination of natural sources like lightning, volcanic eruptions, and human forces (Vasilakos et al. 2009). Fire is one of the main causes of forest degradation, which leads to biodiversity loss and global warming (Rudel et al. 2005). Conversely, climate change influences forest fire frequency, intensity, and expansion of the forest fire season (Flannigan et al. 2009). The spread of anthropogenic forest fires across Asia is aided by an increase in temperature and a decrease in rainfall linked to land-use change (IPCC 2007). Elevation, slope, and aspect, as well as the presence of combustible materials, the terrain's incline, excessive wind velocity, the distance to roadways (Arima et al. 2005), and scorching summer temperatures (Roy et al. 2006) influence fire behavior. The frequency of fire incidences is found to be inversely related to the distance between the forest boundary and the center (Nepstad et al. 1999). Fire is a common occurrence in tropical deciduous ecosystems during the summer months because of greater levels of water stress (Chandra and Bhardwaj 2015). Uncontrolled forest fires, particularly in boreal, temperate, and tropical biomes, may result in massive losses of forest resources and ecosystem services (Nasi et al. 2002). This may also contribute to an increase in weed spread, soil erosion, regeneration constraints, landslides, and habitat destruction (ISFR 2019). Many plant species are affected by fires in terms of natural regeneration, recruitment, establishment, and by facilitating the spread of invasive species in natural forests, contributing to forest deterioration (Reddy et al. 2014). Wildfires are regarded as a hazard that destroys terrestrial habitats and causes economic harm to humans, such as loss of revenue due to land usage, property destruction, agricultural damage, biodiversity loss, deforestation, and desertification (Hernandez-Leal et al. 2006).

The majority of the forest fires in Asia, Africa, Europe, North America, South America, and Australia that occurred in 1997 and 1998 were connected to a drier environment and related to occurrences like El Nino southern oscillation (Narendran 2001). The first forest fire in India occurred ~200 million years ago, and fire affects 3.73 Mha of forests in India each year (Bahuguna and Singh 1999). Of all the fire-prone forest areas, approximately 4% of the country's forest cover is extremely prone to fire, while 6% of forest cover is deemed extremely prone to fire (ISFR 2019). Forest fires are expected to occur with more frequency in more than 36% of the country's forest cover (Kumari et al. 2019). The proportion of fire prone forest areas annually ranges from 33% in some Indian states to over 90% in others (Roy 2003). Adequate planning is required to accomplish fire risk reduction, restoration of the fire regime, and maintenance to effectively control wildfires (Keifer et al. 2000). Forest fire management entails four phases of analysis and evaluation for an efficient fire response; namely mapping possible fire danger and risk, finding hotspots, monitoring current flames, and analyzing post-fire damage (Roy 2003).

Active fire detection, fuel mapping, fire risk mapping, burned area estimations, burn severity assessment, and post-fire vegetation recovery monitoring all benefited from remote sensing (Szpakowski and Jensen 2019). Remote sensing allows for the analysis of circumstances and the tracking of changes over broad geographic areas (Barmpoutis et al. 2020). Coarser resolution MODIS sensors with thermal infrared bands of 3.6–12 m range are efficient enough in detecting real time forest fires (Reddy et al. 2017). Sensors such as Sentinel-2, Landsat series, and ASTER may provide input data for long-term fire risk maps with spatial resolutions ranging from high to moderate (Gibson et al. 2020). Weekly updates of fire risk maps are possible with sensors onboard Sentinel-2, with a five-day revisit interval and medium spatial resolutions (10–20 m) (Roteta et al. 2019). For short-term fire risk mapping, a combination of sensors might be more appropriate, whereas using coarser resolution datasets may provide rapid (daily) updates of dynamic fire risk conditions (Srivastava et al. 2019). A MODIS based forest fire frequency-based hotspot analysis exhibited high forest fire spots with episodic variations during 2001–2020 (Figures 2.5 and 2.6). The high coverage of hot spots (> 95% confidence) was observed in the lower NE regions (comprising states of Nagaland, Mizoram, Manipur, and Tripura) primarily during the period 2011–2015 followed by 2001–2005 and 2006–2010, while least during 2016–2020 because of shifting cultivation (Deb et al. 2013; Kumar 2022).

2.4 Invasive Plant Species (IPS)

Invasive plant species (IPS) are the species that are introduced outside their native ranges intentionally or unintentionally for esthetic (Mukul et al. 2006), or multipurpose values (Rai and Singh 2020). They are known for their threatening behavior toward native plant species and ecosystems (Darwin 1859) and are considered the second largest cause of biodiversity loss (Bellard et al. 2013) and species extinction worldwide (Dueñas et al. 2021). Multiple anthropogenic activities and natural factors (Fuentes et al. 2015) such as biotic factors or appropriate climatic phenomena provide viable conditions for the spread and survival of IPS even in harsh environmental conditions (Wan and Yang 2016). The richness of IPS varies based upon density of human population, level of urbanization, and economic development of a particular zone (Inderjit et al. 2018). However, intensive agricultural practices used to balance food security that elevates the spread of IPS (Rai et al. 2018). Enhanced globalization export and import beyond biogeographical areas, and human-induced environmental modifications lead to establishment of IPS to a new region globally (Early et al. 2016). In comparison to endemic and localized species, IPS exert their dominance over a new area due to their highly competitive ability (Bennett et al. 2011), reproductive rate, fast growth, systematic dispersal capabilities, and resistance to a wide range of environmental conditions (Hellmann et al. 2008). IPS are also known as a barrier to reforestation and regeneration, as their extensive canopy covers impede the availability of light energy and other key resources required for seedling growth and development (D'Antonio and Vitousek 1992). They play a major role in disturbing the forest structure and function by hindering the tree seeds from approaching the soil (Elliott et al. 2013). Forest canopy opening and lesser overstorey magnify the tenure of IPS through supporting the growth of shade tolerant invasive species (Schnitzer et al. 2000). Climatic and atmospheric elements

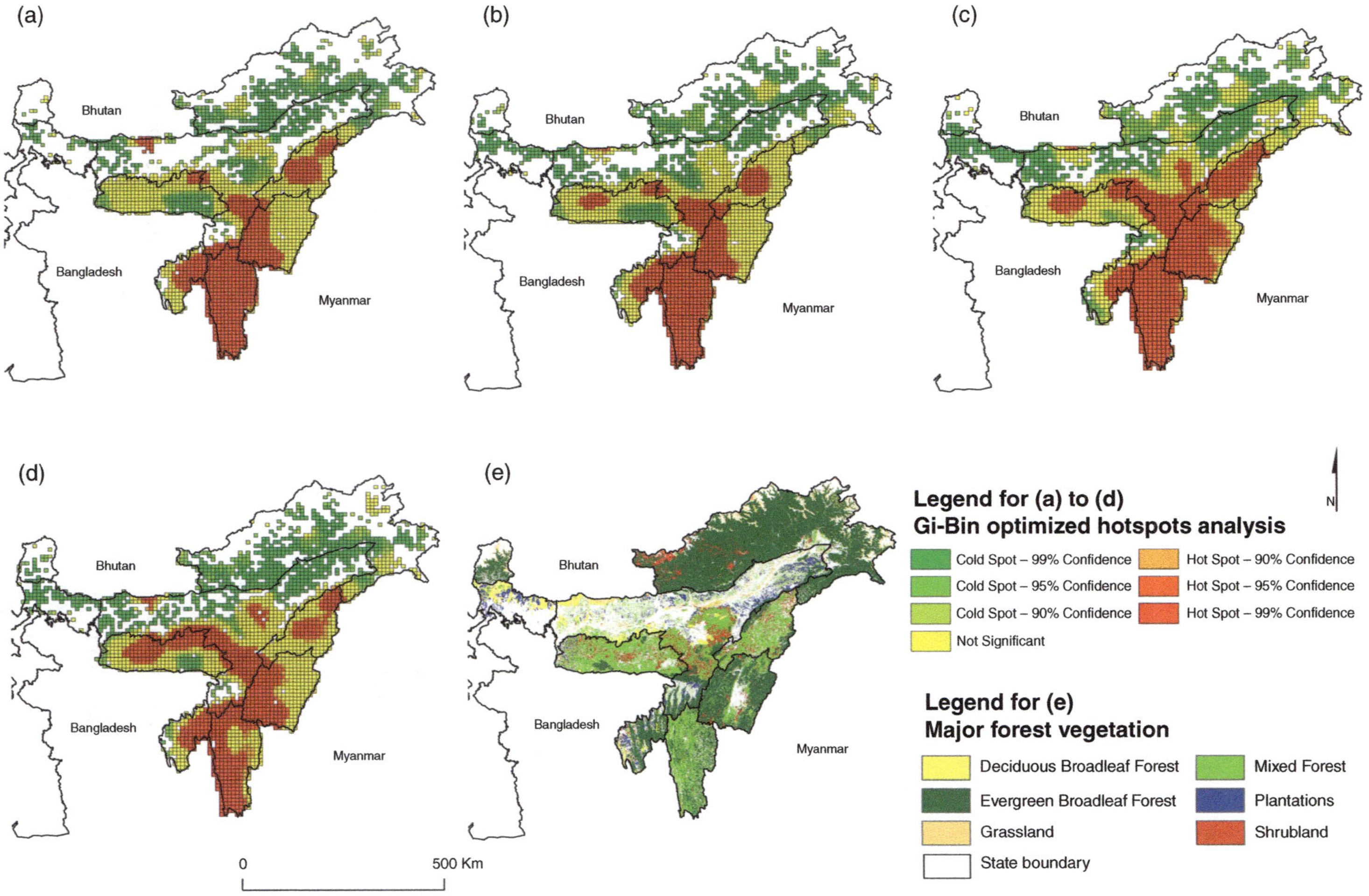

Figure 2.5 Five-year cumulative forest fire frequency based on Gi-Bin optimized hotspots during periods (a) 2001–2005, (b) 2006–2010, (c) 2011–2015, (d) 2016–2020, and (e) forest type. Source: Adapted from Roy et al. 2015 in North East India.

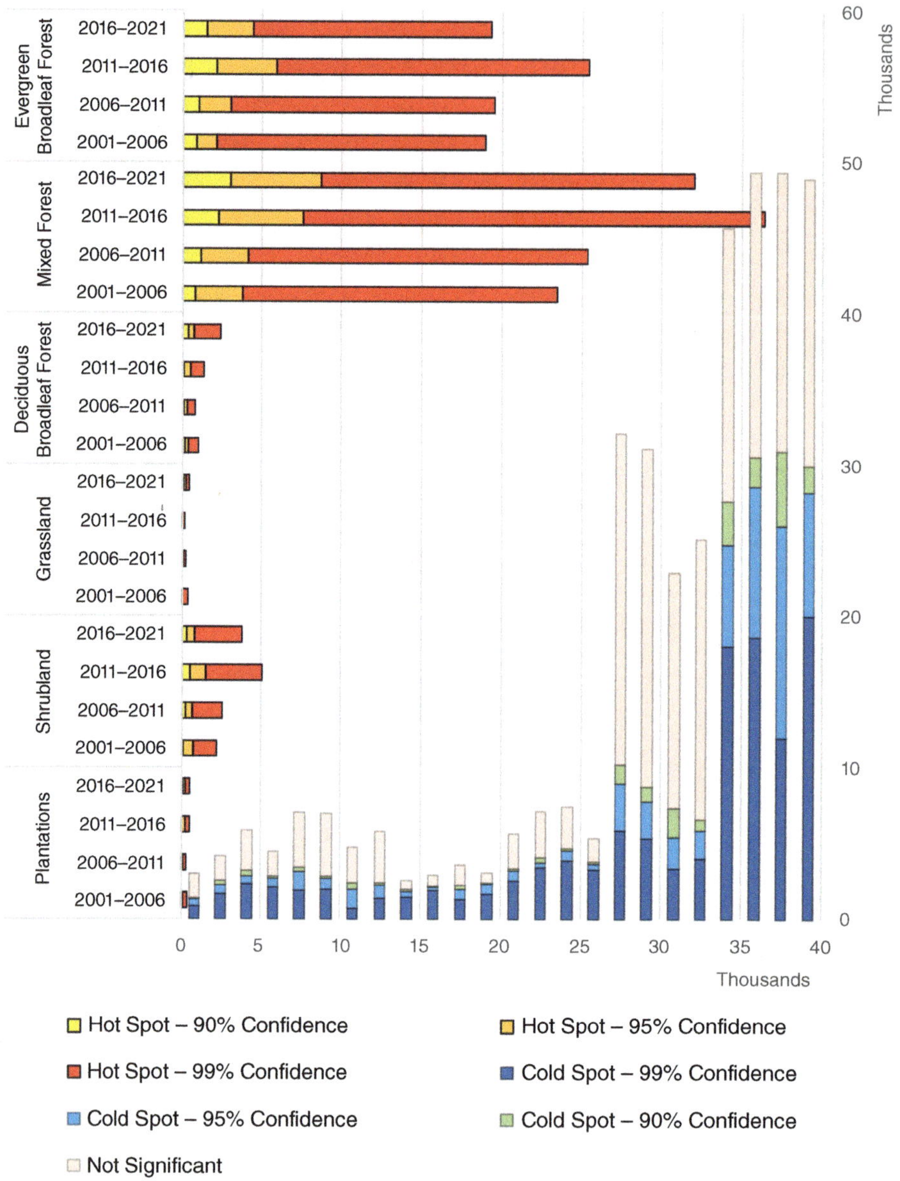

Figure 2.6 Graphical representation of forest fire hotspot during the periods 2001–2005, 2006–2010, 2011–2015, and 2016–2020 over the different vegetation types in North East India.

such as temperature, carbon dioxide, rainfall, and UV radiation levels have the capability to alter the range and distribution of IPS (Shabani et al. 2019). The rise in the spread of IPS has created extreme concern in tropical and subtropical forests (Foxcroft et al. 2017). IPS have the potential to change the forests composition and productivity by increasing or decreasing biomass production, reducing the availability of forage for herbivores, lowering the diversity of native plant species, and degenerating the ecosystem functioning

Figure 2.7 Dominant IPS of Indian Tropical Forests: (a) *Lantana camara*; (b) *Chromolaena odorata*; and (c) *Parthenium hysterophorus*.

(Powell et al. 2011). IPS have played a vital role in human-induced environmental changes (IPBES 2019) and diminishing biodiversity from genetic to landscape or ecosystem levels over the last 30 years (Viciani et al. 2020). *Chromolaena odorata*, *Lantana camara*, *Hyptis suaveolens*, *Mikania micrantha*, and *Parthenium hysterophorus* are dominant invasives of Indian tropical forests (Figure 2.7).

Many remote sensing sensors with moderate to high spatial resolution such as Landsat TM, ETM+, OLI, SPOT, ASTER, IKONOS, Sentinel-2, Quickbird, WorldView, and high spectral resolution sensors such as Hyperion, HyMap, AISA, CASI, and AVIRIS-NG are proven to be compatible in mapping IPS (Kishore et al. 2020; Table 2.1). Fusion of optical, SAR, hyperspectral, and LiDAR data also showed promising results (Peerbhay et al. 2016). Methods such as object-based image analysis, spectral angle mapper, maximum likelihood classification, mahalanobis distance, spectral unmixing, support vector machine, artificial neural network, spectral feature fitting, and mixture tuned matched filtering were popularly used to map IPS (Royimani et al. 2019).

2.5 Climate Change

Globally, climate change is one of the major drivers of terrestrial ecosystem processes that affect the forests by changing forest dynamics (Dale et al. 2000). The rate of photosynthesis, respiration, precipitation, temperature, humidity, and moisture in the forests are typically controlled by climate for longer and shorter durations (Thompson et al. 2009). Thus, climate change has a major impact on timing and occurrence of events in the forest ecosystems both directly and indirectly (Dale et al. 2001). Mainly the severity of incidents, rise in frequency of disturbances, and other altering factors on forests signify the impact of climate change on forests (Loehman et al. 2018). Increased disturbances due to climate-mediated phenomenon enhanced forest resilience, but simultaneously converted forest areas to non-forest areas (Seidl et al. 2016). Due to modification in climate, fire behavior (Lubchenco and Karl 2012), species diversity, and species physiological activities are enhanced, which also altered the biophysical properties of forests (Malhi et al. 2008). Change in temperature of forests also varies the survivability of species with respect to their behavior, predation activities, and habitat which eventually modify their gene pool and leads to loss of biodiversity, misbalance in nutrient cycling, and forest succession (Dale et al. 2001).

Satellite remote sensing provides an essential part of climate system studies through its data acquisition capability of the land, atmosphere, and its interactions (Yang et al. 2013). The climate system of Earth can be better understood using observational data and model simulations (Kumar et al. 2020). By measuring processes and spatiotemporal states of the atmosphere, land, and oceans, remote sensing has made significant contributions to our understanding of the climate system and its variations (Li et al. 2011). Twenty-six out of 50 essential climatic variables (ECVs) are significantly dependent on satellite data, according to the Global Climate Observing System (GCOS) (Houghton et al. 2012). Climate system observations include periodic monitoring of various climate variables such as air temperature, GHGs, surface radiation budget, precipitation, water vapor, wind speed and direction, ozone, air pressure, and aerosols as well as oceanic (sea ice, sea surface temperature, wind velocities, ocean color, bathymetry, coastal processes, and sea level rise), and terrestrial variables (LULC, forest fire, snow cover, river morphometrics, ground water, glacier mass balance, above-ground biomass, fraction of absorbed photosynthetically active radiation, gross primary productivity, leaf area index, soil moisture, forest desertification, and habitat features) (NRC 2008; Lal et al. 2021). Nevertheless, uncertainties in satellite data are frequently induced by sensor biases, retrieval techniques, and inconsistencies between satellite missions using the same sensors (Qu et al. 2013). Many climatic variables still have difficulty in capturing significant long-term trends due to the limited duration of observation series and their uncertainties (Karl et al. 2010).

2.6 Forest Ecosystem Services (FESs)

Forests provide different beneficial services to humankind including provisioning, regulating, cultural, and supporting services (Pavan and Ometto 2018). Forest ecosystem services (FESs) are supposed to contribute to alleviating poverty, specifically in rural areas of developing nations, and deprivation of ecosystem services resulting in negative impact on humans (Tallis et al. 2008). A Millennium Assessment Report (MEA 2005a, 2005b) recognized four categories of FESs (Figure 2.8). Provisioning ecosystem services comprises products derived from ecosystems such as food, which encompasses a diverse range of foodstuffs derived from plants, animals, microbes, and as a result, a range of microbial activities as well as fiber materials like wood, jute, hemp, silk, and grass (Oguh et al. 2021). Fuelwood, dung, dry leaf, biomass, algae, and bioenergy crops are used for energy generation, of which biomass accounts for 10–14% of total world energy demand (Ramachandra and Shruthi 2007). In underdeveloped nations, biomass accounts for nearly 90% of household energy use (Surendra et al. 2014). While air quality management, water regulation and purification, pollination, storm protection, regulation of human diseases, biological control of pests and diseases, and waste treatment are all examples of regulating ecosystem services (de Groot et al. 2010). Both the local and global climates are influenced by ecosystems (Wardle et al. 2011). Changes in land cover can affect both temperature and precipitation at a local scale, while ecosystems play a vital role at the global scale by either sequestering or releasing GHGs (Daba and Dejene 2018). Wetland conversion, forest

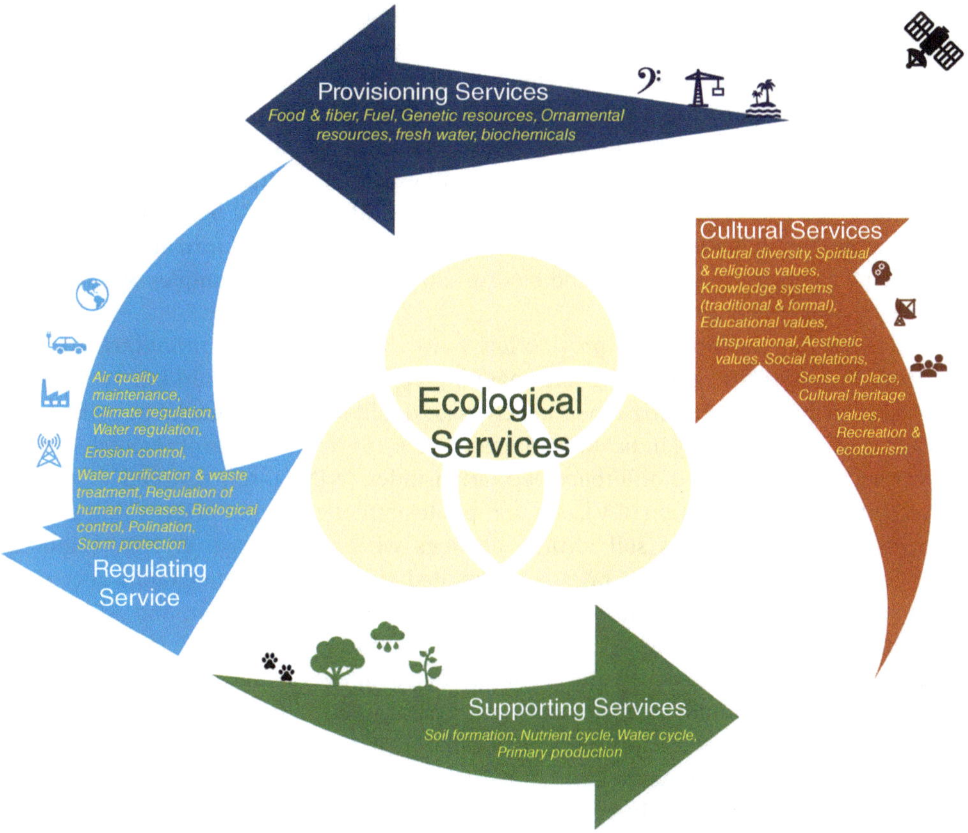

Figure 2.8 Ecosystem services. Source: Adapted from Millennium Ecosystem Assessment 2005a, b.

replacement with agriculture, or agricultural replacement with urban areas, can have a significant impact on the timing and volume of water runoff, flooding, and aquifer recharge (Burkhard et al. 2012). Vegetative cover helps in erosion control by retaining soil and preventing landslides (Lobmann et al. 2020). Forests can help with water filtration, waste treatment, and organic waste breakdown in inland waterways, and coastal and marine ecosystems (Hossain et al. 2016). Changes in ecosystems can have a direct impact on several human diseases as well as on the incidence of pests and diseases in crops and livestock that can be controlled by classical biological control, natural enemy augmentation, and conservation (Gilioli et al. 2020). Insect pollination is an important ecological service that boosts the productivity and quality of about 35% of the world's crops (Melin et al. 2014). Pollinators' distribution, abundance, and efficacy are affected by ecosystem changes, and the main benefits provided by pollinators are their contributions to the growth of a variety of agricultural and horticultural crops (Hanley et al. 2015). Spiritual enrichment, cognitive development, introspection, recreation, and esthetic experiences as well as knowledge systems, social relationships, and a feeling of place are some of the benefits that ecosystems provide to people, and are considered as the cultural ecosystem services, also called nonmaterial benefits (Winthrop 2014). On the other hand, supporting ecosystem services are distinct from other ecosystem services, helping in generating many vital ecosystem functions (Oguh et al. 2021). Services including erosion control are considered as both regulating and supporting, which also depends on the time and its intensity on people (Dominati et al. 2010). Similarly, climate control is included in the regulating services because ecosystem changes have a long-term impact on both local and global climate, which is crucial for human decision-making (MEA 2005c). Photosynthesis, on the other hand, generates oxygen, which is considered as a supporting service (Cao et al. 2017) that helps in maintaining oxygen content in the atmosphere over time. Primary production, soil formation and retention, the nitrogen and water cycle, and biodiversity habitat are all examples of other supporting services (MEA 2005c).

Remote sensing has proven to provide low costs along with fast, continuous, and frequent observations for monitoring and mapping various ecosystem services (de Araujo Barbosa et al. 2015). Regulating services such as air quality can be estimated by using aerosols and trace gases as proxy indicators (Manes et al. 2016). With the use of remotely sensed indices such as normalized difference vegetation index (NDVI), soil adjusted vegetation index (SAVI), green vegetation (GV), and non-photosynthetic vegetation (NPV) supporting services can be mapped as soil fertility declines with an increase in NPV (Numata et al. 2003). Annual soil erosion rates are estimated by integrating topographical components such as elevation, aspect, slope, LULC, vegetation fraction cover, and NDVI (Xue et al. 2018). Changes in forage provisioning services due to restoration efforts are being documented using remote sensing analysis for adaptive management (Malmstrom et al. 2009). Passive remote sensors can detect changes in natural radiation, which are closely linked with water quality parameters like turbidity, chlorophyll-a, and dissolved oxygen (Xie and Ng 2013). Assessment of cultural ecosystem services have benefited with the inclusion of spatial domain by using remote sensing based environmental indicators (Kosanic and Petzold 2020). Combining remotely sensed data with stochastic and simulated models have clear advantages in mapping food production, LULC, vegetation dynamics, soil properties, climatic variabilities, and landscape functioning, which are very complex in nature and provide insights on how each affect the ecosystem services (de Araujo Barbosa et al. 2015).

2.7 Sustainable Uses of Forests and Their Contributions to Livelihood Security

Forests are among the most versatile renewable natural resources, delivering numerous economic, social, environmental, and cultural advantages, and ecosystem services (Aju et al. 2015). They provide various regulatory services and assist in climate regulation, flood management, pollination, and biological disease control, while also playing a vital role in protecting water health and clean air. Forests provide benefits by managing rainfall, reducing biodiversity loss, and giving pollinators, carbon storage, and clean water to almost 1.6 billion individuals on the planet (IUCN 2021). Forest goods provide employment to about 90% of the total population in Indian states like Manipur, with 250 000 women employed in the collection and processing of forest products (Pandey et al. 2016). Similarly, in Chhattisgarh's Bastar district, 75% of the forest-dependent population relies on tubers, flowers, and fruits as a year-round food source (Kujur et al. 2019). Forests are vital for the human future in supporting the livelihood of local people (Langat et al. 2016) by providing food, medicine, fuel, fodder, building materials, employment, and income opportunities through MFPs (Saha and Sundriyal 2012). Collections and processing of MFPs generate employment for ~50 million people in India (Tripathi 2016). Proper management of MFPs is critical for achieving environmental, social, and economic advantages (Pandey et al. 2016). More extensive and thorough studies, as well as the preservation of forests and forest products, utilizing resources efficiently, indigenous knowledge, and abilities, are required to build a conservation strategy (Kumar and Saikia 2020). To protect existing ecosystems and species, a single forest management strategy with surrounding buffer zones should be implemented within protected areas (Bruner et al. 2001). To fulfill growing demand and ensure long-term sustainability, efforts should be made to cultivate and multiply commercially important plant species on a big scale (Kumar and Saikia 2020). Plants for fodder and fuelwood should be planted on terraced terrain, and on community land, as part of an agri-silviculture system (Sharma et al. 2009). Fuelwood should be collected from backyards or gardens, dead tree poles, fallen branches and twigs on the forest floor, and the crop residues (Dao and Hölscher 2018). Adopting sustainable harvesting procedures at the correct time will benefit resource conservation, community socioeconomic status, crop quality, and economic rewards (Pandey et al. 2016). Forest productivity, resilience, and human activities all play a vital role in the long-term viability of the ecosystem and forest services (Awasthi et al. 2003). Increased comprehensive information on the ecology of the species and its socio-ecological impacts, and develop infrastructure for bioprospecting is needed (Uprety et al. 2010). Social forestry in rural areas using multipurpose tree species will provide additional economic benefits that will reduce the pressure of fuelwood on the existing natural forests and enhance ecosystem productive capacity, soil fertility, and local climate control (Wiersum 2004). Plants and ecosystems should be given adequate time to recover by stopping ongoing human pressure (Singh 1998).

2.8 Landscape Based Approach (LbA) and Ecosystem-Based Approach (EbA) of Sustainable Forests Management (SFM)

Landscape based approach (LbA) considers lands that are owned by both government and private sectors for developmental activities to certify sustainable development and strengthen services that are provided by the ecosystem (USAID 2016). The main purpose of LbA is to maintain the integrity of ecosystems for continuous supply of ecological services with sustainable development (Grumbine 1994). These programs restore all the environmental and cultural profits without neglecting the ecological advantages supplied to the economic growth of the nation with respect to local communities (Sedjo et al. 2005). LbA embodies sustainability, multi-functionality, collective planning, placed-based interdisciplinary features, and co-governance (Arts et al. 2017). They create opportunities for different land use practices in different geographic scales (Keskitalo et al. 2016), help in strengthening the ecosystem stewardship (Chapin et al. 2010), and offer schemes for sharing of lands and giving lands on rent basis (Fischer et al. 2014).

LbA combines both the socio-economic and biophysical data (Lindenmayer et al. 2006) as a mitigation measure to climate change (Heller and Zavaleta 2009) and carbon footprint on Mother Nature (Jones et al. 2018). It also helps in regulating the impacts of climate change on humans (Staudinger et al. 2012). In decision-making processes, LbA encompasses a wide range of sections and parties (Turley 2016) that considers policies and practices together for proper functioning of the ecosystem (Cohen-Shacham et al. 2016). National Mission for a Green India has been implemented to deal with the current scenarios of deforestation and forest degradation issues (GIM 2020). Introduction of innovative models such as silviculture modeling and grazing techniques for sustainability, such as sustainable Non-Timber Forest Products (NTFPs) harvesting methods have been put forward (USAID 2016). The provisioning to monitor land surface processes through earth observation datasets aid in better understanding of complex interactions between of anthropogenic activities and natural processes and continuous monitoring of forest, shrubs, grassland. Thus, supporting framing ecosystem and nature based solutions for sustainable forest management (Nagabhatla et al. 2012).

Nature-based solutions (NbSs) are a broad category that encompasses a variety of approaches based on ecosystem services (Cohen et al. 2018), sustainable growth (Maes and Jacobs 2015), risk management of ecosystem disasters, infrastructure, and climatic solutions (Griscom et al. 2017). IUCN describes NbSs for sustainable forest management as an "action to maintain, sustainably manage, and restore natural or modified ecosystems in order to more efficiently and adaptively solve societal concerns while also delivering human well-being and biodiversity advantages" (Cohen-Shacham et al. 2016). The world's 50% GDP mainly depends on nature and development of nature-based strategies in recovery of degraded forests for economic sustainability (Quinney 2020). Restoration of natural resources helps in increasing overall landscape connectivity (Keestra et al. 2018) and assures greater productivity (Maes and Jacobs 2015) by providing numerous ecological and socio-economic advantages (European Commission 2016). Increasing biodiversity, improving ecological goods and services, promoting human health and well-being, and increasing local participation (Kabisch et al. 2016) are some of the major services set by nature-based action plans. These solutions have the potential to mitigate ~30% of the world's climate change phenomenon (Seddon et al. 2019). NbSs accelerate climate modification processes through advanced planning and policymaking (Frantzeskaki 2019). Empowering NbSs help communities to rebuild their environment by decreasing the exposure to flooding, involving locals in capacity building to control local environments, and by strengthening the potential to combat climatic shocks (Lavorel et al. 2019). International bodies initiated various programs in collaboration with different countries to resolve societal challenges through NbSs (Eggermont et al. 2015). REDD+, IUCN, UN Climate Action Summit (NBS for Climate Manifesto 2019) are some of the organizations that deal with climate change scenarios with respect to nature-based action plans and policies. NbSs majorly focused on green and blue infrastructure (Albert et al. 2017), ecosystem-based strategies, and ecological engineering (Borsje et al. 2011). The European Union has created a net zero society by 2050 program based on NbSs under EU Green Deal (European Commission 2016) and Horizon 2020 program for innovation, research, and development (Maes and Jacobs 2015).

2.9 Conclusions

Forests provide immense advantages to all living beings, while indigenous people living in and around forests are inextricably connected to the forest for their survival and livelihood. The unlawful harvesting of forest resources in an unsustainable manner has led to forest degradation all around the world with the increase in population growth and decrease in per capita land availability. Grazing, collection of fuelwoods, MFPs, forest fires, invasive species, and climate change are the major ecological disturbances of forests leading to forest degradation. Climate change affects photosynthesis, respiration, precipitation, water balance, atmospheric CO_2 concentration, species survivability, and predation activities, resulting in changes in gene pools, biodiversity loss, and forest productivity and regeneration. Forest conservation is a current need that can be met by ecosystem-based approaches (EbAs), sustainable forest management, and through comprehensive forest resource management, that addresses ecological, economic, and socio-cultural functions. Ecological engineering, new models, and sustainability strategies in forestry and agriculture can help to reap the full spectrum of economic, social, and environmental advantages while retaining a variety of forest functions in a landscape setting. Integrating geospatial approaches and remote sensing land-use methods can help to support sustainable agricultural and human well-being by improving forest health, accurate evaluation of forest productivity, soil stabilization, climate change, forest fire detection and control, post-fire vegetation recovery monitoring, harvest planning, emissions monitoring, and managing water flows. It can also be achieved by raising public awareness, scientific understanding, and sustainable forest management through government initiatives, and policy intervention, resulting in the preservation and enrichment of forest resources for the benefit of society. Sustainable forest management improves ecological and social resilience, as well as the role of forests in carbon uptake, storage, and other environmental advantages. Forests will help communities and societies to flourish along with conserving biodiversity by providing food, wood energy, shelter, fodder, fiber, and creating revenue and livelihood.

References

Abdollahnejad, A., Panagiotidis, D., Shataee Joybari, S., and Surový, P. (2017). Prediction of dominant forest tree species using Quickbird and environmental data. *Forests* 8: 42).

Ahuja, A. and Bhaskar, S. (2021). *A Forest officer from Jharkhand comes up with an idea to stop forest fires, illegal tree felling and creates jobs*. Available at URL: https://swachhindia.ndtv.com/a-forest-officer-from-jharkhand-comes-up-with-an-idea-to-stop-forest-fires-illegal-tree-felling-and-creates-jobs-58384. [Accessed on 27th December 2021].

Aju, P.C., Iwuchukwu, J.J.I., and Ibe, C.C. (2015). Our forests, our environment, our sustainable livelihoods. *European Journal of Academic Essays* 2 (4): 6–19.

Albert, C., Spangenberg, J., and Schröter, B. (2017). Nature-based solutions: criteria. *Nature* 543: 1–315.

Ali, I., Cawkwell, F., Dwyer, E. et al. (2016). Satellite remote sensing of grasslands: From observation to management. *Journal of Plant Ecology* 9 (6): 649–671. https://doi.org/10.1093/jpe/rtw005.

Allen, J.A., Pimentel, D.P., and Lassoie, J.P. (1988). Fuelwood production and use in rural Swaziland: a case study of two communities. *Forest Ecology and Management* 25 (3,4): 239–254.

Amadi, N.S. and Aleru, P.D. (2021). Influence of open cattle grazing on crop production among rural farmers in rivers state. *Global Journal of Scientific and Research Publications* 1 (4): 1–6.

Andersen, H.E., Reutebuch, S.E., McGaughey, R.J. et al. (2014). Monitoring selective logging in western Amazonia with repeat LiDAR flights. *Remote Sensing of Environment* 151: 157–165.

de Araujo Barbosa, C.C., Atkinson, P.M., and Dearing, J.A. (2015). Remote sensing of ecosystem services: a systematic review. *Ecological Indicators* 52: 430–443.

Arima, E.Y., Walker, R.T., Perz, S.G., and Caldas, M. (2005). Loggers and forest fragmentation: behavioral models of road building in the Amazon Basin. *Annals of the Association of American Geographers* 95 (3): 525–541.

Arts, B., Buizer, M., Horlings, L. et al. (2017). Landscape approaches: a state-of-the-art review. *Annual Review of Environment and Resources* 42: 439–463.

Awasthi, A., Uniyal, S.K., Rawat, G.S., and Rajvanshi, A. (2003). Forest resource availability and its use by the migratory villages of Uttarkashi, Garhwal Himalaya (India). *Forest Ecology and Management* 174: 13–24.

Bahuguna, V.K. and Singh, S. (1999). India: Forest Fire Prevention and Control Strategies (IFFN No. 20)-GFMC. Available at URL: https://gfmc.online/iffn/country/in/in_1.html

Balachander, D. and Ganesan, M. (1993). Extraction of non-timber forest products, including fodder and fuelwood, in Mudumalai. *Economic Botany* 47 (3): 268–274.

Banerjee, A. and Madhurima, C. (2013). Forest degradation and livelihood of local communities in India: a human rights approach. *Journal of Horticulture and Forestry* 5 (8): 122–129.

Banskota, A., Kayastha, N., Falkowski, M.J. et al. (2014). Forest monitoring using landsat time series data: a review. *Canadian Journal of Remote Sensing* 40: 362–384.

Barmpoutis, P., Papaioannou, P., Dimitropoulos, K., and Grammalidis, N. (2020). A review on early forest fire detection systems using optical remote sensing. *Sensors* 20: 6442.

Bellard, C., Thuiller, W., Leroy, B. et al. (2013). Will climate change promote future invasions? *Global Change Biology* 19 (12): 3740–3748.

Belnap, J. and Gillette, D.A. (1998). Vulnerability of desert biological soil crusts to wind erosion: the influences of crust development, soil texture, and disturbance. *Journal of Arid Environments* 39: 133–142.

Bennett, A.E., Thomsen, M., and Strauss, S.Y. (2011). Multiple mechanisms enable invasive species to suppress native species. *American Journal of Botany* 98: 1086–1094.

Borsje, B.W., Wesenbeeck, B.K., Dekker, F. et al. (2011). How ecological engineering can serve in coastal protection. *Ecological Engineering* 37: 113–122.

Brodt, S., Six, J., Feenstra, G. et al. (2011). Sustainable agriculture. *Nature Education Knowledge* 3 (10): 1.

Bruner, A.G., Gullison, R.E., Rice, R.E., and Da Fonseca, G.A. (2001). Effectiveness of parks in protecting tropical biodiversity. *Science* 291: 125–128.

Burkhard, B., Kroll, F., Nedkov, S., and Müller, F. (2012). Mapping ecosystem service supply, demand and budgets. *Ecological Indicators* 21: 17–29.

Cao, W., Li, R., Chi, X. et al. (2017). Island urbanization and its ecological consequences: a case study in Zhoushan Island, East China. *Ecological Indicators* 76: 1–14.

Central Silk Board (2017). *Tasar Value Chain Analysis*. Mahila Kisan Sashaktikaran Pariyojana (MKSP), Central Silk Board, Bangalore and Professional Assistance for Development Action, 1–153.

Chandra, K.K. and Bhardwaj, A.K. (2015). Incidence of forest fire in India and its effect on terrestrial ecosystem dynamics, nutrient and microbial status of soil. *International Journal of Agriculture and Forestry* 5 (2): 69–78.

Chandramolly and Islam, M.A. (2015). Fuelwood, fodder, and timber consumption status in a forest fringe tribal society of Jharkhand. *International Journal of Forestry and Crop Improvement* 6 (1): 71–76.

Chapin, F.S. III, Carpenter, S.R., Kofinas, G.P. et al. (2010). Ecosystem stewardship: sustainability strategies for a rapidly changing planet. *Trends in Ecology Evolution* 25: 241–249.

Chevrel, M., Courtois, M., and Weill, G. (1981). The SPOT satellite remote sensing mission. *Photogrammetric Engineering and Remote Sensing* 47: 1163–1171.

Cohen, W.B., Yang, Z., Healey, S.P. et al. (2018). A LandTrendr multispectral ensemble for forest disturbance detection. *Remote Sensing of Environment* 205: 131–140.

Cohen-Shacham, E., Walters, G., Janzen, C., and Maginnis, S. (ed.) (2016). *Nature-Based Solutions to Address Global Societal Challenges*, 1–115. Gland, Switzerland: IUCN https://portals.iucn.org/library/sites/library/files/documents/2016-036.pdf.

Crotty, F.V., Fychan, R., Theobald, V.J. et al. (2014). The impact of using alternative forages on the nutrient value within slurry and its implications for forage productivity in agricultural systems. *PLoS One* 9 (5): 1–12.

Daba, M.H. and Dejene, S.W. (2018). The role of biodiversity and ecosystem services in carbon sequestration and its implication for climate change mitigation. *International Journal of Environmental Sciences and Natural Resources* 11 (2): 1–10.

Dale, V.H., Joyce, L.A., McNulty, S., and Neilson, R.P. (2000). The interplay between climate change, forests, and disturbances. *Science of the Total Environment* 262: 201–204.

Dale, V.H., Joyce, L.A., McNulty, S. et al. (2001). Climate change and forest disturbances. *BioScience* 51 (9): 723–734.

D'Antonio, C.M. and Vitousek, P.M. (1992). Biological invasions by exotic grasses, the grass/fire cycle, and global change. *Annual Review of Ecology, Evolution, and Systematics* 23: 63–87.

Dao, T.H.H. and Hölscher, D. (2018). Impact of non-timber forest product use on the tree community in North-Western Vietnam. *Forest* 9 (431): 1–15.

Darwin, C. (1859). *On the Origin of Species by Means of Natural Selection, or the Preservation of Favoured Races in the Struggle for Life*, 1–502. London: John Murray.

Deb, S., Lunarah, M., and Tiwari, B.K. (2013). Technological innovations in shifting agricultural practices by three tribal farming communities of Meghalaya, Northeast India. *Tropical Ecology* 54 (2): 133–148.

Diaz, S., Lavorel, S., de Bello, F., and Que'tier, F. (2007). Incorporating plant functional diversity effects in ecosystem service assessments. *Proceedings of the National Academy of Sciences* 104 (52): 20684–20689.

Dominati, E., Patterson, M., and Mackay, A. (2010). A framework for classifying and quantifying the natural capital and ecosystem services of soils. *Ecological Economics* 69 (9): 1858–1868.

Dove, N.C. and Keeton, W.S. (2015). Structural complexity enhancement increases fungal species richness in Northern hardwood forests. *Fungal Ecology* 13: 181–192.

Dueñas, M.A., Hemming, D.J., Roberts, A., and Diaz Soltero, H. (2021). The threat of invasive species to IUCN-listed critically endangered species: a systematic review. *Global Ecology and Conservation* 26: 1–7.

Early, R., Bradley, B.A., Dukes, J.S. et al. (2016). Global threats from invasive alien species in the twenty-first century and national response capacities. *Nature Communications* 7: 1–9.

Eggermont, H., Balian, E., Azevedo, J.M.N. et al. (2015). Nature-based solutions: new influence for environmental management and research in Europe. *GAIA* 24: 243–248.

Elliott, S.D., Blakesley, D., and Hardwick, K. (2013). *Understanding tropical forests.* In: *Restoring Tropical Forests: A Practical Guide* (ed. S. Richmond), 1–344. Royal Botanic Gardens, Kew: Kew Publishing.

European Commission (2016). *Policy Topics: Nature-Based Solutions.* Available at URL https://ec.europa.eu/research/environment/index.cfm?Pg=nbs.

FAO (2014). *Global plan of action for the conservation, sustainable use and development of forest genetic resources.* Commission on Genetic Resources for Food and Agriculture, pp. 1–31.

Filazzola, A., Brown, C., Dettlaff, M.A. et al. (2020). The effects of livestock grazing on biodiversity are multi-trophic: a meta-analysis. *Ecology Letters* 23 (8): 1298–1309.

Fischer, J., Abson, D.J., Butsic, V. et al. (2014). Land sparing versus land sharing: moving forward. *Conservation Letters* 7: 149–157.

Flannigan, M., Stocks, B., Turetsky, M., and Wotton, M. (2009). Impacts of climate change on fire activity and fire management in the circumboreal forest. *Global Change Biology* 15 (3): 549–560.

Forkuor, G., Zoungrana, J.-B.B., Dimobe, K., and Ouattara, B. (2020). Above-ground biomass mapping in West African dryland forest using Sentinel-1 and -2 datasets – a case study. *Remote Sensing of Environment* 236: 111496.

Foxcroft, L.C., Pyšek, P., Richardson, D.M. et al. (2017). Plant invasion science in protected areas: progress and priorities. *Biological Invasions* 19: 1353–1378.

Franklin, J.F., Spies, T.A., van Pelt, R. et al. (2002). Disturbances and structural development of natural forest ecosystems with silvicultural implications, using Douglas-fir forests as an example. *Forest Ecology and Management* 155 (1–3): 399–423.

Frantzeskaki, N. (2019). Seven lessons for planning nature-based solutions in cities. *Environmental Science and Policy* 93: 101–111.

Frison, P.-L., Fruneau, B., Kmiha, S. et al. (2018). Potential of Sentinel-1 data for monitoring temperate mixed forest phenology. *Remote Sensing* 10: 2049.

Fuentes, N., Saldana, A., Kuhn, I., and Klotz, S. (2015). Climatic and socio-economic factors determine the level of invasion by alien plants in Chile. *Plant Ecology & Diversity* 8: 371–377.

Georgiadis, N.J., Ruess, R.W., McNaughton, S.J., and Western, D. (1989). Ecological conditions that determine when grazing stimulates grass production. *Oecologia* 81: 316–322.

Ghiyamat, A. and Shafri, H.Z. (2010). A review on hyperspectral remote sensing for homogeneous and heterogeneous forest biodiversity assessment. *International Journal of Remote Sensing* 31: 1837–1856.

Gibson, R., Danaher, T., Hehir, W., and Collins, L. (2020). A remote sensing approach to mapping fire severity in South-Eastern Australia using Sentinel-2 and random forest. *Remote Sensing of Environment* 240: 111702.

Gil, J.D.B., Reidsma, P., Giller, K., and Todman, L. (2019). Sustainable development goal 2: improved targets and indicators for agriculture and food security. *Ambio* 48 (7): 685–698.

Gilioli, G., Orlando, F., Ghiglieno, I., and Sperandio, G. (2020). Biodiversity and sustainable agriculture. In Guasta, L.T. (ed.). *Equilibri Rivista per lo sviluppo sostenibile* 24 (1): 69–76.

Glatzel, G. (1999). Historic forest use and its possible implication to recently accelerated tree growth in Central Europe. In: *Causes and Consequences of Accelerated Tree Growth in Europe. EFI Proceedings*, vol. 27 (ed. T. Karjalainen, H. Spieker and O. Laroussine), 65–74.

Griscom, B.W., Adams, J., Ellis, P.W. et al. (2017). Natural climate solutions. *Proceedings of the National Academy of Sciences of the United States of America* 114: 11645–11650.

de Groot, R.S., Alkemade, R., Braat, L. et al. (2010). Challenges in integrating the concept of ecosystem services and values in landscape planning, management and decision making. *Ecological Complexity* 7 (3): 260–272.

Grumbine, R.E. (1994). What is ecosystem management? *Conservation Biology* 8: 27–38.

Hanley, N., Breeze, T.D., Ellis, C., and Goulson, D. (2015). Measuring the economic value of pollination services: principles, evidence and knowledge gaps. *Ecosystem Services* 14: 124–132.

Hansen, M.C., Potapov, P.V., Moore, R. et al. (2013). High-resolution global maps of 21st-century forest cover change. *Science* 342 (6160): 850–853.

Harrison, R. (2002). *Forests: The Shadow of Civilization*. Chicago, IL: The University of Chicago Press.

Hartmann, H. (2011). Will a 385 million year struggle for light become a struggle for water and for carbon? How trees may cope with more frequent climate change type drought events. *Global Change Biology* 17 (1): 642–655.

Heller, N.E. and Zavaleta, E.S. (2009). Biodiversity management in the face of climate change: a review of 22 years of recommendations. *Biological Conservation* 142: 14–32.

Hellmann, J.J., Byers, J.E., Bierwagen, B.G., and Dukes, J.S. (2008). Five potential consequences of climate change for invasive species. *Conservation Biology* 22: 534–543.

Heltberg, R., Arndt, T.C., and Sekha, N.U. (2000). Fuelwood consumption and forest degradation: a household model for domestic energy substitution in rural India. *Land Economics* 76 (2): 213–232.

Hernandez-Leal, P.A., Arbelo, M., and Gonzalez-Calvo, A. (2006). Fire risk assessment using satellite data. *Advances in Space Research* 37 (4): 741–746.

Hossain, M.I., Nabi, M.R., Ansari, M.N.A. et al. (2016). Ecosystem services of the world largest mangrove forest Sundarban in Bangladesh. *International Journal of Innovation and Scientific Research* 27 (1): 9–15.

Houghton, J., Townshend, J., Dawson, K. et al. (2012). The GCOS at 20 years: the origin, achievement and future development of the global climate observing system. *Weather* 67: 227–235.

Huntly, N. (1991). Herbivores and the dynamics of communities and ecosystems. *Annual Review of Ecology and Systematics* 22: 477–503.

Inderjit, P. J., van Kleunen, M. et al. (2018). Naturalized alien flora of the Indian states: biogeographic patterns, taxonomic structure and drivers of species richness. *Biological Invasions* 20: 1625–1638.

IPBES (2019). Global Assessment Report on Biodiversity and Ecosystem Services of the Intergovernmental Science-policy Platform on Biodiversity and Ecosystem Services. In: (ed. E.S. Brondizio, J. Settele, S. Diaz and H.T. Ngo), 1–30. Bonn, Germany: *IPBES Secretariat* Available at URL: https://www.intosaicommunity.net/document/articlelibrary/IPBES_Global_Assessment_Report_on_Biodiversity_and_Ecosystem_Services_2019_Link_to_contents.pdf.

IPCC (2007). *Fourth assessment report of the intergovernmental panel on climate change* (pp. 374). Available at URL: https://www.ipcc.ch/report/ar4/wg1

ISFR (2019). Welcome To Forest Survey of India (Technical Information Series) Vol No 1 & 2, Edition 16. *Forest Survey of India*. https://fsi.nic.in/forest-report-2019

IUCN (2021). *Issues Briefs: Forests and Climate Change*. Available at URL: https://www.iucn.org/sites/dev/files/forests_and_climate_change_issues_brief_2021.pdf [Accessed on 26th December 2021

Jang, W., Keyes, C.R., and Page-Dumroese, D.S. (2015). Long-term effects on distribution of forest biomass following different harvesting levels in the Northern rocky mountains. *Forest Ecology and Management* 358: 281–290.

Jansen, V.S., Kolden, C.A., Schmalz, H.J. et al. (2021). Using satellite-based vegetation data for short-term grazing monitoring to inform adaptive management. *Rangeland Ecology & Management* 76: 30–42.

Jones, K.R., Venter, O., Fuller, R.A. et al. (2018). One-third of global protected land is under intense human pressure. *Science* 360: 788–791.

Justice, C.O., Giglio, L., Korontzi, S. et al. (2002). The MODIS fire products. *Remote Sensing of Environment* 83 (1–2): 244–262.

Kabisch, N., Qureshi, S., and Haase, D. (2016). Human–environment interactions in urban green spaces–a systematic review of contemporary issues and prospects for future research. *Environmental Impact Assessment Review* 50: 25–34.

Karl, T.R., Diamond, H.J., Bojinski, S. et al. (2010). Observation needs for climate information, prediction and application: capabilities of existing and future observing systems. *Procedia Environmental Sciences* 1: 192–205.

Kayitakire, F., Hamel, C., and Defourny, P. (2006). Retrieving forest structure variables based on image texture analysis and IKONOS-2 imagery. *Remote Sensing of Environment* 102: 390–401.

Keestra, S., Nunes, J., Novara, A. et al. (2018). The superior effect of nature based solutions in land management for enhancing ecosystem services. *Science of the Total Environment* 610, 611: 997–1009.

Keifer, M., Caprio, E.A., and Lineback, P. (2000). *Incorporating a GIS Model of Ecological Need into Fire Management Planning*. University of Idaho.

Kerketta, J.K., Singh, S., and Kumar, B. (2018). Effect of silvicultural treatments on quantity and quality assessment of Tendu (*Diospyros melanoxylon* Roxb.) leaves. *Journal of Pharmacognosy and Phytochemistry* 2018: 1317–1322.

Keskitalo, E.C.H., Horstkotte, T., Kivinen, S. et al. (2016). Generality of mis-fit? The real-life difficulty of matching scales in an interconnected world. *Ambio* 45: 742–752.

King, R.B. (2002). Land cover mapping principles: a return to interpretation fundamentals. *International Journal of Remote Sensing* 23 (18): 3525–3545.

Kishore, B.S.P.C., Kumar, A., Saikia, P. et al. (2020). Major forests and plant species discrimination in Mudumalai forest region using airborne hyperspectral sensing. *Journal of Asia-Pacific Biodiversity* 13: 637–651.

Kosanic, A. and Petzold, J. (2020). A systematic review of cultural ecosystem services and human wellbeing. *Ecosystem Services* 45: 101168.

Kujur, D., Dutta, J., and Ranga, M.M. (2019). Non wood forest products for rural livelihood and sustainable forestry with special reference to Chhattisgarh. *India. Pacific International Journal* 2 (3): 28–32.

Kumar, A. (2022). Impact of shifting cultivation on green infrastructure: a remote sensing perspective. *International Journal of Ecology and Environmental Sciences.* (In press).

Kumar, D. (2011). Monitoring forest cover changes using remote sensing and GIS: a global perspective. *Research Journal of Environmental Sciences* 5 (2): 105–123.

Kumar, R. and Saikia, P. (2020). *Forests resources of Jharkhand, Eastern India: socio-economic and bioecological perspectives.* In: *Socio-Economic and Eco-Biological Dimensions in Resource Use and Conservation-Strategies for Sustainability* (ed. N. Roy, S. Roychoudhury, S. Nautiyal, et al.), 61–101. Switzerland: Springer International Publishing.

Kumar, A., Pandey, A.C., and Jeyaseelan, A.T. (2012). Built-up and vegetation extraction and density mapping using WorldView-II. *Geocarto International* 27 (7): 557–568.

Kumar, A., Kishore, B., Saikia, P. et al. (2019). Tree diversity assessment and above ground forests biomass estimation using SAR remote sensing: a case study of higher altitude vegetation of North-East Himalayas, India. *Physics and Chemistry of the Earth, Parts A/B/C* 111: 53–64.

Kumar, G., Kumari, R., Kishore, B., and Saikia, P. (2020). *Climate change impacts and implications: an Indian perspective.* In: *Socio-Economic and Eco-Biological Dimensions in Resource Use and Conservation-Strategies for Sustainability* (ed. N. Roy, S. Roychoudhury, S. Nautiyal, et al.), 11–30. Switzerland: Springer International Publishing.

Kumari, R., Banerjee, A., Kumar, R. et al. (2019). *Deforestation in India: consequences and sustainable solutions.* In: *Forest Degradation Around the World* (ed. M.N. Suratman, Z.A. Latif, G.D. Oliveira, et al.), 1–18. London, UK: IntechOpen.

Laamanen, R. and Kangas, A. (2011). Large-scale forest owner's information needs in operational planning of timber harvesting-some practical views in Metsähallitus, Finnish state-owned enterprise. *Silva Fennica* 45 (4): 711–727.

Lal, P., Kumar, A., Saikia, P. et al. (2021a). Effect of vegetation structure on above ground biomass in tropical deciduous forests of central India. *Geocarto International* https://doi.org/10.1080/10106049.2021.1936213.

Lal, P., Shekhar, A., and Kumar, A. (2021b). Quantifying temperature and precipitation change caused by land cover change: a case study of India using WRF model. *Frontiers in Environmental Science.* https://doi.org/10.3389/fenvs.2021.766328.

Lam, N.L., Smith, K.R., Gauthier, A., and Bates, M.N. (2012). Kerosene: a review of household uses and their hazards in low- and middle-income countries. *Journal of Toxicology and Environmental Health: Part B, Critical Reviews* 15 (6): 396–432.

Langat, D.K., Maranga, E.K., Aboud, A.A., and Cheboiwo, J.K. (2016, 2016). Role of forest resources to local livelihoods: the case of East Mau forest ecosystem, Kenya. *International Journal of Forestry Research* (4): 1–10.

Lattimore, B., Smith, C.T., Titus, B.D. et al. (2009). Environmental factors in woodfuel production: opportunities, risks, and criteria and indicators for sustainable practices. *Biomass and Bioenergy* 33 (10): 1321–1342.

Lavorel, S., Colloff, M.J., Locatelli, B. et al. (2019). Mustering the power of ecosystems for adaptation to climate change. *Environmental Science & Policy* 92: 87–97.

Li, J., Wang, M.-H., and Ho, Y.-S. (2011). Trends in research on global climate change: a science citation index expanded-based analysis. *Global and Planetary Change* 77: 13–20.

Lindenmayer, D.B., Franklin, J.F., and Fischer, J. (2006). General management principles and a checklist of strategies to guide forest biodiversity conservation. *Biological Conservation* 131: 433–445.

Linser, S., Wolfslehner, B., Asmar, F. et al. (2018). 25 years of criteria and indicators for sustainable forest management: why some intergovernmental C&I processes flourished while others faded. *Forests* 9 (9): 515.

Lobmann, M.T., Tonin, R., Stegemann, J. et al. (2020). Towards a better understanding of shallow erosion resistance of subalpine grasslands. *Journal of Environmental Management* 276: 111267.

Loehman, R.A., Bentz, B.J., DeNitto, G.A. et al. (2018). Effects of climate change on ecological disturbance in the Northern rockies. In: *Climate Change and Rocky Mountain Ecosystems, Advances in Global Change Research, 63* (ed. J. Halofsky and D. Peterson), 115–141.

Lovric, M., Martinic, I., Lovric, N. et al. (2010). *Assessment of progress of forests in Croatia towards sustainable forest management. SEEFOR, South-east European Forestry* 1 (2): 51–59.

Lubchenco, J. and Karl, T.R. (2012). Predicting and managing extreme weather events. *Physics Today* 65: 1–31.

Lucas, R., Bunting, P., Paterson, M., and Chisholm, L. (2008a). Classification of Australian forest communities using aerial photography, CASI and HyMap data. *Remote Sensing of Environment* 112: 2088–2103.

Lucas, R.M., Lee, A.C., and Bunting, P.J. (2008b). Retrieving forest biomass through integration of CASI and LiDAR data. *International Journal of Remote Sensing* 29: 1553–1577.

MacDicken, K.G., Sola, P., Hall, J.E. et al. (2015). Global progress toward sustainable forest management. *Forest Ecology and Management* 352: 47–56.

Maes, J. and Jacobs, S. (2015). Nature-based solutions for Europe's sustainable development. *Conservation Letters* 10: 121–124.

Malhi, Y., Roberts, J.T., Betts, R.A. et al. (2008). Climate change, deforestation, and the fate of the Amazon. *Science* 319: 169–172.

Malmstrom, C.M., Butterfield, H.S., Barber, C. et al. (2009). Using remote sensing to evaluate the influence of grassland restoration activities on ecosystem forage provisioning services. *Restoration Ecology* 17: 526–538.

Manes, F., Marando, F., Capotorti, G. et al. (2016). Regulating ecosystem services of forests in ten Italian metropolitan cities: air quality improvement by PM10 and O3 removal. *Ecological Indicators* 67: 425–440.

Marshall, E., Newton, A.C., and Schreckenberg, K. (2003). Commercialisation of non-timber forest products: first steps in analysing the factors influencing success. *International Forestry Review* 5 (2): 128–137.

Martin, M.E. and Aber, J.D. (1997). High spectral resolution remote sensing of forest canopy lignin, nitrogen, and ecosystem processes. *Ecological Applications* 7: 431–443.

Melin, A., Rouget, M., Midgley, J.J., and Donaldson, J.S. (2014). Pollination ecosystem services in South African agricultural systems. *South African Journal of Science* 110 (11, 12): 1–9.

Millennium Ecosystem Assessment (MEA) (2005a). *Ecosystem and Human Well-Being: Synthesis*. Washington, DC: Island Press.

Millennium Ecosystem Assessment (MEA) (2005b). *Ecosystem and Human Well-Being: A Manual for Practitioners*. Washington, DC: Island Press.

Millennium Ecosystem Assessment (MEA). (2005c). *Ecosystems and Human Well-being: A framework for assessment, Chapter 2: Ecosystems and their services*. Washington, DC, Island Press.

Ministry of Petroleum and Natural Gas, Government of India (MPNG, GoI) (2021). *Pradhan Mantri Ujjwala Yojana 2.0*. Available at URL: www.pmuy.gov.in [Accessed on 27th December 2021].

Misra, G., Cawkwell, F., and Wingler, A. (2020). Status of phenological research using Sentinel-2 data: a review. *Remote Sensing* 12: 2760.

Mukul, S.A., Uddin, M.B., and Tito, M.R. (2006). Study on the status and various uses of invasive alien plant species in and around Satchari National Park, Sylhet, Bangladesh. *Tigerpaper* 33: 28–32.

Munang, R.T., Thiaw, I., and Rivington, M. (2011). Ecosystem management: tomorrow's approach to enhancing food security under a changing climate. *Sustainability* 3 (7): 937–954.

Muukkonen, P. and Heiskanen, J. (2005). Estimating biomass for boreal forests using ASTER satellite data combined with standwise forest inventory data. *Remote Sensing of Environment* 99: 434–447.

Nagabhatla, N., Sellamuttu, S.S., Bobba, A.G. et al. (2012). Insight to ecosystem based approach (EBA) at landscape level using a geospatial medium. *Journal of the Indian Society of Remote Sensing* 40: 47–64.

Narendran, K. (2001). Forest fires: origins and ecological paradoxes. *Resonance* 6 (11): 34–41.

Nasi, R., Wunder, S., and Campos, J.J. (2002). *Forest ecosystem services: Can they pay our way out of deforestation?* [Working Paper]. CIFOR for the Global Environmental Facility (GEF). https://cgspace.cgiar.org/handle/10568/18673

National Mission for Green India (GIM) (2020). Available at URL: www.indiascienceandtechnology.gov.in/st-visions/national-mission/national-mission-green-india-gim

National Research Council (2008). *Earth Observations from Space: The First 50 Years of Scientific Achievements*. Washington, DC: The National Academies Press https://doi.org/10.17226/11991.

Nautiyal, S. and Kaechele, H. (2008). Fuel switching from wood to LPG can benefit the environment. *Environmental Impact Assessment Review* 28 (8): 523–532.

Nepstad, D., Verissimo, A., Alencar, A. et al. (1999). Large-scale impoverishment of Amazonian forests by logging and fire. *Nature* 398: 505–508.

North, M.P. and Keeton, W.S. (2008). Emulating natural disturbance regimes: an emerging approach for sustainable forest management. In: *Patterns and Processes in Forest Landscapes* (ed. R. Lafortezza, J. Chen, G. Sanesi and T. Crow), 341–372. Dordrecht: Springer.

Numata, I., Soares, J.V., Roberts, D.A. et al. (2003). Relationships among soil fertility dynamics and remotely sensed measures across pasture chronosequences in Rondônia, Brazil. *Remote Sensing of Environment* 87: 446–455.

Oguh, C.E., Obiwulu, E.N.O., Umezinwa, O.J. et al. (2021). Ecosystem and ecological services; need for biodiversity conservation-a critical review. *Asian Journal of Biology* 11 (4): 1–14.

Öllerer, K., Varga, A., Kirby, K. et al. (2019). Beyond the obvious impact of domestic livestock grazing on temperate forest vegetation–a global review. *Biological Conservation* 237: 209–219.

Oregon State University (2021). *Define forages and differentiate between forage types*. Available at URL: https://forages.oregonstate.edu/nfgc/eo/onlineforagecurriculum/instructormaterials/availabletopics/introduction/defineforages [Accessed on 30th November 2021].

Pandey, A.K., Tripathi, Y.C., and Kumar, A. (2016). Non timber forest products (NTFPs) for sustained livelihood: challenges and strategies. *Research Journal of Forestry* 10: 1–7.

Parrotta, J.A., Fui, H.L., Jinlong, L. et al. (2009). Traditional forest-related knowledge and sustainable forest management in Asia. *Forest Ecology and Management* 257: 1987–1988.

Pavan, A.L.R. and Ometto, A.R. (2018). Ecosystem services in life cycle assessment: a novel conceptual framework for soil. *Science of the Total Environment* 643: 1337–1347.

Peerbhay, K., Mutanga, O., Lottering, R. et al. (2016). Detecting bugweed (*Solanum mauritianum*) abundance in plantation forestry using multisource remote sensing. *ISPRS Journal of Photogrammetry and Remote Sensing* 121: 167–176.

Peprah, S., Darambazar, E., Biligetu, B. et al. (2021). Harvest date effect on forage yield, botanical composition, and nutritive value of novel legume-grass mixtures. *Agronomy* 11 (2184): 1–14.

Phiri, D., Simwanda, M., Salekin, S. et al. (2020). Sentinel-2 data for land cover/use mapping: a review. *Remote Sensing* 12: 2291.

Pietikäinen, V. (2006). *Measures to Prevent Overstocking and Overgrazing in Woodlands: A Case Study in Babati, Northern Tanzania*, 1–22. Södertörn University College.

Powell, K.I., Chase, J.M., and Knight, T.M. (2011). A synthesis of plant invasion effects on biodiversity across spatial scales. *American Journal of Botany* 98: 539–548.

Qu, J.J., Powell, A.M., and Sivakumar, M.V.K. (ed.) (2013). *Satellite-Based Applications on Climate Change*. Dordrecht: Springer.

Quinney, M. (2020). *COVID-19 and nature are linked. So should be the recovery*. World Economic Forum. Available at URL: https://www.weforum.org/agenda/2020/04/covid-19-nature-deforestation-recovery

Rai, P.K. and Singh, J.S. (2020). Invasive alien plant species: their impact on environment, ecosystem services and human health. *Ecological Indicators* 111: 1–20.

Rai, P.K., Kumar, V., Lee, S. et al. (2018). Nanoparticle-plant interaction: implications in energy, the environment, and agriculture. *Environment International* 119: 1–19.

Ramachandra, T.V. (2010). Mapping of fuelwood trees using geoinformatics. *Renewable and Sustainable Energy Reviews* 14 (2): 642–654.

Ramachandra, T.V. and Shruthi, B.V. (2007). Spatial mapping of renewable energy potential. *Renewable and Sustainable Energy Reviews* 11 (7): 1460–1480.

Reddy, C.S., Khuroo, A.A., Krishna, P.H. et al. (2014). Threat evaluation for biodiversity conservation of forest ecosystems using geospatial techniques: a case study of Odisha, India. *Ecological Engineering* 69: 287–303.

Reddy, C.S., Padma Alekhya, V.V.L., Saranya, K.R.L. et al. (2017). Monitoring of fire incidences in vegetation types and protected areas of India: implications on carbon emissions. *Journal of Earth System Science* 126 (1): 11.

Reiche, J., Mullissa, A., Slagter, B. et al. (2021). Forest disturbance alerts for the Congo Basin using Sentinel-1. *Environmental Research Letters* 16: 024005.

Roteta, E., Bastarrika, A., Padilla, M. et al. (2019). Development of a Sentinel-2 burned area algorithm: generation of a small fire database for sub-Saharan Africa. *Remote Sensing of Environment* 222: 1–17.

Roy, P.S. (2003). Forest fire and degradation assessment using satellite remote sensing and geographic information system. In: *Satellite Remote Sensing and GIS Applications in Agricultural Meteorology* (ed. M.V.K. Sivakumar, P.S. Roy, K. Harmsen and S.K. Saha), 361–400. World Meteorological Organisation https://library.wmo.int/doc_num.php?explnum_id=5030.

Roy, D.P., Boschetti, L., and Trigg, S.N. (2006). Remote sensing of fire severity: assessing the performance of the normalized burn ratio. *IEEE Geoscience and Remote Sensing Letters* 3 (1): 112–116.

Roy, P.S., Roy, A., Joshi, P.K. et al. (2015). Development of decadal (1985–1995–2005) land use and land cover database for India. *Remote Sensing* 7 (3): 2401–2430. https://doi.org/10.3390/rs70302401.

Royimani, L., Mutanga, O., Odindi, J. et al. (2019). Advancements in satellite remote sensing for mapping and monitoring of alien invasive plant species (AIPs). *Physics and Chemistry of the Earth, Parts A/B/C* 112: 237–245.

Rudel, T.K., Coomes, O.T., Moran, E. et al. (2005). Forest transitions: towards a global understanding of land use change. *Global Environmental Change* 15 (1): 23–31.

Saha, D. and Sundriyal, R.C. (2012). Utilization of non-timber forest products in humid tropics: implications for management and livelihood. *Forest Policy and Economics* 14: 28–40.

Sampson, P.H., Zarco-Tejada, P.J., Mohammed, G.H. et al. (2003). Hyperspectral remote sensing of forest condition: estimating chlorophyll content in tolerant hardwoods. *Forest Science* 49: 381–391.

Sánchez, D.G.R., Viramontes, U.F., Corona, J.S.S. et al. (2015). Forage potential and water productivity in alternative cropping patterns. *Revista Mexicana de Ciencias Pecuarias* 6 (2): 153–170.

Santoro, M., Beaudoin, A., Beer, C. et al. (2015). Forest growing stock volume of the northern hemisphere: spatially explicit estimates for 2010 derived from Envisat ASAR. *Remote Sensing of Environment* 168: 316–334.

Savtchenko, A., Ouzounov, D., Ahmad, S. et al. (2004). Terra and Aqua MODIS products available from NASA GES DAAC. *Advances in Space Research* 34 (4): 710–714.

Schieltz, J.M. and Rubenstein, D.I. (2016). Evidence based review: positive versus negative effects of livestock grazing on wildlife. What do we really know? *Environmental Research Letters* 11: 1–18.

Schlerf, M., Atzberger, C., and Hill, J. (2005). Remote sensing of forest biophysical variables using HyMap imaging spectrometer data. *Remote Sensing of Environment* 95: 177–194.

Schnitzer, S.A., Dalling, J.W., and Carson, W.P. (2000). The impact of lianas on tree regeneration in tropical forest canopy gaps: evidence for an alternative pathway of gap-phase regeneration. *Journal of Ecology* 88: 655–666.

Seddon, N., Sengupta, S., García-Espinosa, M. et al. (2019). *Nature-Based Solutions in Nationally Determined Contributions: Synthesis and Recommendations for Enhancing Climate Ambition and Action by 2020*, 1–62. Gland, Switzerland/Oxford, UK: IUCN and University of Oxford.

Sedjo, R., Bishop, J., and Sayer, J.A. (2005). *Economics issues in ecosystem approaches to forest management*. In: *Forests in Landscapes: Ecosystem Approaches to Sustainability* (ed. J.A. Sayer and S. Maginnis), 1–268. The Earthscan Forestry Library.

Seidl, R., Spies, T.A., Peterson, D.L. et al. (2016). Searching for resilience: addressing the impacts of changing disturbance regimes on forest ecosystem services. *Journal of Applied Ecology* 53: 120–129.

Seymour, R.S., White, A.S., and Philip, G.D. (2002). Natural disturbance regimes in Northeastern North America-evaluating silvicultural systems using natural scales and frequencies. *Forest Ecology and Management* 155 (1–3): 357–367.

Shabani, F., Ahmadi, M., Peters, K. et al. (2019). Climate-driven shifts in the distribution of koala browse species from the last interglacial to the near future. *Ecography* 42: 1587–1599.

Sharma, C.M., Gairola, S., Ghildiyal, S.K., and Suyal, S. (2009). Forest resource use patterns in relation to socioeconomic status: a case study in four temperate villages of Garhwal Himalaya, India. *Mountain Research and Development* 29 (4): 308–319.

Shimada, M., Itoh, T., Motooka, T. et al. (2014). New global forest/non-forest maps from ALOS PALSAR data (2007–2010). *Remote Sensing of Environment* 155: 13–31.

Singh, S.P. (1998). Chronic disturbance, a principal cause of environmental degradation in developing countries. *Environmental Conservation* 25: 1–2.

Singh, J.S., Rawat, Y.S., and Chaturvedi, O.P. (1984). Replacement of oak forest with pine in the Himalaya affects the nitrogen cycle. *Nature* 311: 54–56.

Smith, C.W., Panda, S.K., Bhatt, U.S., and Meyer, F.J. (2021). Improved boreal forest wildfire fuel type mapping in interior Alaska using AVIRIS-NG hyperspectral data. *Remote Sensing* 13: 897.

Solano-Correa, Y., Bovolo, F., and Bruzzone, L. (2018). An approach for unsupervised change detection in multitemporal VHR images acquired by different multispectral sensors. *Remote Sensing* 10 (4): 533.

Srivastava, P.K., Petropoulos, G.P., Gupta, M. et al. (2019). Deriving forest fire probability maps from the fusion of visible/infrared satellite data and geospatial data mining. *Modeling Earth Systems and Environment* 5: 627–643.

Staudinger, M.D., Grimm, M.B., Staudt, A., Carter, S.L. et al. (2012). *Impacts of Climate Change on Biodiversity, Ecosystems, and Ecosystem Services: Technical Input to the 2013 National Climate Assessment*. Cooperative Report to the 2013 National Climate Assessment. National Climate Assessment, pp. 1–296. Available at URL: http://assessment.globalchange.gov

Surendra, K.C., Takara, D., Hashimoto, A.G., and Khanal, S.K. (2014). Biogas as a sustainable energy source for developing countries: opportunities and challenges. *Renewable and Sustainable Energy Reviews* 31: 846–859.

Szpakowski, D.M. and Jensen, J.L. (2019). A review of the applications of remote sensing in fire ecology. *Remote Sensing* 11: 2638.

Tallis, H., Kareiva, P., Marvier, M., and Chang, A. (2008). An ecosystem services framework to support both practical conservation and economic development. *Proceedings of the National Academy of Sciences* 105 (28): 9457–9464.

The Nature-Based Solutions (NbS) for Climate Manifesto (2019). UN Climate Action Summit: New York, NY, USA. Available at URL: http://www.unenvironment.org/nature-based-solutions-climate

Thompson, I., Mackey, B., McNulty, S., and Mosseler, A. (2009). Forest resilience, biodiversity, and climate change. A synthesis of the biodiversity/resilience/stability relationship in forest ecosystems. *Secretariat of the Convention on Biological Diversity, Montreal* 43: 1–67.

Tribal Cooperative Marketing Development Federation of India Limited Ministry of Tribal Affairs, Govt. of India (TRIFED) (2021). *Introduction to Minimum Support* Price *(MSP) for Minor Forest Produces (MFP)*. Available at URL: https://trifed.tribal.gov.in/program/non-timber [Accessed on 27th December 2021].

Tripathi, S. (2016). Livelihood dependency of Garasia tribes utilizing non timber forest products in Abu Road area of Sirohi district in Rajasthan, India. *International Journal of Advanced Research* 4 (7): 498–504.

Turley, L. (2016). *The Landscape Approach Moving Towards Sustainable Land Use Patterns.* State of Sustainability Initiatives, 1–5. Available at URL: http://www.iisd.org/ssi

U.S. Fish and Wildlife Service (2021). *Impacts of grazing*. Available at URL: https://www.fws.gov/invasives/stafftrainingmodule/methods/grazing/impacts.htm [Accessed on 7th December 2021].

United Nations Conference on Environment and Development (UNCED) (1992). *Agenda 21, Rio Declaration, Forest Principles.* New York: United Nations.

United States Agency for International Development (USAID) (2016). *Sustainable Forest Ecosystem Management: A Strategy for India*. Forest-Plus-Partnership for Land Use Science, 1–37. Available at URL: https://www.climatelinks.org/sites/default/files/asset/document/2016_Tetra-Tech_USAID-India-Forest-Plus_Sustainable-Forest-Ecosystem-Management.pdf

Uprety, Y., Boon, E.K., Poudel, R.C. et al. (2010). Non-timber forest products in Bardiya district of Nepal: indigenous use, trade and conservation. *Journal of Human Ecology* 30 (3): 143–158.

Varga, A., Demeter, L., Ulicsni, V. et al. (2020). Prohibited, but still present: local and traditional knowledge about the practice and impact of forest grazing by domestic livestock in Hungary. *Journal of Ethnobiology and Ethnomedicine* 16 (51): 1–12.

Vasilakos, C., Kalabokidis, K., Hatzopoulos, J., and Matsinos, I. (2009). Identifying wildland fire ignition factors through sensitivity analysis of a neural network. *Natural Hazards* 50 (1): 125–143.

Vasquez, R. and Gentry, A.H. (1989). Use and misuse of forest-harvested fruits in the Iquitos area. *Conservation Biology* 3 (4): 350–361.

Verstraete, M.M., Pinty, B., and Curran, P.J. (1999). MERIS potential for land applications. *International Journal of Remote Sensing* 20: 1747–1756.

Viciani, D., Vidali, M., Gigante, D. et al. (2020). A first checklist of the alien dominated vegetation in Italy. *Plant Sociology* 57 (1): 29–54.

Wan, F.H. and Yang, N.W. (2016). Invasion and management of agricultural alien insects in China. *Annual Review of Entomology* 61: 77–98.

Wardle, D.A., Bardgett, R.D., Callaway, R.M., and van der Putten, W.H. (2011). Terrestrial ecosystem responses to species gains and losses. *Science* 332 (6035): 1273–1277.

Wiersum, K.F. (2004). *Social and community forestry*. In: *Encyclopedia of Forest Sciences*. (ed. J. Burley, J. Evans and J.A. Youngquist), 1136–1143. Oxford, UK: Elsevier.

Winthrop, R.H. (2014). The strange case of cultural services: limits of the ecosystem services paradigm. *Ecological Economics* 108: 208–214.

Witkowski, E.T.F., Lamont, B.B., and Obbens, F. (1994). Commercial picking of Banksia hookeriana in the wild reduces subsequent shoot, flower and seed production. *Journal of Applied Ecology* 31 (3): 508–520.

Xie, Y.J. and Ng, C.N. (2013). Exploring spatio-temporal variations of habitat loss and its causal factors in the Shenzhen River cross-border watershed. *Applied Geography* 39: 140–150.

Xue, J., Lyu, D., Wang, D. et al. (2018). Assessment of soil erosion dynamics using the GIS-based RUSLE model: a case study of Wangjianglou watershed from the three gorges reservoir region, Southwestern China. *Water* 10: 1817.

Yadav, A.S. and Gupta, S.K. (2006). Effect of micro-environment and human disturbance on the diversity of woody species in the Sariska Tiger project in India. *Forest Ecology and Management* 225 (1–3): 178–189.

Yadav, R., Jangid, M., Rajpurohit, S., and Kamboj, R.D. (2019). Estimated quantification and valuation of fuelwood collected from forest areas of Narmada forest division of Gujarat state, India. *Plant Archives* 19 (2): 2459–2464.

Yang, J., Gong, P., Fu, R. et al. (2013). The role of satellite remote sensing in climate change studies. *Nature Climate Change* 3: 875–883.

Zhang, Y., Ling, F., Foody, G.M. et al. (2019). Mapping annual forest cover by fusing PALSAR/PALSAR-2 and MODIS NDVI during 2007–2016. *Remote Sensing of Environment* 224: 74–91.

Zhu, J. and Liu, Z. (2004). A review on disturbance ecology of forest. *Chinese Journal of Applied Ecology* 15 (10): 1703–1710.

Zink, M.V., Bloetscher, H., Boerner, T., Kietzmann, H. et al. (1996). *Evaluating the Potential of ERS SAR in Forest Monitoring*. Proc.: PIERS, Progress in Electromagnetics Research Symposium, University of Innsbruck, Austria, 8–12.

Section II

Forest Parameters – Biochemical and Biophysical Parameters

3

Distinguishing Carotene and Xanthophyll Contents in the Leaves of Riparian Forest Species by Applying Machine Learning Algorithms to Field Reflectance Data

*Guillaume Lassalle** and *Sophie Fabre*

Office National d'Études et de Recherches Aérospatiales (ONERA), Toulouse, France

3.1 Introduction

3.1.1 Chapter Overview

This chapter is intended to outline the recent improvements in monitoring the response of vegetation to stressful conditions by remote sensing through the determination of leaf carotenoid composition. Significant advances in the processing of hyperspectral reflectance data have been achieved over the last decade, opening promising outlooks in this domain. The chapter focuses on riparian forest vegetation, which is widely affected by stress exposure worldwide. The first section describes the importance of carotenoids in the assessment of forests' response to stress, as well as the main advances and pitfalls in applying hyperspectral reflectance spectroscopy for determining leaf carotenoids. Then a case study of carotenoid monitoring in a temperate riparian forest is proposed. The study compares the performance of several (ML) algorithms for determining the carotenoid pool of leaves from riparian forest species using field spectroscopy. The applications to satellite spectrometers are then discussed as a perspective toward the broad-scale monitoring of riparian forests.

3.1.2 Threats to Riparian Forests

Riparian forests, also known as riparian corridors, are tree–shrub mixed areas bordering headwaters, lower-order streams, and diverse waterbodies worldwide (Craft 2016). They ensure many ecological services, such as providing habitats for wildlife, enriching streams in organic matter, maintaining water quality, and filtering terrestrial contaminants (Broadmeadow and Nisbet 2004; Cole et al. 2020; Gundersen et al. 2010). Preserving riparian forests has been a global concern over the past decades, and will remain a key challenge in the future, as riparian forests are subject to many stressors of natural and anthropogenic origin (Poff et al. 2011; Stella and Bendix 2019). As part of climate change and increasing

*Corresponding author's email: guillaumelassalle.pro@gmail.com

0000-0002-3281-6464 (Guillaume Lassalle) and 0000-0001-8153-1740 (Sophie Fabre)

Advances in Remote Sensing for Forest Monitoring, First Edition. Edited by Prem C. Pandey and Paul Arellano.

human activity, environmental stressors are expected to increase in frequency, duration, and intensity over the coming decades, affecting vegetation worldwide (Im et al. 2017; Lelieveld et al. 2016; Santini et al. 2018). This includes biotic (pests, diseases, invasive species) and abiotic (heat waves, contaminants, waterlogging, drought) stressors, as well as their combinatory effects (Dreesen et al. 2012; Lassalle et al. 2020; Suzuki et al. 2014; Teskey et al. 2015). Therefore, understanding the biological mechanisms involved in the response of riparian forests to stressful conditions is crucial for preserving these ecosystems. Additionally, broad-scale remote sensing solutions are needed for detecting stress at the early stages of exposure.

3.1.3 Remote Sensing of Riparian Forests

The management of riparian forests includes survey actions (Barker et al. 2002; Magdaleno and Martinez 2014), restoration or rehabilitation (Dufour et al. 2019), repeated maintenance operations (deadwood removal), and consideration of the trade-off between economic and environmental benefits (Huylenbroeck et al. 2020; Sonesson et al. 2020). Historically, active and passive optical remote sensing has proven useful in mapping, characterizing, and monitoring vegetation by covering large areas while giving quick access to the diversity, health, and phenology of species, unlike time-consuming ground surveys and photo-interpretation of aerial images (Dufour et al. 2013; Goetz 2006; Huylenbroeck et al. 2020; Klemas 2014). For example, hyperspectral images combining high spatial and spectral resolution may be used to retrieve the biophysical and biochemical attributes of vegetation related to the species composition and health status (Fassnacht et al. 2016; Lassalle et al. 2019a; Lausch et al. 2016). Multispectral, multitemporal imaging may serve to monitor the phenology of species across seasons and years, thus detecting abnormal changes caused by stressful conditions (Fabre et al. 2020; Phiri et al. 2020). LiDAR data can also be used in a complementary way to obtain information about vegetation structure (e.g. crown or trunk sizes, tree height, foliar density, and woody debris) (Jeronimo et al. 2018; Listopad et al. 2015; Zhen et al. 2016). Hence, many applications have emerged in optical remote sensing of vegetation over the last decade, including on riparian forests (Ba et al. 2020; Furuya et al. 2020; Huylenbroeck et al. 2020; Klemas 2014). However, studies on assessing the response of riparian forests to stressful conditions by imagery remain scarce as compared to other ecosystems such as tropical rainforests and mangroves (Coates et al. 2015; Michez et al. 2016; Pace et al. 2021). So, important progress is needed in developing remote sensing methods specifically suited for the monitoring of riparian trees and shrubs. In that respect, approaches based on retrieving leaf pigment contents from reflectance data proved to be reliable for detecting biotic and abiotic stress in many vegetation types (Lassalle et al. 2019b; Morel et al. 2018). While most of them involve tracking alterations in chlorophylls, carotenoid pigments have sparked a growing interest because of their implication in plant stress.

3.1.4 Implication of Carotenoids in Plant Stress

Many environmental stressors damage leaf tissues owing to the accumulation of reactive oxygen species (Choudhury et al. 2013; Haldimann et al. 2008; Juvany et al. 2013; Shahid et al. 2014). As part of plants' response, carotenoids play a critical role in the photoprotection of light-harvesting structures in leaves (Demmig-Adams and Adams 1996; Demmig et al. 1987). These pigments absorb light mostly in the blue-green region of the visible spectrum (400–500 nm). Based on their chemical structure, carotenoids are classified as carotenes or as xanthophylls. These pigments prevent damage to photosystems through the xanthophyll cycle (Figure 3.1a) (Feret et al. 2008; Zhang et al. 2017). Depending on light availability, carotenoids forming the xanthophyll cycle interconvert by undergoing

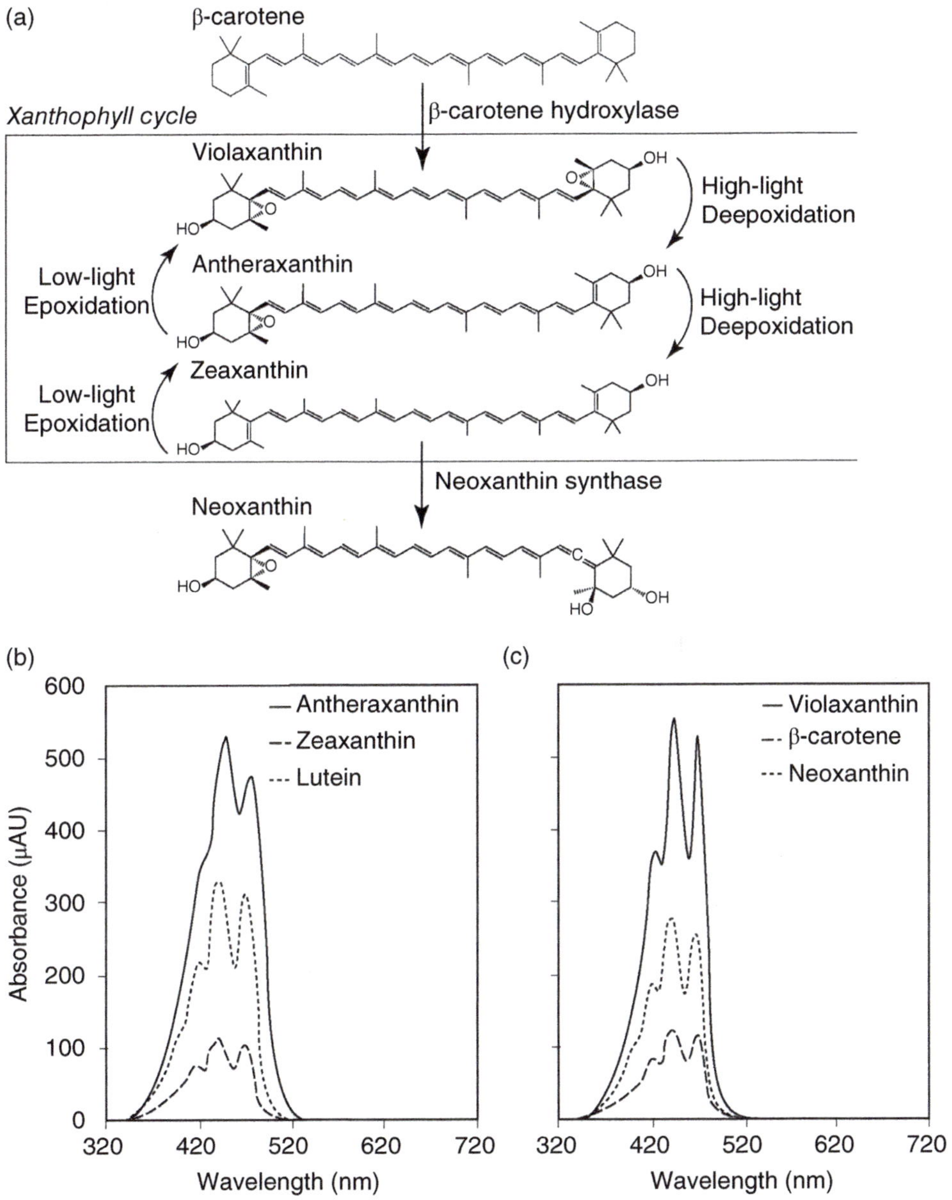

Figure 3.1 (a) β-carotene and xanthophyll transformation pathways. Violaxanthin, antheraxanthin, and zeaxanthin interconvert in response to changes in light availability and plant stress. They form the xanthophyll cycle. (b and c) Absorption spectrum of the main carotenoid pigments in higher plant leaves in the visible region.

epoxidation and deepoxidation reactions. More importantly, carotenoid pigments are involved in a wide range of plant stressors, including heat waves, drought, and soil contamination (Lassalle et al. 2019c; Othman et al. 2014; Young 1991). Hence, plants cope with stressful conditions by modifying the composition of their carotenoid pool. Therefore, the latter is a relevant indicator of plants' response to a changing environment (Demmig-Adams and Adams 1996; Niyogi et al., 2002; Wong et al. 2019), but implies distinguishing carotenes and xanthophylls. In this respect, hyperspectral reflectance spectroscopy offers a non-destructive, time-saving solution for determining leaf pigment contents (Feret et al. 2008; Fu et al. 2020; Lassalle et al. 2019c; Stylinski et al. 2002).

3.1.5 Advances in Carotenoid Retrieval Using Reflectance Spectroscopy

Hyperspectral reflectance spectroscopy exploits leaf optical properties over the continuous visible spectrum (400–750 nm) to retrieve pigment contents (Ustin et al. 2009). While estimating total carotenoid content can be easily achieved, distinguishing carotenes and xanthophylls remains challenging since these pigments share similar light absorption features between them (Figure 3.1b and c) and with chlorophyll, the latter having a stronger influence on leaf reflectance (Junker and Ensminger 2016; Stylinski et al. 2002; Zhang et al. 2017). Two main approaches are commonly used for retrieving total carotenoid content using reflectance data; empirical models and radiative transfer model inversion (Van Cleemput et al. 2018). However, none proved suitable for distinguishing carotenes and xanthophylls until now. Traditional empirical models involve calibrating regression models with leaf reflectance at single or multiple wavelengths or reflectance ratios (i.e. vegetation indices) as input variables. For example, the Photochemical Reflectance Index (PRI)—which is linked to the epoxidation state of the xanthophyll cycle—proved suitable for tracking plants' response to changing light conditions, but is unsuited for determining leaf carotenoid composition precisely (Hikosaka and Noda 2019; Hmimina et al. 2015; Wong and Gamon 2015). Physically-based radiative transfer models aim at modeling the fate of light at the leaf surface and in parenchyma (Jacquemoud et al. 2009). Inversions of these models usually estimate total carotenoid pigment accurately, but still fail in determining leaf carotenoid composition (Feret et al. 2008; Miraglio et al. 2019; Zhang et al. 2017). As an alternative, ML algorithms (Hastie et al. 2009), often referred to as semi-empirical models, might overcome the limits of these two approaches.

3.1.6 Applying Machine Learning to Reflectance Spectroscopy

ML combines explanatory variables—also termed predictors or features—to predict a discrete, categorical, or continuous response variable. Most ML algorithms involve projecting predictors in a high-dimension space (Support Vector Machine [SVM], Neural Network [NN]), or on reducing the predictor space (Partial Least Square Regression [PLSR]), or even combining predictors consecutively (Random Forest [RF]). Some of them deal with multicollinearity among both predictors and response variables, which occurs when using high-dimension reflectance data for distinguishing carotenoid contents (Hastie et al. 2009; Hughes 1968; Lassalle et al. 2020). Therefore, ML algorithms show promising potential for determining leaf carotenoid composition toward monitoring the response of riparian vegetation to environmental stressors. In this regard, the following case study evaluates the feasibility of distinguishing carotenoid pigments in the leaves of riparian plant species by applying ML algorithms to field spectroscopy. Sections 3.2–3.5 describe the dataset and the main results of the study, and Section 3.6 discusses the perspectives of application to satellite imaging spectroscopy.

3.2 Study Area

The site identified for the study is a riparian forest located under a temperate region, in the southwest of France. This forest is made of trees and shrubby vegetation established on sandy loam soil with good water and nutrient supply due to its proximity to rivers. Five species dominate the site, namely *Rubus fruticosus* L. (bramble), a woody perennial shrub commonly found under temperate climate; *Quercus pubescens* Willd. (pubescent oak), a medium-sized marcescent tree that flourishes in sub mediterranean regions; *Populus x canadensis* Moench. (Canadian poplar), a deciduous hybrid of *Populus nigra* L. (black poplar) and *Populus deltoides* (eastern cottonwood); *Cornus sanguinea* L. (bloody dogwood), a deciduous shrub; and *Buddleja davidii* Franch. (butterfly bush), an invasive shrubby species. These five species were studied here.

3.3 Data

3.3.1 Leaf Sampling and Analysis

Young, sun-exposed leaves, which appeared in spring, were sampled on various locations of the site in early July (summer) and mid-October (autumn) 2017, at midday. A total of 151 leaves were sampled, of which 71 came from *R. fruticosus*, the predominant species. Twenty leaves were sampled on each of the four other species. Once sampled, all the leaves were stored in a cooler and brought directly to the laboratory for pigment extraction and analysis. On each leaf, 15-mm wide disks were collected, and leaf pigments were extracted in methanol using the protocol of Diepens et al. (2017) and analyzed by high-pressure liquid chromatography (HPLC) following the method of Barlow et al. (1996). The analysis included β-carotene and common plant xanthophylls; Violaxanthin, neoxanthin, antheraxanthin, zeaxanthin, and lutein, the concentrations of which were expressed in $\mu g\ cm^{-2}$ (Table 3.1) (Feret et al. 2008; Jacquemoud et al. 1996). To illustrate the above-mentioned challenge in distinguishing the carotenoids contents in leaves, we computed the variance inflation Factor (VIF), defined as:

$$VIF_i = \frac{1}{1 - R_i^2} \tag{3.1}$$

where R^2_i is the coefficient of determination of the multiple regression between the content of the carotenoid i as the response variable and all other carotenoid contents as input variables. It is commonly admitted that VIF values above 10 indicate high collinearity among variables (Zarco-Tejada et al., 2018). The VIF was computed for the six carotenoids analyzed by HPLC.

Table 3.1 Statistics of the six carotenoid contents analyzed in plant leaves ($\mu g\,cm^{-2}$).

	Min	Max	Median	Mean	SD	CV (%)
Violaxanthin	0.09	4.98	0.48	0.97	1.03	106.85
Neoxanthin	2.09	7.86	4.91	4.89	1.62	33.14
Antheraxanthin	0.04	1.14	0.43	0.43	0.27	63.07
Zeaxanthin	0.03	1.17	0.37	0.39	0.21	54.18
Lutein	0.32	6.49	1.13	1.71	1.31	76.53
β-carotene	1.1	8.88	4.52	4.83	2.14	44.24

CV: Coefficient of Variation

3.3.2 Reflectance Measurements

The reflectance of the 151 leaves was measured at the exact location of the disks sampled for carotenoid analysis, using a leaf-clip attached to an ASD FieldSpec 4 Hi-Res spectroradiometer (Malvern Panalytical, Malvern, UK). For this purpose, leaf radiance was measured on a black background panel using an internal light source covering the [400:2500] nm optical domain and then converted to reflectance using a white reference calibration panel (Spectralon, Labsphere Inc., North Sutton, USA) (Lassalle et al. 2019c; Milton 1987). Since pigments mainly influence reflectance in the VIS region, only the [400:750] nm reflectance data were exploited in further processing.

3.4 Methodology

3.4.1 Preprocessing of Reflectance Data

The spectral signatures were first smoothed using the Savitzky–Golay algorithm to enhance the signal-to-noise ratio (Savitzky and Golay 1964). Then, the signatures were transformed using the *Continuum* Removal (CR) algorithm proposed by Clark et al. (1985). The CR normalizes spectral signatures by defining a convex hull over a spectral region. Then, each reflectance value from the original spectral signature is divided by that of the convex hull at the corresponding wavelength, resulting in a *continuum*-removed reflectance (CRR) spectrum. The general formula of the CRR is as follows:

$$CRR_{\lambda i} = \frac{\rho_{\lambda i}}{\text{convex hull}_{\lambda i}} \tag{3.2}$$

where $\rho\lambda i$ and convex $\text{hull}_{\lambda i}$ are the reflectance values from the original spectral signature and the convex hull at wavelength $_{\lambda i}$, respectively. The CR algorithm has been widely used to enhance the light absorption features of pigments, especially chlorophylls, in remote sensing studies (Kokaly, 1999; Stimson et al. 2005). Here, we applied the CR separately to the [400:550] and [550:750] nm spectral regions, as described in Noomen et al. (2006) and Kong et al. (2017). An example is illustrated in the Supporting Information Figure S3.1 at the end of the chapter, and the original reflectance and CRR spectral signatures for the five species are presented in Figure 3.2. Applying the CR this

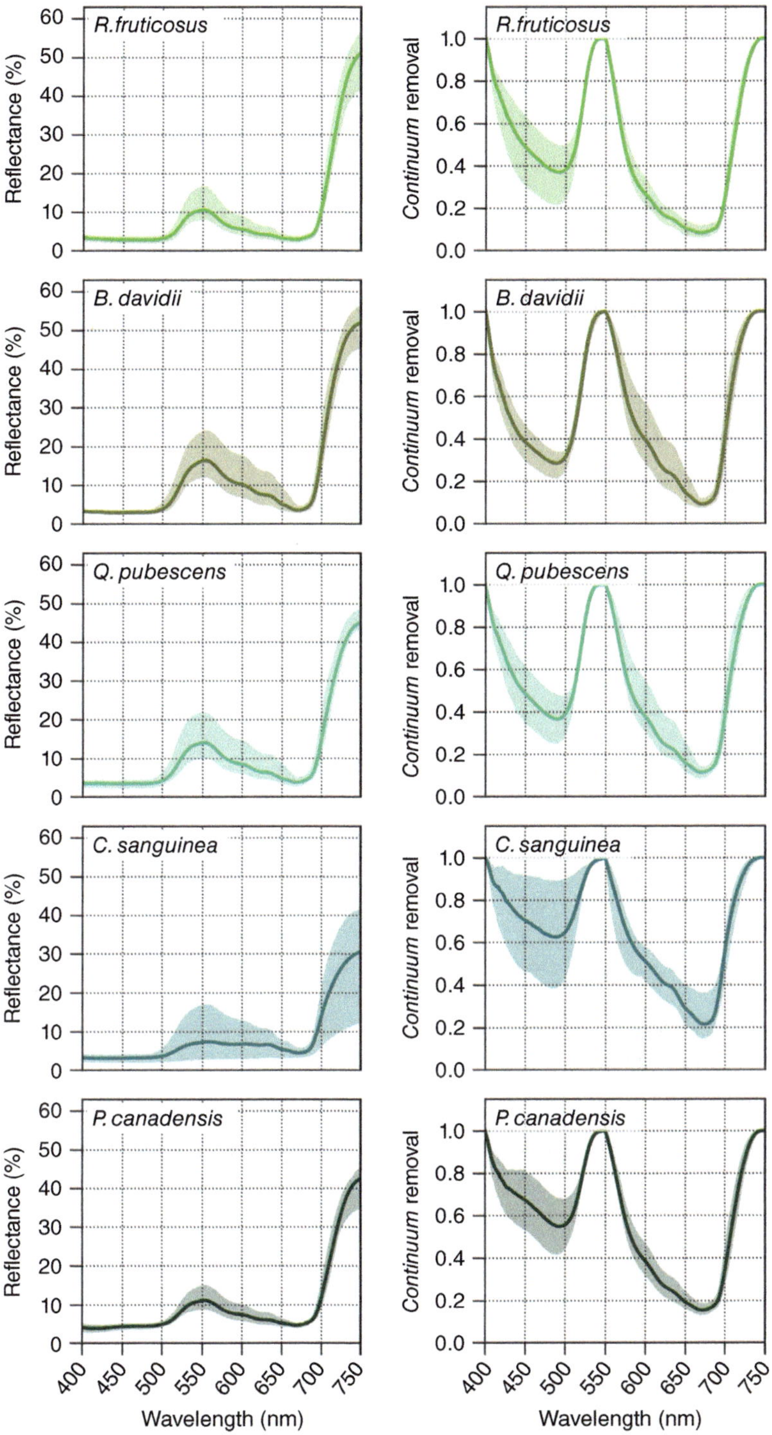

Figure 3.2 (left figures) Original spectral signatures of leaves from the five species measured in the field in 2017, in the visible region (mean±SD). (right figures) *Continuum*-removed spectral signatures obtained after application of the *Continuum* Removal algorithm (mean±SD).

way minimizes the influence of the chlorophyll absorption feature around 550 nm and enhances that of carotenoids in the blue and red regions (Kováč et al. 2013; Sanches et al. 2013). It also modifies the collinearity of reflectance data, as shown in the Supporting Information Figures S3.2 and S3.3.

3.4.2 ML Algorithms

The CRR data from the five species formed the dataset exploited in ML algorithms to retrieve leaf carotenoid contents. As a preliminary step, the linear relationship between each CRR wavelength and the six carotenoid contents in leaves was assessed using the Pearson correlation of coefficient (r), after verifying that the data followed a normal distribution (Shapiro–Wilk test, $p > 0.05$). This helped to identify the potentially relevant spectral regions for estimating each carotenoid. Then the CRR data from the [400:750] nm region were used as predictors (i.e. input variables) in ML algorithms to retrieve the six carotenoid contents (i.e. response variables). Five ML algorithms widely used in remote sensing studies were compared (see Hastie et al. 2009 for the mathematical formulation of these algorithms):

- PLSR (Haenlein and Kaplan 2004)—which combines elements of the *Principal Component Analysis* (PCA) and multiple regression—involves building a set of linear combinations of the predictors, weighted by their univariate influence on the response variable. That is, PLSR produces a series of principal components, formerly known as *latent vectors*, which are used to predict the response variable. PLSR is highly suitable when the set of predictors is very large, such as reflectance data (350 wavelengths in our case), because the orthogonality of the components overcomes the problem of multicollinearity among predictors. However, PLSR does not perform predictor selection. Instead, the relative contribution of each predictor can be estimated through the Variable Importance Projection (VIP) (Tran et al. 2014), defined as:

$$\mathrm{VIP}_k = \sqrt{\frac{K\sum_{a=1}^{A} W_{a,k}^2 SS_a}{\sum_{a=1}^{A} SS_a}} \tag{3.3}$$

where K is the total number of predictors, $w_{a,k}$ is the weight of the variable k for the PLSR component a, A is the total number of PLSR components, and SS_a is the sum of squares explained by the PLSR component a.

- *Support Vector Regression* (SVR) (Smola and Schölkopf 2004), which extends the use of SVM (Cortes and Vapnik, 1995) to regression problems. SVR estimates the linear relationship between pairs of n-dimensional input vectors and a one-dimensional response variable by constructing an optimal hyperplane. The latter involves defining a linear function that minimizes a cost function, which implies maximizing the hyperplane margins and minimizing the approximation error (Okujeni et al. 2013). The width of the margins is guided by a loss function, and the influence of samples located outside is determined by a parameter C. Unlike many regression algorithms, kernel-based SVR handles nonlinear relationships between the predictors and the response variable by varying the kernel function used to project the data. It is particularly suitable when the number of predictors is much greater than the number of samples.

- *Random Forest (RF) Regression (RFR)* (Breiman 2001), an ensemble method that constructs and averages a series of independent decision trees to model the relationship between predictors and a response variable (Hutengs and Vohland 2016). The trees are built from a set of decision rules resulting from the recursive fragmentation of the original feature space into successively smaller sub-regions. The latter are defined by applying binary splits to each predictor. A simple model is then adjusted between the predictors and the response variable, and the splits minimizing the mean squared error between the measured and predicted response variable are selected. The succession of the selected predictor splits embodies the branches of the decision tree, that is, each set of decision rules leads to an estimation of the response variable. RF accounts for non-linear relationships between the predictors and the response variable, and informs on the relative contribution of each predictor to the response variables (Grömping 2009).
- *Elastic net* (ENET) regression (Zou and Hastie 2005), a penalized least square method that benefits from the Ridge and Lasso regressions. Ridge regression shrinks the regression coefficients by penalizing the regression model with a penalty term—called the *L2-norm* (i.e. the sum of squared coefficients)—tuned by a regularization parameter λ, so that predictors with a minor contribution to the response variable are assigned a coefficient close to zero. One major advantage of the Ridge regression is that it handles multicollinearity among predictors. Lasso regression differs from the Ridge one by its penalty term—called the *L1-norm* (i.e. the sum of the absolute coefficients)—attributing exactly zero coefficients to predictors with a minor contribution. By doing this, Lasso regression performs variable selection. ENET combines the advantages of Ridge and Lasso regression to perform efficient predictor selection under multicollinearity (Lassalle et al. 2019c). The contribution of a predictor to the model corresponds to its absolute coefficient value.
- NN (LeCun et al. 2015) is a self-adaptive algorithm that can solve complex regression problems. The most widely used NN model, the Multi-Layer Perceptron (MLP) (Hu and Weng 2009), was used in this study. The MLP consists of multiple layers, each being made of several nodes (i.e. the neurons), which relate to those of a different layer. Here, the input layer is composed of the CRR data, where each node is a wavelength (i.e. a predictor). The hidden layer comprises an activation function that performs regression. The output layer corresponds to the response variable (i.e. the leaf carotenoid content). Following a back-propagation procedure, initial weights are assigned to each node to process the data. Once predictions of the response variable have been made, they are compared to the measured values. Then, the weights are adjusted, and the operation is repeated until reaching the desired accuracy.

3.4.3 Carotenoid Prediction

To compare the performance of the five ML algorithms, the CRR and carotenoid data were divided into 50% training ($n = 75$) and 50% test ($n = 76$) sets using the Kennard–Stone algorithm (Kennard and Stone 1969; Lassalle et al. 2019b). The latter ensures a selection of representative samples in each set. The same proportion of samples from the different species were present in each set. An illustration of the Kennard–Stone sampling output is provided in the Supporting Information Figure S3.4. The regression models were then adjusted on the training set between the CRR data as predictors and each of the six carotenoid contents from all species as the response variable. For each ML algorithm, the optimal parameters were calibrated using a five-fold cross-validation strategy. Before model calibration, the predictors (i.e. the CRR data) were standardized according to the recommendations of Hastie et al. (2009). Then, the resulting regression models were applied to the test set for predicting the carotenoid contents. For each carotenoid, the quality of predictions was assessed using the Pearson's R^2 and the Normalized Root Mean Square Error (NRMSE, in %):

$$NRMSE = \frac{RMSE}{y_{\max} - y_{\min}} \tag{3.4}$$

where $y_{\max}$ and $y_{\min}$ denote the maximum and minimum observed carotenoid concentrations, respectively, and RMSE is the Root Mean Square Error, defined as:

$$\mathrm{RMSE} = \sqrt{\frac{\sum_{j=0}^{n}\left(y_j' - y_j\right)^2}{n}} \tag{3.5}$$

with y'_j and y_j being the predicted and measured carotenoid contents in leaves, respectively. In complement to the NRMSE, we used the Residual Predictive Deviation (RPD), a common metric of prediction accuracy, defined as:

$$RPD = \frac{\sigma}{RMSE} \tag{3.6}$$

where σ is the standard deviation of the observed carotenoid content. RPD values above two were considered as accurate predictions (Lassalle et al. 2019b; Shi et al. 2014).

Finally, the contribution of each predictor (i.e. CRR wavelength) to the prediction of carotenoid contents was analyzed using the metrics described in Section 4.2 for the PLSR, RF, and ENET algorithms. Since SVR and NN are uninformative on the contribution of the predictors, we implemented a recursive feature elimination procedure in these algorithms. This consisted of recursively removing the least relevant predictors from the calibrated regression models until the model accuracy (i.e. the NRMSE) decreased significantly. All data analysis was performed under Python language using the Scikit-Learn package (Pedregosa et al. 2011).

3.5 Results

3.5.1 Leaf Carotenoid Contents

Of the six carotenoids analyzed by HPLC, four were detected in the leaves of all species, in varying concentrations. Neoxanthin was not found in the leaves of *B. davidii*, *Q. pubescens*, and *C. sanguinea*, and zeaxanthin was not found in those of *P. canadensis*. β-carotene showed the highest content (up to 8.88 μg cm^{-2}), followed by neoxanthin (7.86 μg cm^{-2}), lutein (6.49 μg cm^{-2}), violaxanthin (4.98 μg cm^{-2}), zeaxanthin (1.17 μg cm^{-2}), and antheraxanthin

(1.14 µg cm^{-2}) (Table 3.1). However, contrasted carotenoid compositions were observed among species (Figure 3.3). For example, *R. fruticosus* exhibited β-carotene contents higher than all other species, whereas violaxanthin, antheraxanthin, and lutein contents remained low in the leaves of *B. davidii*. As confirmed by VIF values greater than 37 for all carotenoids, the latter were highly correlated with each other, regardless of the species. This highlighted the challenge in distinguishing these pigments.

3.5.2 Predictions of Carotenoid Contents Using ML Algorithms

Carotenoid contents were moderately correlated to single-wavelength CRR data ($r < 0.90$, see Figure S3.5 in Supporting Information), the green (500–550 nm), and red (~700 nm) regions showing the highest correlation coefficients. Hence, CRR data were unsuited for estimating carotenoid contents when considering univariate (*i.e.*

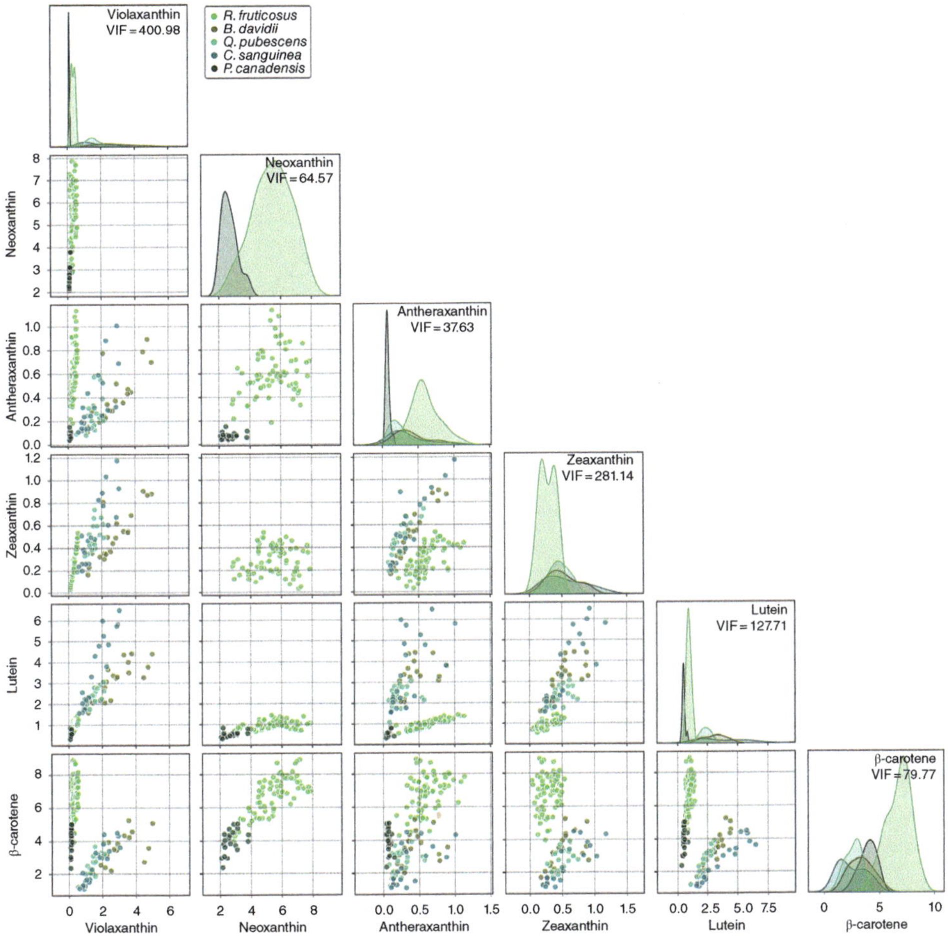

Figure 3.3 Relationship between the six carotenoid pigments (in µg cm^{-2}) analyzed in the leaves of the five species. The top figures show the distribution of contents for each pigment for the five species based on kernel density estimation. [VIF: Variance Inflation Factor, see Eq. (3.1)].

single-wavelength) linear regression. Conversely, ML algorithms—handling non-linear relationships in multiple regression problems—proved suitable in that case. Exploiting CRR data in ML algorithms led to accurate estimations of the carotenoid contents. These results are presented in Figure 3.4 and summarized in the Supporting Information Figure S3.6. For each carotenoid, R^2 and RPD values greater than 0.7 and 1.70 were

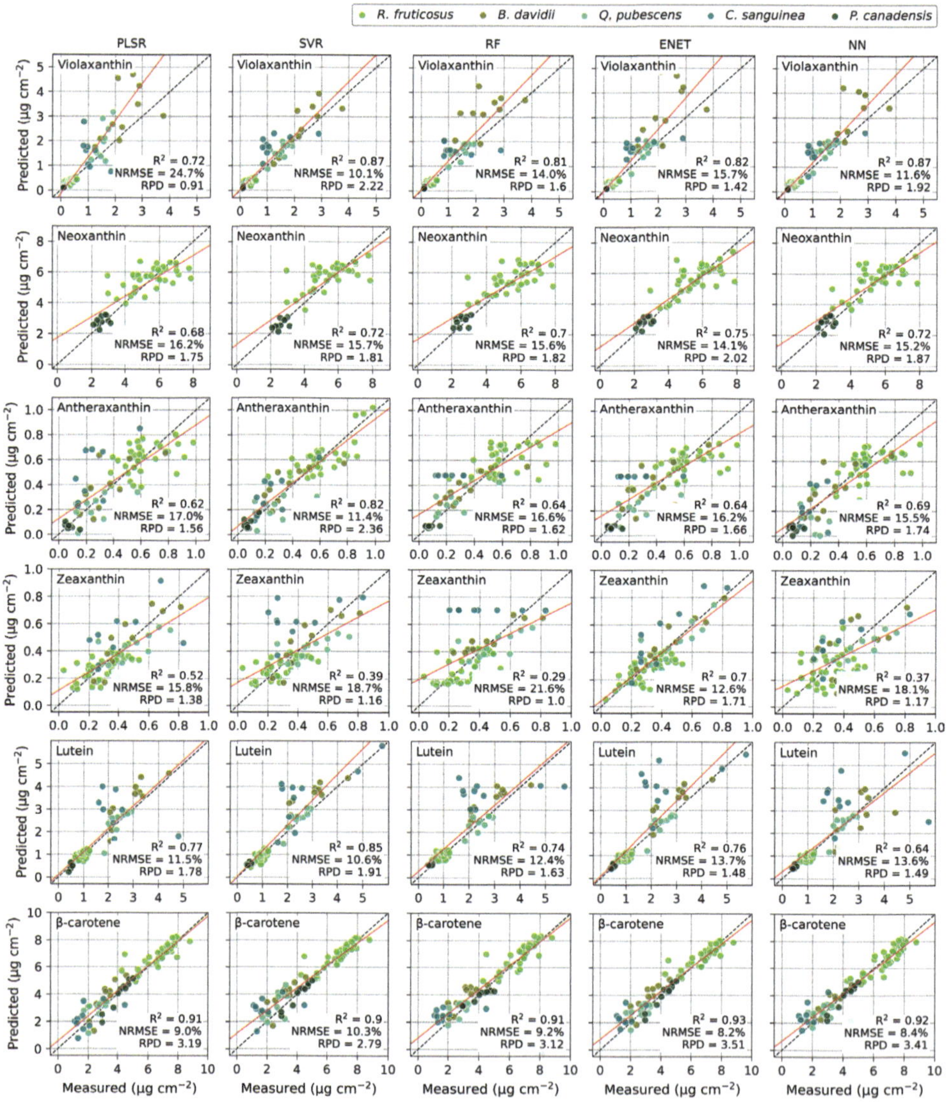

Figure 3.4 Comparison between the measured and predicted carotenoid contents in leaves obtained using the five ML algorithms (in columns, PLSR: Partial Least Square Regression; SVR: Support Vector Regression; RF: Random Forest regression; ENET: Elastic Net regression; NN: Neural Network). For each combination of carotenoid and ML algorithm, the overall R^2, the Normalized Root Mean Square Error (NRMSE), and the Residual Predictive Deviation (RPD) are presented. The dashed line represents the 1:1 line (i.e. perfect predictions).

obtained, respectively, indicating good agreement between the measured and predicted contents. These results were confirmed by NRMSE ≤ 14.1% for at least one ML algorithm. SVR and ENET generally outperformed all other algorithms, especially for low-content carotenoids of the xanthophyll cycle (violaxanthin, antheraxanthin, and zeaxanthin). So, they were retained for further analysis. However, PLSR, RF, and NN were also suitable for estimating β-carotene accurately ($R^2 > 0.9$, RPD > 3). The results obtained using ML algorithms were far better than those of the widely-used PRI on the same dataset, emphasizing the interest of our approach (see the Supporting Information Figure S3.7).

Based on RPD values, the zeaxanthin contents were more difficult to predict than those of other carotenoids (RPD ≤ 1.71), along with Lutein (RPD ≤ 1.91). The analysis of regression residuals revealed that overestimations of intermediary contents (1.5–3 $\mu g\,cm^{-2}$) were largely responsible for inaccuracies (see residuals in the Supporting Information Figure S3.8). Regarding the other carotenoids, the highest violaxanthin contents (> 1 $\mu g\,cm^{-2}$) were slightly overestimated on average, while those of neoxanthin (> 6 $\mu g\,cm^{-2}$), antheraxanthin, and zeaxanthin (> 0.6 g cm^{-2}) stayed below the measured contents. All other concentration ranges were faithfully retrieved by at least one ML algorithm, highlighting the potential of ML in tracking subtle variations in leaf carotene and xanthophyll contents using reflectance spectroscopy.

The contribution of each wavelength to the carotenoid estimation is presented in Figure 3.5. Similarities were observed in the most contributing wavelengths among ML algorithms for the six carotenoids. The 500–550 (green) and 650–700 (red) nm regions were largely responsible for the estimation of neoxanthin, antheraxanthin, and lutein contents. A similar pattern was observed for the other carotenoids, in addition to several wavelengths from the 550–650 nm region. Focusing on β-carotene, only a few wavelengths were needed to obtain accurate estimations using the ENET algorithm. CRR data from the blue region (400–450 nm) remained of lesser importance as compared to the rest of the spectrum.

Several approaches based on reflectance data, such as the PRI and radiative transfer models, are commonly used for determining total carotenoid content, but fail in distinguishing carotenoid pigments individually (Feret et al. 2008; Garbulsky et al. 2011; Garrity et al. 2011; Hernández-Clemente et al. 2012; Hmimina et al. 2015; Wong et al. 2019) (see also the Supporting Information Figure S3.7). Conversely, ML algorithms have been widely applied to reflectance data for assessing vegetation health, mapping species, and estimating chlorophylls (Dronova et al. 2012; Lassalle et al. 2019c; Rumpf et al. 2010; Verrelst et al. 2012). Here, ML algorithms achieved an accurate estimation of the main carotenoids in leaves across various species and over a wide range of concentrations, overcoming the limitations of previous approaches. Our study also emphasizes the importance of choosing the best-suited ML algorithm according to the target carotenoid, to obtain the most accurate estimation. When coupled to the CR algorithm, ML overcomes the problem of high collinearity among both reflectance data and carotenoid contents and remains little influenced by chlorophylls.

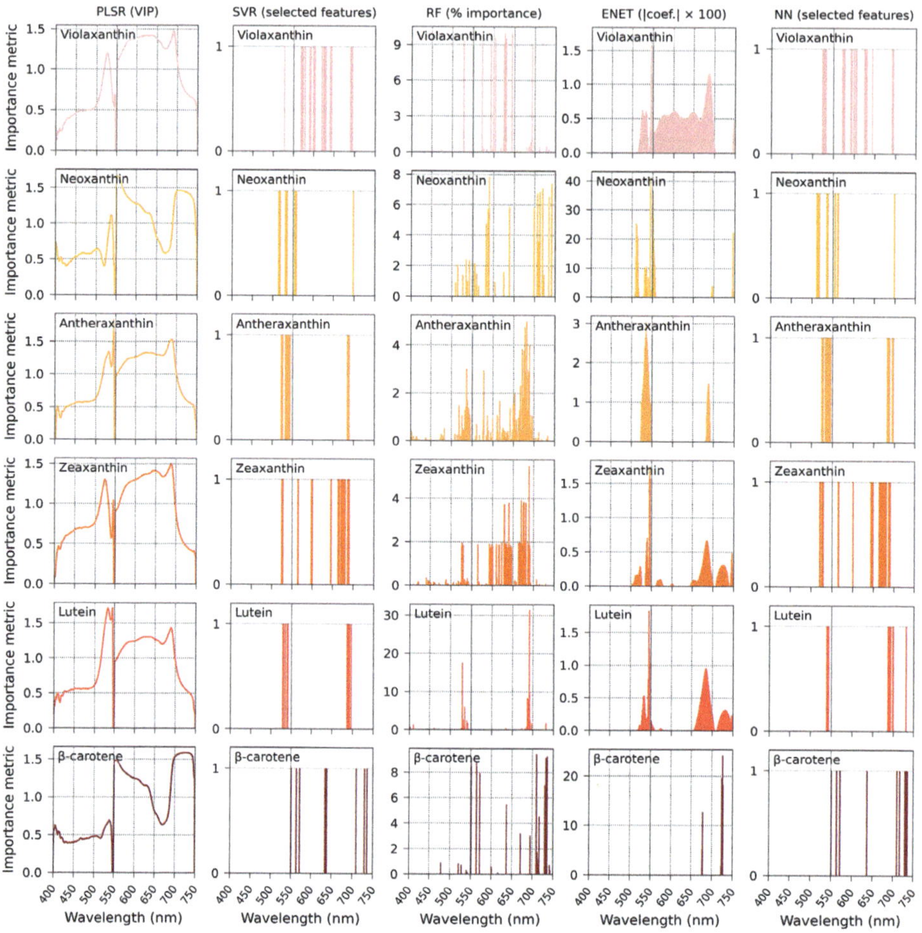

Figure 3.5 Variable importance metrics obtained for the different machine learning algorithms used for predicting the six leaf carotenoid contents (in columns, PLSR: Partial Least Square Regression; SVR: Support Vector Regression; RF: Random Forest regression; ENET: Elastic Net regression; NN: Neural Network). For the SVR and NN algorithms, a value of one means that the variable was retained during the recursive feature elimination procedure. For the other algorithms, the higher the metric, the greater the importance of the variable for predicting leaf carotenoid content. The ENET coefficients were converted to absolute values and multiplied by 100 for graphical convenience. The vertical black line at 550 nm shows the separation of the two spectral regions where the *continuum* removal algorithm was applied. (VIP: Variable Importance Projection.)

3.6 Discussion

3.6.1 Sources of Variability in the Carotenoid Pool among Species

The differences in concentration ranges among species were consistent with that described for other leaf datasets, highlighting the natural variability in leaf pigment composition (Arellano et al. 2017; Feret et al. 2008; Jiang et al. 2018). Neoxanthin was not found in the analysis for three out of the five species studied. Similar observations have been reported previously for other species, for which neoxanthin is replaced by lutein-5,6-epoxide (Bungard et al. 1999; García-Plazaola et al. 2002). Apart from replacing neoxanthin, the presence of lutein epoxide also results from the conversion of lutein, which may result from stress exposure (Matsubara et al. 2007, 2008). Like many carotenoids, lutein contributes to dissipating the exceeding light absorbed, preventing photooxidative damages to plant tissues (Niyogi et al. 2002). Its contribution to the total xanthophylls usually ranges from 30 to 60% (Bungard et al. 1999), but did not exceed 25% in our case. Based on our findings, reflectance spectroscopy could serve as a reliable tool for tracking changes in leaf carotenoid contents in response to environmental stressors. However, it should be able to perform it continuously, ideally on a daily to monthly basis, and on a larger scale. Unlike field measurements, imaging spectrometer may fulfill these requirements (Gamon et al., 2016; Garbulsky et al. 2011; Milton et al. 2009), so they represent a promising solution for mapping plant health before, during, and after stressful events.

3.6.2 Toward a Broad-Scale Monitoring of Carotenoids?

Our study was the first to succeed in distinguishing carotenoid pigments in leaves individually using field reflectance spectroscopy. In line with this work, further research should focus on (i) assessing the performance of ML algorithms in other contexts, and (ii) on upscaling their use to imaging spectrometer. To date, only a few datasets including both reflectance measurements and detailed leaf carotenoid analysis have been made available. To this extent, we encourage researchers to enrich existing datasets by sharing field data upon request in the future, as initiated in recent studies (Meerdink et al. 2019). This would help to extend the scope of ML algorithms to a wider range of plant species, carotenoid pools, and concentration ranges.

Upscaling the use of ML for leaf carotenoid determination is a key step toward continuous, broad-scale monitoring of stressors' impacts on vegetation. Satellite spectrometers cover very large areas with a monthly to weekly revisit time, so they could be well-suited for this purpose. Several multispectral satellite spectrometers currently operate in space, providing reflectance images over a few spectral bands (i.e. wavelengths) with metric to decametric spatial resolution. Sentinel-2 and -3 satellites are of major interest for upscaling our approach since they provide freely-available images of 10- to 20-m spatial resolution in the VIS, while the commercial satellite WorldView-3 offers a resolution lower than 2 m (Frampton et al. 2013; Phiri et al. 2020; Wagner et al. 2018). As highlighted in our study, only a few spectral bands are necessary for determining leaf carotenoid composition accurately using the SVR, ENET, or NN algorithms (Figure 3.5). For some carotenoids, however, hyperspectral spectrometers, which offer a much higher spectral coverage (> 400 wavelengths), might be best suited. In that sense, the upcoming launch of new-generation spectrometers coupling high spatial (≤ 30 m) and spectral resolution (~10 nm) to monthly revisit time are sparking a growing interest in the remote sensing community. For example, in 2019, the Italian Space Agency (*Agenzia Spaziale Italiana,* ASI) and the Japan Ministry of Economy, Trade and Industry (METI) launched the PRISMA and HISUI hyperspectral satellite spectrometer, respectively (20- to 30-m spatial resolution and ~10-nm spectral resolution the VIS) (Loizzo et al. 2019; Matsunaga et al. 2018). Although still in the project stage, the National Aeronautics and Space Administration's (NASA) spectrometer HyspIRI will show close specifications (NASA 2018). Likewise, the French Space Agency (*Centre national d'études spatiales*, CNES) is specifying on a high spatial (8 m) and spectral (~10 nm) resolution spectrometer called HYPXIM-BIODIVERSITY (Michel et al. 2011).

3.6.3 Sensitivity Analysis

To demonstrate the potential of these spectrometers to track changes in the carotenoid pool of leaves, we applied our approach on the same reflectance dataset resampled to 10-nm spectral resolution. The predictions of carotenoid contents obtained with ML algorithms showed good agreement with the laboratory analysis (Table 3.2). That is, at least one algorithm led to NRMSE values ≤ 18.4% for each carotenoid. In line with these results, further study should focus on determining how 8- and 30-m spatial resolutions would affect these performances. Studies based on operating multispectral satellite data are already feasible, and despite most of the abovementioned hyperspectral spectrometers are not yet operating, their suitability for estimating carotenoids could be still evaluated thanks to airborne imaging. Airborne hyperspectral images are frequently used for simulating images with the specifications of future satellite spectrometers (e.g. spatial and spectral resolutions, signal-to-noise ratio, etc.) in vegetation studies (Clark 2017; Clark et al. 2018). Therefore, a similar approach could be conducted for upscaling the use of ML algorithms to the broad-scale tracking of plants' changes in carotenoid contents.

Multi- and hyperspectral spectrometers embedded on Unmanned Aerial Vehicles (UAV) and drones operating daily are also of great interest. Although it does not cover very large areas, drone imaging provides very high spatial resolution images (up to cm-scale). Therefore, it might overcome the limitations of satellite imaging, although important progress is still needed in that field (Barbedo 2019). Hence, the complementarity of drone, airborne, and satellite remote sensing could help to upscale the use of ML algorithms for monitoring plants' response to environmental stresses involving changes in leaf carotenoid composition.

Table 3.2 Normalized Root Mean Square Error (NRMSE) obtained by comparing the measured and predicted carotenoid contents using the five ML algorithms, at 10-nm spectral resolution. This resolution corresponds to that of current and upcoming satellite hyperspectral spectrometers. For each algorithm, the best prediction results are marked bold.

	PLSR	SVR	RF	ENET	NN
Violaxanthin	32.3%	**13.7%**	18.2%	21.6%	14.9%
Neoxanthin	24.6%	24.4%	23.8%	**18.4%**	24.0%
Antheraxanthin	26.9%	**15.1%**	23.9%	23.7%	22.6%
Zeaxanthin	21.3%	24.2%	26.7%	**17.6%**	24.1%
Lutein	19.8%	**15.9%**	18.8%	17.4%	20.2%
β-carotene	13.8%	15.1%	13.3%	**12.1%**	**12.1%**

PLSR: Partial Least Square Regression; SVR: Support Vector Regression; RF: Random Forest regression; ENET: Elastic Net regression; NN: Neural Network

3.7 Conclusion

This chapter outlines the recent advances and future challenges in monitoring riparian forests by remote sensing. Riparian vegetation has sparked a growing interest in the remote sensing community over the last decades (Huylenbroeck et al. 2020), and the tracking of plant stress has become essential for the management and conservation of these ecosystems. By combining high to very-high spatial and spectral resolution, hyperspectral imaging has opened up promising outlooks toward accurate monitoring of tree and understory vegetation in riparian forests. In particular, this technology can give access to a wide range of biochemical parameters related to plant stress. In this chapter, emphasis was made on carotenoid pigments. The case study presented here aimed at evaluating the feasibility of distinguishing carotene and xanthophyll contents in the leaves of riparian plant species from hyperspectral data. The proposed approach highlights the potential of field spectroscopy to quantify these pigments in a non-invasive and non-destructive way, by applying ML algorithms to CRR spectra. Of the five ML algorithms tested, SVR and ENET showed the best performance, regardless of the carotenoid pigment (R^2 and RPD values greater than 0.7 and 1.70, respectively). The weighting or selection of features identified the 500–550 and 650–700 nm intervals as the most contributing regions. In the perspective of broad-scale monitoring, the proposed approach showed promising performance for application to current and upcoming satellite spectrometers when applied to 10-nm spectral resolution spectra (NRMSE $\leq$ 18.4%). Hence, the case study presented in this chapter provides a good basis for monitoring the response of riparian forests to stressful conditions at a broad scale. Attention should now be given to determining the separability of carotenoid pigments at varying spatial resolutions over mixed canopies undergoing stressful conditions. This would take a step toward improving the management and conservation of riparian forests in response to climate change and increasing human activities.

Acknowledgments

The authors gratefully acknowledge A. Credoz, D. Dubucq, and R. Hédacq for their implication in leaf sampling and reflectance measurements, and A. Elger, and D. Lambrigot, and E. Buffan-Dubau for their assistance in leaf carotenoid extraction and analysis.

Funding

Financial support of this work was provided by the ONERA.

References

Arellano, P., Tansey, K., Balzter, H., and Boyd, D.S. (2017). Field spectroscopy and radiative transfer modelling to assess impacts of petroleum pollution on biophysical and biochemical parameters of the Amazon rainforest. *Environment and Earth Science* 76: 1–14. https://doi.org/10.1007/s12665-017-6536-6.

Ba, A., Laslier, M., Dufour, S., and Hubert-Moy, L. (2020). Riparian trees genera identification based on leaf-on/leaf-off airborne laser scanner data and machine learning classifiers in northern France. *International Journal of Remote Sensing* 41: 1645–1667. https://doi.org/10.1080/01431161.2019.1674457.

Barbedo, J.G.A. (2019). A review on the use of unmanned aerial vehicles and imaging sensors for monitoring and assessing plant stresses. *Drones* 3: 40. https://doi.org/10.3390/drones3020040.

Barker, J.R., Bollman, M., Ringold, P.L. et al. (2002). Evaluation of metric precision for a riparian forest survey. *Environmental Monitoring and Assessment* 75: 51–72. https://doi.org/10.1023/A:1014259902449.

Barlow, R.G., Cummings, D.G., and Gibb, S.W. (1996). Improved resolution of mono- and divinyl chlorophylls a and b and zeaxanthin and lutein in phytoplankton extracts using reverse phase C-8 HPLC. *Marine Ecology Progress Series* 161: 303–307.

Breiman, L. (2001). Random forests. *Machine Learning* 45: 5–32. https://doi.org/10.1023/A:1010933404324.

Broadmeadow, S. and Nisbet, T.R. (2004). The effects of riparian forest management on the freshwater environment: a literature review of best management practice. *Hydrology and Earth System Sciences* 8: 286–305. https://doi.org/10.5194/hess-8-286-2004.

Bungard, R.A., Ruban, A.V., Hibberd, J.M. et al. (1999). Unusual carotenoid composition and a new type of xanthophyll cycle in plants. *Proceedings of the National Academy of Sciences* 96: 1135–1139. https://doi.org/10.1073/pnas.96.3.1135.

Choudhury, S., Panda, P., Sahoo, L., and Panda, S.K. (2013). Reactive oxygen species signaling in plants under abiotic stress. *Plant Signaling & Behavior* 8: e23681. https://doi.org/10.4161/psb.23681.

Clark, M.L. (2017). Comparison of simulated hyperspectral HyspIRI and multispectral Landsat 8 and Sentinel-2 imagery for multi-seasonal, regional land-cover mapping. *Remote Sensing of Environment* 200: 311–325. https://doi.org/10.1016/j.rse.2017.08.028.

Clark, R.N., King, T.V.V, and Gorelick, N.S. (1985). *Automatic Continuum Analysis of Reflectance Spectra*. Airborne Imaging Spectrometer Data Analysis Workshop.

Clark, M.L., Buck-Diaz, J., and Evens, J. (2018). Mapping of forest alliances with simulated multi-seasonal hyperspectral satellite imagery. *Remote Sensing of Environment* 210: 490–507. https://doi.org/10.1016/j.rse.2018.03.021.

Coates, A., Dennison, P., Roberts, D., and Roth, K. (2015). Monitoring the impacts of severe drought on Southern California Chaparral species using hyperspectral and thermal infrared imagery. *Remote Sensing* 7: 14276–14291. https://doi.org/10.3390/rs71114276.

Cole, L.J., Stockan, J., and Helliwell, R. (2020). Managing riparian buffer strips to optimize ecosystem services: a review. *Agriculture, Ecosystems and Environment* 296: 106891. https://doi.org/10.1016/j.agee.2020.106891.

Cortes, C. and Vapnik, V. (1995). Support-vector networks. *Machine Learning* 20: 273–297. https://doi.org/10.1007/BF00994018.

Craft, C. (2016). Forested wetlands. In: *Creating and Restoring Wetlands*, 129–160. Elsevier https://doi.org/10.1016/B978-0-12-407232-9.00006-3.

Demmig, B., Winter, K., Krüger, A., and Czygan, F.-C. (1987). Photoinhibition and zeaxanthin formation in intact leaves. *Plant Physiology* 84: 218–224. https://doi.org/10.1104/pp.84.2.218.

Demmig-Adams, B. and Adams, W.W. (1996). The role of xanthophyll cycle carotenoids in the protection of photosynthesis. *Trends in Plant Science* 1: 21–26. https://doi.org/10.1016/S1360-1385(96)80019-7.

Diepens, N.J., Buffan-Dubau, E., Budzinski, H. et al. (2017). Toxicity effects of an environmental realistic herbicide mixture on the seagrass Zostera noltei. *Environmental Pollution* 222: 393–403. https://doi.org/10.1016/j.envpol.2016.12.021.

Dreesen, F.E., De Boeck, H.J., Janssens, I.A., and Nijs, I. (2012). Summer heat and drought extremes trigger unexpected changes in productivity of a temperate annual/biannual plant community. *Environmental and Experimental Botany* 79: 21–30. https://doi.org/10.1016/j.envexpbot.2012.01.005.

Dronova, I., Gong, P., Clinton, N.E. et al. (2012). Landscape analysis of wetland plant functional types: the effects of image segmentation scale, vegetation classes and classification methods. *Remote Sensing of Environment* 127: 357–369. https://doi.org/10.1016/j.rse.2012.09.018.

Dufour, S., Bernez, I., Betbeder, J. et al. (2013). Monitoring restored riparian vegetation: how can recent developments in remote sensing sciences help? *Knowledge and Management of Aquatic Ecosystems* 410 (10): 15. https://doi.org/10.1051/kmae/2013068.

Dufour, S., Rodríguez-González, P.M., and Laslier, M. (2019). Tracing the scientific trajectory of riparian vegetation studies: main topics, approaches and needs in a globally changing world. *Science of Total Environment* 653: 1168–1185. https://doi.org/10.1016/j.scitotenv.2018.10.383.

Fabre, S., Gimenez, R., Elger, A., and Rivière, T. (2020). Unsupervised monitoring vegetation after the closure of an ore processing site with multi-temporal optical remote sensing. *Sensors (Switzerland)* 20 (17): 4800. https://doi.org/10.3390/s20174800.

Fassnacht, F.E., Latifi, H., Stereńczak, K. et al. (2016). Review of studies on tree species classification from remotely sensed data. *Remote Sensing of Environment* 186: 64–87. https://doi.org/10.1016/j.rse.2016.08.013.

Feret, J.B., François, C., Asner, G.P. et al. (2008). PROSPECT-4 and 5: advances in the leaf optical properties model separating photosynthetic pigments. *Remote Sensing of Environment* 112: 3030–3043. https://doi.org/10.1016/j.rse.2008.02.012.

Frampton, W.J., Dash, J., Watmough, G., and Milton, E.J. (2013). Evaluating the capabilities of Sentinel-2 for quantitative estimation of biophysical variables in vegetation. *ISPRS Journal of Photogrammetry and Remote Sensing* 82: 83–92. https://doi.org/10.1016/j.isprsjprs.2013.04.007.

Fu, P., Meacham-Hensold, K., Guan, K. et al. (2020). Estimating photosynthetic traits from reflectance spectra: a synthesis of spectral indices, numerical inversion, and partial least square regression. *Plant, Cell & Environment* 43: 1241–1258. https://doi.org/10.1111/pce.13718.

Furuya, D.E.G., Aguiar, J.A.F., Estrabis, N.V. et al. (2020). A machine learning approach for mapping forest vegetation in riparian zones in an Atlantic biome environment using Sentinel-2 imagery. *Remote Sensing* 12: 4086. https://doi.org/10.3390/rs12244086.

Gamon, J.A., Huemmrich, K.F., Wong, C.Y.S. et al. (2016). A remotely sensed pigment index reveals photosynthetic phenology in evergreen conifers. *Proceedings of the National Academy of Sciences* 113: 13087–13092. https://doi.org/10.1073/pnas.1606162113.

Garbulsky, M.F., Peñuelas, J., Gamon, J. et al. (2011). The photochemical reflectance index (PRI) and the remote sensing of leaf, canopy and ecosystem radiation use efficiencies: a review and meta-analysis. *Remote Sensing of Environment* 115: 281–297. https://doi.org/10.1016/j.rse.2010.08.023.

García-Plazaola, J.I., Hernández, A., Errasti, E., and Becerril, J.M. (2002). Occurrence and operation of the lutein epoxide cycle in Quercus species. *Functional Plant Biology* 29: 1075. https://doi.org/10.1071/FP02002.

Garrity, S.R., Eitel, J.U.H., and Vierling, L.A. (2011). Disentangling the relationships between plant pigments and the photochemical reflectance index reveals a new approach for remote estimation of carotenoid content. *Remote Sensing of Environment* 115: 628–635. https://doi.org/10.1016/j.rse.2010.10.007.

Goetz, S.J. (2006). Remote sensing of riparian buffers: past progress and future prospects. *Journal of the American Water Resources Association* 42: 133–143. https://doi.org/10.1111/j.1752-1688.2006.tb03829.x.

Grömping, U. (2009). Variable importance assessment in regression: linear regression versus random forest. *The American Statistician* 63: 308–319. https://doi.org/10.1198/tast.2009.08199.

Gundersen, P., Laurén, A., Finér, L. et al. (2010). Environmental services provided from riparian forests in the Nordic countries. *Ambio* 39: 555–566. https://doi.org/10.1007/s13280-010-0073-9.

Haenlein, M. and Kaplan, A.M. (2004). A beginner's guide to partial least squares analysis. *Understanding Statistics* 3: 283–297. https://doi.org/10.1207/s15328031us0304_4.

Haldimann, P., Galle, A., and Feller, U. (2008). Impact of an exceptionally hot dry summer on photosynthetic traits in oak (*Quercus pubescens*) leaves. *Tree Physiology* 28: 785–795. https://doi.org/10.1093/treephys/28.5.785.

Hastie, T., Tibshirani, R., and Friedman, J. (2009). *The Elements of Statistical Learning, Springer Series in Statistics*. New York, NY: Springer https://doi.org/10.1007/978-0-387-84858-7.

Hernández-Clemente, R., Navarro-Cerrillo, R.M., and Zarco-Tejada, P.J. (2012). Carotenoid content estimation in a heterogeneous conifer forest using narrow-band indices and PROSPECT+DART simulations. *Remote Sensing of Environment* 127: 298–315. https://doi.org/10.1016/j.rse.2012.09.014.

Hikosaka, K. and Noda, H.M. (2019). Modeling leaf CO_2 assimilation and photosystem II photochemistry from chlorophyll fluorescence and the photochemical reflectance index. *Plant, Cell & Environment* 42: 730–739. https://doi.org/10.1111/pce.13461.

Hmimina, G., Merlier, E., Dufrêne, E., and Soudani, K. (2015). Deconvolution of pigment and physiologically related photochemical reflectance index variability at the canopy scale over an entire growing season. *Plant, Cell & Environment* 38: 1578–1590. https://doi.org/10.1111/pce.12509.

Hu, X. and Weng, Q. (2009). Estimating impervious surfaces from medium spatial resolution imagery using the self-organizing map and multi-layer perceptron neural networks. *Remote Sensing of Environment* 113: 2089–2102. https://doi.org/10.1016/j.rse.2009.05.014.

Hughes, G. (1968). On the mean accuracy of statistical pattern recognizers. *IEEE Transactions on Information Theory* 14: 55–63. https://doi.org/10.1109/TIT.1968.1054102.

Hutengs, C. and Vohland, M. (2016). Downscaling land surface temperatures at regional scales with random forest regression. *Remote Sensing of Environment* 178: 127–141. https://doi.org/10.1016/j.rse.2016.03.006.

Huylenbroeck, L., Laslier, M., Dufour, S. et al. (2020). Using remote sensing to characterize riparian vegetation: a review of available tools and perspectives for managers. *Journal of Environmental Management* 267: 110652. https://doi.org/10.1016/j.jenvman.2020.110652.

Im, E.-S., Pal, J.S., and Eltahir, E.A.B. (2017). Deadly heat waves projected in the densely populated agricultural regions of South Asia. *Science Advances* 3: e1603322. https://doi.org/10.1126/sciadv.1603322.

Jacquemoud, S., Ustin, S.L., Verdebout, J. et al. (1996). Estimating leaf biochemistry using the PROSPECT leaf optical properties model. *Remote Sensing of Environment* 56: 194–202. https://doi.org/10.1016/0034-4257(95)00238-3.

Jacquemoud, S., Verhoef, W., Baret, F. et al. (2009). PROSPECT+SAIL models: a review of use for vegetation characterization. *Remote Sensing of Environment* 113: S56–S66. https://doi.org/10.1016/j.rse.2008.01.026.

Jeronimo, S.M.A., Kane, V.R., Churchill, D.J. et al. (2018). Applying LiDAR individual tree detection to management of structurally diverse forest landscapes. *Journal of Forestry* 116: 336–346. https://doi.org/10.1093/jofore/fvy023.

Jiang, J., Comar, A., Burger, P. et al. (2018). Estimation of leaf traits from reflectance measurements: comparison between methods based on vegetation indices and several versions of the PROSPECT model. *Plant Methods* 14: https://doi.org/10.1186/s13007-018-0291-x.

Junker, L.V. and Ensminger, I. (2016). Relationship between leaf optical properties, chlorophyll fluorescence and pigment changes in senescing *Acer saccharum* leaves. *Tree Physiology* 36: 694–711. https://doi.org/10.1093/treephys/tpv148.

Juvany, M., Müller, M., and Munné-Bosch, S. (2013). Photo-oxidative stress in emerging and senescing leaves: a mirror image. *Journal of Experimental Botany* 64: 3087–3098. https://doi.org/10.1093/jxb/ert174.

Kennard, R.W. and Stone, L.A. (1969). *Computer Aided Design of Experiments, in: Technometrics*, 137–148. London, UK: Palgrave Macmillan https://doi.org/10.1007/978-1-349-95810-8_109.

Klemas, V. (2014). Remote sensing of riparian and wetland buffers: an overview. *Journal of Coastal Research* 297: 869–880. https://doi.org/10.2112/jcoastres-d-14-00013.1.

Kokaly, R. (1999). Spectroscopic determination of leaf biochemistry using band-depth analysis of absorption features and stepwise multiple linear regression. *Remote Sensing of Environment* 67: 267–287. https://doi.org/10.1016/S0034-4257(98)00084-4.

Kong, W., Huang, W., Liu, J. et al. (2017). Estimation of canopy carotenoid content of winter wheat using multi-angle hyperspectral data. *Advances in Space Research* 60: 1988–2000. https://doi.org/10.1016/j.asr.2017.07.015.

Kováč, D., Malenovský, Z., Urban, O. et al. (2013). Response of green reflectance continuum removal index to the xanthophyll de-epoxidation cycle in Norway spruce needles. *Journal of Experimental Botany* 64: 1817–1827. https://doi.org/10.1093/jxb/ert069.

Lassalle, G., Elger, A., Credoz, A. et al. (2019a). Toward quantifying oil contamination in vegetated areas using very high spatial and spectral resolution imagery. *Remote Sensing* 11: 2241. https://doi.org/10.3390/rs11192241.

Lassalle, G., Fabre, S., Credoz, A. et al. (2019b). Application of PROSPECT for estimating total petroleum hydrocarbons in contaminated soils from leaf optical properties. *Journal of Hazardous Materials* 377: 409–417. https://doi.org/10.1016/j.jhazmat.2019.05.093.

Lassalle, G., Fabre, S., Credoz, A. et al. (2019c). Detection and discrimination of various oil-contaminated soils using vegetation reflectance. *Science of Total Environment* 655: 1113–1124. https://doi.org/10.1016/j.scitotenv.2018.11.314.

Lassalle, G., Fabre, S., Credoz, A. et al. (2020). Monitoring oil contamination in vegetated areas with optical remote sensing: a comprehensive review. *Journal of Hazardous Materials* 393: 122427. https://doi.org/10.1016/j.jhazmat.2020.122427.

Lausch, A., Erasmi, S., King, D. et al. (2016). Understanding forest health with remote sensing – part I – a review of spectral traits, processes and remote-sensing characteristics. *Remote Sensing* 8: 1029. https://doi.org/10.3390/rs8121029.

LeCun, Y., Bengio, Y., and Hinton, G. (2015). Deep learning. *Nature* 521: 436–444. https://doi.org/10.1038/nature14539.

Lelieveld, J., Proestos, Y., Hadjinicolaou, P. et al. (2016). Strongly increasing heat extremes in the Middle East and North Africa (MENA) in the 21st century. *Climatic Change* 137: 245–260. https://doi.org/10.1007/s10584-016-1665-6.

Listopad, C.M.C.S., Masters, R.E., Drake, J. et al. (2015). Structural diversity indices based on airborne LiDAR as ecological indicators for managing highly dynamic landscapes. *Ecological Indicators* 57: 268–279. https://doi.org/10.1016/j.ecolind.2015.04.017.

Loizzo, R., Daraio, M., Guarini, R. et al. (2019). Prisma Mission Status and Perspective. In: *IGARSS 2019–2019; IEEE International Geoscience and Remote Sensing Symposium*, 4503–4506. IEEE https://doi.org/10.1109/IGARSS.2019.8899272.

Magdaleno, F. and Martinez, R. (2014). Evaluating the quality of riparian forest vegetation: the riparian forest evaluation (RFV) index. *Forest Systems* 23: 259–272. https://doi.org/10.5424/fs/2014232-04488.

Matsubara, S., Morosinotto, T., Osmond, C.B., and Bassi, R. (2007). Short- and long-term operation of the lutein-epoxide cycle in light-harvesting antenna complexes. *Plant Physiology* 144: 926–941. https://doi.org/10.1104/pp.107.099077.

Matsubara, S., Krause, G.H., Seltmann, M. et al. (2008). Lutein epoxide cycle, light harvesting and photoprotection in species of the tropical tree genus Inga. *Plant, Cell & Environment* 31: 548–561. https://doi.org/10.1111/j.1365-3040.2008.01788.x.

Matsunaga, T., Iwasaki, A., Tsuchida, S. et al. (2018). Hisui Status Toward FY2019 Launch. In: *IGARSS 2018–2018; IEEE International Geoscience and Remote Sensing Symposium*, 160–163. *IEEE*. doi: 10.1109/IGARSS.2018.8518639.

Meerdink, S.K., Hook, S.J., Roberts, D.A., and Abbott, E.A. (2019). The ECOSTRESS spectral library version 1.0. *Remote Sensing of Environment* 230: 111196. https://doi.org/10.1016/j.rse.2019.05.015.

Michel, S., Gamet, P., and Lefevre-Fonollosa, M.-J. (2011). HYPXIM - A hyperspectral satellite defined for science, security and defence users. In: *2011 3rd Workshop on Hyperspectral Image and Signal Processing: Evolution in Remote Sensing (WHISPERS)*, 1–4. *IEEE*. doi: 10.1109/WHISPERS.2011.6080864.

Michez, A., Piégay, H., Lisein, J. et al. (2016). Classification of riparian forest species and health condition using multi-temporal and hyperspatial imagery from unmanned aerial system. *Environmental Monitoring and Assessment* 188: 146. https://doi.org/10.1007/s10661-015-4996-2.

Milton, E.J. (1987). Principles of field spectroscopy. *International Journal of Remote Sensing* 8: 1807–1827. https://doi.org/10.1080/01431168708954818.

Milton, E.J., Schaepman, M.E., Anderson, K. et al. (2009). Progress in field spectroscopy. *Remote Sensing of Environment* 113: S92–S109. https://doi.org/10.1016/j.rse.2007.08.001.

Miraglio, T., Adeline, K., Huesca, M. et al. (2019). Monitoring LAI, chlorophylls, and carotenoids content of a woodland savanna using hyperspectral imagery and 3D radiative transfer modeling. *Remote Sensing* 12: 28. https://doi.org/10.3390/rs12010028.

Morel, J., Jay, S., Féret, J.B. et al. (2018). Exploring the potential of PROCOSINE and close-range hyperspectral imaging to study the effects of fungal diseases on leaf physiology. *Scientific Reports* 8: 1–13. https://doi.org/10.1038/s41598-018-34429-0.

NASA, (2018). HyspIRI Final Report.

Niyogi, K.K., Bjorkman, O., and Grossman, A.R. (2002). The roles of specific xanthophylls in photoprotection. *Proceedings of the National Academy of Sciences* 94: 14162–14167. https://doi.org/10.1073/pnas.94.25.14162.

Noomen, M.F., Skidmore, A.K., van der Meer, F.D., and Prins, H.H.T. (2006). Continuum removed band depth analysis for detecting the effects of natural gas, methane and ethane on maize reflectance. *Remote Sensing of Environment* 105: 262–270. https://doi.org/10.1016/j.rse.2006.07.009.

Okujeni, A., van der Linden, S., Tits, L. et al. (2013). Support vector regression and synthetically mixed training data for quantifying urban land cover. *Remote Sensing of Environment* 137: 184–197. https://doi.org/10.1016/j.rse.2013.06.007.

Othman, R., Mohd Zaifuddin, F.A., and Hassan, N.M. (2014). Carotenoid biosynthesis regulatory mechanisms in plants. *Journal of Oleo Science* 63: 753–760. https://doi.org/10.5650/jos.ess13183.

Pace, G., Gutiérrez-Cánovas, C., Henriques, R. et al. (2021). Remote sensing depicts riparian vegetation responses to water stress in a humid Atlantic region. *Science of the Total Environment* 772: 145526. https://doi.org/10.1016/j.scitotenv.2021.145526.

Pedregosa, F., Varoquaux, G., Gramfort, A. et al. (2011). Scikit-learn: machine learning in Python. *Journal of Machine Learning Research* 12: 2825–2830.

Phiri, D., Simwanda, M., Salekin, S. et al. (2020). Sentinel-2 data for land cover/use mapping: a review. *Remote Sensing* 12: 2291. https://doi.org/10.3390/rs12142291.

Poff, B., Koestner, K.A., Neary, D.G., and Henderson, V. (2011). Threats to riparian ecosystems in western North America: an analysis of existing Literature1. *JAWRA; Journal of the American Water Resoures Association* 47: 1241–1254. https://doi.org/10.1111/j.1752-1688.2011.00571.x.

Rumpf, T., Mahlein, A.-K., Steiner, U. et al. (2010). Early detection and classification of plant diseases with support vector machines based on hyperspectral reflectance. *Computers and Electronics in Agriculture* 74: 91–99. https://doi.org/10.1016/j.compag.2010.06.009.

Sanches, I.D., Souza Filho, C.R., Magalhães, L.A. et al. (2013). Assessing the impact of hydrocarbon leakages on vegetation using reflectance spectroscopy. *ISPRS Journal of Photogrammetry and Remote Sensing* 78: 85–101. https://doi.org/10.1016/j.isprsjprs.2013.01.007.

Santini, A., Liebhold, A., Migliorini, D., and Woodward, S. (2018). Tracing the role of human civilization in the globalization of plant pathogens. *The ISME Journal* 12: 647–652. https://doi.org/10.1038/s41396-017-0013-9.

Savitzky, A. and Golay, M.J.E. (1964). Smoothing and differentiation of data by simplified least squares procedures. *Analytical Chemistry* 36: 1627–1639. https://doi.org/10.1021/ac60214a047.

Shahid, M., Pourrut, B., Dumat, C. et al. (2014). Heavy-metal-induced reactive oxygen species: phytotoxicity and physicochemical changes in plants. In: *Reviews of Environmental Contamination and Toxicology, Reviews of Environmental Contamination and Toxicology* (ed. D.M. Whitacre), 1–44. Cham: Springer International Publishing https://doi.org/10.1007/978-3-319-06746-9_1.

Shi, T., Liu, H., Wang, J. et al. (2014). Monitoring arsenic contamination in agricultural soils with reflectance spectroscopy of rice plants. *Environmental Science & Technology* 48: 6264–6272. https://doi.org/10.1021/es405361n.

Smola, A.J. and Schölkopf, B. (2004). A tutorial on support vector regression. *Statistics and Computing* 14: 199–222. https://doi.org/10.1023/B:STCO.0000035301.49549.88.

Sonesson, J., Ring, E., Högbom, L. et al. (2020). Costs and benefits of seven alternatives for riparian forest buffer management. *Scandinavian Journal of Forest Research* 36 (2–3): 1–9. https://doi.org/10.1080/02827581.2020.1858955.

Stella, J.C. and Bendix, J. (2019). Multiple stressors in riparian ecosystems. In: *Multiple Stressors in River Ecosystems*, 81–110. Elsevier https://doi.org/10.1016/B978-0-12-811713-2.00005-4.

Stimson, H.C., Breshears, D.D., Ustin, S.L., and Kefauver, S.C. (2005). Spectral sensing of foliar water conditions in two co-occurring conifer species: *Pinus edulis and Juniperus monosperma*. *Remote Sensing of Environment* 96: 108–118. https://doi.org/10.1016/j.rse.2004.12.007.

Stylinski, C.D., Gamon, J.A., and Oechel, W.C. (2002). Seasonal patterns of reflectance indices, carotenoid pigments and photosynthesis of evergreen chaparral species. *Oecologia* 131: 366–374. https://doi.org/10.1007/s00442-002-0905-9.

Suzuki, N., Rivero, R.M., Shulaev, V. et al. (2014). Abiotic and biotic stress combinations. *The New Phytologist* 203: 32–43. https://doi.org/10.1111/nph.12797.

Teskey, R., Wertin, T., Bauweraerts, I. et al. (2015). Responses of tree species to heat waves and extreme heat events. *Plant, Cell & Environment* 38: 1699–1712. https://doi.org/10.1111/pce.12417.

Tran, T.N., Afanador, N.L., Buydens, L.M.C., and Blanchet, L. (2014). Interpretation of variable importance in partial least squares with significance multivariate correlation (SMC). *Chemometrics and Intelligent Laboratory Systems* 138: 153–160. https://doi.org/10.1016/j.chemolab.2014.08.005.

Ustin, S.L., Schaepman, M.E., Gitelson, A.A. et al. (2009). Retrieval of foliar information about plant pigment systems from high resolution spectroscopy. *Remote Sensing of Environment* 113: S67–S77. https://doi.org/10.1016/j.rse.2008.10.019.

Van Cleemput, E., Vanierschot, L., Fernández-Castilla, B. et al. (2018). The functional characterization of grass- and shrubland ecosystems using hyperspectral remote sensing: trends, accuracy and moderating variables. *Remote Sensing of Environment* 209: 747–763. https://doi.org/10.1016/j.rse.2018.02.030.

Verrelst, J., Muñoz, J., Alonso, L. et al. (2012). Machine learning regression algorithms for biophysical parameter retrieval: opportunities for Sentinel-2 and -3. *Remote Sensing of Environment* 118: 127–139. https://doi.org/10.1016/j.rse.2011.11.002.

Wagner, F.H., Ferreira, M.P., Sanchez, A. et al. (2018). Individual tree crown delineation in a highly diverse tropical forest using very high resolution satellite images. *ISPRS Journal of Photogrammetry and Remote Sensing* 145: 362–377. https://doi.org/10.1016/j.isprsjprs.2018.09.013.

Wong, C.Y.S. and Gamon, J.A. (2015). The photochemical reflectance index provides an optical indicator of spring photosynthetic activation in evergreen conifers. *The New Phytologist* 206: 196–208. https://doi.org/10.1111/nph.13251.

Wong, C.Y.S., D'Odorico, P., Bhathena, Y. et al. (2019). Carotenoid based vegetation indices for accurate monitoring of the phenology of photosynthesis at the leaf-scale in deciduous and evergreen trees. *Remote Sensing of Environment* 233: 111407. https://doi.org/10.1016/j.rse.2019.111407.

Young, A.J. (1991). The photoprotective role of carotenoids in higher plants. *Physiologia Plantarum* 83: 702–708. https://doi.org/10.1111/j.1399-3054.1991.tb02490.x.

Zarco-Tejada, P.J., Camino, C., Beck, P.S.A. et al. (2018). Previsual symptoms of *Xylella fastidiosa* infection revealed in spectral plant-trait alterations. *Nature Plants* 4: 432–439. https://doi.org/10.1038/s41477-018-0189-7.

Zhang, Y., Huang, J., Wang, F. et al. (2017). An extended PROSPECT: advance in the leaf optical properties model separating total chlorophylls into chlorophyll a and b. *Scientific Reports* 7: 6429. https://doi.org/10.1038/s41598-017-06694-y.

Zhen, Z., Quackenbush, L.J., and Zhang, L. (2016). Trends in automatic individual tree crown detection and delineation-evolution of LiDAR data. *Remote Sensing* 8: 1–26. https://doi.org/10.3390/rs8040333.

Zou, H. and Hastie, T. (2005). Regression and variable selection via the elastic net. *Journal of the Royal Statistical Society. Series B (Statistical Methodology)* 67: 301–320. https://doi.org/10.1111/j.1467-9868.2005.00503.x.

Supporting Information

Distinguishing carotene and xanthophyll contents in the leaves of riparian forest species by applying machine learning algorithms to field reflectance data.

Guillaume Lassalle[a],*
Sophie Fabre[a]

[a]*Office National d'Études et de Recherches Aérospatiales (ONERA), Toulouse, France*
**Corresponding author: Guillaume Lassalle, Office National d'Études et de Recherches Aérospatiales, 2 Avenue Edouard Belin, 31055 Toulouse, France; E-mail: guillaumelassalle.pro@gmail.com*

Table S3.1 Calibrated parameters of the machine learning algorithms retained for carotenoid estimation. (PLSR: Partial Least Square Regression; SVR: Support Vector Regression; RF: Random Forest regression; ENET: Elastic Net regression; NN: Neural Network.)

	Violaxanthin	Neoxanthin	Antheraxanthin	Zeaxanthin	Lutein	β-carotene
PLSR:						
components	2	2	2	2	2	2
SVR:						
kernel	linear	linear	linear	linear	linear	linear
RF:						
number of trees	100	70	10	50	50	10
min samples leaf	10	2	2	5	2	2
min samples split	2	2	5	2	10	2
ENET:						
alpha	0.71	0.13	0.01	0.3	0.62	0.16
L1:L2 ratio	0.3	0.8	1	0.6	0.6	1
NN:						
Neurons (hidden layer)	90	110	50	110	50	150
activation function	rectified linear unit	logistic	logistic	rectified linear unit	logistic	logistic

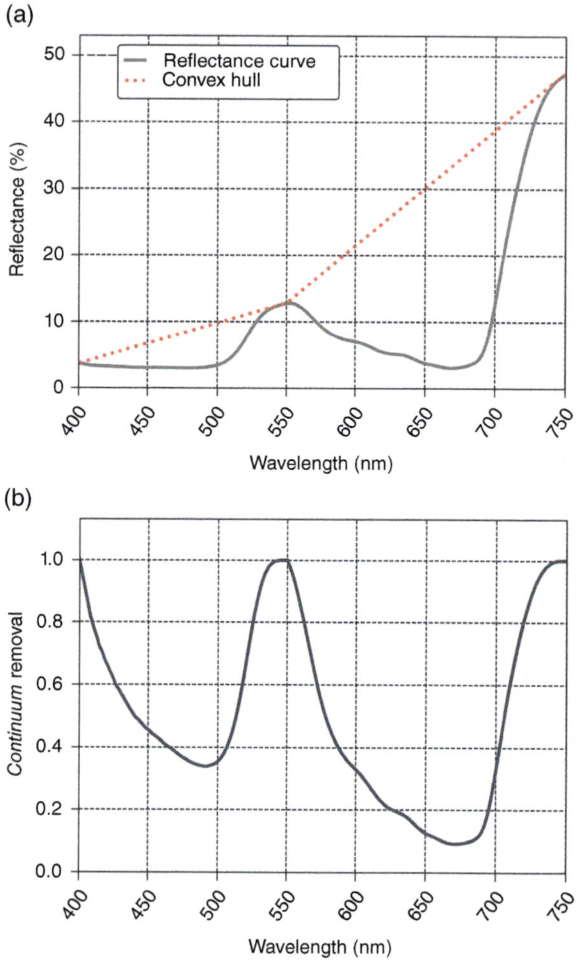

Figure S3.1 Example of the *continuum* removal algorithm applied to the spectral signature of a leaf in the visible region. (a) Original spectral signature (reflectance curve) and convex hull, and (b) resulting *continuum* removed spectral signature.

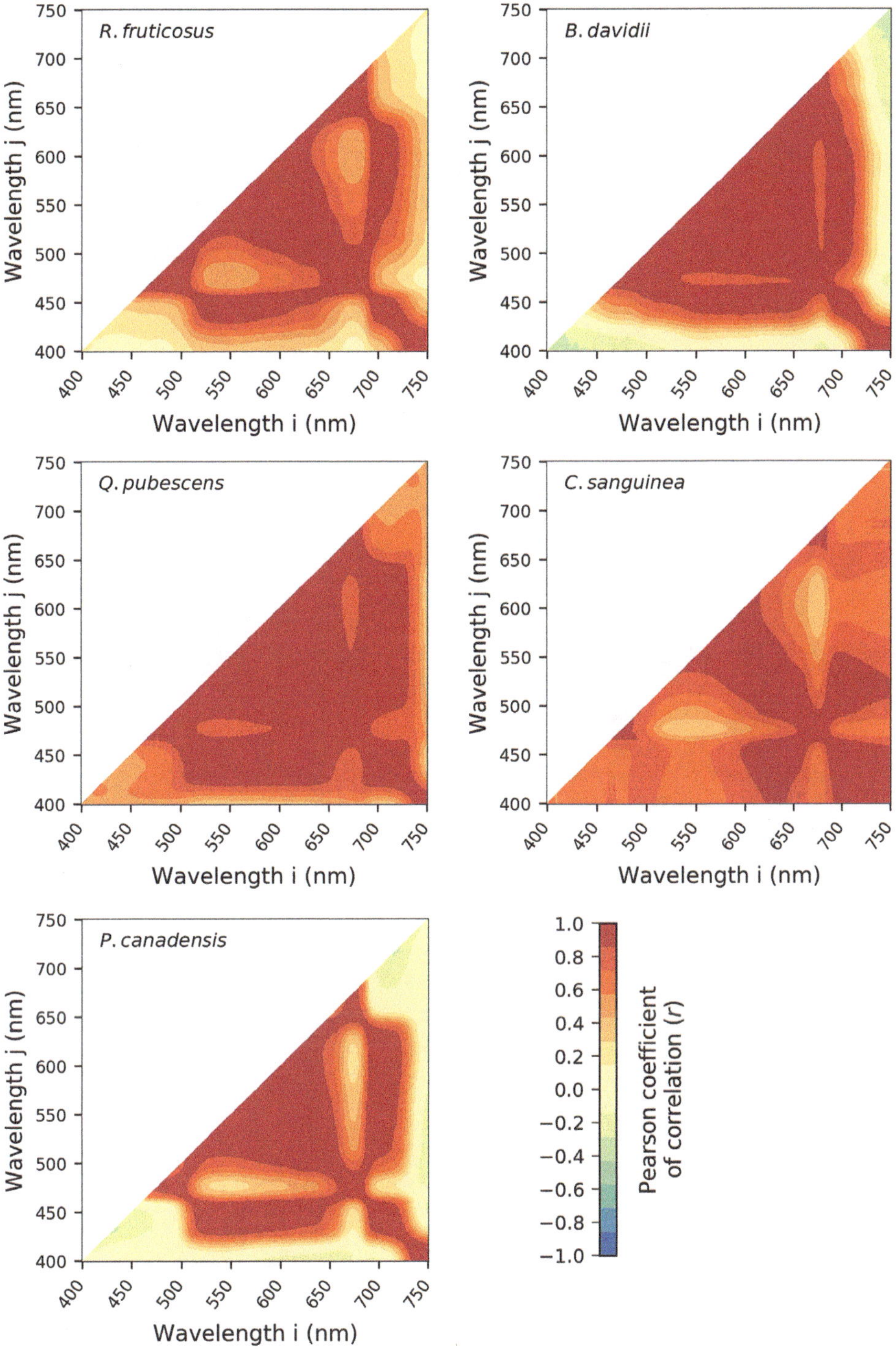

Figure S3.2 Correlation maps obtained between all possible pairs of wavelengths using the leaf reflectance data from the five species. These heat maps were obtained by computing the Pearson correlation coefficient (r) between the reflectance values at each pair of wavelengths i and j at the significance level of $p < 0.05$, after verifying that the data followed normal distribution (Shapiro–Wilk test, $p < 0.05$). The deep blue and red regions show the highest correlations.

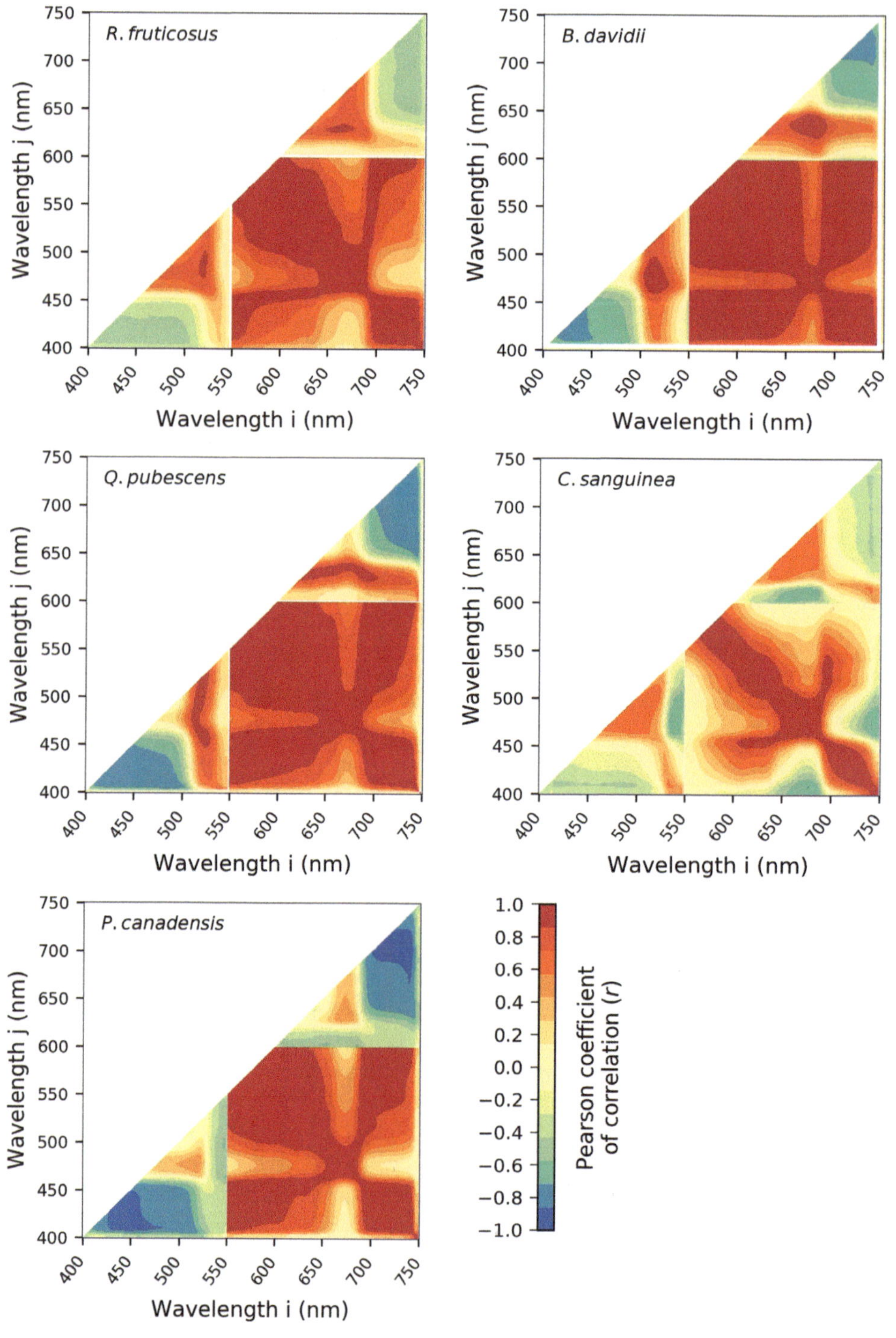

Figure S3.3 Correlation maps obtained between all possible pairs of wavelengths using the leaf *continuum* removed reflectance data from the five species. These heat maps were obtained by computing the Pearson correlation coefficient (r) between the *continuum* removed reflectance values at each pair of wavelengths i and j at the significance level of $p < 0.05$, after verifying that the data followed normal distribution (Shapiro–Wilk test, $p < 0.05$). The deep blue and red regions show the highest correlations.

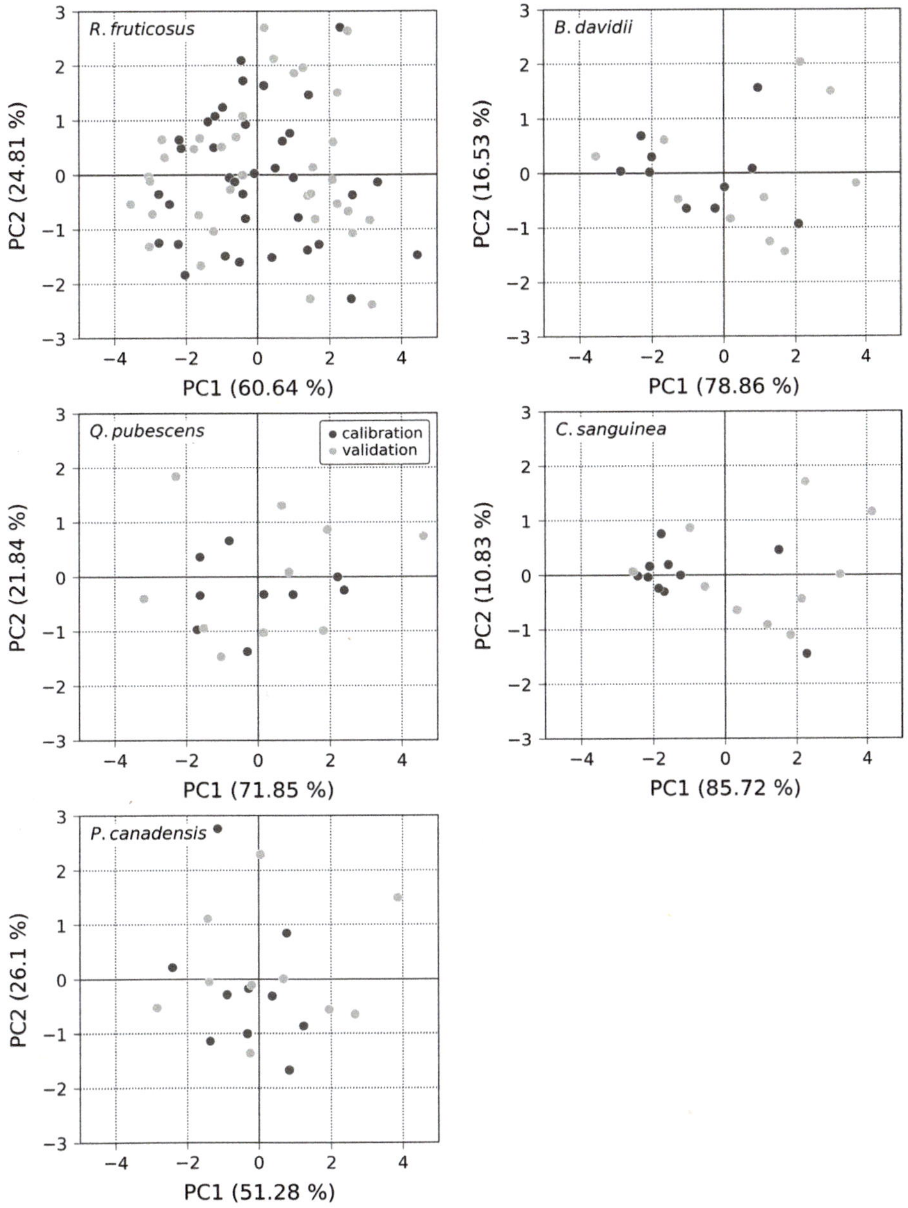

Figure S3.4 Samples selected for the calibration and validation datasets using the Kennard–Stone algorithm. For graphical convenience, the data were projected in the two first components of the PCA and computed using all the carotenoids as input variables, for each species.

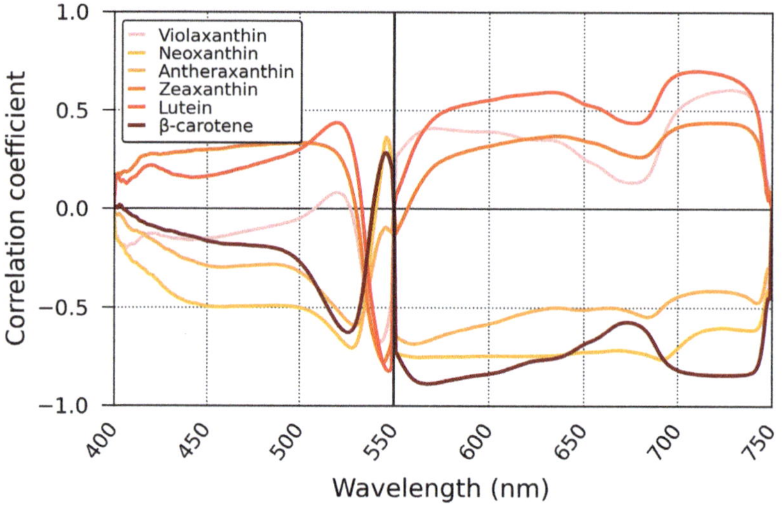

Figure S3.5 Pearson coefficient of correlation (*r*) obtained between continuum removed reflectance values at every wavelength and the six carotenoid contents in the leaves of the five species. The vertical black line at 550 nm shows the separation of the two spectral regions were the continuum removal algorithm was applied.

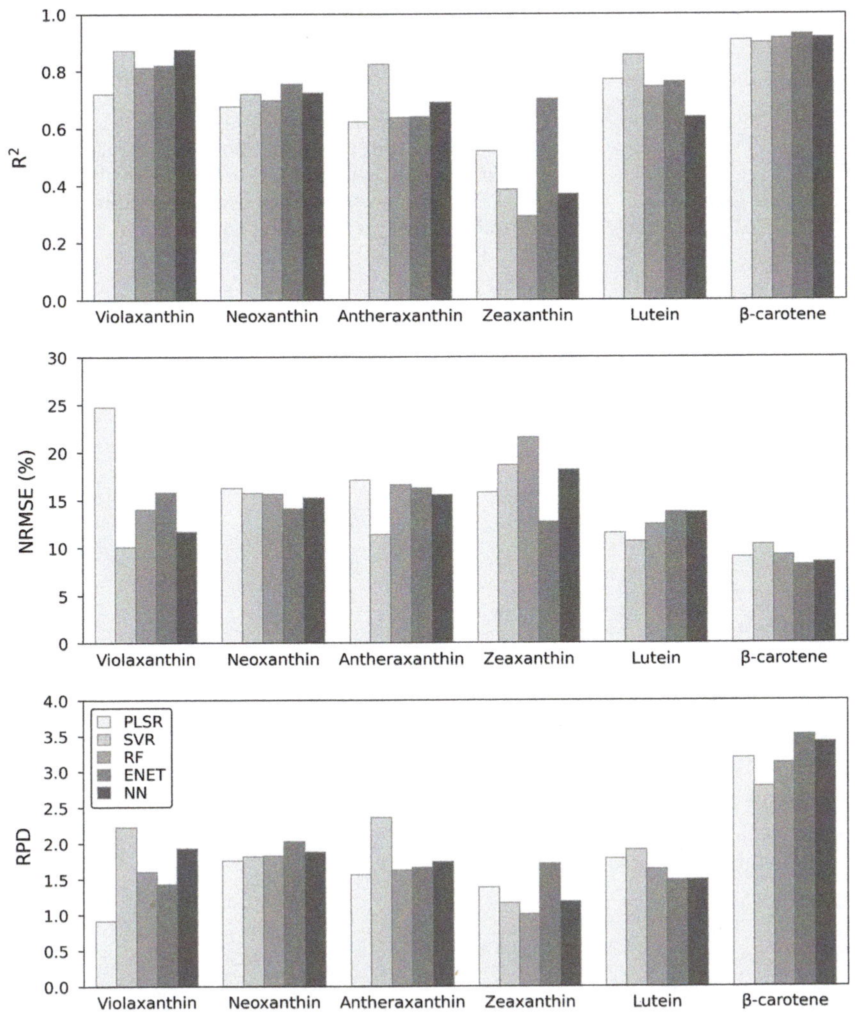

Figure S3.6 Suitability of the different machine learning algorithms to predict the six leaf carotenoid contents (PLSR: Partial Least Square Regression; SVR: Support Vector Regression; RF: Random Forest regression; ENET: Elastic Net regression; NN: Neural Network). (top figure) R^2 obtained between the measured and predicted carotenoid concentrations; (middle figure) Normalized Root Mean Square Errors (NRMSEs) of the predictions; and (bottom figures) Residual Predictive Deviation (RPD) of the predictions. For each carotenoid, the best algorithm is the one that satisfies the highest R^2 and RPD and the lower NRMSE.

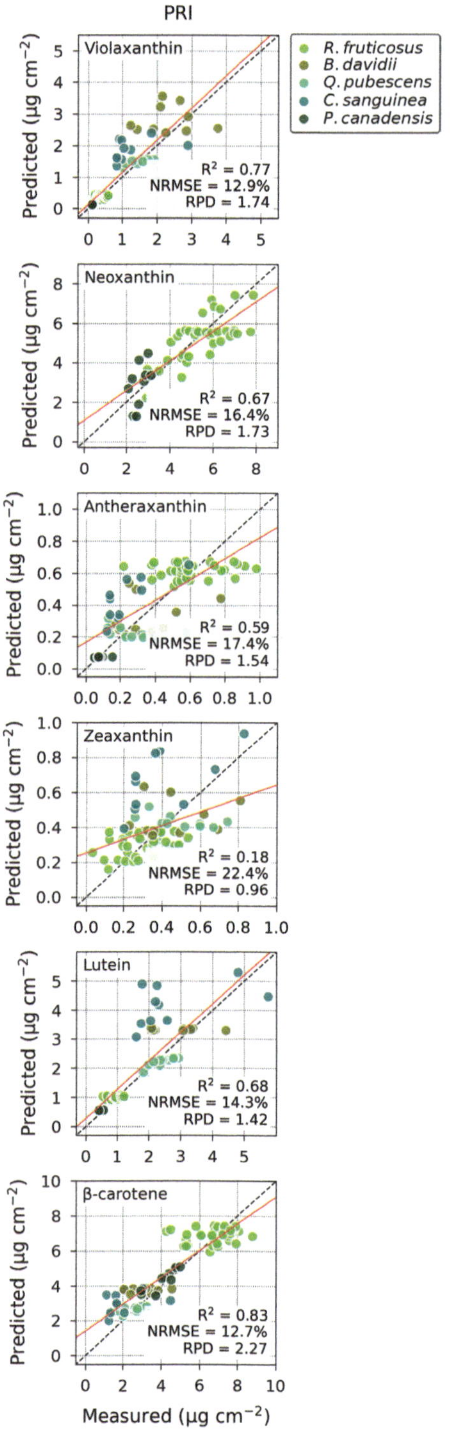

Figure S3.7 Comparison between the measured and predicted carotenoid contents in leaves obtained using the Photochemical Reflectance Index (PRI) (see Wong and Gamon 2015 for the original index formula). For each carotenoid, the overall R^2, the Normalized Root Mean Square Error (NRMSE), and the Residual Predictive Deviation (RPD) are presented. The dashed line represents the 1:1 line (perfect predictions).

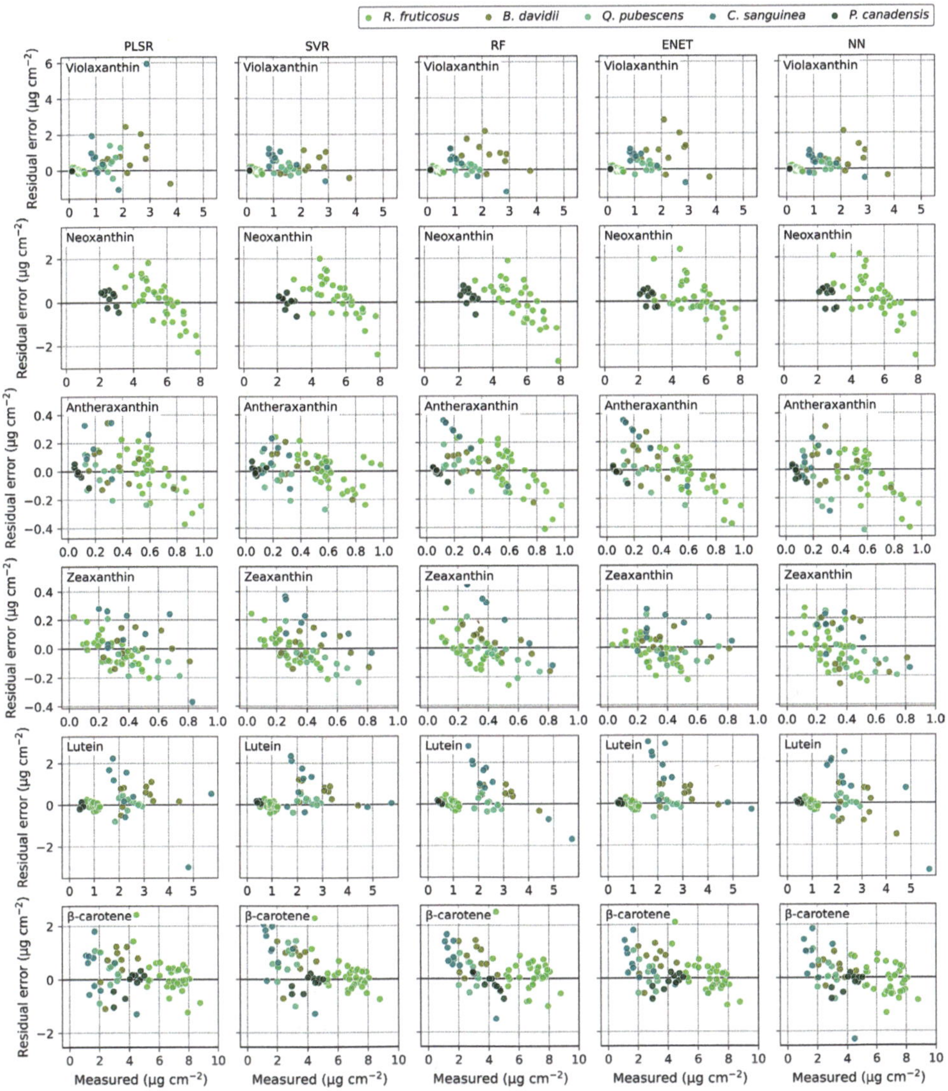

Figure S3.8 Regression residuals for the six predicted carotenoid contents in leaves obtained using the five ML algorithms (in columns, PLSR: Partial Least Square Regression; SVR: Support Vector Regression; RF: Random Forest regression; ENET: Elastic Net regression; NN: Neural Network). Values above zero indicate overestimations of the carotenoid content, whereas values below zero correspond to underestimations

4

Modeling of Abiotic Stress of Conifers with Remote Sensing Data

*Lachezar Filchev**

Space Research and Technology Institute, Bulgarian Academy of Sciences, Bl.1 Acad. G. Bonchev Street, 1113, Sofia, Bulgaria

4.1 Introduction

Over the past four decades, the number of testimonials regarding stressed and dysfunctional forest ecosystems has increased, impacted by human activities at local, regional, and global levels. The importance of monitoring for abiotic stress for the sustainable use and conservation of natural resources and forests in particular is a priority in a number of international and national documents. For example, the list of criteria and indicators for sustainable forest management prepared by the Food and Agriculture Organization (FAO) at The United Nations (UN) as well as in a number of national criteria; Canada, Criteria 2 – Element 2.1 "Occurrence of forest disturbances and stress" is present with eight indicators, Franklin (2001). The protection of forests—in particular conifers—is of special interest in decisions of the Ministerial Conference on the Protection of Forests in Europe (FOREST EUROPE), Oslo 2011; Improved pan-European criteria and indicators for sustainable forest management (2002); The EC Green Paper on Forest Protection and Information in the EU: "Preparing Forests for Climate Change" (2010); The EC White Paper "Adapting to Climate Change: Toward a European Framework for Action" (2009); The European Biodiversity Conservation Strategy (2011) (see Haslett et al. 2010; New EU Forest Strategy 2012; the New EU Forest Strategy 2020) which had a new improved release with the EU 2030 Biodiversity Strategy (2021). In addition, the forest condition is monitored EU-wide by ICP Forest jointly by European Commission (EC) and the United Nations Economic Commission for Europe and published on a regular basis since 1990s, i.e. EC (2012), Forest Condition in Europe (2010, 2012).

Significant efforts are also being made to determine the permitted concentrations/loads of heavy metals, metalloids, and radionuclides for certain plant and animal species that are bioindicators of chemical pollution of ecosystems. Efforts are also made by a number of interdisciplinary scientific fields, such as plant physiology, phytopathology, phytocenology,

*Corresponding author's email: lachezarhf@space.bas.bg

0000-0002-6248-0148 (Lachezar Filchev)

Advances in Remote Sensing for Forest Monitoring, First Edition. Edited by Prem C. Pandey and Paul Arellano.

stress ecology, geochemical ecology, geochemistry of landscapes, ecotoxicology, radioecology, and radiobiology to detect the factors and causative agents of various types of stress.

Internationally, organizations such as the International Atomic Energy Agency (IAEA) and the European Commission (EC) deal with the definition of the element loads as a reflection of the guidelines set out in their recommendations, acts, and directives. Some countries have already adopted nonhuman protection standards for biota, such as England and Wales (UK), whose requirements for environmental impact assessment (EIA) affecting NATURA 2000 habitats include ionizing radiation. In the US, biota protection guidelines and minimum toxicity levels are given by USDOE Orders 5400.5 and 450.1. Major joint projects, such as impacts of elevated CO_2 levels and air pollutants on tree physiology (ICAT) developed in the EU are examples of the specific efforts and interest in the field of air basin pollution with harmful substances and chemical compounds and their impact on the physiology of forests and conifers in particular.

Atmospheric pollution and acid rainfall continue to be a problem for some regions of Bulgaria, as shown by observations under the International Co-operative Programme on Assessment and Monitoring of Air Pollution Effects on Forests (ICP "Forest"). In some coniferous ecosystems, for example, *Pinus ponderosa* L., the so-called "coniferous ecosystems" are observed with Ecosystem Syndrome Distress syndrome (ESD), accompanied by dramatic changes throughout the ecosystem (Rapport 1989; Rapport et al. 1998). In the late 1990s, problems with soil acidification and acid rain led to the emergence of the so-called "acid rains" and "new forest decline." For example, because of soil acidification in the Sudets' Ore Mountain, an area of about 8000 ha has developed Mg deficiency (Omasa et al. 2005).

The concept of *stress* was first introduced in 1935 by H. Selye and it defined as a state of critical load, which is expressed by a specific syndrome—stress syndrome—consisting of all nonspecific changes occurring in the biological system (Selye 1976). The reverse state of stress is *ex-stress* or *eu-stress*, or a positive non-specific reaction of the body to any impact on it ("good" stress) (Reimers 1980). Contrary to ex-tress is *distress* or a "bad" stress, and which covers all negative changes in excessive irritant force that the system cannot overcome (Dedew 1989). There are also degrees of manifestation of the stress situation, such as *eu*-stress, where it is understood as a stress in which adaptation occurs and protective mechanisms are developed under the action of the stressor; while *di*-stress means stress which, under the collective action of the stressor, culminates in a fatal end. Physiological stress means all nonspecific protective physiological reactions and manifestations causing an abnormal state of the body as a result of various adverse factors. Environmental stress is understood to disrupt the internal structure of the population, reduce the fertility of species, increase mortality, etc. (Dedew 1989).

A similar concept of stress but on a community level is the *digression* known by ecosystem ecology; it is a deterioration of the state of biotic communities under the action of internal or external factors, which differs from: *exodynamic digression*—with long-acting external factors—flooding, salinization; *anthropodynamic digression*— under the action of human factors; and *endodynamic digression*—under the action of biogenic internal factors (Kutov and Kostova 1988). Both stress and digression end in degradation of the biotic environment or ecosystem; *catacenosis*. Stressful situations in vegetation are usually accompanied by vegetation disturbances, which are the result of the occurrence of the stress situation itself and consists of a change in the biomass of the plant or co-society due to a process or phenomenon (Odum 1986).

The *disturbance of vegetation* in plant ecology, unlike stress, does not lead to the death of the individual or co-society, and is thus perceived as a narrower concept than the stressful situation. Another related definition of *forest health* is given by stress ecology, such as systemic organization, sustainability, and viability, as well as the absence of signs of ecosystem distress (Schaeffer, Henricks et al. 1988; Rapport et al. 1998). Forest health indicators include indicators such as tree density and productivity, nutrient stocks and cycles, predominance of communicable diseases, and destructive fire potential (Rapport et al. 1998).

The German physiologist Stocker (1947), cited later by Lichtenthaler (1998) in his notorious work on the stress concept in plants, determines the following phases of the stressful state of vegetation:

- *Reaction phase* or *alarm phase* is an initial action of the stressor leading to functional changes, which lead to reduced vitality of the system;
- *Restorative phase* in which normal functional processes are restored and the stability of the system is increased;
- *Resistance phase* in which adaptation mechanisms are developed;
- *Final phase* in which discontinuation of the effects of irritants and restoration of the original state of the system is observed.

If the impact of stressors continues or their intensity increases, the system breaks and death occurs; an *exhausting phase* occurs with too long of a load on the stressor side and exhaustion of its adaptive capabilities. It is expressed in chronic changes (imbalance of metabolic processes, as catabolism prevails, and rapid aging and death processes occur).

Stressful situations in coniferous species, as in other plant species, are characterized by their *factors* (stressors, causative agents) and *response* (external expressions; strain, exhaustion). The factors that determine the occurrence and development of a stressful situation are divided into three groups; *biotic*, *abiotic*, and *anthropogenic* (Sukhih and Sinicina 1979). Changes occurring in vegetation status are autogenic—driven by characteristics of plant species and allogeneic ones —and caused by external factors (Mulligan and Wainwright 2004). According to other studies, factors and stressful situations are grouped into two more general groups: *abiotic* and *biotic* (Jones and Schofield 2008). To the first group, the authors refer diseases and pests to forest plantations, as well as mechanical damage caused by various environmental factors. The main cause of the appearance and development of one or another abiotic stress factor is the lowering of the biological resistance of the individual, which is caused by other environmental factors. For example, in drought or forest damage conditions, a number of pests of coniferous plantations develop. Acute contamination and defeat may be caused by high concentrations of toxicants acting in short periods of time; minutes and hours. It is expressed in irreversible phenotypic changes of assimilation tissues, leading to a violation of gas exchange and the death of the plant. The second group, biotic, consists of chronic damage resulting from the systematic action of one or a group of pollutants occurring as drywall, cutting down the crowns of trees, lowering of growth, reduction of dimensions, necrosis, staining change, deflation, and cessation of fertility. Hidden or physiological stressful situations occur in negligible concentrations of phytotoxin. Physiological changes are expressed in violation of the growth and functioning of the body, the lowering of chlorophyll content, and chlorosis.

4.2 Natural Factors

Ingestion of chemical element by plants is influenced by *natural* and *anthropogenic* factors. Natural factors include the amount of falling solar radiation (insolation), fluctuations (24/7, seasonal and annual) of air temperature, and the amount of precipitation.

4.2.1 Soils

4.2.1.1 Oxidation–Reduction Conditions

Depending on the pH^- and eH+ of soils, heavy metals are caught in the soil exchange or non-exchange and form insoluble salts, which affects their mobility and hence the retention of plant growth, which leads to depigmentation and yellowing—chlorosis and subsequent necrosis of the root system. It is also known that when soil acidification and pH^- changes from 4 to 5.5 it leads to Zn sorption on Fe and Al hydroxides, while the ions of Cu^{2+} and Ca^{+2} are not sorb due to the lack of free active centers—ions of Al. At 7.5 pH, the solubility of zinc increases again due to the formation of organo-metal compounds, and the addition of Cu in soils in turn lowers the sorption of Zn. With increased pH^- levels, it reduces the solubility and mobility of Hg, Pb, Zn, Cu, and Ni, and vice versa increases the mobility of Ni and Se. Dangerous toxicants, such as As, Sr, and Cd retain their mobility within wide limits of changing pH^- values (Bliznakov and Ilieva 2004).

The *nutritional balance of plants* is just as important as the concentration of nutrients in them. It is defined as the ratio of 100 units of nitrogen to the concentration of micro- and macronutrients. From the results of numerous studies, we see that the relative optimal concentration of macronutrients for seven coniferous species of the temperate continental climatic region are: K – 54.2; P – 18.9; Mg – 5.4; Ca – 5.3. Deficiencies of a chemical element of either group lead to certain stressful situations in conifers. For example, iron deficiency, which occurs with a strong basic reaction of the soil, leads to iron deficiency, which some authors refer to as the age structure of coniferous plants rather than to its absence in the soil; Mn deficiency leads to chlorosis with a slightly yellow tint of leaves, while the absence of Mg in the leaves leads to a golden-yellow tint of the needles. Deficiencies of Cu in the soil lead to a sharp fall of the leaf mass in young plants, or to turning the top of the branches of the annual growth in the form of a "shepherd's pole.

4.3 Anthropogenic Factors

The harmful effects of *anthropogenic environmental factors* depend on human activities. The anthropogenic effects of the environment are transmitted to the plant cover through the abiotic components of the natural environment or ecosystem, such as atmosphere, water, and soils. One of the most pronounced anthropogenic factors for stress is pollution. As early as the 1960s, the *order of toxicity of chemical elements in plants* was created, according to which toxic elements were subdivided into three classes: very toxic elements—detrimental to plants in concentrations lower than 1 ppm^{-1}; moderately toxic at concentrations 1–100 ppm^{-1}; and slightly toxic at concentrations higher than 100 ppm^{-1} (Bowen 1966; Fortescue 1971). A short time later, Kowalsky (Ermakov and Kovalsky 2019) also proposed the *theory of biogeochemical thresholds*, which is essentially a transfer of the *law on the critical load of the factor*, which states that if one factor reaches its critical thresholds, despite the optimal combination of the other factors, the species is at risk of death.

The theory of biogeochemical thresholds, in turn, was successfully developed in the USSR and other countries by the former Warsaw Treaty, thanks to the development of ecology theory in the 1950s, and especially in the works of H. Ellenberg (1979) (see Hill et al. 2000), and its derived curves of the physiological and environmental response to the development of the species depending on a given abiotic factor of the environment (Dedew 1989). The theory of *biogeochemical thresholds* and the *law on the critical load of the factor* are private cases of the *law of the minimum (law of the limiting factor)* of J. Liebig (1840) (Sinclair and Park 1993) and the *law of environmental tolerance (plasticity/maximum)* formulated by V. E. Shelford (1931) with regard to chemical elements, according to which, at lower and higher than normal threshold concentrations of a certain chemical element, the development of the plant deteriorates. These "laws" in turn are a description of private cases of the *law of physiological interactions* (law of collective action of factors) found and named *law of the effectiveness of factors* by the German agrochemist and plant physiologist A. Mitscherlich (1909), and renamed by B. Baule (1917) in the *law of cumulative action* (Harmsen, 2000).

4.3.1 Atmospheric Pollution

One of the most dynamic components of the landscape is the atmosphere, with the main pollutants transmitted by air being grouped by particle size. For example, aerosol pollutants formed by gaseous reactions have a particle size of 0.1–1 μm. If the aerosol is formed as a result of the breakdown of a substance—from Aeolic weathering or particles as a result of forest fires—their dimensions are in the order of 1–10 μm.

In terms of toxicity, aerosol compounds are divided into toxic and non-toxic. *Non-toxic aerosol compounds* are usually deposited on the crowns of trees without going into chemical reactions with the leaf mass, but hinder photosynthetic processes, and clog the mouth of the leaves, which also disrupts breathing processes. *Toxic aerosols*, and in particular those of heavy metals (Zn, Cu, Pb, and Ni), fall into plants directly from the air or from the soil through the root system (Stoyanov 1999). Heavy metals are commonly accepted as all metals with an atomic weight of more than 40–50 or a density of more than (7000–8000 kg/m^3) except for the group of precious and rare metals (Orlov et al. 2002). In terms of some heavy metals in precipitation, such as (Zn, Cu, Pb, and Co), Zn concentrations are highest and vary widely, with the highest occurring during the warmer months and lowest during the coldest. By their significance (in a declining order of toxicity), regional pollutants having a direct effect on the state of coniferous forests are ranked: SO, NOx and fluorine compounds, ozone (O_3), Cl, NH_4 (Agess 1982; Vinogradov 1981). Depending on the depth of the damage and the intensity of the pollutants in coniferous landscapes stressed under the influence of air pollutants, Vinogradov (1981) separates the following zones; acute, chronic, and hidden (physiological) pollution. This zoning may be applied to both atmospheric and other types of pollutants acting on vegetation.

4.3.1.1 Non-organic Gaseous Pollutants

As early as the 1930s, V.N. Sharpinski began intensified research into perishable coniferous forests near aluminum, porcelain, and coal plants due to fluorine emissions from aluminum production in dust (NaF, CaF_2) and gas fractions (HF, SiF_4) (Ilkun 1978). The various stages of coniferous forest pollution with fluorine compounds are reflected differently in the coniferous vegetation spectra. For example, three stages of vegetation change spectra are separated in the case of fluorine contamination; (i) initial whitening of conifers, (ii) subsequent darkening, and (iii) defoliation (Vinogradov 1981).

Physiogo-biochemical disorders in plants occur after the gases on their way to the plant cell, and especially in SO_2, occur with the lipoproteid cell membrane—plasmalem. Gas ions have the properties of free radicals, and in lipids cause an increase in free-range chain oxidation processes. Additionally, the penetration of this ionic complex into the plasmalem of the cells of plants leads to a violation of its permeability and structure. The study of induced glowing caused by this process was first studied in Russia by Rubin (1976) and Markarova et al. (1976). The results of three years of research on the permeability of cell membranes in various coniferous species polluted with various atmospheric pollutants, Nikolaevski (1983), concluded that the permeability of cell membranes increases and a general decrease in photosynthesis is observed.

The degree of damage to plants from different gases has been investigated against a number of environmental factors, such as temperature, illumination, 24-hour cycles and weather conditions, duration of action, and gas concentration. From the studies of Nikolaevski (1983) with matched atoms and molecules of SO, NOx, and NH_4, it was concluded that conifers, as well as other types of vegetation, have the property of absorbing harmful gases and returning sulfur to the soil in the form of sulfates, and the ability to translocate by separating in the form of hydrogen sulfide into the atmosphere. Even more important are the conclusions regarding nitrogen oxide pollution (NOx), which are converted into ammonia and nitrogen oxides in even higher quantities. Some studies also indicate the compaction of wood and the sewing of annual coniferous circles resulting from increased CO_2 content (Yazaki et al. 2005).

For the first time, Nikolaevskiy and Miroshnikova (1974) also set maximum permissible air pollution concentrations for plants, which must be equal to or lower than sulfur anhydride, 0.02 mg/m^3; nitrogen oxides, 0.05 mg/m^3; ammonia, 0.1 mg/m^3; sulfuric acid vapor, 0.1 mg/m^3; formaldehyde, 0.02 mg/m^3; methanol, 0.2 mg/m^3; benzene, 0.1 mg/m^3; cyclohexane, 0.2 mg/m^3; ammonia and sulfur powders, 3 mg/m^3. According to the intensity of damage to coniferous vegetation, acute lesions occur at single concentrations of SO_x; 1–2 mg/m^3. Chronic disorders resulting from the action of moderate concentrations, 0.4–0.5 mg/m^3, and low concentrations, 0.1 mg/m^3 of SO_x result in fragmentation, spotting, and contrasting texture of vegetation. The weakest physiological changes in vegetation are recorded at concentrations of SO_x in the order of 0.05–0.08 mg/m^3 (Vinogradov 1981).

4.3.1.2 Ozone Stress

Ozone stress is observed in places prone to the creation of photochemical smog; a mixture of gases and aerosol particles of primary origin (NO, C_mN_n, SO_2) and secondary origin (Runeckles and Resh 1975; Vinogradov 1981). For example, the methods of chemical interaction and conversion of hydrocarbons with ozone, which form peroxyacetyl nitrate (PAN), peroxy propionyltrate (PPP), and propyl-peroxybutylenetrate (PB) have been established. The order of toxicity of these compounds (phyto-oxidants) relative to vegetation is as follows; PBS, PPP, and PAN (Ilykun 1978). The visible effects of ozone on trees produced in some laboratory experiments have, to date, not been confirmed in field experiments. Only the Ponderosa pine forests in the San Bernardino Mountains of southern California have exhibited confirmed visual damage from ambient ozone exposures. New hypotheses, based on field case studies in the eastern United States, suggest that ozone may be contributing to productivity declines in forests without visible damage symptoms [Hakkarinen 1987 (Forest Health and Ozone)].

4.3.1.3 Dust and Aerosols Pollutants

Aerosol particles fall on the crowns of trees and cause a corresponding physiological reaction and stress if the particles encounter atmospheric precipitation. There are several mechanisms to meet this condition: (i) diffusioresis, for particles of a size of 0.1 μm; (ii) brown diffusion; (iii) collision and interception; (iv) dissolving the gases; and (v) formation of drops of condensation nuclei. In addition to precipitation, polluting substances are also deposited in contact with the sub-limiting surface, the so-called "dry deposition." Dry deposition is determined by three factors; the speed of turbulence of the atmosphere, the physico-chemical properties of the sub-cooling surface, and the physico-chemical properties of polluting substances. With the lowest dry absorption rate are particles measuring 0.1–0.5 μm, which is due to the repulsion forces in the laminar layer around the vegetation, at the expense of brown diffusion and foresis.

Nitrate particles are characterized by dimensions several times larger than sulfates, and with the greatest intensity of "dry ingestion" are heavy metals and hydrocarbons. It is known that coniferous crops are the most affected by dust atmospheric pollution. Because coniferous species are year-round evergreens, they are also preferred for afforestation in areas with a highly developed heavy industry. For example, it has been found that pine forests with an area of 1 acre retain up to 3.6 t of dust, while beech forests retain up to 6.8 t. During vegetation, dust reduces photosynthetic activity, clogs the cyst, retains stature, and causes other damage to coniferous vegetation. For example, the cement dust found on primroses and soil leads to a targeted boiling effect, which can worsen the nutrient availability in soils with a basic reaction and lead to stress. TPP gas ash has a variable composition, and lignite ash has been shown to have a high CaO content (up to 20%), which harms some coniferous species. On the other hand, lower-calcium oxide TPP ash contains more heavy metals. Apart from TPP ash, waste repositories resulting from coal burning are highly toxic due to the trace elements contained therein. Dust on the green tissues of plants increased the integral brightness factor and smoothed the maximum reflection coefficient in the green part of the spectrum (λλ 0.52–0.54 μm), strongly increased the coefficient values (λλ 0.62–0.68 μm), and lowered its values in the nearby IC region (λλ 0.8–1.1 μm) (Vinogradov 1981).

Dust pollution is also caused by the waste of the chemical and metallurgical industry, such as slurry and slags, which are rich in heavy metals and trace elements. Transport is a straight source of dust and aerosol pollution, with the main pollutants other than Pb being Cd, Ni, Zn, Cu, Cr, Co, which are obtained from mechanical wear of tires (Zn, Cd), metal parts of cars (Ni, Cr, Cu) or additives in lubricants (Zn, Cd). However, when rain falls, the accumulated dust on the crowns of the trees is washed off and a regeneration from a sort of air filter is carried out. Dust and aerosol pollutants washed away from rainwater and falling into soils often include some of the 16 most important chemical elements for the growth of green plants. These elements are divided into two groups according to the relative need of plants; *micro-* and *macronutrients*.

These are the nine macronutrients described in decreasing order according to their importance and quantity in plants and foliar mass ≥ 1000 ppm; carbon (C), oxygen (O), hydrogen (H), nitrogen (N), potassium (K), calcium (Ca), magnesium (Mg), phosphorus (P), and sulfur (S). There are seven micronutrients in quantities ≤ 100 ppm; chlorine (Cl), iron (Fe), boron (B), manganese (Mn), zinc (Zn), copper (Cu), and molybdenum (Mo) (Kutov and Kostova 1988). Moreover, both micro- and macronutrients are unevenly distributed on the organs of plants. A significant part accumulates in the ground parts of plants (leaves, stems); Mn, Mo, Sr, Cu, Ti, and Ni, and to a lesser extent Fe, A1, and Co. In the roots of plants are accumulated mainly Ag, Pb, Sn, W, Cr, V, and U.

Evenly distributed in the organs of plants are Zn, Pb, and Cr. Natural radionuclides such as $2^{22}Rn$ and ^{238}U and their products from radioactive families and radionuclides, which are constantly formed under the action of cosmic radiation, such as ^{3}H, ^{7}Be, ^{10}Be, ^{14}C, ^{22}Na (Poppytz and Penchev 1989), are carried through the air.

Uranium and uranium compounds have a proven combined effect of low stress and delay in greening in conifers in the presence of iron. This is mainly because iron is an essential photochemical element in the formation of chlorophyll in green plants, however, it is not physiologically active, since manganese ions are present in the soil and plants, because manganese has an oxidative effect (Kabata-Pendias and Pendias 1984; Aronoff et al. 1985). Like natural radionuclides, artificial radionuclides derived from mining, atomic energy, and nuclear tests are also carried over plants. Such are the isotopes ^{131}I, ^{90}Sr, and ^{137}Cs, which get on the plant and are washed away from the rainfall into the soil, and fall into the food chain of the plant itself.

Products obtained from nuclear tests and incidents with nuclear plants have a short half-life, ^{238}Pu, ^{239}Pu, ^{240}Pu, ^{14}C, ^{55}Fe, and ^{60}Co, the most durable being the isotope tritium (^{3}H). The uptake of natural or artificial radionuclide by the plant depends, as with other contaminants, on the acidity of the soil, pH^-, and on the organic matter in the soil. The isotopes of radionuclides, such as Neptunium, Plutonium, and relatively slower Ameritium and Curium, were most intensely absorbed. The behavior and ingestion of ^{90}Sr is like that of Ca's, while the ingestion of ^{137}Cs is like that of K. In nature, and more precisely aqueous solutions, ^{137}Cs is presented as a simple single-digit ion, and uranium is in the form of $UO_2(CO_3)^{2-}$ or $UO_2\left(CO_3\right)_3^{2-}$.

In soils, ^{137}Cs is retained and accumulated depending on the humus content. For example, the order of retaining ability of soils in relation to Cs is subsolists (gray forest) < brown forest < black earth resins (WSC). A big problem is the bioaccumulation of this radionuclide in mosses, lichens, and fungi, since in the northern regions of Europe, this is the main food of deer, as a result of which there are increased concentrations in the tissues and bone system of deer and people inhabiting these regions, for example Lapland (Onianwa 2001). Radioactivity also disrupts chromosomes, causing chromosomal aberration and mutations, which in turn can lead to a violation of the metabolism of plants and changes in their characteristics in subsequent generations (Orlov et al. 2002).

4.3.2 Soil and Groundwater Pollution

In addition to the atmosphere and soils, coniferous landscapes are also *polluted by surface water and groundwater.* The technogenic contamination of soils with heavy metals is of a regressive accumulative type, i.e. the concentrations are highest in the surface soil layer (0–10 cm) and decrease in depth, and secondary minerals, organic substances, and hydroxides of Fe and Al (Bliznakov and Ilieva 2004) play an important role in their accumulation. This fact protects the deep-rooted vegetation from pollution to some extent. However, it is a well-known fact that the root system of many coniferous species is shallow and widespread, which allows for a higher impact of technogenic contamination on conifers than other species. For the deep roots, which reach for the groundwater-table, the main contaminants carried through groundwater are: cyanides—typical enzymatic poisons suppressing the active mechanisms of root nutrition; phenols—ingested in the soil by hemsorption and subsequent humification; nitrogen compounds—are as diverse in chemical composition and therefore different in harmfulness, synthetic nitrogen compounds are distinguished by great toxicity; sulfur compounds—directly affected by atmospheric rainfall or groundwater are highly biologically active, with some exhibiting bactericidal and pesticide action, and in good air mode on soils all mineral forms of sulfur are oxidized to stable sulfates; chlorine compounds— some chlorine and hypochlorine compounds are a prerequisite for secondary soil salting and therefore salt stress. Organochlorine compounds, in turn, act as a pesticide and destroy the microflora in the root system of coniferous species, their high biological resistance makes them difficult to remove from soils; synthetic oils, oil, and petroleum products—an integral part of wastewater—violate the water, physical and structural properties, and air regime of soils, as well as inactivate microorganisms, and suppress its self-cleaning ability.

The high concentration of one element in soils often suppresses the uptake of another similar element of plants. For example, the high concentration of Mg in the soil suppresses the uptake of Ca. Under the influence of aluminum, the entry of Cl, Ca, and Fe into plants is reduced; nitrogen slows down or ceases the absorption of Mn; and potassium fertilizers reduce the absorption of Fe and Co (Orlov et al. 2002). Soil pollution is particularly strong when saturated with heavy metals and radionuclides, as well as organic pollutants from domestic and industrial wastewater (Pavlotskaya 1974; Badyaev et al. 1990). For example, the relationship between oxidative stress in coniferous and increased Cr content in soil (Panda and Choudhury 2005) has been clarified. The content of this and other heavy metals in the soil cause the manifestation of chlorosis, and in the presence of natural and artificial radionuclides in soils and groundwater, cause the appearance of physiological and morphological changes of vegetation (Anisova and Yakushev 2008). Some inorganic pesticides contain As and S, which cause symptoms of toxicity, such as chlorosis, necrosis, holes in the branches of trees, premature needle wrapping, etc. Some of the strongest pollutants with cumulative action from heavy metals in soils are Hg, Cd, Pb, and Ni, with some heavy metals able to form the so-called "heavy metals synergistic mixtures," the toxic action of which exceeds at times the cumulative action of each pollutant individually (Kutov and Kostova 1988). The multiplication of the negative effects of the heavy metal content on the main physiological processes leads to various structural-functional disorders in the plant organism, combined under the common name metal phytotoxicity (Vasilev et al. 2010). As indicators of phytotoxicity in most tests, seed germination, biometric parameters of plants, the activity of stress-sensitive enzymes, etc. are used (Vangronsveld and Clijsters 1992). One of the most used plant tests is the bean test of Vangronsveld and Clijsters (1992), in which the main indicators are the leafy area and fresh mass of plants, the activity of stress enzyme,s and the isoenzyme spectrum of guayacol peroxidase.

Plant tests classify the phytotoxicity of the substrate into different groups (classes). The Vangronsveld and Clijsters (1992) test used the following phytotoxic classes; non-toxic (I class), slightly toxic (II class), moderately toxic (III class), and highly toxic (IV class) soil. Heavy metals are protoplasmic poisons, the toxicity of which increases with an increase in atomic weight (Orlov et al. 2002). The phytotoxicity of heavy metals manifests itself differently, and an attempt to summarize toxicity is the order of phytotoxicity; Ni < Cu < Co < Mn < Zn (Alekseev 1987).

4.4 Thresholds and Critical Loads

Conifers, and pines in particular, are known to perish at very low doses of radiation 20–30 R/day, while most plants perish at 200 R/day. The lowest known lethal dose of 0.8 Gy/yr was noted on the Douglas Fir (*Pseudotsuga douglasii*); in the eastern white pine (*Pinus strobus*), 2.7 Gy/yr; and in red pine 6 Gy/yr (*Pinus resinosa*); while 3 Gy/yr in resinous pine caused only visible disturbances (*Pinus rigida*) (Poschl and Nollet 2007). The best overview of the minimum doses of organisms' sensitiveness with regard to natural and artificial radionuclides, excluding man, was given by Rose (1992) and Barescut (2005).

The maximum permissible concentrations of discarded radionuclides in the environment during normal operation of the NPP are: ^{131}I – 1*10^{-4} Bq/kg; ^{137}In – 2*10^{-7} Bq/kg; and ^{90}Sr – 4*10^{-9} Bq/kg. As a result of uranium mining from lignite and brown coal, the concentrations of the following isotopes in some regions reach: ^{238}U – 37 Bq/kg; ^{226}Ra – 18.5 Bq/kg; ^{210}Pb – 25 Bq/kg; ^{210}Po – 30 Bq/kg; ^{232}To – 18.5 Bq/kg; ^{228}To – 7.5 Bq/kg; and ^{40}K – 130 Bq/kg.

Visible changes in the condition of coniferous vegetation are not observed until the content of some heavy metals in the needles is reached, such as: Zn – 80–90 μg/kg; and at Cu content – 40 μg/kg (Orlov et al. 2002).

4.4.1 Satellite Multi-Band Remote Methods for Detecting Abiotic Stress

The use and application of multiband remote sensing (RS) methods in forestry and landscape research have a long history of application for detecting damage to forest plantations (Sukhih and Sinicina 1979; Vygodskaya and Gorshkova 1987; Franklin 2001). Stressful situations in forests, in particular coniferous vegetation, and more broadly coniferous landscapes, have been the subject of research since the 1920s, with the earliest applications associated with the use of panchromatic aerophotos to delineate them by diseases and pests, forest fires, and natural disaster forest plantations in Ontario, Canada (Sukhih and Sinicina 1979).

Later, in the 1950s, methodologies were drawn up in the United States and the USSR to decipher forests by aerial photography, and in addition, work appeared concerning the selection and evaluation of materials, aerophotos of various photo materials, and scales for forest decryption. With the advent of infrared (IR) photography, opportunities for detecting stressful situations in forest plantations expanded, yet most authors consider ordinary color photography is informative enough to separate the affected forest areas (Sukhih and Sinicina 1979).

On October 3, 1962, the American cosmonaut W.M. Shira took the first 12 panchromatic images of the Earth's surface in the range (λλ 320–720 nm), using six different filters, thus launching multi-band satellite RS (Vinogradov 1981). Later, with the advent of manned orbital stations, such as "Skylab," "Salut," and the "Apollo" program, the application of satellite multi-band data in the assessment of stress situations in coniferous forests began (Warren 1969). The first multi-channel satellite images obtained aboard the Soviet Pilotable SpaceCraft (PKK) "Soyuz-12" using the 9-channel KN-3 camera were taken in 1973.

The first similar multi-band images were obtained from the US Skylab orbiter station, from the 8-channel S190B, as well as the spectrophotometric S191 aerophoto cameras. Since the first satellites of the Landsat series were launched in 1972, originally known as ERTS, some of the largest anthropogenic pollution has been identified around the metallurgical center in the Province of Ontario, Canada, with a length of the affected area 50–60 km and a width of 20–30 km in the opposite direction, to the direction of prevailing winds in the area from the south–southeast (Murtha and Watson 1973; Vinogradov 1981).

With these studies, the optical dependence of an increase in the reflectivity of vegetation in the orange-red region of the spectrum (λλ 0.6–0.7 μm) has been established, and on this basis three pollution zones around the metallurgical compound have been identified. At the same time, using multi-band satellite images, similar studies were done in the USSR in areas around metallurgical enterprises of the aluminum industry, where mass drying of pine plantations was noted. In the area around a nickel enrichment complex in the tundra, a paragenetic zone of ecosystem change has been established, traceable in television multi-band photos with a 1 km resolution (Kupriyanov and Prokacheva 1976). With the help of multi-band images around a copper plant, an area of 28 completely disturbed ecosystems and an area with severely disturbed ecosystems of 40 km^2 (Makunina 1978) has been established.

The filming by US reconnaissance satellites in connection with the Chernobyl NPP incident, when Keyhole-11-6 (international number 84 122-1) captured the damage done to the environment in and around Chernobyl. Subsequently, evaluations of the impact of the Chernobyl Nuclear Power Plant accident on biota and in particular coniferous vegetation, with the so-called "red-leaf vegetation" gained notoriety, were subsequently prepared and expanded. The "red forest" withered as a result of radioactive cloud deposits (Richelson 1990). There were some attempts to study the vegetation condition with milti-band imagery from Landsat TM (Modzelewska et al. 2013) and spaceborne SAR data (Zakharov et al. 2003). In

Figure 4.1 An extract of a burned area mapping product generated by CIMA Foundation and Fadeout using the WASDI processing environment based on images acquired by Copernicus Sentinel-2 on 26 March and 10 April 2020. It shows the burned area around Chernobyl in the Ukraine on 10 April. Source: © contains Copernicus Sentinel data [2020], processed by CIMA Foundation and Fadeout srl. Credits by CIMA Research Foundation – Center of competence of the Italian Civil Protection (https://www.cimafoundation.org).

2020, the wildfire in the Red Forest created a great concern around Europe, which lead to the activation of the Copernicus Emergency Mapping Service, and the Copernicus Sentinel-2 satellite mission has imaged the fires and smoke. It also mapped the resulting area of burned ground, see Figure 4.1 (Mapping Chernobyl fires from space 2020). The emergency situation was called upon because of the well-known state of the Red Forest, where combustion could potentially cause raising of the radiation background in the neighboring countries.

The assessment of the consequences resulting from the Chernobyl accident also includes satellite platforms for civil use, such as SPOT-1, Landsat TM, etc., on the basis of which the scientific community analyzes the state of the environment in the area of the accident. Satellites such as SPOT-1 launched just 20 days before the Chernobyl nuclear accident, and were able to give the first images on May 6, 1986, beginning EIAs. As a result, the measures taken to cleanse radioactive products from the environment, the first coniferous forests directly affected by the accident, as well as the surface soil horizon (10–30 cm) in the decontamination area, have been seized. The impaired normal processes of self-recovery (cone formation) of coniferous forests were only resumed 10 years after the accident in the alienated area from the site of the accident within a radius of 30 km (25 years of satellite imagery over Chernobyl 2011).

The multi-channel satellite radiometers with high spatial resolution (HSR) with the greatest notoriety and applicability in detecting stressful situations in forest plantations are the French SPOT HRV and HRG, American Landsat 5 TM, Lansat 7 enchanced thematic mapper plus (ETM+), and Landsat 8 operational land imager (OLI), and advanced space-borne thermal emission and reflectance radiometer (ASTER) aboard EOS Moderate resolution imaging spectroradiometer (MODIS) – TERRA, DMS series satellites, Japanese ALOS-PALSAR, Indian IRS-1D and 1C/LISS-3, Russian Resource-DK, and RapidEye; as well as radiometers with ultra-high HRS: GeoEye-1, IKONOS-2, QuickBird, OrbView-3, WorldView-2 and -3, etc. (Van Der Meer et al. 2002).

The low SR satellites most commonly used to assess stressful situations at regional level, country, continent, biome are the US TERRA and AQUA spectrometers under the EOS-MODIS program, SeaWiFS aboard SeaStar, AVHRR aboard NOAA, Russian Meteor, MSU-SKI aboard Resource-Ol, Medium Resolution Imaging Spectrometer (MERIS) aboard Envisat-1, as well as SPOT VEGETATION 1 and 2 aboard the French SPOT 4 and SPOT 5 subsequently replaced by SPOT 6 and SPOT 7. Unlike spectrometric remote methods that study stressful situations along selected absorption lines in the plant spectrum, multispectral sensors do not have these options.

The limitation of modern satellite EO systems in the capture and transmission of information to the terrestrial segment is mainly due to limitations in contemporary memory, power supply, and telemetry systems, which allows, on the other hand, to develop other technical solutions in multispectral systems, such as multi-angle imaging, high temporal resolution, stereo capture, etc., which allow for other types of analyses and assessments of the state of forest plantations. For example, the US MISR radiometer captures the same territory from nine positional angles in order to recreate the object reflectance, and the European CHRIS radiometer on board PROBA captures up to 62 bands simultaneously from four observation angles.

4.4.2 Satellite Infrared Remote Sensing Methods for Detecting Abiotic Stress

A detailed overview of the theory and applications of thermal imaging for the study of interactions between plants and water, and for the diagnosis and monitoring of diseases in vegetation, is presented by Jones and Schofield (2008). Infrared (IR) images are mainly used to study the interactions between vegetation and water, and in particular leaf temperature. In rare cases, the temperature of the leaves may be affected by other physiological processes, for example, the heat generated in exothermic processes or the freezing of water in the leaves, while in some exceptional cases of high levels of breathing, for example, with Arum spadix, increased temperatures are generated by these elevated respiratory levels (Chaerle and Van Der Straeten 2000).

For water stress, the definition of the Crop Water Stress Index (CWSI) (see Table 4.1) by Idso et al. (1981) is often assumed:

$$\mathrm{CWSI} = \left(\mathrm{T_{canopy}} - \mathrm{T_{nws}}\right) / \left(\mathrm{T_{max}} - \mathrm{T_{nws}}\right) \tag{4.1}$$

where T_{max} is the temperature of the dry surface and T_{nws} is the empirical temperature of "not water stressed base" (this is the plant breathing at potential levels in the same environment at optimal irrigation levels; good soil moisture retention). The value of CWSI – 0 is

Table 4.1 List of the used vegetation indices in the chapter.

Index name	Formula	Source
Crop water stress index (CWSI)	CWSI = (Tcanopy − Tnws)/(Tmax − Tnws)	(Idso et al. 1981)
Temperature condition index (TCI)	TCI = 100 * (Tmax − T)/(Tmax − Tmin)	(Gutman et al. 1995)
Modified chlorophyll absorption ratio index (MCARI)/ MTVI2	MCARI = (λ700 − λ670 nm) − 0.2(λ700 − λ550 nm) * (λ700 − λ670 nm) MTVI2 = (1.5*(1.2(λ800 − λ550 nm) − 2.5(λ670 − λ550 nm)/ (λ2800 + 1)2 − (λ6800 − λ5670 nm√) − 0.5)	(Rondeaux et al. 1996)
Transformed chlorophyll absorption ratio (TCARI)	TCARI = 3[(λ700 − λ670 nm) − 0.2(λ700 − λ550 nm) (λ700 nm/λ670 nm)]	(Haboudane et al. 2002)
Photochemical reflectance index (PRI)	PRI = (λ528 − λ567 nm)/(λ528 + λ567 nm)	(Gamon et al. 1992)

indicative of the absence of water stress, and a value of the order of 1 is indicative of maximum water stress (Thermal Crop Water Stress Indices 2001). This index is an effective measure of the openness of the leaf cyst. The vegetation temperature status index – Temperature Condition Index (TCI) was developed to aproximate the importance of thermal bands when assessing the state of vegetation. The TCI formula has the following appearance:

$$TCI = 100 * \left(T_{max} - T\right) / \left(T_{max} - T_{min}\right) \tag{4.2}$$

where T is the actual adjusted value of weekly temperature amounts, and T_{max} and T_{min} are the maximum and minimum smoothed temperatures for the same weeks for a multiannual observation period. For example, based on experimental data, it has been found that the brightness temperatures of AVHRR bands 4 and 5 (K°) depend mainly on the temperature of the Earth's surface, absolute humidity, and vertical temperature gradient "earth surface-atmosphere" (Gutman et al. 1995). The first satellite spectrometers and radiometers capturing in the thermal IR area of the spectrum are the weather satellites of the Nimbus series – Nimbus-III and the Russian Cosmos – Cosmos-243 (Henel and Conrath 1969; Bashirinov et al. 1969).

The most famous operational satellite radiometers, capturing in Infra-Red (IR) and with a significant archive of accumulated information over a long period of time—from the 1980s and 1990s to the present day—are the Multispectral Scanner System (MSS) aboard Landsat 4, Thematic Mapper (TM) aboard Landsat 4 and 5, and ETM+ aboard Landsat 7. From ETM+ bands, fifth band captures in the range 1.55–1.75 μm and has 30 m SR, sixth band in 10.4–12.5 μm with 120 m SR, and seventh band in 2.08–2.35 with 30 m SR. Soon after Landsat 8 OLI (originally called Landsat Data Continuity Mission) was launched in

February 2013, the Landsat program started preparing for the successor; Landsat 9. Landsat 9 will carry two science instruments, i.e. OLI 2 and Thermal Infrared Sensor (TIS 2), which are sensors with moderate spatial resolution—15, 30, and 100 m depending on spectral band—and the ability to detect the same range in intensity as Landsat 8, or better. Landsat 9 will be placed in an orbit on 16/09/2021 so that it is eight days out of phase with Landsat 8 to increase temporal coverage of observations (Landsat 9 2021). In this system, the fifth and seventh bands are most often used to detect geological deposits of Cu and skarn carbonate deposits of various mineral deposits, while the sixth band is most often used when examining the earth's own thermal radiation (van der Meer and de Jong 2006). One of the main radiometers capturing in the IR area of the electromagnetic spectrum is ASTER, aboard EOS-AM1 TERRA, developed jointly by the Japanese Space Agency (JAXA) and NASA. This radiometer captures in three bands in the visible and nearby IR area of the spectrum with 15 m spatial resolution, six bands in the near IR with 30 m PRS, since 2003 dysfunctional due to sensor cooling problems, and five bands in the thermal IR area with a PRS of 90 m. Another low PRS satellite sensor aboard NASA's EOS-AM1 TERRA is MODIS, which acquires in 36 bands, in the visible and infrared area of the spectrum 0.4–14.4 μm.

4.4.3 Hyperspectral Satellite Remote Sensing Methods for Detecting Abiotic Stress

Spectrometric RS, unlike other RS systems, register the spectral reflection coefficient of natural sites in a series of narrow (in the order of 10 nm and less), contiguous spectral bands (van der Meer and de Jong 2006). The spectar obtained is a function of the physico-chemical (humidity rate) and biological (phenological state, project coating, etc.) features of the filmed object (van der Meer and de Jong 2006; Charles and Zyl 2006). The value of the brightness of a natural object, L, or the radiation emitted for a spatial angle from a non-point source in a particular direction per unit of projected source area in that direction, depends on the intensity of the incident radiation Eo(λi), and the reflective properties of the natural object. The very brightness of the natural formation is transformed by the influence of the transmission function of the atmosphere, which has the following appearance, Mishev et al. (1987).

$$\mathrm{P}_{\lambda_i}(\mathrm{H}) = \frac{\mathrm{L}(\lambda_i, \theta, \varphi, \mathrm{H})}{\mathrm{L}_0(\lambda_i, \theta, \varphi, H_0)} \tag{4.3}$$

where a certain direction (φ, φ) and a certain spectral range in the spectrum $\lambda_i + \Delta\lambda$, $L_0(\lambda_i$, $(\varphi, \varphi, H)_0)$—spectral brightness recorded on the surface No and $L(\lambda_i, \theta, \varphi, H)H$—is reflected by the natural formation solar radiation at altitude, H. Taking into account the importance of the indicate and albedo of natural and anthropogenic objects, satellite platforms with spectrometers captured from different positional angles have been created in order to study the positional effects on the spectrum. These are the Global Change Observation Mission (GCOM)-C by the Japan Aerospace Exploration Agency's (JAXA) Second Generation Global Imager (SGLI) spectrometer, shooting in three positional angles; −45°, nadir, and +45° along the shooting path (Suzuki et al. 2010). Partially multi-angle and multiband radiometry from the PRS; ATSR, AATSR, MISR, AVHRR, and MODIS (Vogt and Verstraete 2009).

The first hyperspectral satellite systems were launched by subsisting experiments conducted synchronously aboard the Soyuz-6, Soyuz-7, and Soyuz-8, with aircraft experiments with similar equipment (RSS-2) in the late 1970s (Vinogradov 1981). In the 1990s, NASA and TRW developed a spectrometry sensor (HSI) for the LEWIS mission, which was scheduled to capture 128 bands in the range of 0.4–1 μm and another 256 bands in the 0.9–2.5 μm area with a band width of 5 and 6.5 nm, respectively. Another similar spectrometry radiometer (spectrometer) is MightySat II's FTHSI, developed by Air Force Research Lab with 256 bands in the 0.35–1.05 μm area. It was only with the advent of NASA's New Millennium Program/Earth Observer-1 (NMP/EO-1) satellite platforms with the Hyperion spectrometer on board and ESA's CHRIS/PROBA orbiting in 1999 and 2001, respectively that spectrometric satellite RS entered the real civil and scientific applications (van der Meer and de Jong 2006). Despite the long planning and postponement, several spectrometric missions are at an advanced stage of their realization already. These are the German Aerospace Center's (DLR) EnMap and Modular Optoelectronic Scanner (MOS), NASA's High Resolution Imaging Spectrometer (HIRIS), and the U.S. Navy's Naval Earth Map Observer (NEMO). Unlike existing satellite sensors such as Hyperion and CHRIS/PROBA, NEMO will be dual-purpose for military and civilian use, with the spectrometer on board; Coastal Ocean Imaging Spectroradiometer (COIS) will capture in the spectral range 400–2500 nm and have a spectral resolution of 10 nm. The width of the stage will be 30 km, with a pixel size of 60–30 m, as well as an improved signal-to-noise ratio (S/N ratio) compared to previous similar systems.

The HIRIS spectrometer is planned to capture 192 bands, with spectral resolution from 9.4 to 11.7 nm in different areas of the electromagnetic spectrum; 0.4–2.5 μm. The planned Australian Resource Information and Environment Satellite (ARIES) will have full coverage in the visible and nearby IR spectral range (Visible Near InfraRed [VNIR]) 400–1050 nm, with a distance between bands of 20 nm, and subsequent continuation in the short wave Infra-Red (SWIR-2) spectral range 2000–2500 nm with a minimum distance of 16 nm between bands, or a total of 105 bands (van der Meer and de Jong 2006). The OrbView-4 satellite, also known as Warfighter, developed by Orbimage, is scheduled to have one panchromatic channel with 1 m GRDs and a multi-channel system with 4 m PRPs, as well as 200 bands in the spectral region 0.4–2.5 μm with 8 m PRPs. Unlike other spectrometers, OrbView-4 has the ability to capture in different angles, with a deviation of up to 45° from the zenith, but for civilian and scientific purposes, information will only be provided with 24 m spatial resolution. One of the current studies on the capabilities of the future Hyperspectral Environment and Resource Observer (HERO) satellite mission for the study of signals of stress and disturbance in the forests of the Canadian Aeronautics and Space Institute (CASI) shows the weaknesses of such a mission, and outlines the possibilities for improving the parameters of the mission with a view to future research (Peddle et al. 2008).

A study utilizing EO-1/Hyperion data over coniferous landscapes of decommissioned uranium ore extraction area in Teyna watershed, Iskar River left tributary (Bulgaria) was carried out by Filchev (2013), see Figure 4.2.

Four vegetation indices (VIs), such as Modified Chlorophyll Absorption Ratio Index (MCARI), Transformed Chlorophyll Absorbtion Ratio (TCARI) (Haboudane et al. 2002), MTVI 2 (Rondeaux et al. 1996), and photochemical reflectance index (PRI) (Gamon et al. 1992), as well as the shape and area attributes of the red-edge spectral features, have been studied in order to detect changes caused by the ore-extraction. The analysis was performed on four test plots, each 30 × 30 m in size, to match the EO-1/Hyperion pixel size, in

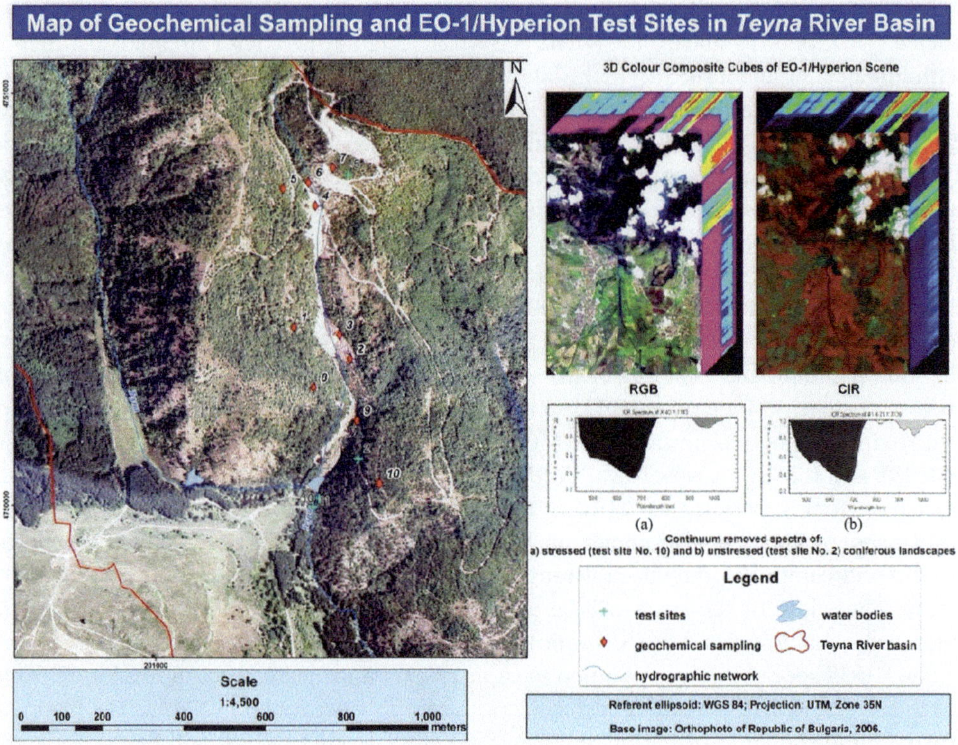

Figure 4.2 Map of geochemical test sites at Teyna watershed. Source: Filchev, L., 2013, The SAO/NASA Astrophysics Data System.

homogenous coniferous landscapes in the Teyna River. In effect, the red-edge position index for unstressed coniferous landscapes (Black pine) was found to be at about $\lambda = 683$ nm, whereas the stressed ones' red-edge position was at $\lambda = 671$ nm. The tested MTVI 2 and TCARI/MCARI values match the in-situ collected geochemical data, but PRI does not reflect the pollution pattern. The conclusion was that the present-day satellite hyperspectral systems are not sensitive enough to map the abiotic stress of uranium ore extraction and decommisioning (Filchev 2013).

4.4.4 Fluorescent Satellite Remote Sensing Methods for Detecting Vegetation Stress

More information on the response to stress on leaves in tree crowns can be obtained from chlorophyll fluorescence emissions. The main wavelengths included in fluorescence and green leaf emission when irradiated with UV-A radiation are in the blue wavelength 440 nm, in the green 520 nm, in the red 690 nm, and the distant red wavelength 740 nm. These peaks in the red and distant red regions are primarily associated with chlorophyll-a fluorescence, which is directly related to photosynthetic processes, while shorter emissions in short wavelengths are predominantly associated with fluorescence from other, predominantly phenolic, leaf ingredients, in particular ferrous and chlorogenic acids associated with cell walls (Buschmann et al. 2000). At each wavelength, the intensity of fluorescence is influenced by the concentration of the emitting substance, the internal optics of the leaf, including factors that affect the partial re-absorption of fluorescence, in particular photosynthetically related fluorescence, energy distribution between photosystems and damping and energy reduction processes in chloroplasts.

In practice, the raw amount of fluorescence at any time is a relatively useless signal, since it is too dependent on lighting and on the structural characteristics of the leaf. It is therefore useful to normalize the data and extract analogues of the classical parameters of chlorophyll fluorescence, which are usually obtained both from fluorescent transitions at illumination (fluorescence transients on illumination) or those obtained using modulating fluorescent techniques (Maxwell and Johnson 2000; Malenovsky et al. 2009).

The most useful fluorescence parameters are generally considered to be the ratio of the variable of maximum fluorescence after assimilation (acclimation) in the dark (Fv/Fm), indicating the potential maximum efficiency of photo system II and quantum levels (transitions) of photosystem II, which is given through the Fv/fm ratio during stable state photosynthesis (Maxwell and Johnson 2000). These relationships are obtained from a pair of fluorescent images in indication of the variation in these characteristics above the surface of the leaves.

More research on chlorophyll fluorescence has focused on the use of single point measurements, although in recent years numerous fluorescent imaging systems have become available to the general public (Lichtenthaler and Miehe 1997). These systems have become extremely useful not only for the study of photosynthetic response to stress, but also for their applications for studying a wide range of biotic stress, with multi-color and chlorophyll fluorescence data being used to demonstrate and distinguish the early stages of infection with fungi, viruses, and bacteria before symptoms become visible in standard reflective images (Chaerle and Van Der Straeten 2000). The first multi-channel satellite mission based on fluorescence was JAXA's IBUKI (GOSAT), which was launched into orbit on 23 January 2009 under international number 2009-002A. The first results of fluorescent composite images globally for 2009, after a two-year validation of the satellite data, have been made public by NASA's Goddart Space Flight Center research team (Joiner et al. 2011).

4.4.5 Modeling in Geoscience

An early classification of models in geoscience is related to the works of Chorley (1969), which separated three main types of models and several subtypes:

- I) Models – analogies with natural systems
  - a) historical analogies
  - b) spatial analogies
- II) Models – specialized systems
  - a) constructions
    - 1) likenesses
    - 2) analogues
  - b) deterministic
    - 3) stochastic
  - c) schemes of the experiment

III) Models – common systems
   a) synthetic
   b) partial
   c) black boxes

By definition given by Armand (1975), the model (from lat. modulus – measure, specimen) is "a system that is similar in something to the original." Armand classification models (1971) are ideal (concepts), descriptive, cartographic, graphic (graphics, columns, and block charts, etc.); physical – static, dynamic, and symbolic; logical; and mathematical. Patterns in geography are still divided into material and thoughtful or ideal. According to the form (way) of reflecting reality they differ (i) verbal (verbal) patterns – any description of an object or process; (ii) graphic patterns – with the help of geometric shapes and arrows, the connections between the elements of the natural environment are transmitted; (iii) cartographic models (belong to the group of graphic models) – maps or series of maps on different scales and for different purposes; mathematical models – depicting connections and processes using mathematical symbols (Landscape Preservation 1986). The landscapes, of which vegetation is an integral part, are characterized by their structure, functioning, composition, set of conditions – dynamics (so-called repertoire), and behavior. The model can mimic the original in relation to any of these properties. If the model mimics the system completely, it is called isomorphic. Homomorphic systems occur when we can establish unambiguous conformity of the elements of a system to the second, but not in the opposite case. In the practice of computer-assisted modeling of complex models and systems, it is necessary to use Unified Modeling Language (UML) as a standard object-oriented–object-oriented (O–O) modeling language (Trujillo and Parsons 2007).

At present, in the development of natural sciences for simulation of geosystems, the following modeling languages are widely used: STELLA (http://hpsinc.com); VENSIM (www.vensim.com); Power Sim (www.powersim.com); ModelMaker (www.cherwell.com); and SIMULINK, MATLAB or SIMILE (www.simulistics.com) modeling environment, as well as procedures and functions in high-level programming languages such as BASIC, FORTRAN, Pascal, C++ or Java, or the specialized geospatial information modeling language of the University of Utrecht, Netherlands – PCRASTER (www.pcraster.nl). Similar geospatial data-oriented tools are ModelBuilder, distributed with ESRI Inc.'s ArcInfo/ArcGIS license, and the bitmap-based modeling system – LANDLORD, which expands by scripting high-level languages such as Python 3.x. Most of the applications or graphical environments for modeling different processes and phenomena are coupled by connecting programs to geographical information systems (GISs), to use the functionality of GIS software packages for analysis and evaluation of results. Similar is the SIMARC (http://www.ecoap.unina.it/software.htm) tool developed as a SIMILE interface to ArcView (ESRI Inc.) or LSD (Landlord Simile Dynamic link) to connect LANDLORD and SIMILE (Mulligan and Wainwright 2004). The connection capabilities of ENVI image processing and classification system and the GIS ArcGIS of ESRI Inc. enable the use of ModelBuilder with data and classification results from the earth observation.

The first works giving the principles of work and modeling of geochemical anomalies in GIS environment were the works of Bonham-Carter (1988). At the modern stage, a number of biogeochemical models have been developed, such as JABOWA, FORECAST, TREEGRO, FOREST-BGC/PnET, NuChem, FORSANA – models of a particular pollutant in vegetation, as well as models modeling simple biogeochemical cycles (C, N2, Hg, Mg, etc.) (GOUDRIAN, DECOMP, SOIL, SOIL-N, COUP, PROFILE), and biomass growth models in coniferous forests (EFIMOD 2) are used to simulate the relevant biogeochemical cycle of a particular pollutant (Komarov et al. 2003). Other similar commercial software packages, which involve geochemical modeling of groundwater are those of Schlumberger Water Services' division with software, such as Visual MODFLOW, Hydro GeoAnalyst, AquaChem, Aquifer Test Pro, and GW Contour, designed to solve closely specialized tasks of assessing pollution and quality of surface water and groundwater with heavy metals and radionuclides, drilling management, and cartographic model creation. However, all these models duplicate the cycle of certain pollutants, one of which requires that input data have a repeatable monitoring character over a certain time period, which is not feasible within the scope of this study.

4.4.6 Models of Geosystems and Abiotic Stress in Ecology and Radioecology

Modeling in the ecology of plants is another approach of modeling the processes and phenomena taking place in the landscape, which puts at the center of the study of functions, dynamics, and conditions at the ecosystem level; plants (conifers in particular). The characteristic of plant ecology is the spatial and time scale of the models—from microscopic to planetary level, as well as the speed of processing from centuries and commensurate with geological eras to the order of seconds—in photosynthetic reactions (Mulligan and Wainwright 2004).

Interactions between plant functions and abiotic and biotic environmental factors are studied by plant physiology, and in most cases, such as streams of substances and energy, these interactions are considered at the individual or community level. The structure of plants is directly related to their physiology and functioning. For example, the state of the crowns of trees, trunks, and root system are directly dependent on the water consumption of the plant. This link is highlighted in numerous works by both scientists involved in both ecology and plant physiology and EO (Osborne and Woodward 2001). The dynamics of plants, or certain parts of plants, are generally an essential part of understanding the structure of plants at individual or community levels. Therefore, as Mulligan and Wainwright (2004) point out; the structure of vegetation is determined by the history of the development of the particular plant or society, population dynamics and abiotic and biotic stressors, Figure 4.3. In forest ecology, according to the objectives and tasks of research, vegetation growth, self-recovery, habitat patterns, and mortality patterns (Hawkes 2000) are examined and modeled.

The beginning of mathematical modeling of ecosystems, population dynamics, began in the 1920s and 1930s, with the works of A. Lotki and V. Volterra. Later began the application of a number of diversity indices (Shannon index, etc.) and curves of dominance-diversity to assess the influence of anthropogenic abiotic stress (chemical contamination) (Odum 1986). With the advent of cybernetics and modern computing, science began to build on increasingly complex ecosystem models (Gvishiani et al. 1986). Some examples are the compartment (parts of the plant) migration patterns and biogenic accumulation of chemical elements such as C, Ca, Na, K, P, etc., primary ecosystem productivity models,

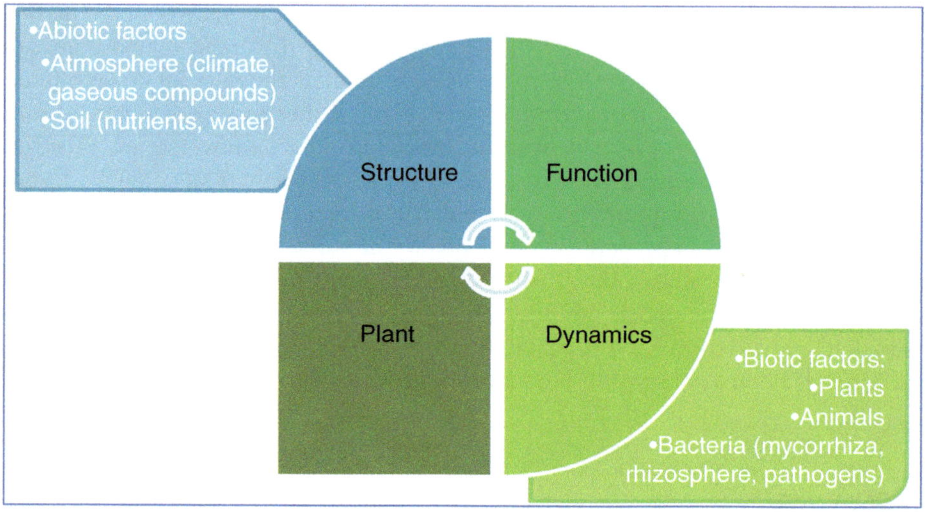

Figure 4.3 Environmental interactions between an individual plant and other plants, animals, atmosphere, and soil. Source: Mulligan, Wainwright, 2004 / John Wiley & Sons.

photosynthesis, migration and redistribution of substances, and energy and thermal balance, etc., most modern graphics systems use the principles of system compartmentation and the determination of flows between departments (compartments), as well as the direction of influence of certain parameters and/or variables on these flows.

In ecology, mostly ordinary differential, algebraic, and patterns of differences are accepted, and differentiation of private derivatives. In order to solve biological and environmental tasks, great attention is paid to the concept of simple differential equations, in which the rate of changes in system state variables is expressed as the sum of the factors affecting them. Linear dynamic systems, such as an ecosystem, are presented as linear dynamic systems of the n-th order with a number of linear differential equations of the n-th order, which in turn are converted into n number of linear differential equations of the first order. The main disadvantages of this type of models is the lack of variability characteristic of living nature, therefore more and more attention is paid to probability patterns or models of determinism (Jeffers 1988). Another weakness of simple differential equations is that they present the state of the system as a point in time, which in fact contradicts the continuality of geosystems.

The group of bioenergy models is joined by the model of the authors Koehn and Bayne (1989), which modeled the effects of stressors on individuals. The model has the following appearance in Eq. (4.4):

$$P_g + P_r = (C * AE) - (R_m + R_r) \tag{4.4}$$

where, the left side is the sum of somatic (P_g) and reproductive (P_r) production, and the right side is the difference between the product of energy consumption (C) and energy assimilation efficiency (AE), and the sum of the metabolic costs of maintenance of the soma (R) and the cost of maintaining all processes, such as nutrition, growth and reproduction (R). The right side of the equation reflects the output produced taking into account the energy taken (C * AE) and the energy costs of maintaining somatic, nutrition, growth and reproduction ($R_m + R_r$). Based on this model, Koehn and Bayne (1989) built graphs of tolerance zones and stress zones in terms of stress intensity and energy balance (Newman and Clemens 2008).

With the advent of mathematical and stochastic methods in radioecology, a number of chamber static and dynamic models of the distribution and compilation of a number of heavy metal pollutants, natural and artificial radionuclides in the soil–water–plant system are gradually created. One of the most versatile models developed for the purposes of migration, differentiation, and response of plant organisms, populations, and ecosystems of radioactive contamination gives Polikarpov (1998). In the case of composite transport patterns and the exchange of natural and artificial radionuclides in the environment and in the plant, and conifers in particular, radioecology takes into account the properties of each radionuclide, a period of half-life T1/2, type and energy of radiation, chemical compound, age and gender of the individual, anatomical and physiological characteristics of individual organs, tissues and systems, such as mass and form of the organ, the situation of the authorities vis-à-vis other authorities, etc., effective half-out period (semi-detachment), it is possible to calculate the doses in the individual compartments (departments or parts) of the plant. The effective half-life is a combination of the radionuclide half-life – $T_{1/2}$ and $T_б$ (4.5):

$$T_{ef} = \frac{(T_{1/2} . T_б)}{(T_{1/2} + T_б)} \tag{4.5}$$

The effective half-life, together with the physico-chemical properties of radionuclide, determines the radiotoxicity of the particular radionuclide. It is also essential for the biokinetics (metabolism) of radionuclide whether it enters certain chemical compounds.

The accepted unit of measurement in which the absorbed dose for plant, as well as the animal organisms, is determined is Gray (Gy), which expresses the amount of energy per unit mass of organ or tissue of the plant. A fundamental unit of measurement, to measure internal irradiation is the absorbed part (AF). If the activity and density of the medium and the body are considered equal for both external (D_{ext}) and internal dosimetric levels (D_{int}), then the simplest model of radiation levels for α-, β-, and γ radiation in units of measurement (Gy/s) are described as a function of the absorbed part:

$$D_{int} = 1.6 * 10^{-13} * q * E_i * AF(E_i) \tag{4.6}$$

$$D_{ext} = 1.6 * 10^{-13} * q * E_i * (1 - AF(E_i)) \tag{4.7}$$

where, $1.6 * 10^{-13}$ is a conversion factor (J MeV^{-1}), i – denotes radiation type (α-, β-, γ rays or spontaneous products of radioactive decay), q is the radionudal activity concentration in the body or in the environment (Bq kg^{-1}), and E_i is the energy (MeV). Therefore, the absorbed dose may never exceed that of the infinite media (environment) (D_∞), for which any monoenergy particle is given of the type:

$$D_\infty = 1.6 * 10^{-13} * q * E_i \tag{4.8}$$

If the body's dimensions are much smaller than the radiation in the media, especially in long-wave radiation (high-energy electrons and protonins), then $AF \rightarrow 0$, $D_{int} \rightarrow 0$, and

$D_{ext} \to D_\infty$. Conversely, when the size of the organisms is greater than the radiation in the media (specifically for α-particles and low-energy electrons with a range of less than 50–100 μm), then $AF \to 1$, $D_{int} \to D_\infty$, and $D_{ext} \to 0$ (Modeling Radiation Exposure and Radionuclide Transfer..., 2007). In order to synchronize and compare the results of the different distribution patterns of radionuclides in the water–soil–plant system in 2004, the Biota Working Group (BWG) was established at the IAEA's Environmental Modeling for Radiation Safety (EMRAS) program. As a result of the first phase of the work of the Task Force, 15 models from countries such as the U.S., Canada, Belgium, Russia, Lithuania, England, and Wales were compared, as well as models developed under international programs. These are the models applied by Atomic Energy Canada Limited (AECL); RESRAD-BIOTA of the U.S. Department of Energy, Framework for the Assessment of Environmental Impact (FASSET) developed during the fifth Framework Program EC; Environmental Risk from Ionizing Contaminants – Assessment and Management (ERICA Tool) developed under the sixth EU Framework Programme; Environment Agencies of England and Wales – R&D128 (EA R&D128), LIETDOS-BIO, DosDiMEco; Russian ECOMOD, Elementary Dose Evaluation for Natural Environment (EDEN 2); French CASTEAUR (simplified calculation of the transfer levels of radionuclides in receiving waterways); EPIC DOSES3D, FASTer, D-Max University of Portsmouth, UK; and the Dutch LAKECO-B etc. (Modeling Radiation Exposure and Radionuclide Transfer..., 2007).

4.5 Conclusions

As the authors of the recent report entitled "Monitoring of Forests through Remote Sensing" point out, the EO can contribute to the cost-efficient mapping and monitoring of abiotic and biotic factors by detecting the resulting disturbances as manifested in modified spectral/temporal signatures. Unfortunately, many other factors may also alter the biochemical and structural characteristics of trees and forest stands, as well as their spectral/thermal properties. Thus, making such changes both relatively unspecific and often buried in "background noise" (Atzberger et al. 2020).

The processes of transport and distribution of technogenic heavy metals and natural and artificial radionuclides in plants are partially modeled after the stress situations themselves in a particular type of vegetation, and in the combination of a certain set of parameters at the current stage of the development of scientific knowledge, so it is not possible to create them. Models have been created that try to fully model ecosystems at structural, physical, and chemical (material) levels. However, stressful situations and the resulting baselines of the system to be detected by EO methods have not been studied in their entirety due to the different "toxicity thresholds" of heavy metals and radionuclides in the soil in terms of development of both soil microflora and the development, anatomy, and physiology of the studied coniferous species.

In conclusion, the following more important problems for recognizing abiotic stress situations in coniferous landscapes are formed at this stage of development of satellite systems for EO (i) creation of additional and better spatial resolution (SR) satellite systems for remote research based on fluorescence; (ii) improvement of the spectral and SR of infrared satellite RS and systems in order to better identify stressful situations caused by the abiotic environmental factors; and (iii) establishment of expert systems based on databases of knowledge and data from satellite RS in order to define and resolve a specific stressful situation more quickly and clearly.

In summary, the study attempts to (i) reveal the importance of modeling coniferous vegetation stress using EO to the community of stakeholders; (ii) provide an overview of the numerous uncorrelated attempts to study the abiotic stress, and underline the importance to streamline the process in view of the global challenges such as climate change and deforestation; (iii) through an interdisciplinary issue to bridge the gap between the EO community of practitioners and ecophysiologists, meteorologists and climatologists, geochemists, and landscape ecologists to name a few, with the modelers; either working in physical, geochemical, physiological, or mathematical modeling.

References

25 years of satellite imagery over Chernobyl (2011). link – ASTRIUM an EADS company: 21.

Ages, P. (1982). *Keys to Ecology*. Leningrad: Hydrometeoizdat (in Russian).

Alekseev, Y.V. (1987). *Heavy Metals in Soils and Plants*. Leningrad: (in Russian).

Anisova, Z.M. and Yakushev, B.I. (2008). *Natural Isotopes of Uranium in Soils and Plants of Pine Forests of the Minsk Upland*. Belarusian Science: Minsk (in Russian).

Armand, D.L. (1975). *Science of Landscape – Fundamentals of Theory and Logical-Mathematical Methods*. Moscow: Mysl (in Russian).

Aronoff, S., Collin, E.D. et al. (1985). Biogeochemical Anomallies and Landsat Imagery: A Comparison on the Wollaston Lake Area, Saskatchewan. In: *Miscellaneous Report, Saskatchewan Geology Survey, Saskatchewan Energy and Miners*, vol. 85-4, 116–124.

Atzberger, C., Zeug, G. et al. (2020). *Monitoring of Forests through Remote Sensing*. ENV.D.1/ETU/2018/0022MV, Final Report.

Badyaev, V.V., Egorov, U.A. et al. (1990). *Environmental Protection During the Operation of AES*. Energoatomizdat Moscow (in Russian).

Barescut, J., Gariel, J., Péres, J., and Sazykina, T. (2005). A system of dose-effects relationships for the northern wildlife: Radiation protection criteria. *Radioprotection* 40 (S1): S889–S892. https://doi.org/10.1051/radiopro:2005s1-130.

Bashirinov, A.E., Gurvich, A.S. et al. (1969). Determination of geophysical parameters by measurements of thermal radio emission at the Kosmos-243 equipment. *DAN USSR* 188: 6. (in Russian).

Baule, B. (1917). Mitscherlich's law of physiological relations. *Landwirtschafliche Jahrbuecher* 51 (1916–1917): 363–385. (in German).

Bliznakov, A. and Ilieva, V. (2004). Heavy metals in Bulgarian soils. *Science* XIV: 3–6. (in Bulgarian).

Bonham-Carter, G.F. (1988). Numerical procedures and computer program for fitting an inverted Gaussian model to vegetation reflectance data. *Computers & Geosciences* 14 (3): 339–356.

Bowen, H.J.M. (1966). *Trace Elements in Biochemistry*. Academic Press.

Buschmann, C., Langsdorf, G. et al. (2000). Imaging of the blue, green, and red fluorescence emission of plants: an overview. *Photosynthetica* 38: 483–491.

Chaerle, L. and Van Der Straeten, D. (2000). Imaging techniques and the early detection of plant stress. *Trends in Plant Science* 5 (11): 495–501.

Charles, E. and Zyl, J.V. (2006). *Introduction to the Physics and Techniques of Remote Sensing*. NY: Wiley.

Chorley, R.J. (1969). Geography and analogue theory. *Annals of the Association of American Geographers* 54: 127–137.

Dedew, I.I. (1989). *Ecological encyclopedic dictionary*. In: *Moldavian Soviet Encyclopedia*, vol. 408 (ed. V.V. Kishinev). German: (in Russian).

Ellenberg, H. (1979). *Vegetation Ecology of Central Europe*, 4e. Cambridge, UK: Cambridge University Press (Translated from Russian in 1988).

EMRAS. (2007). Modelling Radiation Exposure and Radionuclide Transfer for Non-human Species – Report of the Biota Working Group of EMRAS Theme 3, Environmental Modelling for RAdiation Safety (EMRAS) Programme; IAEA.

Ermakov, V.V. and Kovalsky, Y.V. (2019). V.V. Kovalsky's Contribution in the Development of Biogeochemistry and Geochemical Ecology. In: *Biogeochemistry – The Scientific Basis for Sustainable Development and Protection of Human Health*, vol. 1, 8–14. Tula: TSPU 288 p. ISBN 978-5-6042450-0-2. http://www.spsl.nsc.ru/FullText/konfe/%D0%91%D0%B8%D0%BE%D0%B3%D0%B5%D0%BE%D1%85%D0%B8%D0%BC%D0%A3%D1%81%D1%82%D0%A0%D0%B0%D0%B7%D0%B220191.pdf (In Russian).

EU 2030 Biodiversity Strategy. (2021). *The EU 2030 biodiversity strategy*. EC https://op.europa.eu/s/s3mX

Filchev, L. (2013). Detection and assessment of abiotic stress of coniferous landscapes caused by uranium mining (using hyperspectral EO-1/Hyperion data). *Aerospace Research in Bulgaria* 25: 138–153.

Forest Condition in Europe (2010). *Forest Condition in Europe: 2010*. Technical report of ICP Forests, Thünen-Institute, Institute for World Forestry, Germany; Hamburg, June 2010.

Forest Condition in Europe (2012). *Forest Condition in Europe: 2012*. Technical report of ICP Forests, Lorenz, M. and Becher, G. (eds.). Thünen-Institute, Institute for World Forestry, Germany; Hamburg, June 2010.

Forest Health and Ozone (1987). *Forest health and ozone*. Hakkarinen, C. (ed.). (Electric Power Research Inst., Palo Alto, CA (USA)). Apr 1987. 87p. (EPRI-EA-5135-SR). Research Reports Center, Box 50490, Palo Alto, CA 94.303.

Fortescue, J.A.C. (1971). Biogeochemistry, plant growth and the environment. *CIM Bulletin* 64 (August): 77–82.

Franklin, S.E. (2001). *Remote Sensing for Sustainable Forest Management*. Lewis Publishers; CRC Press LLC.

Gamon, J.A., Penuelas, J., and Field, C.B. (1992). A narrow-waveband spectral index that tracks diurnal changes in photosynthetic efficiency. *Remote Sensing of Environment* 41 (1): 35–44.

Gutman, G., Tarpley, D. et al. (1995). The enhanced NOAA global land data set from the advanced very high resolution radiometer. *The Bulletin of the American Meteorological Society* 76 (7): 1141–1156.

Gvishiani, D.M., Novika, I.B. et al. (ed.) (1986). *The Nature of Models and Models of Nature*. Moscow: Mysl (in Russian).

Haboudane, D., Miller, J.R. et al. (2002). Integrated narrow-band vegetation indices for prediction of crop chlorophyll content for application to precision agriculture. *Remote Sensing of Environment* 81 (2, 3): 416–426.

Harmsen, K. (2000). A modified mitscherlich equation for rainfed crop production in semi-arid areas: 1. *Theory. NJAS - Wageningen Journal of Life Sciences* 48 (3): 237–250, ISSN 1573-5214. https://doi.org/10.1016/S1573-5214(00)80016-0.

Haslett, J., Berry, P., Bela, G. et al. (2010). Changing conservation strategies in Europe: a framework integrating ecosystem services and dynamics. *Biodiversity and Conservation* 19: 2963–2977. https://doi.org/10.1007/s10531-009-9743-y.

Hawkes, C. (2000). Woody plant mortality algorithms: description, problems, and progress. *Ecological Modelling* 126: 225–248.

Henel, R. and Conrath, B. (1969). Preliminary results from the interferometer experiment on Nimbus-III. *Science* 165 (3899): 1258–1260. https://doi.org/10.1126/science.165.3899.1258. PMID: 17813600.

Hill, M.O., Roy, D.B., Mountford, J.O., and Bunce, R.G.H. (2000). Extending Ellenberg's indicator values to a new area: an algorithmic approach. *Journal of Applied Ecology* 37: 3–15. https://doi.org/10.1046/j.1365-2664.2000.00466.x.

Idso, S.B., Jackson, R.D. et al. (1981). Normalizing the stress-degree-day parameter for environmental variability. *Journal of Agricultural Meteorology* 24: 45–55.

Ilkun, G.M. (1978). *Atmospheric Pollution and Plants*. Kiev: Naukova dumka (in Russian).

Ilykun, G.M. (1978). *Pollutants of the Atmosphere and Plants*, 247. Naukova Dumka, Kyiv: Academy of Sciences of the Ukrainian SSR, Central Republican Botanical Garden.

Jeffers, J.N.R. (ed.) (1988). *Practitioner's Handbook on the Modelling of Dynamic Change in Ecosystems*. Chichester, UK: *SCOPE*. John Wiley & Sons.

Joiner, J., Yoshida, Y. et al. (2011). First observations of global and seasonal terrestrial chlorophyll fluorescence from space. *Biogeosciences* 8: 637–651.

Jones, H.G. and Schofield, P. (2008). Thermal and other remote sensing of plant stress. *General and Applied Plant Physiology* 34 (Special Issue [1, 2]): 19–32.

Kabata-Pendias, A. and Pendias, H. (1984). *Trace Elements in Soils and Plants*. Boca Raton, Florida: CRC Press.

Koehn, R.K. and Bayne, B.L. (1989). Towards a physiological and genetical understanding of the energetics of the stress response. *Biological Journal of the Linnean Society* 37: 157–171.

Komarov, A., Chertov, O. et al. (2003). EFIMOD 2- a model of growth and cycling of elements in boreal forest ecosystems. *Ecological Modelling* 170 (2, 3): 373–392.

Kupriyanov, V.V. and Prokacheva, V.G., (eds.). (1976). Satellite information and land study (introduction to satellite hydrology). *Proceedings of the GYI*. Leningrad (in Russian).

Kutov, A. and Kostova, Z. (ed.) (1988). *Dictionary of Ecology and Environment Protection*, vol. 288. Partizdat, Sofia, Bulgaria (in Bulgarian).

Landsat 9. (2021). *Overview, 2021*. https://landsat.gsfc.nasa.gov/landsat-9/landsat-9-overview

Landscape Preservation (1986). Dictionary. Sofia, Science and Art, 248: (in Bulgarian).

Lichtenthaler, H.K. (1998). The stress concept in plants: an introduction. *Annals of the New York Academy of Sciences* 30 ((851)): 187–198. https://doi.org/10.1111/j.1749-6632.1998.tb08993.x. PMID: 9668620.

Lichtenthaler, H.K. and Miehe, J.A. (1997). Fluorescence imaging as a diagnostic tool for plant stress. *Trends in Plant Science* 2 (8): 316–320.

Makunina, G.S. (1978). Anthropogenic modification of the low-mountain southern secret landscape in the sphere of influence of copper smelting production. *MSU Bulletin Geographical Series* 3: (in Russian).

Malenovsky, Z., Mishra, K.B. et al. (2009). Scientific and technical challenges in remote sensing of plant canopy reflectance and fluorescence. *Journal of Experimental Botany* 60 (11): 2987–3004.

Mapping Chernobyl fires from space. (2020). link - ESA Applications. Партиздат.

Markarova, E.N. et al. (1976). *On the use of chemiluminescent methods in works on plant pathogenesis. Pathological Physiology and Immunity of the Plant*: MGU Publishing House, Moscow.

Maxwell, K. and Johnson, G.N. (2000). Chlorophyll fluorescence – a practical guide. *Journal of Experimental Botany* 51: 659–668.

van der Meer, F., Schmidt, K.S. et al. (2002). New environmental remote sensing systems. In: *Environmental Modelling with GIS and Remote Sensing* (ed. A.K. Skidmore), 26–51. London.: Taylor & Francis.

Mishev, D.N., Dobrev, T.B. et al. (1987). *Remote Sensing Methods in Geophysics and Geology – Aerospace Methods for the Study of Earth*. "Technica" Publishing House, Sofia (in Bulgarian).

Mitscherlich, E.A. (1909). The law of the minimum and the law of diminishing soil productivity. *Landwirtschafliche Jahrbuecher* 38: 537–552. (in German).

Modelling Radiation Exposure and Radionuclide Transfer for Non-human Species: Report of the Biota Working Group of EMRAS Theme 3. (2007). *Environmental Modelling for RAdiation Safety (EMRAS) Programme - IAEA*.

Modzelewska, A., Jarocinska, A., Pochrybniak, P., and Mostowski, M. (2013). *The vegetation condition changes near Chernobyl based on Landsat TM*. EARSeL Symposium.

Mulligan, M. and Wainwright, J. (ed.) (2004). *Environmental Modelling – Finding Simplicity in Complexity*. Chippenham, Wiltshire: Wiley.

Murtha, P.A. and Watson, E.K. (1973). SO2 damage of forest recorded by ERTS-1. In: *Proceedings of Third ERTS Symposium*, 10–14 December, 1973, NASA, Washington DC, USA.

Natura (2000). Environment - European Commission (europa.eu).

New EU Forest Strategy (2012). New EU Forest Strategy – possible options and instruments Maria Gafo-Gomez Zamaloa, (ed.). *EC*, June 2012.

Newman, M.C. and Clemens, W.H. (2008). *Ecotoxicology - A Comprehensive Approach*. Boca Raton: Taylor & Francis.

Nikolaevski, V.S. (1983). The effect of industrial gases on vegetation. In: *Regional Environmental Monitoring: (On the Example of the Upper Oka Basin)* (ed. V.A. Kovda and A.S. Kerzhencev), 264. Moskow: Nauka (in Russian).

Nikolaevski, V.S. and Miroshnikova, A.T. (1974). Permissible norms of air pollution for plants. *Hygiene and Sanitation* 4: 16–18. (in Russian).

Odum, E. (1986). *Ecology*, vol. 2, 376. Moskow: Mir (in Russian).

Omasa, K., Nouchi, I., and Kok, L.J.D. (ed.) (2005). *Plant Responses to Air Pollution and Global Change*. Tokyo: Springer-Verlag.

Onianwa, P.C. (2001). Monitoring atmospheric metal pollution: a review of the use of mosses as indicators. *Environmental Monitoring and Assessment* 71 (1): 13–50.

Orlov, A.S., Sadovnikova, L.K. et al. (2002). *Ecology and environment protection from chemical pollution*. Higher education, Moskow (in Russian).

Osborne, C.P. and Woodward, F.I. (2001). Biological mechanisms underlying recent increases in the NDVI of Mediterranean shrublands. *International Journal of Remote Sensing* 22: 1895–1907.

Panda, S.K. and Choudhury, S. (2005). Chromium stress in plants. *Brasilian Journal of Plant Physiology* 17 (1): 95–102.

Pavlotskaya, F.I. (1974). *Migration of Radioactive Products of Global Fallout in Soils*. Atomizdat Moscow (in Russian).

Peddle, D.R., Boulton, R.B. et al. (2008). Hyperspectral detection of chemical vegetation stress: evaluation for the Canadian HERO satellite mission. *Canadian Journal of Remote Sensing* 34 (Suppl. 1): S198–S216.

Polikarpov, G.G. (1998). Conceptual model of responses of organisms, populations and ecosystems to all possible dose rates of ionising radiation in the environment. *Radiation Protection Dosimetry* 75: 181.

Poppytz, R. and Penchev, B. (1989). *Life with Radiation – Doses, Risk*. National Education, Sofia (in Bulgarian): *Protection*.

Poschl, M. and Nollet, L.M.L. (ed.) (2007). *Radionuclide Concentrations in Food and the Environment*. Boca Raton, FL: Taylor & Francis.

Rapport, D.J. (1989). What constitutes ecosystem health? *Perspectives in Biology and Medicine* 33: 120–132.

Rapport, D.J., Costanza, R. et al. (1998). Assessing ecosystem health. *TREE* 13: 10.

Reimers, N.F. (1980). *Alphabet of Nature – Micro-Encyclopedia of Biosphere*, 208. Moskow: Znanie (in Russian).

The EC Green Paper on Forest Protection and Information in the EU: Preparing Forests for Climate Change, EC. (2010). https://op.europa.eu/s/s3mQ

EC (2012). *EU Forestry Strategy and the EU Forest Action Plan* (ed. T. Szedlac). EC.

Richelson, J.T. (1990). *America's Secret Eyes in Space*, vol. 383. Harper & Row.

Rondeaux, G., Steven, M. et al. (1996). Optimization of soil-adjusted vegetation indices. *Remote Sensing of Environment* 55 (2): 95–107.

Rose, K.S.B. (1992). Lower limits of radiosensitivity in organisms, excluding man. *Journal of Environmental Radioactivity* 15 (2): 113–133.

Rubin, B.A. (1976). *Physiology of plants*. Moscow: Vysshaya Shkola.

Runeckles, V.C. and Resh, H.M. (1975). The effect of chronic ozone injury to leaves by reflectance spectrophotometry. *Atmospheric Environment* 9 (4): 447–452.

Schaeffer, D.J., Henricks, E.E. et al. (1988). Ecosystem health: 1 measuring ecosystem health. *Environmental Management* 12: 445–455.

Selye, H. (1976). *Stress without distress*. In: *Psychopathology of Human Adaptation* (ed. G. Serban). Boston, MA: Springer https://doi.org/10.1007/978-1-4684-2238-2_9.

Shelford, V.E. (1931). Some concepts of bioecology. *Ecology*. 12 (3): 455–467. https://doi.org/10.2307/1928991 ISSN 1939-9170. JSTOR 1928991.

Sinclair, T.R. and Park, W.R. (1993). Inadequacy of the Liebig limiting-factor paradigm for explaining varying crop yields. *Agronomy Journal* 85 (3): 472–476. https://doi.org/10.2134/agronj1993.00021962008500030040x.

Stocker, O. (1947). Probleme der pflanzlichen. *Dürreresistenz. Die Naturwissenschaften* 34 (12): 362–371. https://doi.org/10.1007/BF00623626.

Stoyanov, S. (1999). *Heavy Metals in Environment and Food Products - Toxicity, Clinics, Healing and Prophylactic*. Sofia: Pensoft - PublishSciSet-Agri (in Bulgarian).

Sukhih, V.I. and Sinicina, S.G. (1979). *Aerospace Methods in Environment Protection and Forestry*. Lesnaya promishlenost: Moskow (in Russian).

Suzuki, R., Ishii, R. et al. (2010). Forest Vegetation Analyses by GCOM-C/SGLI Accompanied with Field Data With Field Data. *International Archives of the Photogrammetry, Remote Sensing and Spatial Information Science* 38 (8): 3.

The New EU Forest Strategy (2020). *The new EU Forest Strategy 2020* – Report to the SFC, draft 2, 20 March 2012.

Thermal Crop Water Stress Indices (2001). *Thermal Crop Water Stress Indices*, 12.

Trujillo, J. and Parsons, J. (2007). Preface to FP-UML 2007. In: *Advances in Conceptual Modeling – Foundations and Applications* - ER 2007 Workshops CMLSA, FP-UML, ONISW, QoIS, RIGiM, SeCoGIS, Auckland, New Zealand, 2007.

Van der Meer, F. and Jong, S.M.D. (ed.) (2006). *Imaging Spectrometry - Basic Principles and Prospective Applications. Remote Sensing and Digital Image Processing*. Springer.

Vangronsveld, J. and Clijsters, H. (1992). A biological test system for the evaluation of metal phytotoxicity and immobilisation by additives in metal contaminated soils. In: *Metal Compounds in Environment and Life, 4. Special Supplement to Chemical Speciation and Bioavailability* (ed. E. Merian and W. Haedi), 117–125. Wilmington: Science Reviews Inc.

Vasilev, A., Berova, M. et al. (2010). Metal phytotoxicity - appropriate indicators and tests for ecotoxicological evaluation of contaminated soils. *Agricultural Sciences II* (4): 9.

Vinogradov, B.V. (1981). *Transformed Earth – Aerospace Research*. Moskow: Mysl (in Russian).

Vogt, P. and Verstraete, M. M. (2009). *Physics of AnisView*, 5.

Vygodskaya, N.N. and Gorshkova, I.I. (1987). *Theory and Expriment in Remote Sensing Studies of Vegetation*. Leningrad: Gidrometeoizdat (in Russian).

Warren, W.B. (1969). *Apollo Mission G Spacecraft to Spacecraft VHF Coverage*. TRW Report Number 11176-H272-RO-00. 26 June, 1969.

White Paper "Adapting to Climate Change: Towards a European Framework for Action", EC (2009). EC. http://eur-lex.europa.eu/LexUriServ/LexUriServ.do?uri=CELEX:52009 DC0147:EN:NOT

Yazaki, K., Maruyama, Y. et al. (2005). Effects of elevated carbon dioxide concentration on wood structure and formation in trees. In: *Plant Responses to Air Pollution and Global Change* (ed. K. Omasa, I. Nouchi and L.J.D.(e.). Kok). Tokyo: Springer-Verlag.

Zakharov, A., Zakharova, L., Sinilo, V.P., and Sorochinsky, M.V. (2003, 2003. IGARSS '03. 2003). *Monitoring of the forests state in the Chernobyl area using remote sensing data*. In: *Geoscience and Remote Sensing Symposium*, vol. 4, 2474–2476.

5

Retrieval of Mangrove Forest Properties Using Synthetic Aperture Radar

Jigarkumar B. Solanki[1], *Rina Kumari*[*,1], *Anup Kumar Das*[2], *Parul Maurya*[1], *and Nikhil Lele*[2]

[1] *School of Environment Science and Sustainable Development, Central University of Gujarat, Gandhinagar, Gujarat, 382030, India*
[2] *Space Applications Centre, ISRO, Ahmedabad, Gujarat, 380015, India*

5.1 Introduction

Worldwide, changes in land use and land cover is a major concern, and it contributes to approximately 20% of global anthropogenic CO_2 emissions (Avtar et al. 2012). A number of initiatives have been taken to mitigate the concentration of CO_2 from the atmosphere and maintain the balance of carbon (Trexler and Kosloff 1998). For instance, Reducing Emission from Deforestation and forest Degradation (REDD) is a United Nations-adopted scheme to mitigate climate change. It helps the developing countries by assisting in making strategies to support afforestation, reforestation, reduce deforestation, agroforestry, and management of the forest. For the implementation of REDD policies an accurate assessment of (i) place of deforestation with forest types, (ii) aerial extent (hectares), (iii) biomass lost (percentage), (iv) carbon contents in each forest type (metric tons of carbon per hectare), and (v) the process of forest loss is needed (Ramakutty et al. 2007; Avtar et al. 2012; Nyamari and Cabral 2021). Forests sequester and store more carbon compared to any other terrestrial ecosystem, so one effective way to reduce the CO_2 from the atmosphere is carbon sequestration by plants or autotrophs (Gibbs et al. 2007). Mangroves cover only 0.7% of the area of total tropical forest, although each hectare of mangroves has the potential to sequestrate more carbon than any other terrestrial forest of tropics (Duke et al. 2007).

Mangroves are groups of trees, palms, and shrubs that are concentrated at the land–ocean interfaces and river–ocean interfaces (estuary) (Duke 2006; FAO 2007). Mangrove plants have the ability to grow with salty water and low oxygenated soil due to their unique root (pneumatophores and prop) adaptation (Simard 2019). Globally, mangroves forests are found in 118 countries, having the areal extent of 137 760 km^2 until 2000 (Giri et al. 2011). The world's most diverse and extensive mangrove forest is found in South and Southeast Asia, encompassing 40%

*Corresponding author's email: rina.sesd@cug.ac.in

0000-0002-5848-8374 (Jigarkumar B. Solanki), 0000-0003-1824-6351 (Rina Kumari), 0000-0002-5024-5159 (Anup Kumar Das), 0000-0002-8217-1422 (Parul Maurya), and 0000-0002-1412-6646 (Nikhil Lele)

Advances in Remote Sensing for Forest Monitoring, First Edition. Edited by Prem C. Pandey and Paul Arellano.

of world mangroves (FSI 2019). Mangroves serve a wide range of ecological services, such as nurseries for marine mammals, fish, crustaceans, and birds; medicines; and construction materials; and they act as shields from storms, tsunamis, and floods (Dahdouh-Guebas 2011; Kathiresan 2012; Kripa et al. 2020). Worldwide, mangroves are found to be the most threatened and quickly vanishing ecosystem (Siikamaki et al. 2012). Aquaculture expansion, agriculture, and industrialization are the major threats to the mangrove ecosystem. In Thailand, more than half of the nation's mangroves were converted to aquaculture during the period of 1961 to 1993, while in Vietnam, 5.1 million ha of mangroves were degraded from 1980 to 2000 (Seto and Fragkias 2007). In Myanmar, the expansion of rice fields, and in Malaysia and Indonesia, oil palm farms, were found as causes of degradation of mangrove forests (Richards and Friess 2016). If the current trends in mangroves conversion continue, after a century there will be no mangroves left (Duke et al. 2007). Thus, the continuous monitoring of this vital ecosystem is necessary. The field monitoring of mangroves is difficult due to their locations, where the maze of trunks and roots, muddy substrate, and tides are present (Cougo et al. 2015). Thus, to study the long-term dynamics of inaccessible mangrove ecosystems, remote sensing data with different spectral, spatial, and temporal resolutions have proven to be crucial tools (Cougo et al. 2015).

Optical sensors record the sun's energy, which is reflected from the earth's surface. Red radiation is absorbed by the plant leaves due to the presence of chlorophyll, and then emit NIR radiation due to internal scattering (Allen and Richardson 1968). The quantity of leaf tissue and pigment concentration governs the reflected radiation (Allen and Richardson 1968). Optical remote sensing has been used in mangrove mapping (Held et al. 2003; Giri et al. 2011), change detection (Giri et al. 2015; Jia et al. 2015), health status (Chellamani et al. 2014), biomass estimation (Foody et al. 2003), field survey planning, and emergency data supply for managing this calamity (Kuenzer et al. 2011).

Spectral variations of the canopy depend on several optical properties such as leaf area index (LAI), background reflectance, and leaf inclination. This also depends on the age, vitality, and phenological and physiological characteristics of that particular species. Spectral reflectance is also affected by the climatic condition, which in turn affects leaf dynamics (such as leaf foliation and leaf senescence) (Kuenzer et al. 2011).

Various physiological phenomena of mangroves such as predicting photosynthesis, respiration, carbon and nutrient cycling, transpiration, and rainfall interception are indicated by the LAI. Due to the low spatial resolution of existing satellite RS data (e.g. Landsat TM) before 1990, quantification of these was difficult. However, after 1990 with the emergence of high-resolution satellite image SPOT-1, mangrove LAI inversion became possible. Some authors, such as Ramsey and Jensen (1996) and Green et al. (1998), established the relationship between in-situ canopy spectra and mangrove LAI, and the relationship of normalized differences of vegetation index (NDVI) and ground measured LAI using SPOT and LANDSAT images. The optical sensor capability became limited during the cloud cover, but this cloud cover is significantly more transparent for microwave sensors (Meyer 2019).

5.2 Microwave Remote Sensing

Microwave sensors are RADAR, which is an acronym for the RAdio Detection And Ranging that can operate in both active and passive modes. Active sensors include Side Looking Airborne Radar (SLAR), Synthetic Aperture Radar (SAR), Scatterometer, and Altimeter

Table 5.1 Details of microwave frequency bands and their suitability for typical applications.

Band	Frequency (GHz)	Wavelength (cm)	Application
P	0.3–1	30–100	Forestry mapping, monitoring, and biomass
L	1–2	15–30	Forest mapping and biomass, geophysical monitoring
S	2–4	7.5–15	Agriculture monitoring, earth observation
C	4–8	3.8–7.5	Mapping, change analysis, monitoring of low to moderate vegetation, maritime navigation
X	8–12	2.4–3.8	Urban monitoring, used extensively on airborne systems for military reconnaissance and terrain mapping
Ku	12–18	1.7–2.4	Used in initial airborne radar but now being used mostly for satellite telecommunication
K	18–27	1.1–1.7	
Ka	27–40	0.8–1.1	

Source: Adapted from Meyer 2019.

that transmit their own signals, and operate day and night independent of solar illumination conditions. Passive sensors are radiometers that measure the microwave emissions from a target as brightness temperatures. Active sensors transmit signals toward the target and measure the portion of energy returned back to the sensor as "backscatter" energy represented in decibels (dB). SAR systems operated at different microwave frequency bands suitable for various applications are shown in Table 5.1. RADAR returned signal or backscatter is governed by the various sensor parameters such as frequency, polarization, and incident angle, and target parameters such as dielectric properties, size and distribution of individual scatterers within a resolution cell, and the target roughness (Lee and Pottier 2009; Meyer 2019).

5.2.1 Polarization

SAR signals are polarized electromagnetic waves. Polarization describes the locus of the electric field in the plane perpendicular to the propagating signal (Meyer 2019). Radar systems can transmit and receive signals in either horizontal (H) polarization or in vertical (V) polarization. There are different types of transmit/receive polarization combinations such as co-polarization (HH; horizontal transmit, horizontal receive, or VV; vertical transmit, vertical receive), and cross-polarization (HV;horizontal transmit, vertical receive or VH; vertical transmit, horizontal receive) (Meyer 2019). The HV response is dominated by the volume scattering, which is related to the canopy and overall biomass, while the underlying surface gives a higher response in HH and VV polarization (Thomas et al. 2015). During high tide conditions, the underlying surface of mangroves becomes smooth due to water, so specular scattering dominates, but the presence of vertical stems of mangroves give a high HH polarized response due to the double bounce scattering between woody structure and water (Lucas et al. 2007; Thomas et al. 2015).

5.2.2 Interaction Mechanism of SAR

The interaction of SAR signals with various surface features is described by three major scattering mechanisms such as odd-bounce or surface, even-bounce or dihedral, and volume or diffuse scattering, which depends on the SAR system configuration and the physical and geometrical properties of the surface being scanned (Figure 5.1) (Lucas et al. 2007).

The surface scattering mainly occurs at water body, dry lakebeds, low vegetation fields, and bare rocks. The double-bounce scattering occurs at building sides, tree trunks, light poles, and rice and corn fields. Most natural vegetation such as forests, shrubs, and grasslands produce volume scattering. The backscatter values corresponding to the surface, volume, and double-bounce scattering are low medium and high, respectively, which depend on the frequency, polarization, and local incidence angle of the SAR signal. For example, at L-band HH polarization and at moderate incidence angle (30°–40°) the typical backscatter values associated with the surface, volume, and double-bounce scattering from natural distributed targets are less than −18 dB, −18 to −8 dB, and greater than −8 dB, respectively (Lucas et al. 2007).

Mangroves primarily exhibit volume scattering from the vegetation canopy. The partially inundated vegetation allows a higher contribution of double-bounce scattering resulting in increased backscattering as compared to non-inundated vegetation. During the low tide period, the dead plant material and complex root system of mangroves become exposed above the water surface, leading to an increase in roughness, and hence high diffuse scattering. The backscatter value also depends on the nature of the root system, as some mangroves species have longer aerial woody root (e.g. *Rhizophora*), while some have the smaller pneumatophores (e.g. *Avicennia*) (Lucas et al. 2007).

At a longer wavelength, the potential of radar signals to penetrate the vegetation canopy and interaction with forest components such as branches, trunks, and leaves increases (Lucas et al. 2007). Figure 5.2 illustrates the relationship between signal wavelength and signal penetration into the vegetation. It shows that X-band, with 2.4 to 3.8 cm wavelength, is only able to interact with the canopy or upper portion of the vegetation, while the C-band can interact with canopy as well as some upper branches also, and L-band can penetrate

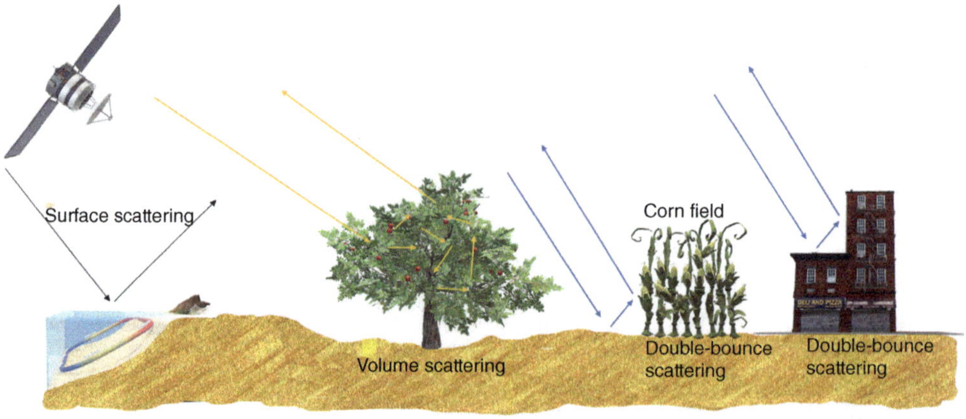

Figure 5.1 Sketch of scattering types.

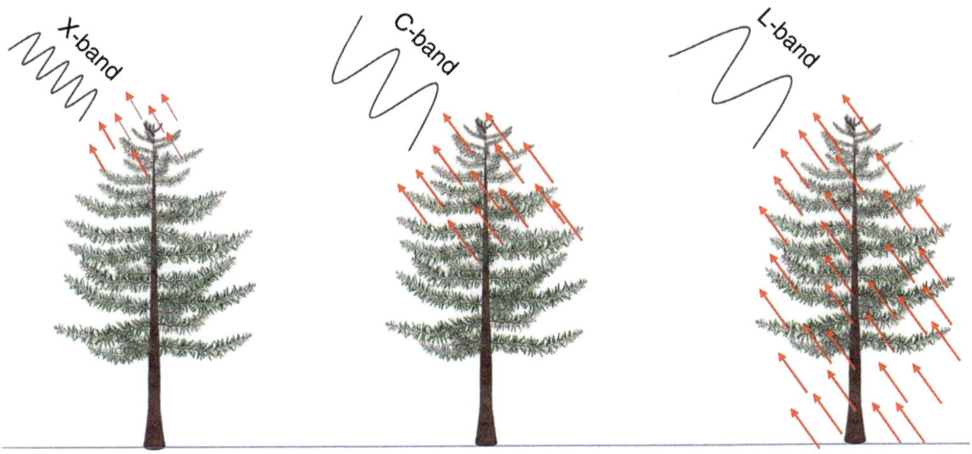

Figure 5.2 Penetration capability of different SAR band into forest.

deeper into the vegetation structure as well as underlying surfaces, and comparatively extract more information than the shorter wavelengths. Moreover, the surface roughness and moisture content of the underlying surface contribute to the overall backscatter. Hence, to characterize the vegetation structure, biomass and other biophysical parameters using SAR data, the use of a longer wavelength radar band is recommended (Imhoff et al. 1986; Lucas et al. 2007; Meyer 2019).

SAR data or data products from a few past and present space-borne missions are freely available, and data from several upcoming missions have been declared available freely for various geoscience applications. Table 5.2 lists some of the space-borne SAR missions that provide or are committed to providing free and open access data to the global community. These datasets hold great potential for space-based monitoring of vegetation and retrieval for vegetation biophysical parameters.

5.2.3 SAR Based Mangroves Studies

In recent years studies related to the estimation of mangrove cover, mangrove classification, and biophysical parameter retrieval were conducted by many using SAR data. Some of these studies are listed in Table 5.3; these studies have demonstrated the potential of SAR data for mangroves structural parameter retrieval and biomass estimation.

5.2.4 SAR Image of the Mangroves

The SAR image of Tapi estuary in Surat, Gujarat acquired in L-band HH polarization is shown in Figure 5.3. Here the brighter backscatter is the result of strong double bounce scattering effect from the Surat city and nearby settlements (blue circle), and medium backscatter with mid-gray tone is the result of volume scattering from the mangroves (red circle). Water bodies are easily identifiable in this image as dark tones due to specular reflection, and other land cover classes like mudflats and saltpan/aquaculture are also easily identifiable in this image.

Table 5.2 Detail of free and open data from past, present, and forthcoming space-borne SAR missions.

Mission	Duration	Wavelength	Modes /Polarization	Resolution	Swath	Data / product access
ALOS-1	2006–2011	L-band (24 cm)	FBS: HH, VVFBD: HH/HV, HH/VHPLR: HH/HV/VH/VVScanSAR: HH, VV	FBS: 10× 10mFBD: 20×10mPLR: 30×10m ScanSAR: 100 m	FBS: 70 kmFBD: 70 kmPLR: 30 km ScanSAR: 250–350 km	Free and open (limited datasets)
ALOS-1/2	ALOS-1; 2006–2011 ALOS-2; 2015 – ongoing	L-band (24 cm)	Dual-pol: HH/HV	25 m and 100 m annual mosaic products	Global coverage	Free and open
Sentinel-1	2014 ongoing	C-band (05.6 cm)	Single: HH, HVDual: HH/HV, VV/VH	Stripmap: 5×5 mIW: 5×0 mEW: 20–40 m	Stripmap: 375 kmIW: 250 kmEW: 400 km	Free and open
NISAR	Planned in 2023	L- and S-band (24 cm/ 9.3 cm)	Single: HH, HV, VH, VVDual: HH/HV, VV/VH, HH/VVquad	3–20 m (mode dependent)	250 km	Free and open
BIOMASS	Planned in 2022	P-band (70 cm)	Quad	≤ 60 × 50 m	160 km	Free and open
TanDEM-L	Planned in 2023	L-band (24 cm)	Single, dual, quad	12 × 12 m	350 km	Free and open

5.2.5 Mapping the Mangrove Area

Mapping of the mangrove areal extent with radar data alone is difficult, particularly when the mangrove area is adjoined with forest or other vegetation. The mapping is also difficult with only optical data, as spectral tone may not be enough to differentiate mangroves from other vegetation. If a mangrove patch is surrounded by mudflats, bare ground, and fallow croplands, it is easier to differentiate the mangroves from the surrounding land-use class, but the discernibility reduces when another vegetation type (e.g. forest or grassland) is present, adjacent to the mangroves. Thomas et al. (2015) suggested two measures to overcome this confusion. They concluded that a baseline map should be prepared from optical data, along with using a combination of SAR and optical data for mangroves extent mapping, as various authors have also done. de Souza et al. (2012) used L-band quad-pol data from Advanced Land Observing Satellite (ALOS) and concluded that HH polarization intensity and radar biomass index were efficient to differentiate fringe and basin type mangroves, while HV polarization data was capable to discriminate mangroves at different growth stages.

Table 5.3 Studies done on the mangroves forest using SAR data.

No.	Authors	Study area	Study related to	Data used
1	Wang and Imhoff 1993	Ganges delta, Bangladesh	Monitoring the mangrove aerial extent	SIR-B
2	Pasqualini et al. 1999	Northwestern Madagascar	Monitoring the mangrove aerial extent	SPOT-XS, SIR-C
3	Mougin et al. 1999	French Guiana	Retrieval of structural parameters and total biomass	AIRSAR (P-, L-, C-band)
4	Mitchell et al. 2005	Kakadu National Park, Australia	Retrieval of forest structure and biomass	AIRSAR (P-, L-, C-band)
5	Lucas et al. 2007	Malaysia, Australia, and French Guiana	Mapping, change detection, retrieval of structural attributes, and biomass	ALOS-PALSAR, AIRSAR, JERS-1 SAR
6	Thomas et al. 2015	Global mangrove watch program	Monitoring the mangrove aerial extent	ALOS-PALSAR, JERS-1, Landsat
7	Cougo et al. 2015	Northern coast of Brazil	Modeling of biophysical parameter	Radarsar-2
8	de Souza Pereira et al. 2016	Southeast Brazil	Physical characteristics of mangroves	ALOS-PALSAR
9	Argamosa et al. 2018	Philippines; San Juan, Masinloc	Modeling above ground biomass	Sentinel-1
10	Lucas et al. 2020	Matang Mangrove Forest Reserve, Malaysia	Retrieval of biophysical properties with optical and SAR data	Landsat, ALOS-PALSAR-1 and -2, SRTM, TanDEM-X, WorldView-2

Land cover classification may be done with radar data also, both with supervised or unsupervised methods. To apply supervised classification, the classifier should be trained and for that, the ground-based information related to the mangroves species and distribution is essential. Held et al. (2003) tried to explore the best possible result using a combination of Compact Airborne Spectrographic Imager (CASI) and Airborne SAR by NASA/Jet Propulsion Laboratory (AIRSAR) data for monitoring and mapping the mangroves in the estuary. They performed maximum likelihood, normal neural network, and hierarchical neural network classification for the individual as well as integrated datasets. They found the highest accuracy (79.8%) using the hierarchical neural network classification method with integrated datasets (CASI + AIRSAR). Turkar et al. (2012) performed artificial neural network (ANN) and Wishart supervised classified on the RADARSAR-2, ALOS-PALSAR, ENVISAT-ASAR, and TerraSAR-X data to evaluate the classification ability of fully versus partially polarimetric SAR data. They found the highest classification accuracy using full polarimetric data. Their results show that when the phase info was involved, the classification accuracy was more than 80%, and with intensity HH, VV data classification accuracy

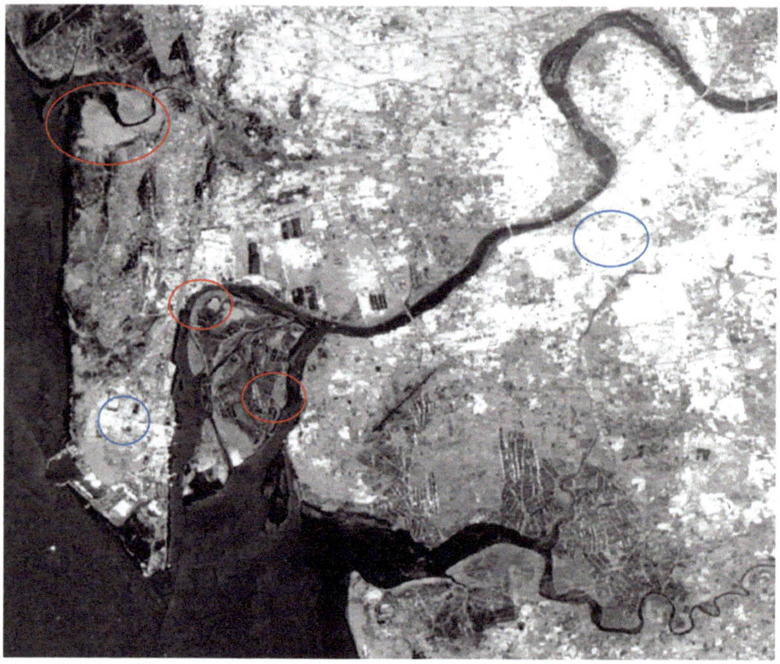

Figure 5.3 Global PALSAR-2 fnf (forest/non-forest) image of L-HH band of the Tapi estuary.

was around 80%. Similar results were also observed by Lee et al. (2001). Turkar et al. (2012) found better classification of the forest with complex HH and HV than complex HH and VV, in the case of single pol data they found good classification accuracy with HV, compare to HH and VV.

5.2.6 Identification of Mangrove Degraded Area Using SAR

Deforestation mapping is important to manage and monitor forest resources. Optical sensors are able to detect the deforested area in mangroves but the combination of optical and SAR gives more accurate results, as demonstrated by several earlier indicated studies. One of the most effective ways to represent SAR dual-polarization data through an RGB image is to put SAR co-polarization (eg. HH or VV) intensity, cross-polarization (e.g. HV or VH) intensity, and the ratio of SAR co- and cross-polarization intensities (HH/HV or VV/VH ratio) in the red, green, and blue layers, respectively. RGB images were constructed from ALOS-2 PALSAR-2 L-band and Sentinel-1 C-band dual-pol data over the region covering Tapi estuary, Gujarat, India (Figure 5.4). In the RGB images, the mangroves vegetation is seen as green color, and the brightness of the green color is dependent on the canopy density and biomass of the vegetation.

In this first case study, the mangrove area of Tapi estuary, Gujarat, India was chosen to map the deforested area using ALOS FnF data, L-band data, and Sentinel-1 C-band data (see Figure 5.5). The ALOS FnF and Sentinel-1 C-band data of 2017 and 2020 were

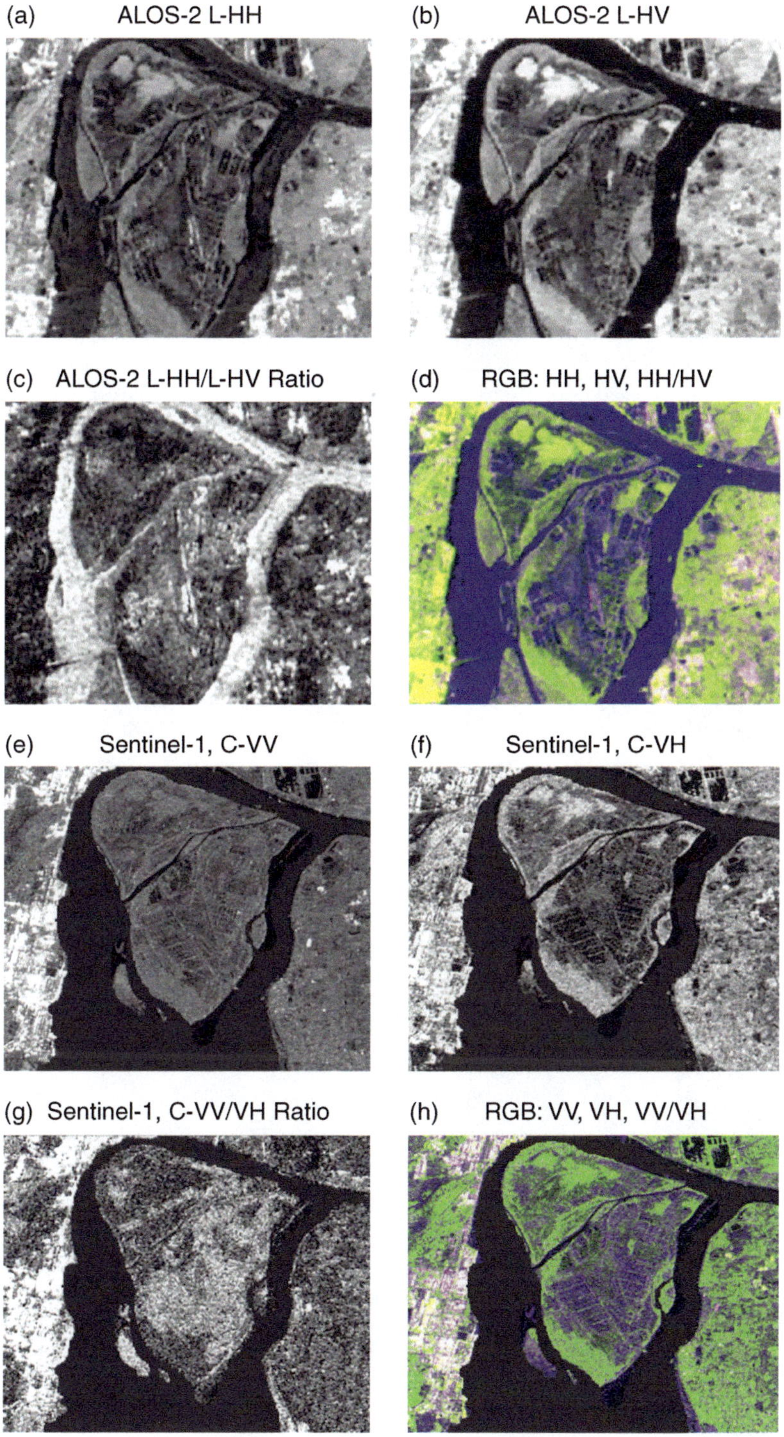

Figure 5.4 Images of the Tapi estuary, Gujarat, India: (a–d) represent ALOS-PALSAR-2 HH, HV, HH/HV ratio, and RGB images, respectively; and (e–h) represent Sentinel-1 VV, VH, VV/VH ratio, and RGB images, respectively.

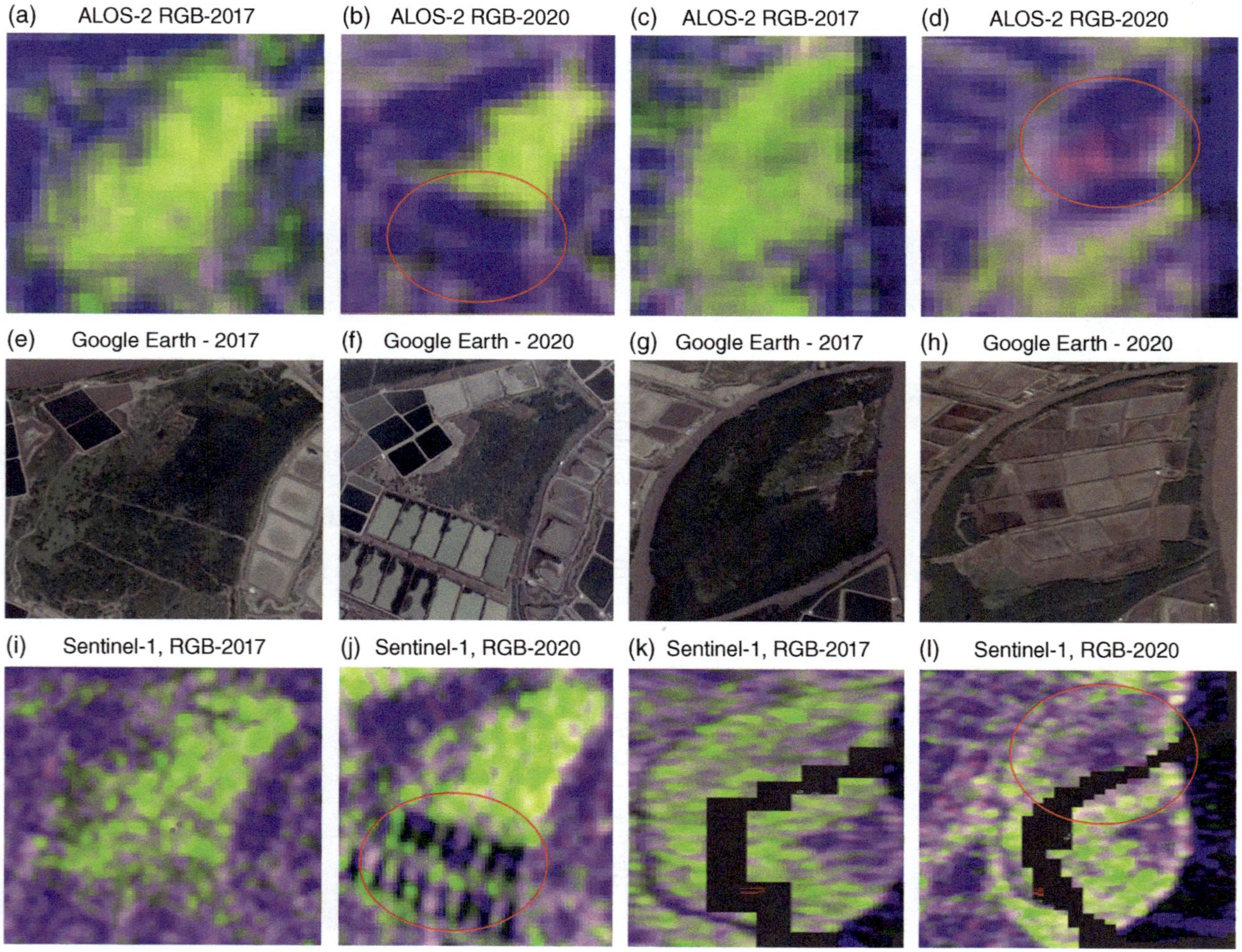

Figure 5.5 (a, c, i, k) shows the mangrove area; (b, d, j, l) shows the mangrove area after degradation with red circle; and (e, f, g, h) is the google earth imagery correspond with that particular SAR data.

downloaded from the JAXA website and Copernicus server, respectively. The processing (backscatter calculation, and application of Lee 3 × 3 filter) of ALOS image were done in SNAP. The preprocessing (apply orbit file, radiometric calibration, geocoding, Lee 3 × 3 filter, linear to dB scale) of Sentinel-1 images were also done in SNAP. The ALOS image was converted into a backscatter image using formula 1. In Figure 5.5, it is clearly visible that the mangrove area was converted into aquaculture/saltpan and it is marked with red circle in 2020 images.

$$\sigma^{\circ} = 10 \times \log 10 \left(DN \right)^{2} - 83.0$$

where, σ° is the backscattering coefficient and −83.0 is the calibration factor.

Figure 5.6 shows the mangrove area near Khajod Village, Gujarat, India, and was taken as a second case study to map the deforested area using Sentinel-1 HV pol data. C-band Sentinel-1, GRDH images of 2019 and 2021 were downloaded from Copernicus server. The dB scale of the Sentinel-1 images were done using SNAP tools. It was observed that after degradation of mangrove the HV-backscatter become lower (Table 5.4). The threshold technique was applied on σ° HV value to delineate mangrove-degraded area; it was observed that 1.5 ha mangrove were degraded and converted into the aquaculture.

5.2.7 Mangrove Forest Structure Parameters and SAR

Forest structure parameters include the areal extent of forest, areal heterogeneity, tree cover, canopy height, and biomass (Simard 2019). Mougin et al. (1999) used multi-frequency and full polarimetric AIRSAR data over French Guiana for estimation of tree density ($N\,ha^{-1}$), basal area ($m^2\ ha^{-1}$), tree height (m), tree DBH (cm), and its correlation with

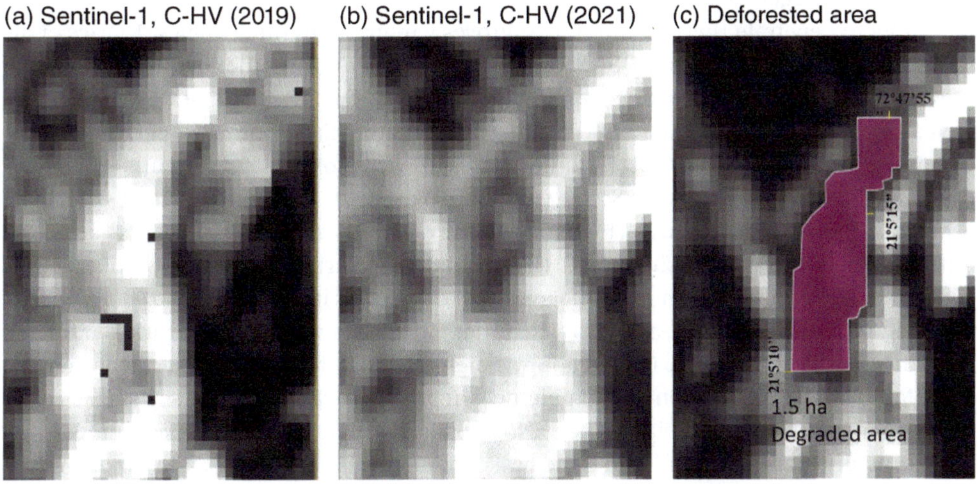

Figure 5.6 Mangrove forest, Khajod village, Gujarat, India; (a) before degradation (b) after degradation, and (c) mapped mangrove degraded area with the location.

Table 5.4 SAR backscatter values before and after degradation.

Radar band	Before degradation	After degradation
L-HH	−9 dB	−21 dB
L-HV	−17 dB	−29 dB
L-HH/HV	9 dB	10 dB
C-VV	−9 dB	−12 dB
C-VH	−14 dB	−19 dB
C-VV/VH	7 dB	8 dB

backscatter. For all the parameters, P-HV and L-HV gave best result. Kovacs et al. (2006) used the fine beam RADARSAT-1 (C-band) data, to examine the relationship between the structural parameter of white mangroves of the Mexican Pacific and the backscatter coefficient. They considered five biophysical parameters such as stem density, basal area, mean DBH, mean stem height, and LAI in the dry and wet seasons. They chose a two-band position, with an incident angle of 40° and 47°. They found a significant correlation between backscatter and the logarithms of LAI (dry and wet) and mean stem height at both incident angles. The result of this study reveals that it is possible to differentiate dead white mangroves from the healthy mangroves. They suggested utilization of SAR images of the dry season, where steeper incident angle may improve the relationship. Cougo et al. (2015) studied the correlation between Radarsat-2 C-band SAR backscatter ($\sigma°$, $\beta°$, γ) and the biophysical attributes of mangroves of the Amazon River. They considered plant height, DBH, basal area, and density for their study. A significant correlation was found between $\sigma°$ VH, with r^{-} value of 0.53 for the DBH, 0.46 for BA, and 0.63 for average height. Imhoff et al. (1986) were used Shuttle Imaging Radar-B (SIR-B) data over southern Bangladesh to determine the effect of incident angle (26°, 46°, and 58°) on the penetration capability of SAR in the mangrove vegetation and information content. They concluded that L-band has the capability to penetrate the 12.5-m-tall mangroves canopy at all three angles. Kovacs et al. (2008) used the multi-polarized SAR backscatter for monitoring the health of the mangrove's ecosystem with other biophysical parameters. They observed that polarization, as well as incident angle, play an important role in estimating the biophysical parameter, and concluded that co-polarized (i.e. HH and VV) images were not able to differentiate between dead and healthy mangroves. de Souza et al. (2012) used full polarimetric ALOS-PALSAR, L-band to study the mangroves forest structure of Southeast Brazil. They extracted 10 attributes from SAR data, such as $\sigma°$ of HH, HV, and VV polarization, the cross-polarization ratio, the parallel polarization ratio, the total power, biomass index, canopy structure index, and volume scattering index, and correlated those attributes with the structural parameter of mangroves such as mean DBH, canopy height, density, basal area, and mean height. They found a high correlation of HH channel with two structural parameters such as canopy height and mean tree height. They also applied multiple linear regression models for each structural parameter and found the best result using biomass index and canopy index for mean tree height, canopy height, and mean DBH.

5.2.8 Mangrove Biomass and SAR

Biomass is organic material present both above and below ground. Forest biomass estimation is important to understand the productivity of the forest ecosystem and carbon budget (Devagiri et al. 2013). The most common and appropriate method for biomass estimation is correlating the SAR backscatter with field measured biomass using a regression technique. Mangrove biomass estimation using SAR data was first published by Mougin et al. (1999), who used C-, L-, and P-band of AIRSAR data over French Guiana for estimation of total above ground dry biomass and its correlation with backscatter. The total biomass of that mangrove's region ranged from approximately 5–437 tons of dry matter per hector. They observed both P-HV and L-HV had a good correlation with the biomass, but couldn't find a significant correlation between P-VV and biomass. The backscatter value of the C-band increases with biomass up to saturation value which was 160 t DM ha^{-1} for the HH and 70 t DM ha^{-1} for HV, the saturation value for the L-band was 30 and 140 t DM ha^{-1}, respectively. The saturation value for the P-band was approximately 70 t DM ha^{-1} for VV, 160 t DM ha^{-1} for HH, and between 160 and 240 t DM ha^{-1} for HV (Mougin et al. 1999). Thomas et al. (2015) also observed a similar result, and conclude that L-band and P-band SAR are suitable for calculating the biomass and three-dimensional woody structures due to its deeper penetration in the canopy. Whereas Das and Patnaik (2017) suggest that the SAR data obtained during the leaf-off period has good sensitivity and it is more suitable for biomass estimation. Proisy et al. (2000) observed that for L-HH and VV, the double bounce scattering was dominant below 100–150 t DM ha^{-1}, whereas mangroves with more than 150 t DM ha^{-1} gives volume scattering. Hamdan et al. (2014) used ALOS L-band FBD with 25 m pixel spacing SAR data, to find out the correlation between the backscattering coefficient and the above ground biomass of the Matang Mangrove Forest Reserve in Malaysia. They concluded that HV backscatter was the best fit to determine the AGB. They also observed that HV backscatter saturated around 100 MG ha^{-1}, and error occurred when more than 150 MG ha^{-1} AGB was observed. The saturation of SAR backscatter at a given polarization is governed by the forest structure, such as plants per unit area, size, the distribution pattern of the leaves and branches (Imhoff 1995; Wolter and Townsend 2011), background surface roughness, and soil moisture (Quinones and Hoekman 2004), in other words the longer the wavelength, the higher the saturation point (Simard 2019). Castillo et al. (2017) used Sentinel-1 C-band data for retrieval of above ground biomass and its prediction of Honda Bay, Philippines' mangrove forest. They used two prediction models, i.e. conventional linear regression and machine learning algorithms, to predict the biomass of mangroves. They conclude that in the case of single date SAR image, significant correlations were observed with VH polarization, while the fusion of multi-temporal data VH and VV channels correlated greater with biomass than single date SAR data. When they included the elevation data with multi-date VV-VH polarization, the correlation was improved. When they compared the result of biomass prediction, machine-learning algorithms had a better result than conventional regression. Proisy et al. (2003) also found a similar result for C-band AIRSAR data where cross-polarization (HV) had significant relation with biomass than co-polarized (VV) channel for the mangroves of French Guiana and northern Australia. Kumar et al. (2012) used Envisat ASAR data to retrieve tropical forest biomass in India and observed that the multi-date SAR images performed better than single-date SAR images. Vaghela et al. (2021) used time-series data of Sentinel-1A C-band for modeling the above ground biomass of mangroves, and its changes of Mundra Taluka in the Kachchh district of western India. To estimate the biomass, tree height, DBH and canopy cover, and wood density were measured and incorporated in an allometric model. A Polarimetric Radar Vegetation Index (PRVI) images were generated and utilized for modeling AGB. They concluded that the PRVI based method was efficient to estimate the AGB through dual polarimetric data.

5.3 Conclusions

The mangrove ecosystem is essential and yet it continues to be threatened. Mangrove forests are problematic for admittance for field research, so remote sensing has proven to be pivotal tools to describe mangroves and their structure and evaluate the services that they deliver. This chapter provides information related to the mangrove ecosystem and the role of SAR to retrieve the mangrove forest properties. SAR data has proved to be very useful to study the mangrove ecosystem due to its attributes such as sensitivity with moisture content, polarization, surface roughness, various incident angles, and penetration potential. SAR is capable of imaging land surfaces irrespective of weather conditions and allows day and night imaging. The combination of SAR data with optical data, improves the discrimination of mangroves from other vegetation, as well as providing good results in the assessment of biophysical parameters such as canopy height, LAI, leaf pigment, and biomass. Before assessment of biophysical parameter with SAR, the researcher must be aware of the extent of mangroves and in the selection of polarization mode. The open-source and free sentinel data of Sentinel-1 and -2, as well as SNAP software, would be very useful for biomass mapping and monitoring in coastal regions. In this chapter, L-band ALOS-2 and C-band Sentinel-1 GRDH images were processed in the SNAP software, and used to identify the mangrove-deforested areas.

Biomass estimation of mangroves is playing an essential role to understand the productivity of the ecosystem, carbon budget, and planning for sustainable use of mangroves services. SAR sensors are sensitive to the volumetric properties of the canopy (Le Toan et al. 2011; Neumann et al. 2012). In various studies, C-, L-, and P-bands were commonly used to estimate the biomass. The estimation of biomass of mangrove vegetation with only radar backscatter has certain limitations due to saturation of SAR backscatter to higher biomass. Instead, some authors have recommended the use of SAR coherence parameters and radar interferometry to study the mangroves' structural parameters. The frequently used method is to correlate the field-based result with backscatter of single frequency and polarization, or with multiple frequencies and polarization using multivariate regression-based modeling (Mougin et al. 1999; Kovacs et al. 2006; Cougo et al. 2015). The use of polarimetric and polarimetric–interferometric SAR based approaches have been reported to have good potential for addressing mangroves biomass (Solberg et al. 2010; Neumann et al. 2012).

References

Allen, W.A. and Richardson, A.J. (1968). Interaction of light with a plant canopy. *Journal of the Optical Society of America* 58 (8): 1023. https://doi.org/10.1364/JOSA.58.001023.

Argamosa, R., Blanco, A., Baloloy, A. et al. (2018). Modelling above ground biomass of mangrove forest using sentinel-1 imagery. In: *ISPRS Annals of the Photogrammetry, Remote Sensing and Spatial Information Sciences*, IV-3, 13–20. https://doi.org/10.5194/isprs-annals-IV-3-13-2018.

Avtar, R., Sawada, H., Takeuchi, W., and Singh, G. (2012). Characterization of forests and deforestation in Cambodia using ALOS/PALSAR observation. *Geocarto International* 27 (2): 119–137. https://doi.org/10.1080/10106049.2011.626081.

Castillo, J., Apan, A., Maraseni, T., and Salmo, S. (2017). Estimation and mapping of above-ground biomass of mangrove forests and their replacement land uses in the Philippines using sentinel imagery. *ISPRS Journal of Photogrammetry and Remote Sensing* 134: 70–85. https://doi.org/10.1016/j.isprsjprs.2017.10.016.

Chellamani, P., Prakash Singh, C., and Panigrahy, S. (2014). Assessment of the health status of Indian mangrove ecosystems using multi temporal remote sensing data. *Tropical Ecology* 55 (2): 245–253.

Cougo, M., Souza-Filho, P., Silva, A. et al. (2015). Radarsat-2 backscattering for the modeling of biophysical parameters of regenerating mangrove forests. *Remote Sensing* 7 (12): 17097–17112. https://doi.org/10.3390/rs71215873.

Dahdouh-Guebas, F. (2011). World atlas of mangroves. Spalding, M., Kainuma, M., and Collins, L. (eds). *Human Ecology* 39 (1): 107–109. https://doi.org/10.1007/s10745-010-9366-7.

Das, A. and Patnaik, C. (2017). Monitoring Forest Above-ground Biomass of Gujarat State Using Multi-temporal Synthetic Aperture RADAR Data. *The 38th Asian Conference on Remote Sensing Theme - Space Applications: Touching Human Lives.* Asian Association on Remote Sensing (AARS). Retrieved from: https://www.researchgate.net/publication/321361018_MONITORING_FOREST_ABOVEGROUND_BIOMASS_OF_GUJARAT_STATE_USING_MULTI-TEMPORAL_SYNTHETIC_APERTURE_RADAR_DATA

Devagiri, G.M., Money, S., Singh, S. et al. (2013). Assessment of above ground biomass and carbon pool in different vegetation types of southwestern part of Karnataka, India using spectral modeling. *Tropical Ecology* 54 (2): 149–165.

Duke, N.C. (2006). Indo-West Pacific stilt mangroves: *Rhizophora apiculata, R. mucronata, R. stylosa, R. X annamalai, R. X lamarckii.* In: *Traditional Trees of Pacific Islands: Their Culture, Environment, and Use* (ed. C. Elevitch and C. Elevitch), 641–660. Holualoa, Hawaii: Permanent Agriculture Resources (PAR).

Duke, N., Meynecke, J.-O., Dittmann, S. et al. (2007). A world without mangroves? *Science* 317 (5834): https://doi.org/10.1126/science.317.5834.41b.

Food and Agriculture Organization (FAO) (2007). *The World's Mangroves 1980–2005.* Rome: FAO.

Foody, G.M., Boyd, D.S., and Cutler, M. (2003). Predictive relations of tropical forest biomass from Landsat TM data and their transferability between regions. *Remote Sensing of Environment* 85 (4): 463–474. https://doi.org/10.1016/S0034-4257(03)00039-7.

Forest Survey of India (FSI). (2019). *India State of Forest Report* 2019.

Gibbs, H., Brown, S., Niles, J., and Foley, J. (2007). Monitoring and estimating tropical forest carbon stocks: making REDD a reality. *Environmental Research Letters* 2 (4): https://doi.org/10.1088/1748-9326/2/4/045023.

Giri, C., Ochieng, E., Tieszen, L. et al. (2011). Status and distribution of mangrove forests of the world using earth observation satellite data. *Global Ecology and Biogeography* 20 (1): 154–159. https://doi.org/10.1111/j.1466-8238.2010.00584.x.

Giri, C., Long, J., Abbas, S. et al. (2015). Distribution and dynamics of mangrove forests of South Asia. *Journal of Environmental Management* 148: https://doi.org/10.1016/j.jenvman.2014.01.020.

Green, E.P., Clark, C.D., Mumby, P.J. et al. (1998). Remote sensing techniques for mangrove mapping. *International Journal of Remote Sensing* 19 (5): 935–956. https://doi.org/10.1080/014311698215801.

Hamdan, O., Khali Aziz, H., and Mohd Hasmadi, I. (2014). L-band ALOS PALSAR for biomass estimation of Matang mangroves, Malaysia. *Remote Sensing of Environment* 155: 69–78. https://doi.org/10.1016/j.rse.2014.04.029.

Held, A., Ticehurst, C., Lymburner, L., and Williams, N. (2003). High resolution mapping of tropical mangrove ecosystems using hyperspectral and radar remote sensing. *International Journal of Remote Sensing* 24 (13): https://doi.org/10.1080/0143116031000066323.

Imhoff, M. (1995). A theoretical analysis of the effect of forest structure on synthetic aperture radar backscatter and the remote sensing of biomass. *IEEE Transactions on Geoscience and Remote Sensing* 33 (2): 341–351. https://doi.org/10.1109/TGRS.1995.8746015.

Imhoff, M., Story, M., Vermillion, C. et al. (1986). Forest canopy characterization and vegetation penetration assessment with space-borne radar. *IEEE Transactions on Geoscience and Remote Sensing* GE-24 (4): 535–542. https://doi.org/10.1109/TGRS.1986.289668.

Jia, M., Wang, Z., Zhang, Y. et al. (2015). Landsat-based estimation of mangrove forest loss and restoration in Guangxi Province, China, influenced by human and natural factors. *IEEE Journal of Selected Topics in Applied Earth Observations and Remote Sensing* 8 (1): https://doi.org/10.1109/JSTARS.2014.2333527.

Kathiresan, K. (2012). Importance of mangrove ecosystem. *International Journal of Marine Science.* https://doi.org/10.5376/ijms.2012.02.0010.

Kovacs, J., Vandenberg, C., and Flores-Verdugo, F. (2006). Assessing fine beam RADARSAT-1 backscatter from a white mangrove (*Laguncularia racemosa* [*Gaertner*]) canopy. *Wetlands Ecology and Management* 14 (5): 401–408. https://doi.org/10.1007/s11273-005-6237-x.

Kovacs, J.M., Vandenberg, C.V., Wang, J., and Flores-Verdugo, F. (2008). The use of multipolarized Spaceborne SAR backscatter for monitoring the health of a degraded mangrove Forest. *Journal of Coastal Research* 24 (1): 248–254. https://doi.org/10.2112/06-0660.1.

Kripa, M., Nivas, A.H., Lele, N. et al. (2020). Seasonal dynamics and light use efficiency of major mangrove species over Indian region. *Proceedings of the National Academy of Sciences, India Section B: Biological Sciences* 90 (2): 267–275. https://doi.org/10.1007/s40011-019-01077-x.

Kuenzer, C., Bluemel, A., Gebhardt, S. et al. (2011). Remote sensing of mangrove ecosystems: a review. *Remote Sensing* 3 (5): 878–928. https://doi.org/10.3390/rs3050878.

Kumar, S., Pandey, U., Kushwaha, S. et al. (2012). Aboveground biomass estimation of tropical forest from Envisat advanced synthetic aperture radar data using modeling approach. *Journal of Applied Remote Sensing* 6 (1): 063588. https://doi.org/10.1117/1.JRS.6.063588.

Le Toan, T., Quegan, S., Davidson, M.W.J. et al. (2011). The BIOMASS mission: mapping global forest biomass to better understand the terrestrial carbon cycle. In: *Remote Sensing of Environment*, vol. 115(11), 2850–2860. https://doi.org/10.1016/j.rse.2011.03.020.

Lee, J.-S. and Pottier, E. (2009). *Polarimetric Radar Imaging from Basics to Application* (ed. B.J. Thompson). New York: CRC Press.

Lee, J.-S., Grunes, M., and Pottier, E. (2001). Quantitative comparison of classification capability: fully polarimetric versus dual and single-polarization SAR. *IEEE Transactions on Geoscience and Remote Sensing* 39 (11): 2343–2351. https://doi.org/10.1109/36.964970.

Lucas, R.M., Mitchell, A.L., Rosenqvist, A. et al. (2007). The potential of L-band SAR for quantifying mangrove characteristics and change: case studies from the tropics. *Aquatic Conservation: Marine and Freshwater Ecosystems* 17 (3): 245–264. https://doi.org/10.1002/aqc.833.

Lucas, R., Van De Kerchove, R., Otero, V. et al. (2020). Structural characterisation of mangrove forests achieved through combining multiple sources of remote sensing data. *Remote Sensing of Environment* 237: 111543. https://doi.org/10.1016/j.rse.2019.111543.

Meyer, F. (2019). Spaceborne synthetic aperture radar – principles, data access, and basic processing techniques. In: *SAR Handbook: Comprehensive Methodologies for Forest Monitoring and Biomass Estimation* (ed. A. Flores, K. Herndon, R. Thapa and E. Cherrington). NASA: https://doi.org/10.25966/nr2c-s697.

Mitchell, A., Lucas, R., Proisy, C., and Melius, A. (2005). *Sensitivity of radar backscatter to mangrove forest structure and AIRSAR imaging parameters*. Proceedings of the 2005 IEEE International Geoscience and Remote Sensing Symposium, 2005 (IGARSS '05), (pp. 2090–2093). *IEEE*. https://doi.org/10.1109/IGARSS.2005.1526428

Mougin, E., Proisy, C., Marty, G. et al. (1999). Multifrequency and multipolarization radar backscattering from mangrove forests. *IEEE Transactions on Geoscience and Remote Sensing* 37 (1): 94–102.

Neumann, M., Saatchi, S., Ulander, L., and Fransson, J. (2012). Assessing performance of L- and P-band Polarimetric interferometric SAR data in estimating boreal Forest above-ground biomass. *IEEE Transactions on Geoscience and Remote Sensing* 50 (3): 714–726. https://doi.org/10.1109/TGRS.2011.2176133.

Nyamari, N. and Cabral, P. (2021). Impact of land cover changes on carbon stock trends in Kenya for spatial implementation of REDD+ policy. *Applied Geography* 133: 102479. https://doi.org/10.1016/j.apgeog.2021.102479.

Pasqualini, V., Iltis, J., Dessay, N. et al. (1999). Mangrove mapping in North-Western Madagascar using SPOT-XS and SIR-C radar data. *Hydrobiologia* 413: 127–133. https://doi.org/10.1023/A:1003807330375.

Proisy, C., Mougin, E., Fromard, F., and Karam, M. (2000). Interpretation of polarimetric radar signatures of mangrove forests. *Remote Sensing of Environment* 71 (1): 56–66.

Proisy, C., Mitchell, A., Lucas, R., Fromard, F. et al. (2003). Estimation of Mangrove Biomass using Multifrequency Radar Data. Application to Mangroves of French Guiana and Northern Australia. *Proceeding of the Mangrove 2003 Conference*. Salvador, Bahia, Brazil.

Quinones, M. and Hoekman, D. (2004). Exploration of factors limiting biomass estimation by Polarimetric radar in tropical forests. *IEEE Transactions on Geoscience and Remote Sensing* 42 (1): 86–104. https://doi.org/10.1109/TGRS.2003.815402.

Ramakutty, N., Gibbs, H., Achard, F. et al. (2007). Challenges to estimating carbon emissions from tropical deforestation. *Global Change Biology* 13 (1): 51–66. https://doi.org/10.1111/j.1365-2486.2006.01272.x.

Ramsey, E. and Jensen, J. (1996). Remote sensing of mangrove wetlands: relating canopy spectra to site-specific data. *Photogrammetric Engineering and Remote Sensing*, 62(8). https://www.asprs.org/wp-content/uploads/pers/1996journal/aug/1996_aug_939-948.pdf

Richards, D.R. and Friess, D.A. (2016). Rates and drivers of mangrove deforestation in Southeast Asia, 2000–2012. *Proceedings of the National Academy of Sciences of the United States of America* 113 (2): 344–349. https://doi.org/10.1073/pnas.1510272113.

Seto, K. and Fragkias, M. (2007). Mangrove conversion and aquaculture development in Vietnam: a remote sensing-based approach for evaluating the Ramsar convention on wetlands. *Global Environmental Change* 17 (3–4): 486–500. https://doi.org/10.1016/j.gloenvcha.2007.03.001.

Siikamaki, J., Sanchirico, J., and Jardine, S. (2012). Global economic potential for reducing carbon dioxide emissions from mangrove loss. *Proceedings of the National Academy of Sciences of the United States of America* 109 (36): 14369–14374. https://doi.org/10.1073/pnas.1200519109.

Simard, M. (2019). Radar remote sensing of mangrove forests. In: *SAR Handbook: Comprehensive Methodologies for Forest Monitoring and Biomass Estimation* (ed. A. Flores, K. Herndon, R. Thapa and E. Cherrington). E.NASA: https://doi.org/10.25966/nr2c-s697.

Solberg, S., Astrup, R., Gobakken, T. et al. (2010). Estimating spruce and pine biomass with interferometric X-band SAR. *Remote Sensing of Environment* 114 (10): 2353–2360. https://doi.org/10.1016/j.rse.2010.05.011.

de Souza Pereira, F., Kampel, M., and Cunha-Lignon, M. (2016). Mangrove vegetation structure in Southeast Brazil from phased array L-band synthetic aperture radar data. *Journal of Applied Remote Sensing* 10 (3): https://doi.org/10.1117/1.JRS.10.036021.

de Souza, R., Pereira, F., Kampel, M., and Cunha-Lignon, M. (2012). Mapping of mangrove forests on the southern coast of São Paulo, Brazil, using synthetic aperture radar data from ALOS/PALSAR. *Remote Sensing Letters* 3 (7): 567–576. https://doi.org/10.1080/01431161.2011.641511.

Thomas, N., Lucas, R., Itoh, T. et al. (2015). An approach to monitoring mangrove extents through time-series comparison of JERS-1 SAR and ALOS PALSAR data. *Wetlands Ecology and Management* 23 (1): 3–17. https://doi.org/10.1007/s11273-014-9370-6.

Trexler, M. and Kosloff, L. (1998). The 1997 Kyoto protocol: what does it mean for project-based climate change mitigation? *Mitigation and Adaptation Strategies for Global Change* 3 (1): 1–58.

Turkar, V., Deo, R., Rao, Y. et al. (2012). Classification accuracy of multi-frequency and multi-polarization SAR images for various land covers. *IEEE Journal of Selected Topics in Applied Earth Observations and Remote Sensing* 5 (3): 936–941. https://doi.org/10.1109/JSTARS.2012.2192915.

Vaghela, B., Chirakkal, S., Putrevu, D., and Solanki, H. (2021). Modelling above ground biomass of Indian mangrove forest using dual-pol SAR data. *Remote Sensing Applications: Society and Environment* 21: 100457. https://doi.org/10.1016/j.rsase.2020.100457.

Wang, Y. and Imhoff, M. (1993). Simulated and observed L-HH radar backscatter from tropical mangrove forests. *International Journal of Remote Sensing* 2819–2828: https://doi.org/10.1080/01431169308904311.

Wolter, P. and Townsend, P. (2011). Multi-sensor data fusion for estimating forest species composition and abundance in northern Minnesota. *Remote Sensing of Environment* 115 (2): 671–691. https://doi.org/10.1016/j.rse.2010.10.010.

6

Photosynthetic Variables Estimation in a Mangrove Forest

Somnath Paramanik[1], *Roma Varghese*[1], *Mukunda D. Behera*[*,1], *Surbhi Barnwal*[1], *Soumit K. Behera*[2], *and Bimal K. Bhattyacharya*[3]

[1] *Centre for Ocean, River, Atmosphere and Land Sciences (CORAL), Indian Institute of Technology Kharagpur, Kharagpur, 721302, India*
[2] *Plant Ecology and Climate Change Science Division, CSIR–National Botanical Research Institute, Rana Pratap Marg, Lucknow, 226001, India*
[3] *Space Applications Centre, ISRO, Ahmedabad, 380015, India*

6.1 Introduction

6.1.1 Mangroves

Mangroves are shrubs and trees of medium height which can sustain in brackish water, seawater, and salty water. They have exceptional physiological, morphological, and biological characteristics that help to adapt to extreme events, e.g. high temperature, salinity, cyclones, high tides, and sedimentation load. They are distributed in the inter-tidal regions between the sea and the land in the tropical and sub-tropical regions between approximately 30°N and 30°S latitude, which covers up to 75% of the tropical and sub-tropical shorelines (Spalding 1997). The total area of mangroves covers between 167 000 and 181 000 km^2 (Spalding 1997; Valiela et al. 2001). Mangrove habitat has been declining considerably in recent decades due to agriculture, aquaculture farming, ecotourism, and urbanization (Alongi 2002). The loss of mangroves can progressively affect sea-level rise that might endanger marine life and terrestrial biological diversity, which in turn affects natural barriers and carbon sequestration. This alarming proportionate rate (0.39% per year globally) can abolish the mangroves in the next 100 years if they are not protected (Duke et al. 2007; Hamilton and Casey 2016).

The largest amount of mangroves can be found in Southeast Asia, with maximum species diversity, including the Indian mangroves (Lovelock et al. 2015b). The Bhitarkanika mangrove wetlands (Figure 6.1), located on Odisha's east coast in the Kendrapara district, is the second-largest mangrove forest in the Indian mainland (Paramanik et al. 2019) and

* Corresponding author's email: mdbehera@coral.iitkgp.ac.in

0000-0002-4509-8801 (Somnath Paramanik), 0000-0002-5347-6040 (Roma Varghese), 0000-0002-9976-6270 (Mukunda D. Behera), 0000-0002-6276-5477 (Surbhi Barnwal), 0000-0002-9181-3263 (Soumit K. Behera), and 0000-0002-8430-4313 (Bimal K. Bhattyacharya)

Advances in Remote Sensing for Forest Monitoring, First Edition. Edited by Prem C. Pandey and Paul Arellano.

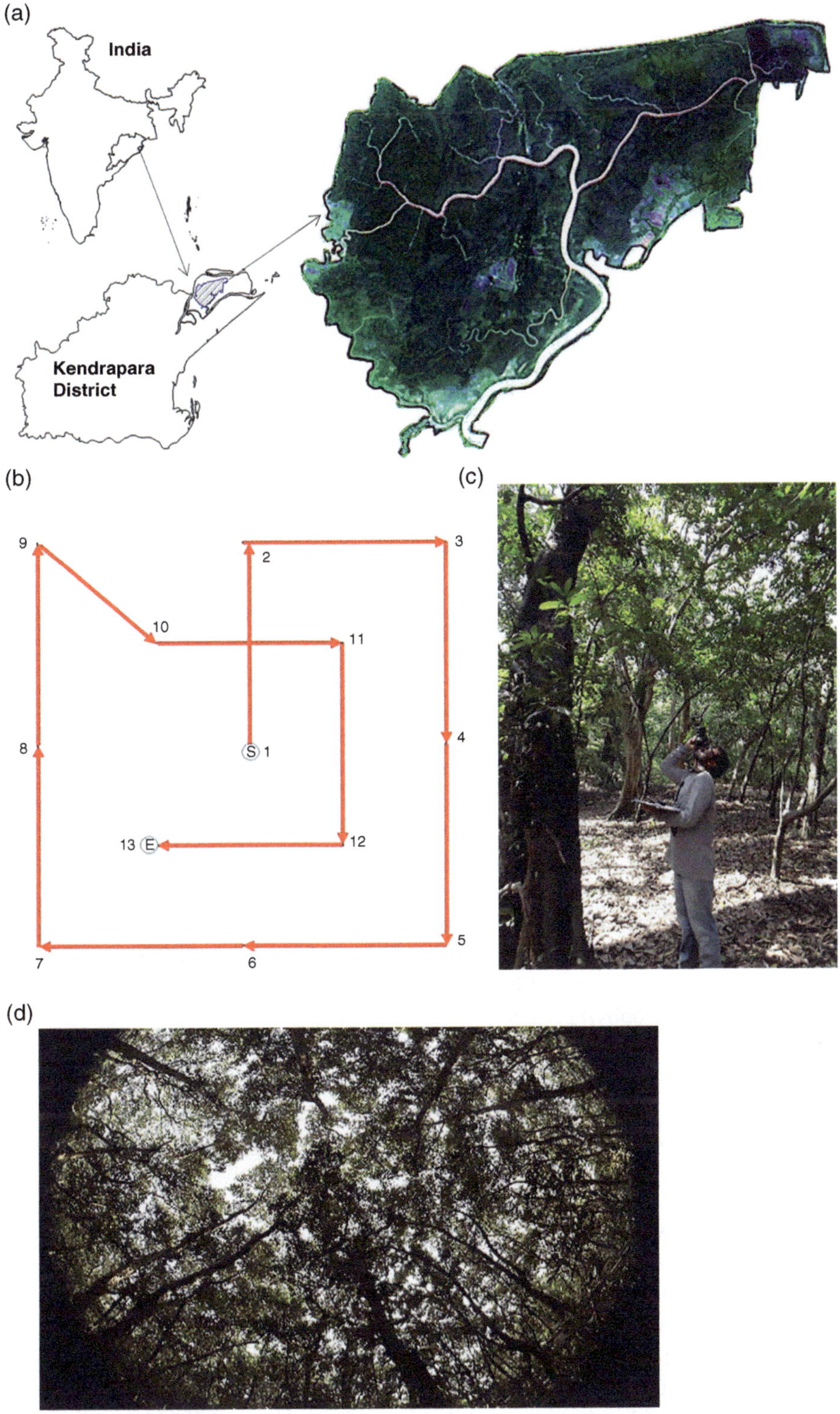

Figure 6.1 (a) A true color composite (TCC, Sentinel-2, March 2019) image of Bhitarkanika Wildlife Sanctuary, the study site. (b) Design of 13 locations within an elementary sampling unit (ESU) of $20 \times 20\,m^2$ used to capture the hemispherical photographs, and SPAD readings, where 'S' and 'E' represent the start and end points of measurement path. (c) An upward looking hemispherical photograph captured on-field using a handheld DSLR camera with a fish eye lens. (d) A representative 2-dimensional RGB hemispherical photograph.

accommodates 62 plant species. This forest area is situated in the estuarine environment created by the Brahmani, Dhamra, and Baitarani rivers, and intersects with the Bay of Bengal by the network of creeks (Ghosh et al. 2021). The region is now protected as the Bhitarkanika Wildlife Sanctuary (BWS), having an area of 165 km^2. BWS lies in the tropical belt and experiences a warm and humid climate with an average annual rainfall of 1670 mm. The average air temperature is in the range of 10–43 °C, and the relative humidity changes between 70 and 84%. The ratio of precipitation to potential evapotranspiration varies between 0.5 and 0.75. The tide is semi-diurnal, it varies in the range between 0.5 and 2.8 m in the winter season, and sometimes it can go up to 4.5 m in the estuary mouth (Behera et al. 2021). Some of the dominant mangrove species found in BWS are *Avicennia officinalis, Ceriops decandra, Cynometra iripa, Exoceria agallocha, Heritiera fomes, Heritiera littoralis, Sonneratia apetala, Bruguiera sexangula, Xylocarpus mekongensis, Xylocarpus granatum, Aglaia cucullata, Pongamia pinnata, Aegiceras corniculatum, Cerbera odollam,* and *Intsia bijuga* (Pattanaik et al. 2008; Pandey et al. 2019). In BWS, the height of trees generally ranges between 8 and 10 m, but a few taller (14–16 m) and smaller trees (2–3 m) are also present (Ghosh et al. 2020).

6.1.2 Photosynthesis/Carbon Sequestration

In the tropics, mangrove forests are one of the most carbon-rich ecosystems, containing an average of 937 t C ha^{-1}; they can sequester almost four times more carbon per unit area than other tropical forests (Alongi 2012). Although mangroves cover only 0.7% of the world's tropical forests, their carbon sequestering capacity indicates the importance of carbon emission mitigation (Giri et al. 2011). The average net primary production for mangroves is 11.1 t dry weight ha^{-1}/$year^{-1}$ (Alongi 2012). The stem regions of mangroves in BWS store around 40.08 t ha^{-1} of CO_2 in a year (Banerjee et al. 2017). In the Bhitarkanika mangroves, the average carbon stock measured from field sampling is 131.07 t C ha^{-1} (Anand et al. 2020). A majority of the mangroves have above ground biomass up to 400 t/ha (Ghosh et al. 2021). Mangroves account for approximately 1% of carbon sequestration by the world's forests, yet they sequester 14% of carbon (Chazdon et al. 2016). Leaf area index (LAI), the amount of chlorophyll present in canopy level, and solar-induced fluorescence (SIF) serve as good indicators of vegetation carbon storage in forests, and those can be quantified as primary productivity.

6.1.3 Leaf Area Index

This biophysical variable measures the one-sided leaf area per unit ground surface area, so it is a dimensionless entity (Watson 1947). The surface of the leaves is the important site for energy and mass (gas) exchange. Therefore, LAI is proportional to canopy interception, evapotranspiration, precipitation interception, and primary productivity. Long-term analysis of LAI dynamics can be helpful to understand the vegetation stress and health condition of the forest. LAI is identified as the primary contributor of carbon uptake, as it indicates the amount of foliage that intercepts light and assimilates carbon. Two methods are used to measure the LAI in the field; a direct approach such as litter trap which is destructive, or an indirect method that uses optical instruments such as plant canopy analyzer LAI 2200, digital hemispherical photography (DHP). An allometric method to estimate LAI is linked to other plant functional parameters such as canopy cover, diameter at breast height (DBH), height, and other climate and topographic variables. These direct and indirect field measurement techniques are laborious and time-consuming, which is relieved with remote sensing data use.

Several global satellite LAI products have been generated over the last two decades with a spatial resolution of 250 m to 7 km. The National Oceanic and Atmospheric Administration (NOAA) Advanced Very High-Resolution Radiometer (AVHRR) global LAI data products are available since the 1980s. Previous research (Blinn et al. 2019) related to LAI estimations focused on developing empirical equations using field observed LAI and different vegetation indices (VIs) based on the correlation or regression analysis. This method is very common for its simplicity, and it can be easily applied to larger areas based on the satellite data coverage. The major limitation of this LAI-VIs relationship is the saturation effect due to higher biomass and LAI. Another issue is reliance on VIs as they are very site- and species-specific, even though they depend on the topography and sun-sensor geometry (Thenkabail et al. 2000). Although NDVI is widely used, it suffers from saturation at a certain level when the concentration of chlorophyll becomes higher, as it enables maximum absorption in the red wavelength. Paramanik et al. (2019) applied the modified Choudhary method and demonstrated LAI variations between 1.54 and 5.18 in BWS (Choudhury et al. 1994).

6.1.4 Chlorophyll Concentration

Chlorophyll is the crucial pigment for photosynthesis and it is directly linked to photosynthetically active radiation (PAR) and the amount of carbon storage. The leaf chlorophyll pigments efficiently indicate the plant's physiological condition and health status. The standard approach of chlorophyll extraction uses organic solvent (acetone, ethanol, etc.) and spectrophotometer analysis following Arnon's equation (Porra et al. 1989). Chlorophyll is also measured by optical instruments such as Opti-Sciences CCM-200 Chlorophyll Content Meter (CCM-200) and Minolta SPAD-502 Chlorophyll Meter (SPAD-502). Remote sensing approaches offer an alternative way to estimate canopy chlorophyll, using regression or radiative transfer models (RTMs) (Pastor-Guzman et al. 2015). The in-situ chlorophyll measurements are linked to the VIs using empirical equations in the regression approach, while the physical-based models are inverted to obtain the chlorophyll estimate by optimizing other model parameters in the RTMs (Zhu et al. 2017).

The leaf chlorophyll has a strong relationship with spectral reflectance (Ali et al. 2020; Sonobe et al. 2021). Limited studies are available on the seasonal variation of chlorophyll from mangrove ecosystems (Zhen et al. 2021). Heenkenda et al. (2015) quantified the distribution of canopy chlorophyll over Rapid Creek mangrove forest in Australia using WorldView-2 satellite data, wherein they revealed that red, near infra-red (NIR), and red-edge bands provide higher importance for chlorophyll estimates. Zhen et al. (2021) estimated the canopy level chlorophyll using the statistical relationship between VIs and soil plant analysis development (SPAD) for a mangrove forest of China using Sentinel-2 optical dataset. Parida and Kumari (2021) reported a chlorophyll concentration of 40–90 $\mu g/cm^2$ in BWS.

6.1.5 Solar Induced Fluorescence

Chlorophyll fluorescence emitted by plants is linked to light reactions and therefore, indicates the actual photosynthetic activity. This radiation flux originating from the photosystems consists of a small fraction of absorbed photons (<5% of absorbed photosynthetically active radiation [APAR]) released at a longer wavelength (650–850 nm) with peaks at 690 and 740 nm (Duveiller and Cescatti 2016). It responds well to environmental conditions, such as water and light (Kalaji et al. 2016). The solar-induced chlorophyll fluorescence (SIF) is an efficient optical signal for tracking terrestrial photosynthetic activity from space to understand its spatial and temporal dynamics (Gamon et al. 2019). Since SIF directly relates to the internal photosynthesis mechanisms, it carries information on the eco-physiological state of vegetation. Therefore, SIF acts as an efficient proxy of plant photosynthesis and can be useful for the assessment of various ecosystem services.

At the ground level, SIF measurements are made possible with hand-held and clip-on devices, and top-of-canopy sensors mounted on the flux towers. With the application of remote sensing techniques, airborne systems carrying high-performance spectrometers can efficiently retrieve SIF. Satellite sensors originally designed for atmospheric chemistry applications such as GOSAT, TANSO-FTS, GOME, OCO-2, ACGS, TROPOMI, MetOp, TanSat, etc., are widely used for measuring SIF signals from space. However, satellite-derived SIF retrievals have a coarser resolution which needs to be downscaled using approaches like Random Forest and Neural Network algorithms (Yu et al. 2019). Global Orbiting Carbon Observatory-2 (OCO-2) SIF dataset (GOSIF) is an example of a downscaled product with spatially continuous data (2000–2020), which helps in spatially explicit studies of ecosystems with heterogeneous land covers (Wei et al. 2019). The temporal variability of SIF and GPP demonstrated a significant correlation for a subtropical mangrove forest in southeastern China (Zhu et al. 2021). SIF simulated using Soil Canopy Observation, Photochemistry, and Energy fluxes (SCOPE) model and field measurements demonstrated a good correlation for the Pichavaram mangrove forest (Kripa et al. 2017).

6.1.6 Gross Primary Productivity (GPP)

The gross primary productivity (GPP) represents the total amount of carbon fixed into organic material through photosynthesis in the terrestrial biosphere for a given period. It plays a critical role in climate dynamics and indicates vegetation response (Varghese and Behera 2019). The biogeochemical module of the ecosystem process models simulates GPP and other functional characteristics of vegetation (Liu et al. 2014). The remote sensing data-driven models estimate GPP at a broader scale using less labor and economic input, and the satellite data continuity helps to understand the dynamics of GPP. The GPP models can be categorized as (i) vegetation index (VI) models, (ii) production efficiency models (PEM), (iii) process-based ecosystem models, and (iv) artificial intelligence (AI) based algorithms (Sun et al. 2019). VI models use their conceptual connection to photosynthetic activity to quantify GPP (Noumonvi and Ferlan 2020), while PEMs apply light use efficiency (LUE) theory, which provides a stronger physical foundation for GPP estimate (Wu et al. 2012). The application of AI in GPP modeling appears to be an advanced technology due to its efficiency in tackling the non-linearity involved. MODIS-based GPP products provided better resolution (500 m, 8-day) with optimum temporal coverage during 2000–2020. GPP estimated using the vegetation photosynthesis model across the BWS demonstrated strong seasonality (Lele et al. 2021).

6.1.7 Vegetation Indices (VIs)

VIs are the combination of reflectance bands that enhance the sensitivity of vegetation characteristics by minimizing the effect of atmospheric conditions, background noise, and solar geometry (Baret and Guyot 1991). The sensitivity of a VI depends on the responsive nature between the spectral bands and targeted variables. In vegetation application, the red and NIR band has maximum use. In addition, the red-edge region at around 700 nm, is crucial to identify senescence, growth stages, or the health condition of different species. VIs are widely used in estimating biophysical parameters such as chlorophyll, LAI, and biomass. The target variables are considered a function of different VIs treated as independent variables, and an equation is established using parametric or non-parametric regression. Estimation of biophysical parameters using the VI approach is a simple method but has some limitations, such as saturation and sensitivity to background variability. Most importantly, VIs are highly dependent on site, vegetation type, species, topography, and sun-sensor geometry. Many modifications were done on normal VIs by replacing the red band with other visible bands or adding other bands to reduce the atmospheric effects or soil background effect, and thus, enhanced vegetation index (EVI) or soil adjusted vegetation index (SAVI) were evolved. The red-edge and SWIR bands are also recognized as important variables for biophysical variable estimation.

6.2 Materials and Methodology

This study analyzed the temporal pattern of LAI, GPP, and SIF, and estimated LAI and chlorophyll concentration using parametric and non-parametric regression from moderate resolution Landsat-8 data.

6.2.1 Dataset

6.2.1.1 In-situ Data

The DHP and soil plant analysis development (SPAD) measurements were collected from the field at $20 \times 20\,m^2$ Elementary Sampling Unit (ESU) for LAI and chlorophyll estimation, and for further linkage to satellite pixels. Thirteen hemispherical photographs were collected using a fisheye lens mounted DSLR camera with a field of view of 178° for each ESU. Similarly, 13 trees were selected for SPAD readings, and from each tree, SPAD readings were measured from the leaf at the maximum upper level of reach. A total of six readings were taken from each leaf by the SPAD from the side of its midrib. The distance between the points in an ESU was maintained as 10 m to take 13 photographs or the select 13 trees for SPAD reading. Any error from field-measured LAI may come from clumping effect, woody materials, slope, sampling strategy, or observation conditions. As *Heritiera fomes* (HF), *Excoecaria agallocha* (EA), and *Avicennia officinalis* (AO) were the most dominant species, 15 leaf samples were collected for each plant to estimate leaf chlorophyll concentration in the laboratory.

Table 6.1 Description of satellite datasets that were used in the study.

Sl. No.	Satellite	Product	Acquisition	Spatial resolution	Temporal resolution (days)	Purpose (analysis/generation)
1	MODIS	MOD15A2H	2003–2018	500 m	8, 4	LAI
2	MODIS	MOD17A2H	2003–2018	500 m	8	GPP
3	GOSIF	OCO-2	2003–2018	0.05°	8	SIF
4	Landsat-8	OLI	10 Mar 2019	30 m	16	VIs

6.2.1.2 Satellite/airborne Data/products

This study used MODIS-based LAI and GPP products, OCO2 derived SIF products, and Landsat-8 data (Table 6.1). The V6 version of MODIS LAI product at 500 m spatial and 8- and 4-days temporal resolutions have been available since 2015. All the MODIS LAI products (e.g. MOD15A2H, MYD15A2H, MCD15A2H, and MCD15A3H) are retrieved from Terra MODIS, Aqua MODIS, and Terra MODIS + Aqua MODIS combined (Yan et al. 2016), and adopt the same LAI retrieval algorithm. The nomenclature starting with "MOD" "MYD" and "MCD" indicate capture by MODIS Terra, MODIS Aqua, and both MODIS/Terra+Aqua, while A2H and A3H represent their compositing methods such as 8-day for A2H and 4-day for A3H. MODIS LAI products provide six scientific data sets (SDS), among them, the second one is the LAI. These products contain the corresponding Quality Control (QC) data in the third and fourth layers, which helps to detect cloud cover, snow coverage, and the presence of aerosol.

The continuous estimate of GPP provided by MODIS multispectral sensor (MOD17A2H) is the first satellite-data-based global product developed to track the vegetation primary productivity in terrestrial ecosystems (Running et al. 2004). Characterized by a moderate spatial resolution (500 m) and high temporal resolution (8-day), the MOD17A2H dataset was generated based on the LUE concept using the following mathematical framework (Tian et al. 2010)

$$\mathrm{GPP} = \varepsilon_{\max} \times \mathrm{PAR} \times \mathrm{FPAR} \times f(\mathrm{VPD}) \times f(\mathrm{T}_{\min}) \tag{6.1}$$

Here PAR was calculated from incident shortwave radiation, the fraction of photosynthetically active radiation (FAPAR) absorbed by the canopy taken from MODIS MOD15 LAI/FPAR product, meteorological data sets (temperature, incoming solar radiation, and vapor pressure deficit) collected from GMAO and NCEP/NCAR Reanalysis II, and LUE (ε_{max}) obtained from biome properties lookup table (Running and Zhao 2015).

Global orbiting carbon observatory-2 SIF (GOSIF) is a discrete OCO-2 SIF data-based estimate of chlorophyll fluorescence emitted from vegetation developed by incorporating multivariate linear regression involving vegetation biophysical properties and meteorological variables derived from MODIS and MERRA-2 (Li and Xiao 2019). GOSIF shows a strong correlation with GPP on the global scale ($R^2 = 0.73$, $p < 0.001$) compared with in-situ measurements of terrestrial carbon fluxes from 91 FLXNET sites. Although other SIF datasets are available, GOSIF provides better SIF retrievals with relatively fine spatial (0.05°) and temporal (8-day) resolution and a continuous global coverage over the period 2000 to present. Landsat-8 Operational Land Imager (OLI) surface reflectance level-3 data products of 10 March 2019 at 30 m spatial resolution were used (refer to Table 6.1). This higher-level data product is already atmospherically, geometrically, and radiometrically corrected. The Landsat-8 scene was acquired from the United States Geological Survey (USGS) Earth Explorer website (https://earthexplorer.usgs.gov). Thus corrected, the cloud-free dataset was used for further analysis, such as VIs generation and LAI and chlorophyll estimation.

6.2.2 Methods

6.2.2.1 Digital Hemispherical Photography (DHP) Processing

The DHP technique uses a digital camera with a fisheye lens, which helps measure canopy gap fraction for different viewing directions (Jonckheere et al. 2004). In this work, a Nikon D5300 DSLR was used with a fisheye mount lens of 8–15 mm focal length. Photographs were taken looking upward. Exposure was set manually to reduce the saturation effect and to achieve good contrast between canopy elements and sky. The DHP image resolution was 6000 × 4000 pixels. Hemispherical photographs were processed using an open-source software package, Can-Eye (Weiss and Baret 2014), which works on the MATLAB platform and measures the plant area index (PAI), but the term LAI is used instead of PAI throughout the text due to its universal use and acceptance. It is well known that indirect methods use the gap fraction to estimate the PAI (Breda 2003). Images were divided into a certain zenith angle (100) and azimuth cell (36) to calculate the gap fraction. Before that, images were classified between two broad classes, namely sky or background and vegetation elements. The gap fraction was calculated that was further inverted to estimate the PAI using Miller's integral formula (Miller 1967),

$$\mathrm{PAI}(\mathrm{LAI}) = 2\int_{0}^{\frac{\pi}{2}} -ln\left(P_0(\theta_v)\cos\theta_v \sin\theta_v \, d\theta_v\right) \tag{6.2}$$

where, Θ_v is the view zenith angle, P_0 is the gap fraction.

6.2.2.2 Laboratory Chlorophyll Extraction

The leaf samples were brought to the laboratory using appropriate ice-packed boxes. In the laboratory, a 3 ml chlorophyll solution was created with 80% aqueous acetone for grinding the leaf samples. A spectrophotometer recorded the absorbance value for all the samples at wavelength 645 and 663 nm. Further, the chlorophyll content (mg/g) was calculated using Arnon's (1949) equation. From each leaf sample, six SPAD readings were recorded prior to the chlorophyll extraction to develop the relationship between them.

6.2.2.3 Vegetation Index Generation

VIs represent an empirical approach for parameter inversion, and are quite straightforward to use. VIs were derived from Landsat-8 data for prediction of LAI and chlorophyll concentration (Table 6.2). The VIs were used as dependent variables in RF regression for LAI and chlorophyll concentration estimation.

Table 6.2 Vegetation indices used to estimate LAI and chlorophyll concentration from Landsat-8 data.

Index	Formula	References
Normalized Difference Vegetation Index (NDVI)	(NIR − Red)/(NIR + Red)	Rouse et al. (1974)
Normalized Difference Water Index (NDWI)	(Green − NIR)/ (Green + NIR)	McFeeters (1996)
Ratio Vegetation index (RVI)	NIR/Red	Birth and McVey (1968)
Soil Adjusted Vegetation Index (SAVI)	(1 + 0.5) × (NIR − Red)/ (NIR + Red + 0.5)	Huete (1988)

6.2.2.4 Machine Learning Approaches

For this study, parametric and non-parametric regression approaches were used to test the performance of the dependent variables (VIs) with independent variables (DHP-based LAI, leaf chlorophyll content [LCC]). RF is an ensemble-based decision tree approach that is used for both classification and regression (Breiman 2001). Nodes or leaves of each decision tree in RF are connected hierarchically with a target prediction for a specific unique feature. RF utilizes the bagging strategy to select training samples and features randomly. It adopts the average prediction from all decision trees and verifies the trained model using the out-of-bag samples, which provide unbiased estimation. Tuning parameters such as the number of trees (ntree) and the number of explanatory variables (mtry) are adjusted to overcome noise and overfitting. This study used the "randomForest" and an associate package "caret" in the R programming platform to predict LAI and LCC. For both the cases, "ntree" was fixed as five hundred, whereas "mtry" was fixed as "2" as it demonstrated minimum RMSE using the grid search method.

6.2.2.5 Time Series Analysis

Time series analysis was carried out to understand the temporal variability of mangrove biophysical properties. Seasonal averages of GPP, SIF, and LAI were generated for the months of January and February (JF), March, April, and May (MAM), and October, November, and December (OND). The time series of area-averaged GPP, SIF, and LAI were plotted. The trend was assessed using the linear regression technique, wherein the seasonal components were highlighted. The linear fit of the time series of LAI, SIF, and GPP was estimated using Pearson's r-value.

6.3 Results

6.3.1 Seasonal Variation of LAI, SIF, and GPP

As per the seasonal trend analysis using MODIS LAI (MOD15A2H) during 2003–2018, the average LAI value varied between 2.71 and 3.27. On a seasonal average JF months hold the highest LAI ($^{avg}LAI_{JF} = 4.00$) range followed by OND months ($^{avg}LAI_{OND} = 3.6$) and MAM

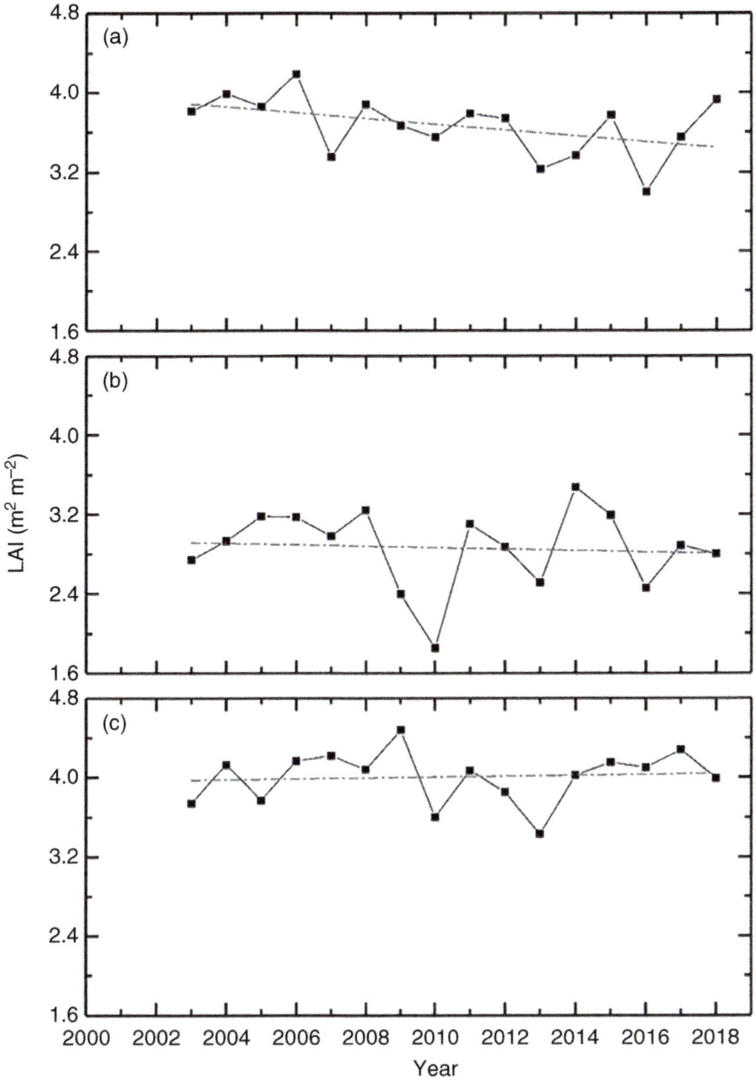

Figure 6.2 Seasonal trend analysis of (a) LAI, (b) SIF, and (c) GPP, during 2000–2003 to 2018 for (a) JF, (b) MAM, and (c) OND months.

months ($^{avg}LAI_{MAM}$ = 2.85). The LAI variability in JF months (R = −0.45) and MAM months (R = −0.08) exhibited a negative trend over the 16 years, and OND months (R = +0.08) demonstrated a weak positive trend (Figure 6.2a).

The temporal dynamics of photosynthetic activity across the study area were depicted in the time series of seasonally averaged SIF (Figure 6.2b). Overall, the mean SIF ranged between 1.46 and 2.7 $W m^{-2}$ μm^{-1} Sr^{-1}. The maximum and minimum SIF observed during the OND months (≈2.33 $W m^{-2}$ μm^{-1} Sr^{-1}) and JF months (≈1.65 $W m^{-2}$ μm^{-1} Sr^{-1}). The SIF depicted during MAM months was similar to the range found during OND months (2.1 $W m^{-2}$ μm^{-1} Sr^{-1}). The SIF variability in JF months (R = −0.60) and MAM months

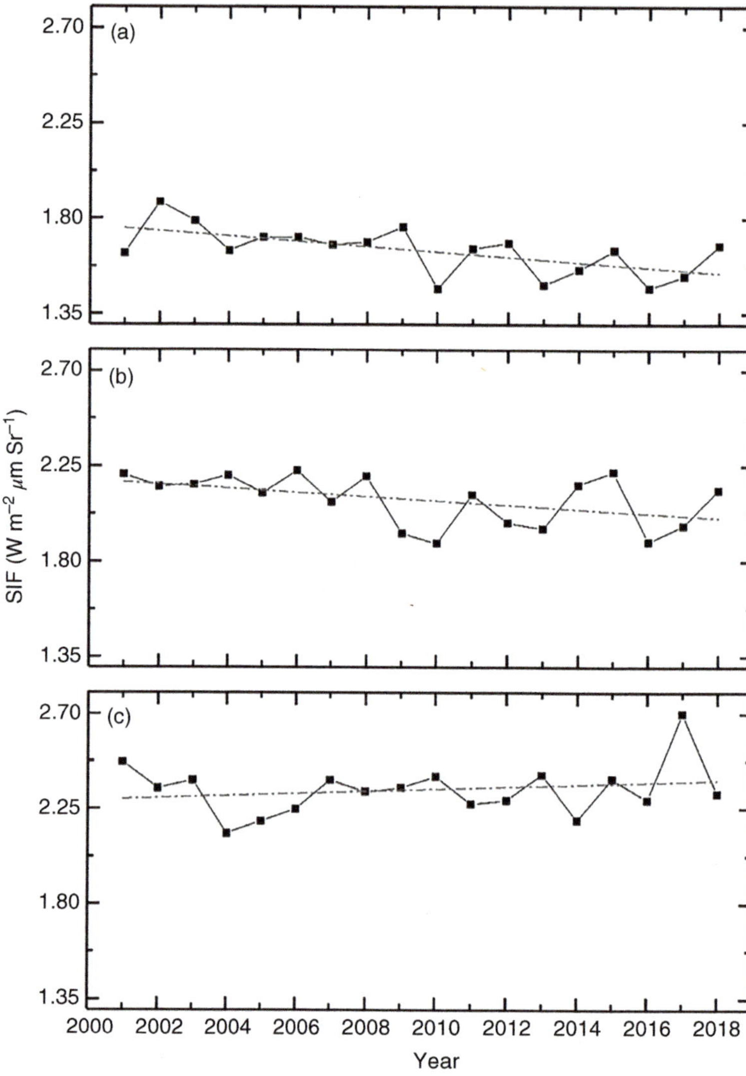

Figure 6.2 (Continued)

(R = −0.44) demonstrated a strong negative trend during 2001–2018, while OND months showed a weak positive trend (R = +0.22).

The time series trend evaluated the variation in primary productivity across seasons, with the seasonal average GPP fluctuation between 52.2 and 95.6 g C m^{-2} mon^{-1} (Figure 6.2c). The highest GPP was observed during OND months (83.09 g C m^{-2} mon^{-1}), while MAM months demonstrated the least productive period (66.03 g C m^{-2} mon^{-1}). Furthermore, GPP variability in both JF and MAM months demonstrated an increasing trend, while a decreasing trend was noticed for the OND GPP months. The variation in GPP can be seen during JF months (R = −0.45), even though no sharp slope was observed in the time series trend. However, the GPP trend was poor during MAM (R = −0.09) and OND (R = +0.08) months.

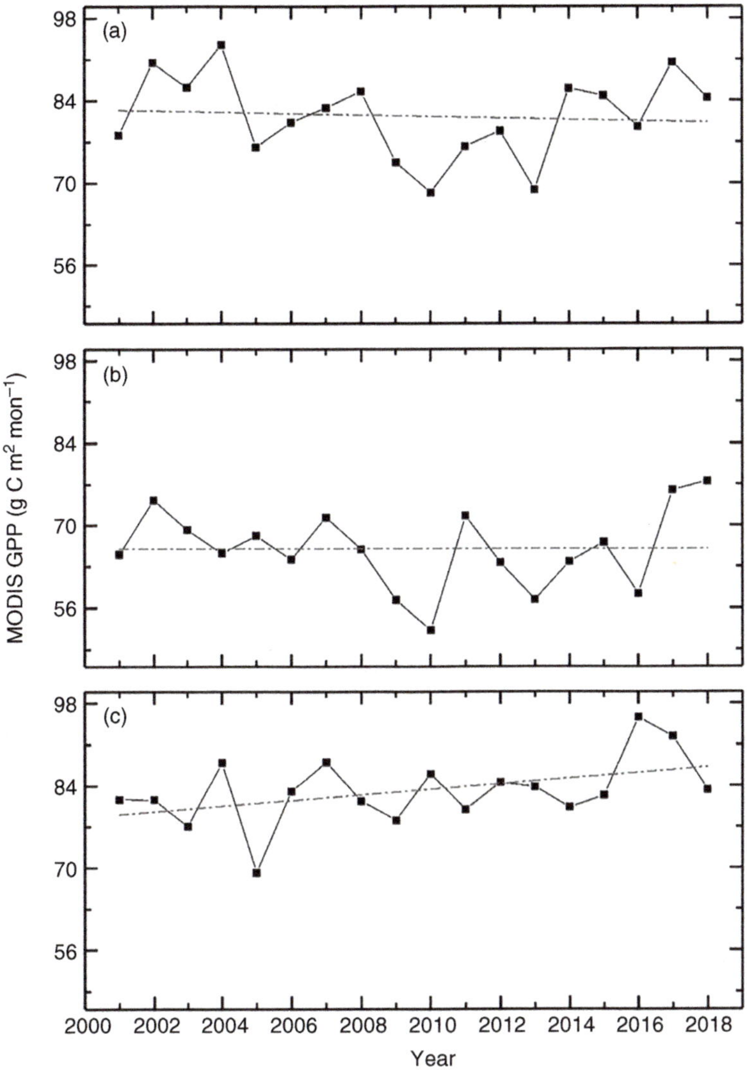

Figure 6.2 (Continued)

Overall, GPP demonstrated better correlation with LAI ($R^2 = 0.59$) than SIF, while LAI and SIF demonstrated poor correlation (Figure 6.3).

6.3.2 Landsat-8 Predicted LAI

The Landsat-8 derived VIs (NDVI, NDWI, SAVI, and RVI) demonstrated a different relationship with DHP based LAI (Figure 6.4a–d), which was further inverted using RF model for LAI prediction for BWS (Figure 6.5a). All the VIs were positively correlated with the DHP based LAI and found statistically significant (p-value < 0.05). NDVI and SAVI showed

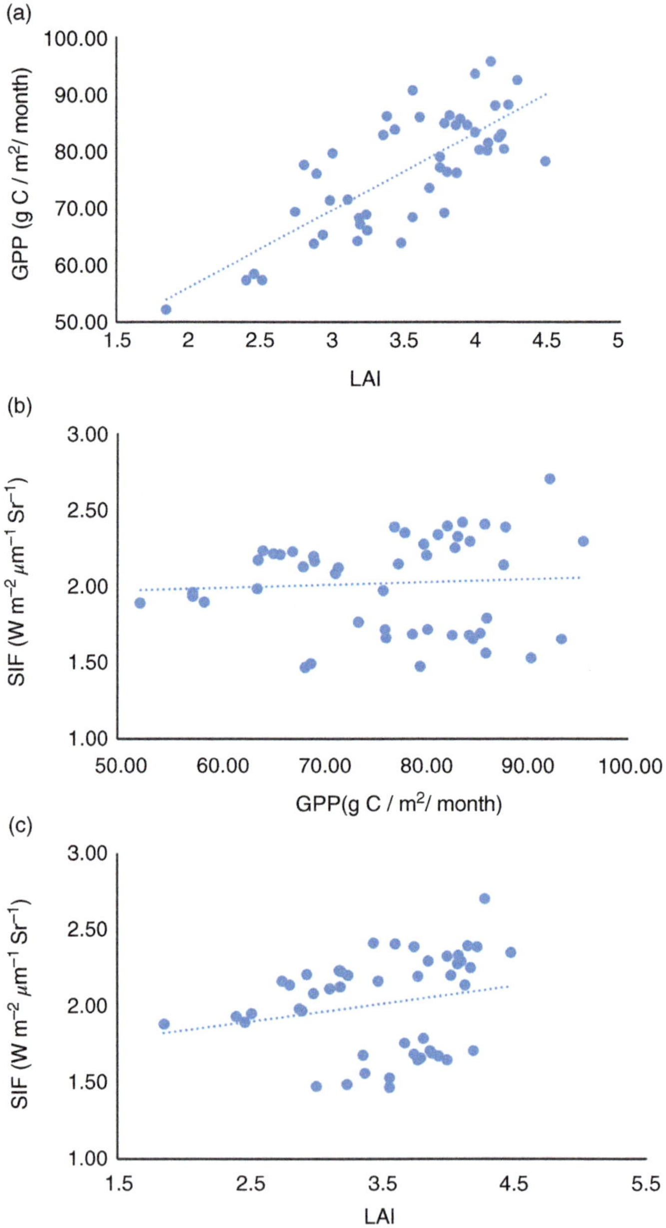

Figure 6.3 Relationship between satellite-derived (a) GPP and LAI, (b) GPP and SIF, (c) SIF and LAI product during 2003–2018. The GPP demonstrated strong positive correlation with LAI.

maximum similarity ($R^2 = 0.29$) followed by RVI ($R^2 = 0.26$), while NDWI has a comparatively weaker correlation ($R^2 = 0.22$). Overall the predicted LAI varied from 1.81 to 4.92 and matched well with the DHP based LAI estimate (R = 0.68 and RMSE = 0.74) with significant goodness of fit (Figure 6.5b).

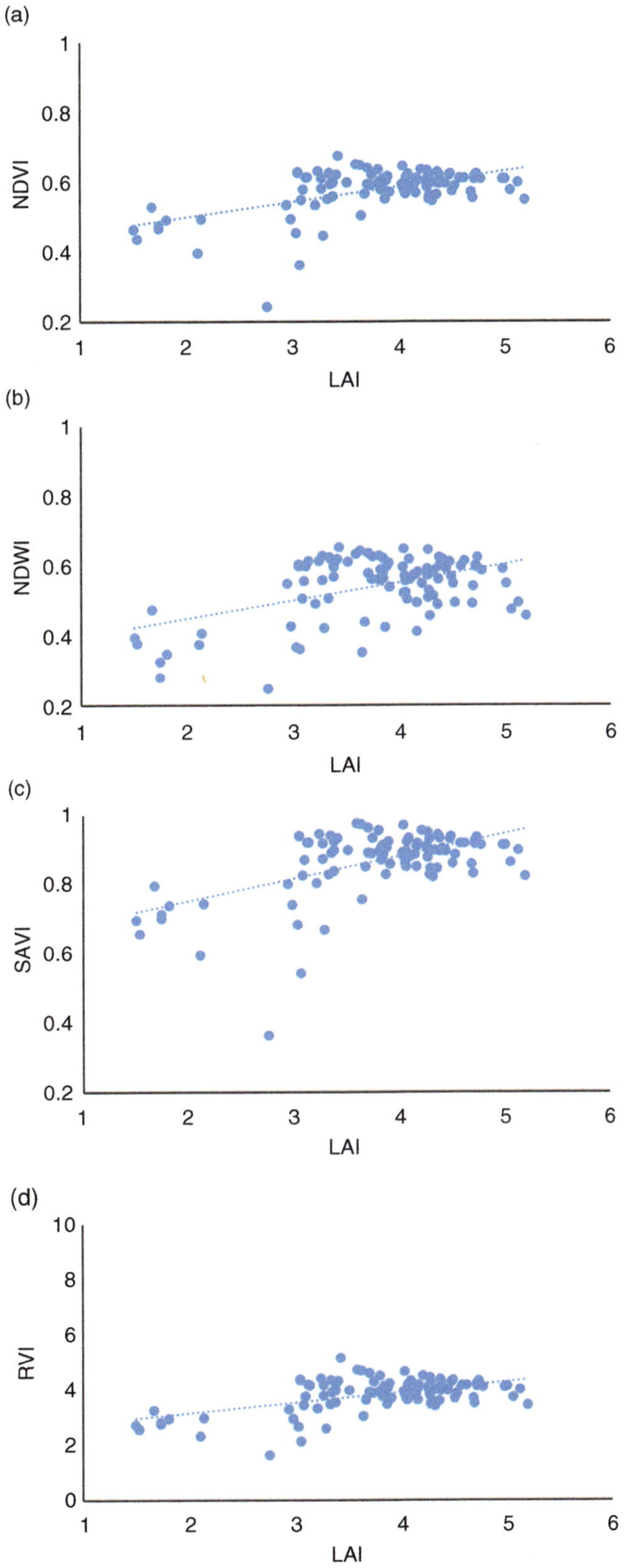

Figure 6.4 Relationship between DHP-based field LAI and VIs; (a) NDVI, (b) NDWI, (c) SAVI, and (d) RVI. These VIs were further used as independent variables for predicting LAI.

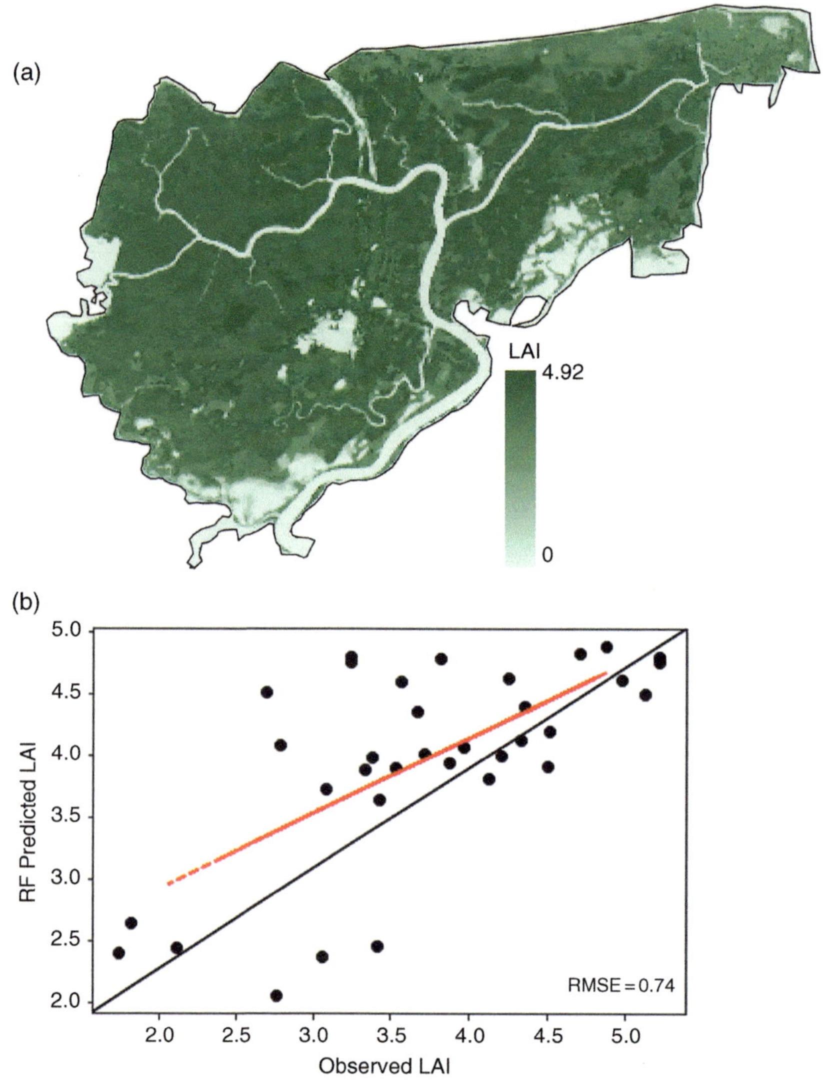

Figure 6.5 (a) Predicted LAI image using RF algorithm wherein four VIs (NDVI, NDWI, SAVI, RVI) were used as independent variables; (b) validation of the predicted LAI using DHP-based observed LAI dataset.

6.3.3 Landsat-8 Predicted Canopy Chlorophyll Content (CCC)

The NDVI, SAVI, RVI, and NDWI showed weak correlations with the laboratory-extracted leaf CCC (Figure 6.6a–d). Comparatively, RVI exhibited a better correlation ($R^2 = 0.022$), followed by SAVI ($R^2 = 0.015$), NDVI ($R^2 = 0.015$), and NDWI ($R^2 = 0.003$). The spatial distribution of canopy canopy chlorophyll content (CCC) was well predicted from the VIs using the RF algorithm that varied from 2.16 to 11.15 mg/g over BWS (Figure 6.7a).

Figure 6.6 Relationship between leaf chlorophyll content and Vis; (a) NDVI, (b) NDWI, (c) RVI, and (d) SAVI. These four VIs were further used as independent variables for predicting chlorophyll concentration from Landsat-8 data.

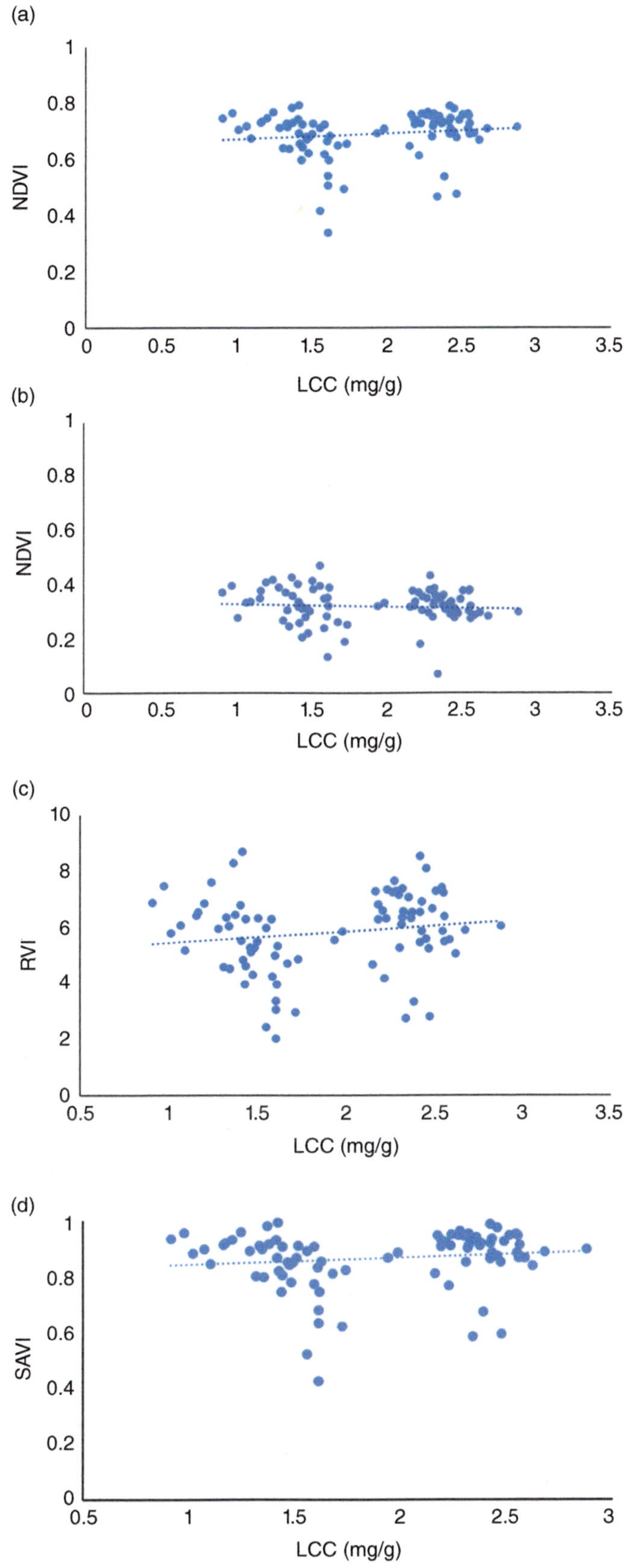

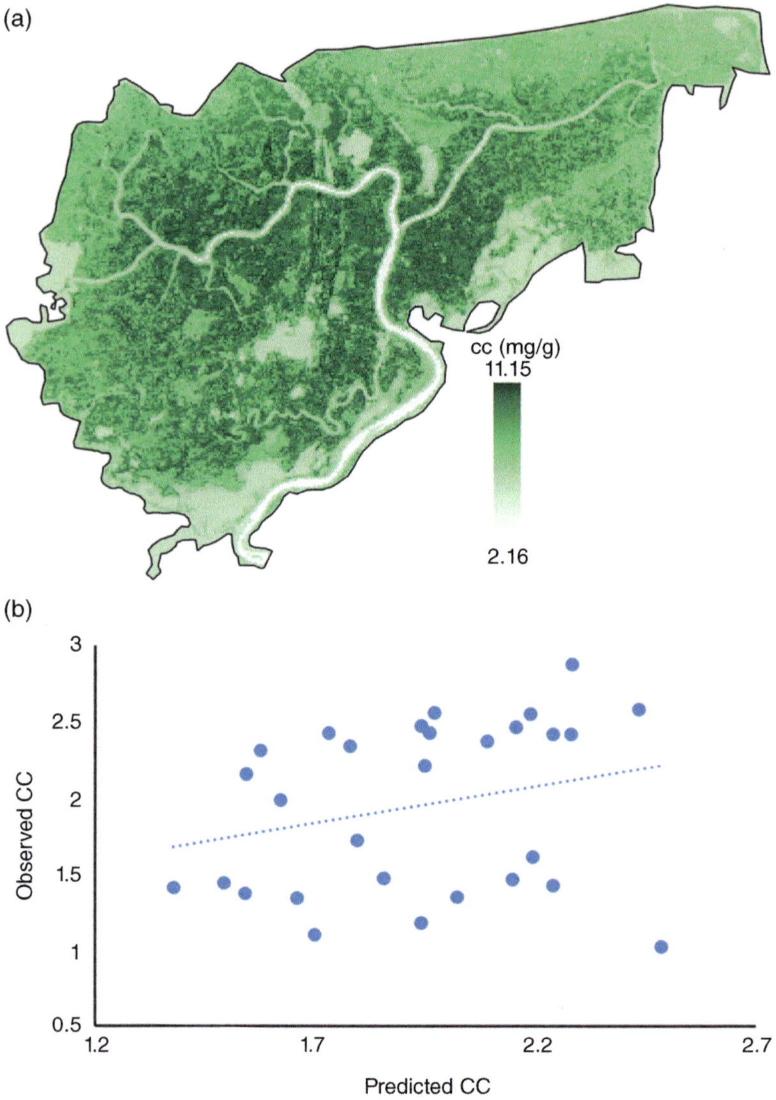

Figure 6.7 (a) Chlorophyll concentration image generated using RF algorithm where four VIs (NDVI, NDWI, RVI, SAVI) were used as independent variables and LAI used as a multiplicative factor; (b) validation between observed and VIs based RF predicted LCC.

The relationship between SPAD reading and leaf CCC demonstrated a positive correlation ($R^2 = 0.5$), establishing the following equation.

$$\text{SPAD} = 14.46 \times \text{LCC} + 23.12 \tag{6.3}$$

The leaf CCC could be transferred to canopy CCC by multiplying Landsat-8 derived VIs and DHP based LAI estimates. The RF model predicted leaf CCC demonstrated RMSE of 0.54 mg/g when tested with another set of observed data (Figure 6.7b).

6.4 Discussion

6.4.1 Seasonal Behavior

The SIF, GPP, and LAI time series analysis demonstrated very marginal variations in their temporal pattern. The GPP and LAI exhibited a good correlation owing to a similar algorithm basis and spatial resolution (Figure 6.8a,b). Chamberlain et al. (2021) observed similar seasonal peaks and troughs for a mangrove forest in Australia due to phenology. From our long term temporal analysis, it is clear that the winter season (JF months) demonstrated maximum LAI than other months, which may be due to influence of tropical conditions such as frequent cloud cover and higher temperature in summer and post-monsoon, and local tidal effects (Lovelock et al. 2015a). It is found that winter is the start of the season (SOS) and post-monsoon is the end of the season (EOS) as far as mangrove productivity is concerned for the BWS (Chamberlain et al. 2021). Some species shed their leaves in winter, such as *Xylocarpus* spp., whereas *Avicennia marina* has lower leaf shedding and growth throughout the year (Duke 2006).

The maximal SIF in the post-monsoon season (OND) may be attributed to the cloudy conditions (Figure 6.8c), when the lower and middle parts of the canopy receive more diffuse PAR, leading to an increase in the SIF emission (Porcar-Castell et al. 2014; Lu et al. 2020). Normally, GPP and SIF are strongly correlated, but PAR brings some uncertainty to the GPP–SIF relationship (Cheng et al. 2021), because of this seasonal fluctuation in PAR (Lu et al. 2020; Cheng et al. 2021). Other than PAR, soil moisture and vapor pressure deficit demonstrated a significant impact on the GPP–SIF relationship

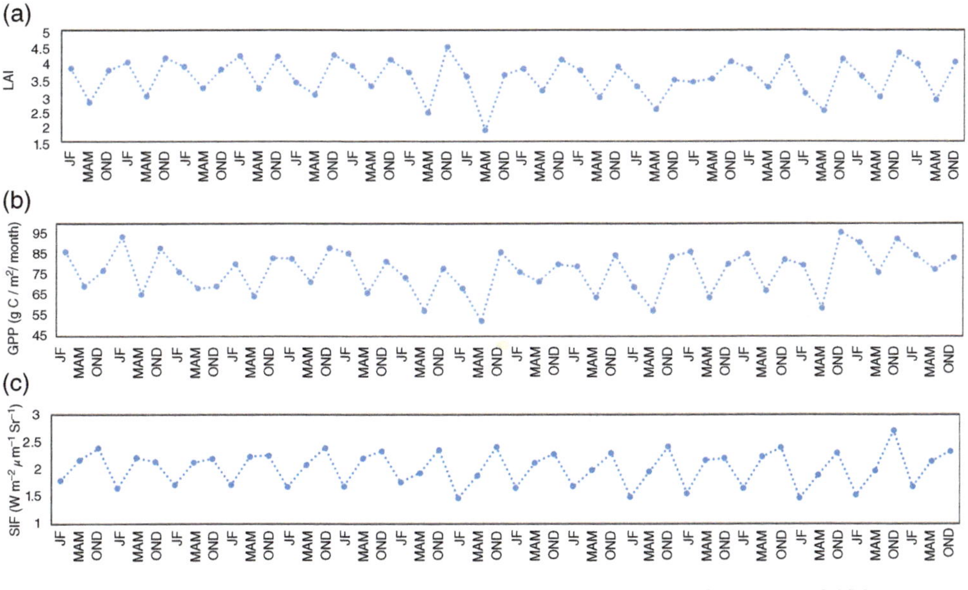

Figure 6.8 Temporal pattern of (a) MODIS-based LAI, (b) MODIS-based GPP, and (c) OCO2-based SIF for the Bhitarkanika Wildlife Sanctuary during 2000–2003 to 2018. Where JF: January and February months; MAM: March, April, and May months; OND: October, November, and December months.

(Zhou et al. 2020; Ma et al. 2022). The soil moisture is comparatively higher in the post-monsoon season, which could trigger a positive relationship with SIF (Liu et al. 2020). High vapor pressure deficit can make the closure of stomata, and reduce the GPP by inhibiting the gaseous exchange (Paul-Limoges et al. 2018), which could be the possible reason behind the poor correlation between SIF and LAI.

6.4.2 Random Forest-based LAI and LCC estimation

The LAI estimated from the Landsat-8 reflectance bands derived VIs well matched with the previous studies (Paramanik et al. 2019; Parida and Kumari 2021), and the RF algorithm performed well (Yao et al. 2017). Only simple univariate linear or parametric regression would be an unwise decision over non-parametric regression (Verrelst et al. 2015) to make a model for biophysical variables, e.g. LAI and chlorophyll estimation in the heterogeneous field such as mangrove forests. The VIs were chosen based on the contrast of NIR and red reflectance bands, considering the effect of moisture and soil background (Blinn et al. 2019). The RMSE of 0.74 for LAI prediction could be due to some uncertainty from DHP image processing or GPS-based positional error of ± 3 m. Three of the four VIs were found suitable for leaf-level chlorophyll using RF regression. However, the accuracy of predicted chlorophyll concentration can be further improved if the sources of uncertainties can be minimized, such as SPAD reading from the top canopy and minimal time lag between leaf collection from the field and CC extraction in the laboratory.

6.5 Conclusions

This study demonstrated that despite the difficulty of conducting a field campaign in the mangrove ecosystem, several photosynthetic variables such as LAI, chlorophyll, and GPP can be estimated efficiently by utilizing remote sensing techniques. The results may be improved by incorporating more field observations of DHP-based LAI maintaining the density compositions, more leaf samples for laboratory extraction from the upper canopy, and corresponding SPAD value by keeping minimal time lag. For estimating the LAI and chlorophyll concentration, more focus should be given on proper VI selection, as they are site and vegetation type-specific and depend on the sun-sensor geometry. Future studies may test other VIs to improve the results with new data such as hyperspectral Sentinel-2, as they have unique bands in the red-edge region with higher spatial resolution. Though it is well documented that RF models perform well for prediction, other advanced machine learning or deep learning approaches may be tested for refinement. On a regional scale, more remote sensing-based models should be made with finer resolution datasets, as coarser-resolution GPP products or SIF products are very challenging in precise analysis or small-scale monitoring. In a climate change scenario, these biophysical or biochemical parameters should be estimated more accurately and directly linked to primary productivity. Accurate estimation of LAI or chlorophyll concentration will enable better assessment of GPP, which will certainly be advantageous for efficient management practices.

References

Ali, A.M., Darvishzadeh, R., Skidmore, A. et al. (2020). Comparing methods for mapping canopy chlorophyll content in a mixed mountain forest using Sentinel-2 data. *International Journal of Applied Earth Observation and Geoinformation* 87: https://doi.org/10.1016/j.jag.2019.102037.

Alongi, D.M. (2002). *Present state and future of the world's mangrove forests. Environmental Conservation* 29: 331–349.

Alongi, D.M. (2012). Carbon sequestration in mangrove forests. *Carbon Management* 3: 313–322.

Anand, A., Pandey, P.C., Petropoulos, G.P. et al. (2020). Use of hyperion for mangrove forest carbon stock assessment in Bhitarkanika forest reserve: a contribution towards blue carbon initiative. *Remote Sensing* 12: 597.

Arnon, D.I. (1949). Copper enzymes in isolated chloroplasts. Polyphenoloxidase in Beta vulgaris. *Plant Physiology* 24: 1.

Banerjee, K., Bal, G., Pal, N. et al. (2017). Carbon sequestration by mangrove vegetation: a case study from Odisha Bhitarkanika wild life sanctuary (BWLS). *Parana Journal of Science and Education* 3: 1–9.

Baret, F. and Guyot, G. (1991). Potentials and limits of vegetation indices for LAI and APAR assessment. *Remote Sensing of Environment* 35: 161–173.

Behera, M.D., Barnwal, S., Paramanik, S. et al. (2021). Species-level classification and mapping of a mangrove forest using random forest—utilisation of AVIRIS-NG and sentinel data. *Remote Sensing* 13 (11): 2027.

Birth, G.S. and McVey, G.R. (1968). Measuring the color of growing turf with a reflectance spectrophotometer 1. *Agronomy Journal* 60: 640–643.

Blinn, C.E., House, M.N., Wynne, R.H. et al. (2019). Landsat-8 based leaf area index estimation in loblolly pine plantations. *Forests* 10: 222.

Breda, N.J. (2003). Ground-based measurements of leaf area index: a review of methods, instruments and current controversies. *Journal of Experimental Botany* 54: 2403–2417.

Breiman, L. (2001). Random forests. *Machine Learning* 45: 5–32.

Chamberlain, D.A., Phinn, S.R., and Possingham, H.P. (2021). Mangrove Forest cover and phenology with Landsat dense time series in Central Queensland, Australia. *Remote Sensing* 13: 3032.

Chazdon, R.L., Broadbent, E.N., Rozendaal, D.M. et al. (2016). Carbon sequestration potential of second-growth forest regeneration in the Latin American tropics. *Science Advances* 2: e1501639.

Cheng, X., Zhou, Y., Hu, M. et al. (2021). The links between canopy solar-induced chlorophyll fluorescence and gross primary production responses to meteorological factors in the growing season in deciduous broadleaf forest. *Remote Sensing* 13: 2363.

Choudhury, B.J., Ahmed, N.U., Idso, S.B. et al. (1994). Relations between evaporation coefficients and vegetation indices studied by model simulations. *Remote Sensing of Environment* 50: 1–17.

Duke, N.C. (2006). *Australia's mangroves: the authoritative guide to Australia's mangrove plants*. MER.

Duke, N.C., Meynecke, J.-O., Dittmann, S. et al. (2007). A world without mangroves? *Science* 317: 41–42.

Duveiller, G. and Cescatti, A. (2016). Spatially downscaling sun-induced chlorophyll fluorescence leads to an improved temporal correlation with gross primary productivity. *Remote Sensing of Environment* 182: 72–89. https://doi.org/10.1016/j.rse.2016.04.027.

Gamon, J.A., Somers, B., Malenovský, Z. et al. (2019). Assessing vegetation function with imaging spectroscopy. *Surveys in Geophysics* 40: 489–513. https://doi.org/10.1007/s10712-019-09511-5.

Ghosh, S.M., Behera, M.D., and Paramanik, S. (2020). Canopy height estimation using sentinel series images through machine learning models in a mangrove forest. *Remote Sensing* 12: 1519.

Ghosh, S., Behera, M., Jagadish, B. et al. (2021). A novel approach for estimation of aboveground biomass of a carbon-rich mangrove site in India. *Journal of Environmental Management* 292: 112816.

Giri, C., Ochieng, E., Tieszen, L.L. et al. (2011). Status and distribution of mangrove forests of the world using earth observation satellite data. *Global Ecology and Biogeography* 20: 154–159.

Hamilton, S.E. and Casey, D. (2016). Creation of a high spatio-temporal resolution global database of continuous mangrove forest cover for the 21st century (CGMFC-21). *Global Ecology and Biogeography* 25: 729–738.

Heenkenda, M.K., Joyce, K.E., Maier, S.W., and de Bruin, S. (2015). Quantifying mangrove chlorophyll from high spatial resolution imagery. *ISPRS Journal of Photogrammetry and Remote Sensing* 108: 234–244.

Huete, A.R. (1988). A soil-adjusted vegetation index (SAVI). *Remote Sensing of Environment* 25: 295–309.

Jonckheere, I., Fleck, S., Nackaerts, K. et al. (2004). Review of methods for in situ leaf area index determination: part I. Theories, sensors and hemispherical photography. *Agricultural and Forest Meteorology* 121: 19–35.

Kalaji, H.M., Jajoo, A., Oukarroum, A. et al. (2016). Chlorophyll a fluorescence as a tool to monitor physiological status of plants under abiotic stress conditions. *Acta Physiologiae Plantarum* 38: https://doi.org/10.1007/s11738-016-2113-y.

Kripa, M.K., Gohel, A., Lele, N., et al. (2017). Modelling chlorophyll fluorescence through scope and in-situ measurements - A case study for mangrove vegetation. *38th Asian Conference on Remote Sensing - Space Applications: Touching Human Lives*, ACRS 2017 2017-Octob.

Lele, N., Kripa, M., Panda, M. et al. (2021). Seasonal variation in photosynthetic rates and satellite-based GPP estimation over mangrove forest. *Environmental Monitoring and Assessment* 193: 1–20.

Li, X. and Xiao, J. (2019). A global, 0.05-degree product of solar-induced chlorophyll fluorescence derived from OCO-2, MODIS, and reanalysis data. *Remote Sensing* 11: 517.

Liu, D., Cai, W., Xia, J. et al. (2014). Global validation of a process-based model on vegetation gross primary production using eddy covariance observations. *PLoS One* 9: https://doi.org/10.1371/journal.pone.0110407.

Liu, L., Gudmundsson, L., Hauser, M. et al. (2020). Soil moisture dominates dryness stress on ecosystem production globally. *Nature Communications* 11: 1–9.

Lovelock, C.E., Adame, M.F., Bennion, V. et al. (2015a). Sea level and turbidity controls on mangrove soil surface elevation change. *Estuarine, Coastal and Shelf Science* 153: 1–9.

Lovelock, C.E., Cahoon, D.R., Friess, D.A. et al. (2015b). The vulnerability of Indo-Pacific mangrove forests to sea-level rise. *Nature* 526: 559–563.

Lu, X., Liu, Z., Zhao, F., and Tang, J. (2020). Comparison of total emitted solar-induced chlorophyll fluorescence (SIF) and top-of-canopy (TOC) SIF in estimating photosynthesis. *Remote Sensing of Environment* 251: 112083.

Ma, L., Sun, L., Wang, S. et al. (2022). Analysis on the relationship between sun-induced chlorophyll fluorescence and gross primary productivity of winter wheat in northern China. *Ecological Indicators* 139: 108905.

McFeeters, S.K. (1996). The use of the Normalized Difference Water Index (NDWI) in the delineation of open water features. *International Journal of Remote Sensing* 17: 1425–1432.

Miller, J.B. (1967). A formula for average foliage density. *Australian Journal of Botany* 15 (1): 141–144.

Noumonvi, K.D. and Ferlan, M. (2020). Empirical vs. light-use efficiency modelling for estimating carbon fluxes in a mid-succession ecosystem developed on abandoned karst grassland. *PLoS One* 15: 1–18. https://doi.org/10.1371/journal.pone.0237351.

Pandey, P.C., Anand, A., and Srivastava, P.K. (2019). Spatial distribution of mangrove forest species and biomass assessment using field inventory and earth observation hyperspectral data. *Biodiversity and Conservation* 28 (8): 2143–2162. https://doi.org/10.1007/s10531-019-01698-8.

Paramanik, S., Behera, M.D., Bhattacharya, B.K., and Tripathi, S. (2019). Evaluation and Validation of the MODIS LAI Algorithm with Digital Hemispherical *Photography at Bhitar Kanika Mangrove Forest, India*. In: *International Geoscience and Remote Sensing Symposium (IGARSS)*, 6558–6561. https://doi.org/10.1109/IGARSS.2019.8899056.

Parida, B.R. and Kumari, A. (2021). Mapping and modeling mangrove biophysical and biochemical parameters using Sentinel-2A satellite data in Bhitarkanika National Park, Odisha. *Modeling Earth Systems and Environment* 7: 2463–2474.

Pastor-Guzman, J., Atkinson, P.M., Dash, J., and Rioja-Nieto, R. (2015). Spatiotemporal variation in mangrove chlorophyll concentration using Landsat-8. *Remote Sensing* 7: 14530–14558. https://doi.org/10.3390/rs71114530.

Pattanaik, C., Reddy, C.S., Dhal, N.K., and Das, R. (2008). Utilisation of mangrove forests in Bhitarkanika wildlife sanctuary, Orissa. *Indian Journal of Traditional Knowledge* 7: 598–603.

Paul-Limoges, E., Damm, A., Hueni, A. et al. (2018). Effect of environmental conditions on sun-induced fluorescence in a mixed forest and a cropland. *Remote Sensing of Environment* 219: 310–323.

Porcar-Castell, A., Tyystjärvi, E., Atherton, J. et al. (2014). Linking chlorophyll a fluorescence to photosynthesis for remote sensing applications: mechanisms and challenges. *Journal of Experimental Botany* 65: 4065–4095.

Porra, R., Thompson, W., and Kriedemann, P. (1989). Determination of accurate extinction coefficients and simultaneous equations for assaying chlorophylls a and b extracted with four different solvents: verification of the concentration of chlorophyll standards by atomic absorption spectroscopy. *Biochimica et Biophysica Acta (BBA)-Bioenergetics* 975: 384–394.

Rouse, J.W., Haas, R.H., Schell, J.A. et al. (1974). *Monitoring vegetation systems in the Great Plains with ERTS*, vol. 351, 309. *NASA special publication*.

Running, S.W. and Zhao, M. (2015). *Daily GPP and annual NPP (MOD17A2/A3) products NASA Earth Observing System MODIS land algorithm*. MOD17 User's Guide 2015, 1–28.

Running, S.W., Nemani, R.R., Heinsch, F.A. et al. (2004). A continuous satellite-derived measure of global terrestrial primary production. *Bioscience* 54: 547–560.

Sonobe, R., Yamashita, H., Mihara, H. et al. (2021). Hyperspectral reflectance sensing for quantifying leaf chlorophyll content in wasabi leaves using spectral pre-processing techniques and machine learning algorithms. *International Journal of Remote Sensing* 42: 1311–1329.

Spalding, M. (1997). The global distribution and status of mangrove ecosystems. *International News Letter of Coastal Management-Intercoast Network* 1: 20–21.

Sun, Z., Wang, X., Zhang, X. et al. (2019). Evaluating and comparing remote sensing terrestrial GPP models for their response to climate variability and CO_2 trends. *Science of the Total Environment* 668: 696–713. https://doi.org/10.1016/j.scitotenv.2019.03.025.

Thenkabail, P.S., Smith, R.B., and De Pauw, E. (2000). Hyperspectral vegetation indices and their relationships with agricultural crop characteristics. *Remote Sensing of Environment* 71 (2): 158–182. https://doi.org/10.1016/S0034-4257(99)00067-X.

Tian, H., Chen, G., Liu, M. et al. (2010). Model estimates of net primary productivity, evapotranspiration, and water use efficiency in the terrestrial ecosystems of the southern United States during 1895–2007. *Forest Ecology and Management* 259: 1311–1327.

Valiela, I., Bowen, J.L., and York, J.K. (2001). Mangrove forests: one of the World's threatened major tropical environments: at least 35% of the area of mangrove forests has been lost in the past two decades, losses that exceed those for tropical rain forests and coral reefs, two other well-known threatened environments. *Bioscience* 51: 807–815.

Varghese, R. and Behera, M. (2019). Annual and seasonal variations in gross primary productivity across the agro-climatic regions in India. *Environmental Monitoring and Assessment* 191: 1–19.

Verrelst, J., Rivera, J.P., Veroustraete, F. et al. (2015). Experimental Sentinel-2 LAI estimation using parametric, non-parametric and physical retrieval methods–a comparison. *ISPRS Journal of Photogrammetry and Remote Sensing* 108: 260–272.

Watson, D.J. (1947). Comparative physiological studies on the growth of field crops: I. Variation in net assimilation rate and leaf area between species and varieties, and within and between years. *Annals of Botany* 11: 41–76.

Wei, J., Tang, X., Gu, Q. et al. (2019). Using solar-induced chlorophyll fluorescence observed by OCO-2 to predict autumn crop production in China. *Remote Sensing* 11: 1715.

Weiss, M. and Baret, F. (2014). *CAN-EYE user manual*. CAN-EYE.

Wu, C., Chen, J., Desai, A. et al. (2012). Remote sensing of canopy light use efficiency in temperate and boreal forests of North America using MODIS imagery. *Remote Sensing of Environment* 118: 60–72. https://doi.org/10.1016/j.rse.2011.11.012.

Yan, K., Park, T., Yan, G. et al. (2016). Evaluation of MODIS LAI/FPAR product collection 6. Part 1: consistency and improvements. *Remote Sensing* 8: 359.

Yao, X., Yu, K., Yang, Y. et al. (2017). Estimation of forest leaf area index based on random forest model and remote sensing data. *Transactions of the Chinese Society for Agricultural Machinery* 48: 159–166.

Yu, L., Wen, J., Chang, C.Y. et al. (2019). High-resolution global contiguous SIF of OCO-2. *Geophysical Research Letters* 46: 1449–1458. https://doi.org/10.1029/2018GL081109.

Zhen, J., Jiang, X., Xu, Y. et al. (2021). Mapping leaf chlorophyll content of mangrove forests with Sentinel-2 images of four periods. *International Journal of Applied Earth Observation and Geoinformation* 102: 102387.

Zhou, H., Wu, D., and Lin, Y. (2020). The relationship between solar-induced fluorescence and gross primary productivity under different growth conditions: global analysis using satellite and biogeochemical model data. *International Journal of Remote Sensing* 41 (19): 7660–7679.

Zhu, Y., Liu, K., Liu, L. et al. (2017). Exploring the potential of worldview-2 red-edge band-based vegetation indices for estimation of mangrove leaf area index with machine learning algorithms. *Remote Sensing* 9: 1060.

Zhu, X., Hou, Y., Zhang, Y., and Lu, X. (2021). Potential of sun-induced chlorophyll fluorescence for indicating mangrove canopy photosynthesis. *Journal of Geophysical Research: Biogeosciences* 126: 1–17. https://doi.org/10.1029/2020JG006159.

7

Quantifying Carbon Stock Variability of Species Within a Reforested Urban Landscape Using Texture Measures Derived from Remotely Sensed Imagery

*Mthembeni Mngadi, John Odindi, and Onisimo Mutanga**

School of Agricultural, Earth and Environmental Sciences, Discipline of Geography, University of KwaZulu-Natal, Pietermaritzburg, South Africa

7.1 Introduction

Continuous transformation of natural into impervious urban landscapes that typifies urbanization has been considered a major driver of environmental change (Sithole et al. 2018; Odindi and Mhangara 2012). Such transformation affects the quality of critical ecological processes and ecosystem services, resulting in among others, an increase in urban thermal heat, and air and noise pollution (Xu et al. 2016; Livesley et al. 2016). Although urban landscapes cover marginal land surface worldwide, they account for approximately 70% of the global carbon emissions due to dense settlement and high energy consumption (Ribeiro et al. 2019, Esch et al. 2017). Literature shows that urbanization is often associated with unprecedented deforestation and forest degradation; resulting in declining ecosystem goods and services and sequestration potential, while increasing greenhouse gas emissions and risk of climate change (Keenan et al. 2015; Payn et al. 2015; Delphin et al. 2016; Mngadi et al. 2021). Hence, the recently launched program for reducing emissions from deforestation and forest degradation (REDD) has identified urban reforestation as the most practical and long-term alternative for assimilating emitted carbon and reducing the impacts of climate change (Mansourian and Vallauri 2005; Livesley et al. 2016; Deo et al. 2017; Curiel-Esparza et al. 2015). The recognition of urban reforestation initiatives as reliable carbon sinks is presumed to contribute greatly to the local, regional, and global carbon budget (Sithole et al. 2018; Curiel-Esparza et al. 2015; Mngadi et al. 2021). However, despite the immediate need to understand their value on the global carbon cycle and climate change mitigation potential, information on carbon stock variability in reforested trees remains largely unknown. In this regard, there is need for carbon quantification in reforested landscapes in order to facilitate timely and accurate information necessary for informed urban ecosystem services regulation, urban environmental sustainability, and carbon and climate

*Corresponding author's email: mutangaO@ukzn.ac.za

0000-0002-9854-203X (Mthembeni Mngadi), 0000-0002-4934-1346 (John Odindi), and 0000-0002-7358-8111 (Onisimo Mutanga)

Advances in Remote Sensing for Forest Monitoring, First Edition. Edited by Prem C. Pandey and Paul Arellano.

change policy formulation. Despite the recent increasing interest in carbon stock assessment in sub-Saharan Africa, especially South Africa, there is paucity in literature that seeks to understand carbon variability within re-forested landscapes.

Quantification of carbon stock variability between reforested tree species is an important step toward understanding the contribution of urban reforestation initiatives to the global carbon balance and climate change (Giardina and Ryan 2002; Dube and Mutanga 2015; Van der Werf and Nagel 1996). Chen et al. (2015) noted that *Eucalyptus urophylla* produced the highest above ground carbon stock compared to *Castanopsis hystrix* and another 10 mixed plantation species; while Dube and Mutanga (2015) showed that *Pinus taeda* and *Eucalyptus grandis* species constitute the highest proportion of above ground carbon stock. Wei et al. (2013) also affirmed that above ground carbon stock varies with different forest species. These studies demonstrate that explicit knowledge on carbon balance necessitates an understanding of variability of carbon stock between and within forest species.

Despite numerous studies estimating carbon variations across different forest species types, most of the work has been done on commercial forests; with limited information on indigenous reforested urban trees (Wei et al. 2013; Dube and Mutanga 2015; Chen et al. 2015). Furthermore, available studies have often relied on conventional quantification approaches such as field measurements and survey models, which are costly, labor intensive, and time consuming. Literature has demonstrated that field surveys or models result in underestimation and uncertainty of the actual magnitude of carbon stored in terrestrial ecosystems (Dube and Mutanga 2015; Guo et al. 2010; Raich et al. 2014; Zhang et al. 2012). Meanwhile, the emergence of remote sensing has shown remarkable potential in providing reliable spectral data necessary for accurate and concise quantification of carbon stock variability (Gara et al. 2016; Hickey et al. 2018). Remote sensing minimizes the costs associated with extensive field survey and data collection for large spatial coverage, thus it is valuable for wall-to-wall quantification of aboveground carbon variability. However, despite the suitability of high spatial resolution remotely sensed datasets (e.g. Worldview series, QuickBird, and RapidEye) in forest carbon quantification, they are costly and not readily available. These limitations impede recurrent quantification and monitoring of carbon stock, especially in resource constrained regions such as Southern Africa. Hence, the adoption of freely and readily available optical sensors remain the most convenient sources of spatial and temporal datasets for forest ecosystems carbon stock estimation. For instance, Sentinel-2 data with improved spatial, spectral, and radiometric properties has shown promising potential in modeling forest biomass and carbon stock (Adamu et al. 2021; Khan et al. 2020). The sensor consists of strategically positioned bands in the red-edge region of the electromagnetic spectrum, which increases vegetation sensitivity and spectral response. However, whereas the utility of Sentinel-2 multispectral image (MSI) is gaining popularity, the rich information contained by the sensor is yet to be fully explored using texture-measures for estimating forest carbon stock, especially in a reforested urban landscape. Literature shows that analysis using spectral vegetation indices have been the most commonly used techniques in remote sensing for biomass and carbon modeling (Pandit et al. 2020; Foody et al. 2003; Dube and Mutanga 2015; Steininger 2000). The strength of spectral vegetation indices relies on the relationship between red and near infrared bands, which are critical for detecting green biomass, including the ability to reduce atmospheric effects, senesced vegetation, and soil background (Lu 2006; Foody et al. 2003; Sarker and

Nichol 2011). Additionally, estimating biomass and carbon stock using spectral indices generated from medium-to-coarse spatial resolution sensors in dense tropical and sub-tropical forests has been challenging, due to saturation problems and complexity of forest structure ecosystems (Lu 2006; Pandit et al. 2020; Dube and Mutanga 2015).

Recently, numerous studies have advocated for integration of texture measures with spectral data to enhance the quality and accuracy of image dataset (Kupidura 2019; Lottering and Mutanga 2012; Dube and Mutanga 2015). Image texture characteristics are considered a valuable source of information capable of detecting forest structural attributes, leaf area index, age, and density using medium or high spatial resolution image data (Dube and Mutanga 2015; Pandit et al. 2020; Lottering and Mutanga 2012; Sarker and Nichol 2011). Texture represents the spatial arrangement of gray levels of pixels in an image based on local variance of spectrally unique dominant objects, as well as their size (Zhu and Yang 1998; Kupidura 2019). Hence, texture measures provide robust information on the spatial distribution of pixel-values. For instance. Dube and Mutanga (2015) showed that texture measures derived from medium spatial resolution remote sensing could be effectively used to estimate carbon stock variations across different forest species on plantation forests. Generally, the value of texture measures used to understand carbon variation in indigenous reforested trees remain largely unknown. It is therefore necessary to evaluate the value of texture measures computed from Sentinel-2 MSI to estimate carbon stock variability within reforested urban landscape. Therefore, this study sought to quantify carbon stock variation across different species of reforested trees within an urban landscape using Sentinel-2 computed texture measures. The study hypothesized that the estimated carbon stock significantly varies among reforested tree species.

7.2 Materials and Methods

7.2.1 The Study Site

The study was conducted at the Buffelsdraai reforestation site situated on the north of Durban city, within the eThekwini Municipality, KwaZulu-Natal province, South Africa (Figure 7.1). The reforestation site is located between 30°58′20.08″E and 29°37′55.17″S, and covers approximately 800 ha. Mean annual temperature of the area ranges between 22 and 27 °C, while the mean annual rainfall ranges from 600–1000 mm. The study area is characterized by uneven topography supported by a resistant underlying dwyka tillite rock and dominated by *Acacia* (i.e. *Caffra and Robusta*), *Syzygium cordatum*, *Bridelia micrantha*, and *Erythrina caffra* vegetation species.

7.2.2 Field Survey and Data Collection

Data collection and field surveys were conducted from April 13–17, 2021 during clear sky and sunny weather conditions that favor maximum biomass productivity. In this study, 120 pre-determined random sample points were generated and loaded into a global positioning system (GPS) and used to navigate to the sites. Thereafter, 10× 10 m plot-size windows were developed around each point, and tree height and circumference (at basal-height of

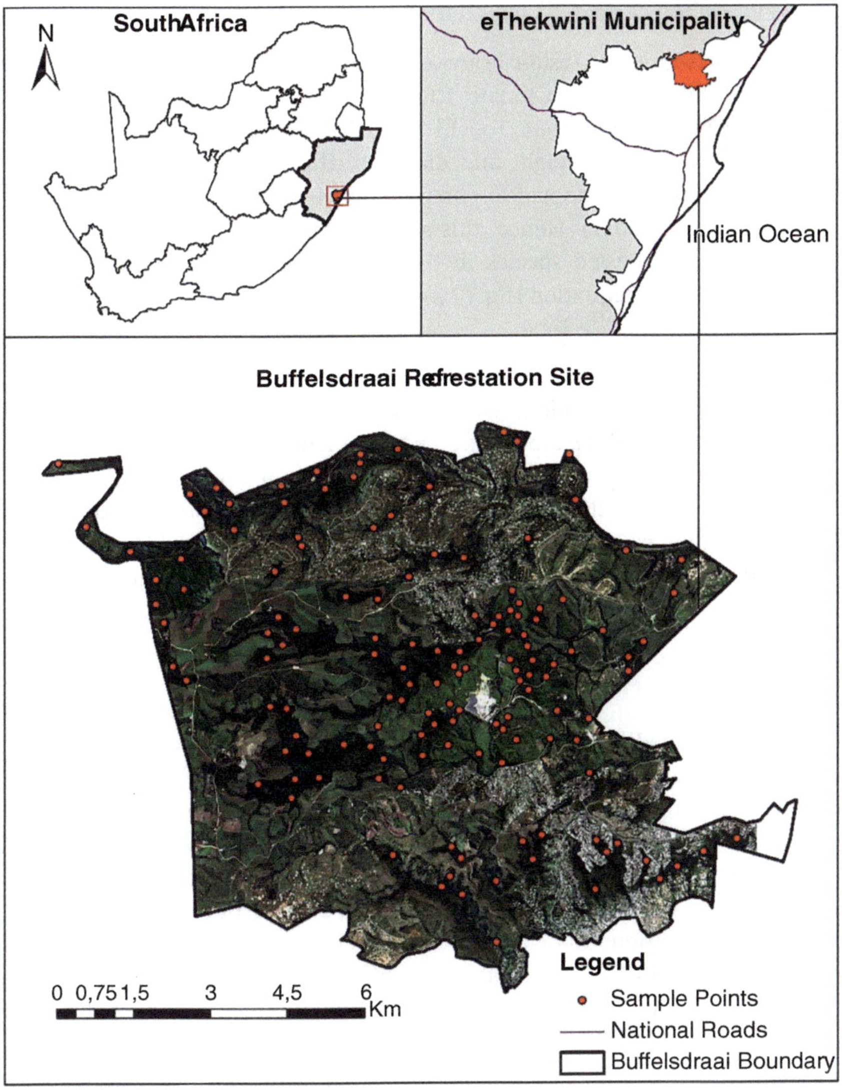

Figure 7.1 The study area and field sample points.

1.3 m) of the dominant tree species measured and recorded onsite. The tree species height was measured using clinometer (Vertex IV Hypsometer), while the circumference was measured using tape measure. The tree diameter (D) was then computed from circumference measurements using the generic diameter equation [Eq. (7.1)]. In addition, the geographic location of each sampled tree species was recorded using Trimble GPS.

$$D = \frac{C}{\pi} \tag{7.1}$$

where D represents diameter, C is the circumference and π is a constant (3.14).

7.2.3 Allometric Modeling of Above Ground Biomass and Carbon Stock

The adoption of an allometric model to estimate above ground biomass has been recommended by the Intergovernmental Panel on Climate Change (IPCC) as the most ideal non-destructive approach (Clark III et al. 1986; Toochi 2018). Literature indicates that allometric relationships between the tree height and diameter affect tree biomass, thus their measurements can be accurately adopted to compute above ground biomass (Dube and Mutanga 2015; Mngadi et al. 2021). Hence, this study consolidated tree height (*H*) and diameter (*D*) of different reforested species in the allometric model to produce above ground biomass using a generic equation [Eq. (7.2)].

$$AGB = a\left(D^2H\right)^b \tag{7.2}$$

where *AGB* represent above ground biomass, *D* is the diameter at breast height (cm), *H* indicates tree height (m), and *a* and *b* are regression coefficients.

Furthermore, literature indicates that dry biomass contains approximately 50% of carbon, thus, the factor of 0.5 is commonly used to convert dry biomass into carbon stock (Hu et al. 2015; Dube and Mutanga 2015; Tang et al. 2016; Mngadi et al. 2021). Hence, in this study, the factor of 0.5 was used to compute the actual carbon stock from above ground biomass.

7.2.4 Image Acquisition and Pre-processing

In this study, a freely and readily available Sentinel-2 MSI captured under clear weather conditions on April 18, 2021 was downloaded on May 5, 2021 from the European Space Agency (ESA) portal. The image consists of 13 spectral wavebands located in the visible (443.9–664.5 nm), near infrared (835.1–864.8 nm), and shortwave infrared (1613.7–2202 nm) regions of the electromagnetic spectrum. The sensor also provides unique and strategically positioned band settings in the red-edge region (i.e. b5, 6, 7, and 8A) of the electromagnetic spectrum; valuable for vegetation modeling. Sentinel-2 MSI captures spectral information at varying spatial resolutions of 10, 20, and 60 m with high (five days) temporal resolution. The sensor's spectral data was atmospherically and radiometrically corrected in the Quantum Geographic Information System (QGIS) software using Dark Object Subtraction (DOS) approach.

7.2.5 Sentinel-2 MSI Texture Metrics Derivation

Texture metrics were statistically derived from Sentinel-2 MSI using gray level co-occurrence matrix (GLCM) technique embedded in Environment for Visualizing Images (ENVI 3.4) software. The texture measures were calculated using a co-occurrence displacement vector (*d*) of 1:1 and the direction (θ) of 45° in GLCM algorithm. However, the effect of angle parameter is considered minimal on the coefficient of determination (Kayitakire et al. 2006; Lottering and Mutanga 2012); hence texture measures were computed at a constant angle. Although the GLCM technique offers many texture metrics, this study selected nine-texture metrics, which includes mean, variance, correlation, contrast, entropy, dissimilarity, homogeneity, and angular second moment (Table 7.1).

Table 7.1 Image-texture metrics derived from Sentinel-2 MSIs and their formulae.

Texture variables	Formulae	References
Mean (ME)	$\sum_{i=1}^{Ng}\sum_{j=1}^{Ng} i * P(i,j)$	Materka and Strzelecki (1998)
Variance (VAR)	$\sum_{i=1}^{Ng}\sum_{j=1}^{Ng} (i-\mu)^2 P(i,j)$	Materka and Strzelecki (1998)
Correlation (COR)	$\frac{\sum_i \sum_j (i,j) P(i,j) - \mu_x \mu_y}{\sigma_x \sigma_y}$	Kayitakire et al. (2006)
Contrast (CON)	$\sum_{i=1}^{Ng}\sum_{j=1}^{Ng} P(i,j)(i-j)^2$	Kayitakire et al. (2006)
Entropy (ENT)	$-\sum_{i=1}^{Ng}\sum_{j=1}^{Ng} P(i,j)\log(P(i,j))$	Yuan et al. (1991)
Dissimilarity (DI)	$\sum_{i=1}^{Ng}\sum_{j=1}^{Ng} P(i,j)\lvert i-j \rvert$	Puzicha et al. (1999)
Homogeneity (HO)	$\sum_{i=1}^{Ng}\sum_{j=1}^{Ng} \frac{1}{1+(i-j)^2} P(i,j)$	Tuttle et al. (2006)
Angular second moment (ASM)	$\sum_{i=1}^{Ng}\sum_{j=1}^{Ng} \{P(i,j)\}^2$	Yuan et al. (1991)

Note: $P(i, j)$ is the frequency in which two compared pixels occur (e.g. one with gray level i and the other with gray level j).

These texture measures are considered valuable in remote sensing image analysis (Baraldi and Panniggiani 1995; Rao et al. 2002). The content of these texture metrics depends on the sensor's spectral domain, spatial resolution, and characteristics of sensed features (i.e. shape, dimension, and spatial distribution). Image-texture measures are robust in detecting forests structure, leaf area index, biomass, density, and age (Sarker and Nichol 2011; Dube and Mutanga 2015). These ecosystem parameters are strongly related to remote sensing spatial and spectral information often used to estimate above ground biomass and carbon stock. Furthermore, one of the key parameters of the GLCM technique is the moving window size. Thus, in this study, the window size of 3×3 was established to compute texture metrics. In addition, texture values from individual band-metric were extracted using ArcMap (version 10.6).

7.2.6 Statistical Analysis

Above ground carbon stock estimates across different tree species was performed using random forest model implemented in RStudio (version 3.6.3.) software. Random forest is an ensemble machine learning algorithm developed to improve classification and regression trees approach by combining multiple sets of decision trees. In a regression, random forest builds individual trees by randomly selecting a subset of variables from the input dataset. The optimal random forest model is determined using three parameters: *ntree*, which is based on a large set of regression trees developed from a bootstrap sample of the observations (with default value set to 500); *mtry*, which is the number of predictor variables tested at each tree node (*mtry* takes the square-root of the total number of variables as a default value in the classification, whereas in the regression it divides all predictors with a default value of three); and *node-size*, which is the smallest value of terminal node sizes of the trees (with a default value of 1). Commonly, the out-of-bag error is used to determine optimal *ntree* and *mtry* values for the best prediction model. In this study, *ntree* was evaluated between 100 and 500 at an interval of 100, while the *mtry* values were evaluated between 1 and 12 using interval value of 1. The *node-size* was accepted at a default value of 1 and used throughout the prediction analysis. Furthermore, backward elimination methods embedded in random forest algorithm were used to determine a subset of predictor variables, which were ideal for the final prediction model. Backward elimination is necessary for removing predictor variables that suffer from multicollinearity, while remaining with optimal predictors, which provide better estimation performance. In this study, the prediction output of random forest model was integrated into the statistical analysis of variance (ANOVA) embedded in OriginPro (version 9.0) software for testing significant difference ($\alpha \leq 0.05$) in carbon stock variability between different tree species.

7.2.7 Model Accuracy Assessment

The prediction performance of random forest was tested based on the10-fold cross validation approach. The total dataset ($N = 82$) was initially separated into training ($n = 58$) and testing ($n = 26$) datasets. Coefficient of determination (R^2) and root mean square error (RMSE) values were used for rating the prediction performance of random forest model. The RMSE was performed based on the following equation:

$$RMSE = \sqrt{\frac{\sum_{i=1}^{n} \left(\hat{y}_i - y_i\right)^2}{n}}$$

where y_i represents observed values, $\hat{y}_i$ represents predicted values, and n is the number of data points.

7.3 Results

7.3.1 Carbon Stock of Reforested Tree Species

Figure 7.2 illustrates the statistical summary of measured carbon stock variability between five different species. The *Acacia caffra* have the highest mean carbon stock of 6.91 t/ha, compared to *E. caffra* and *S. cordatum*, which have the lowest carbon stock of 3.98 and 4.25 t/ha, respectively.

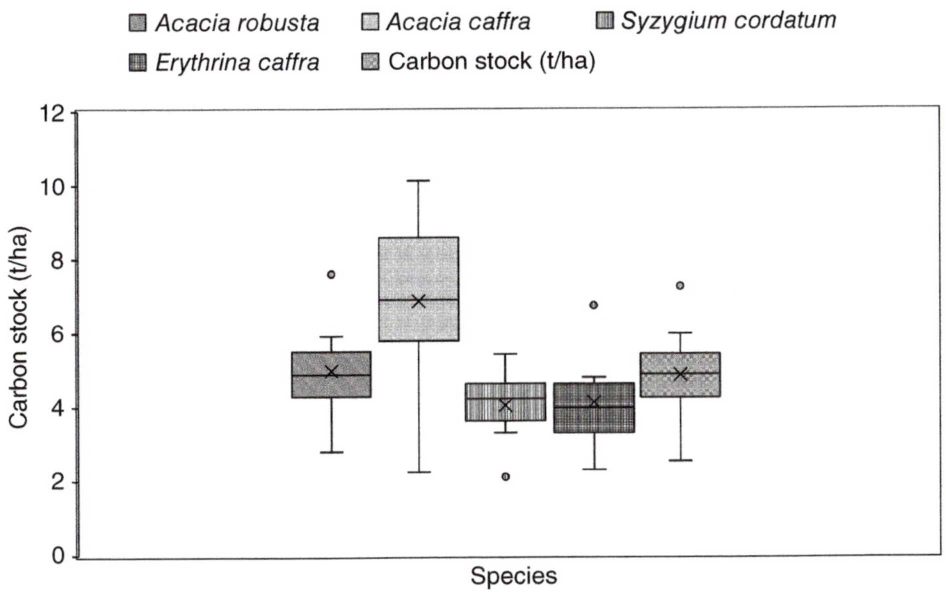

Figure 7.2 Descriptive statistics of the aboveground measured carbon stock variability between tree species.

Results in Table 7.2 show the optimal Sentinel-2 bands derived texture measures for estimating carbon stock across different forest species using cross-validation and backward elimination approaches. Optimal texture measures for *A. caffra* and *Acacia robusta* species carbon stock estimations were mean, variance, homogeneity, contrast, and correlation, while optimal texture measures for *Briderliar microntha* and *S. codartum* species were dissimilarity, second moment, mean, contrast, and correlation. Contrast, second moment, and correlation were selected as the most optimal texture predictors for *E. caffra*. The majority of these texture measures were commonly derived from the wavebands located between the visible (e.g. b2, 3, and 4), red-edge (e.g. b5, 6, and 7), and near infrared (e.g. b8 and 8A) regions of electromagnetic spectrum. The selected texture measures were used to build final models for predicting carbon stock variability across different species.

7.3.2 Prediction Performance of Carbon Stock Using Remotely Sensed Data and the Random Forest Model

Results in Table 7.3 show the performance of remote sensing derived texture measures and random forest model in estimating carbon stock in the reforested tree species. The consolidation of optimal texture measures in the model produced a reasonable prediction performance (R^2) of between 0.56 and 0.88 and error rate (RMSE) of 0.80–0.31 t/ha for estimating carbon stock among reforested tree species. The relationships between texture predicted carbon against field-measured carbon stock for individual species and combined datasets are shown in Figure 7.3a–f.

Table 7.2 Selection of optimal bands texture measures at the best moving window size (3 × 3) using random forest model for estimating carbon stock across different tree species.

Species	Image bands	Texture measures	R^2	RMSE (t/ha)
	b2 (Blue)	Mean	0.29	1.62
Acacia robusta	b8 (NIR)	Variance	0.43	1.03
	b6 (Red-edge)	Homogeneity	0.46	1.01
	b11 (SWIR)	Contrast	0.38	1.27
	b6 (Red-edge)	Variance, correlation	0.49	1.00
Acacia caffra	b7 (Red-edge)	Homogeneity	0.55	0.81
	b8A (Red-edge)	Correlation	0.51	0.87
	b4 (Red)	Contrast	0.31	1.45
Bridelia micrantha	b2 (Blue)	Dissimilarity	0.52	0.84
	b5 (Red-edge)	Second moment	0.45	1.01
	b3 (Green)	Correlation	0.40	1.16
	b11 (SWIR)	Mean	0.32	1.41
Syzygium cordatum	b7 (Red-edge)	Variance, dissimilarity	0.56	0.79
	b2 (Blue)	Dissimilarity	0.39	1.21
	b4 (Red)	Contrast, correlation	0.36	1.33
Erythrina caffra	b3 (Green)	Contrast	0.39	1.22
	b8A (Red-edge)	Second moment	0.45	1.01

Table 7.3 Performance of optimal texture measures in predicting carbon stock variability among different tree species.

Species	R^2	RMSE (t/ha)
Acacia robusta	0.58	0.78
Acacia caffra	0.73	0.48
Bridelia micrantha	0.67	0.59
Syzygium cordatum	0.59	0.73
Erythrina caffra	0.56	0.80
All data	0.88	0.31

7.3.3 Carbon Stock Estimates and Variability Between Reforested Tree Species

The estimates of aboveground carbon stock in reforested urban landscapes for all tree species studied are shown in Figure 7.4. The results of this study demonstrate that *A. robusta*, *B. micrantha*, and *A. cacia caffra* contain the largest proportion of carbon stock estimates, compared to *S. cordatum* and *E. caffra*, which contain lower carbon stock. Based on the statistical analysis of variance (ANOVA), the variation in total mean carbon stock

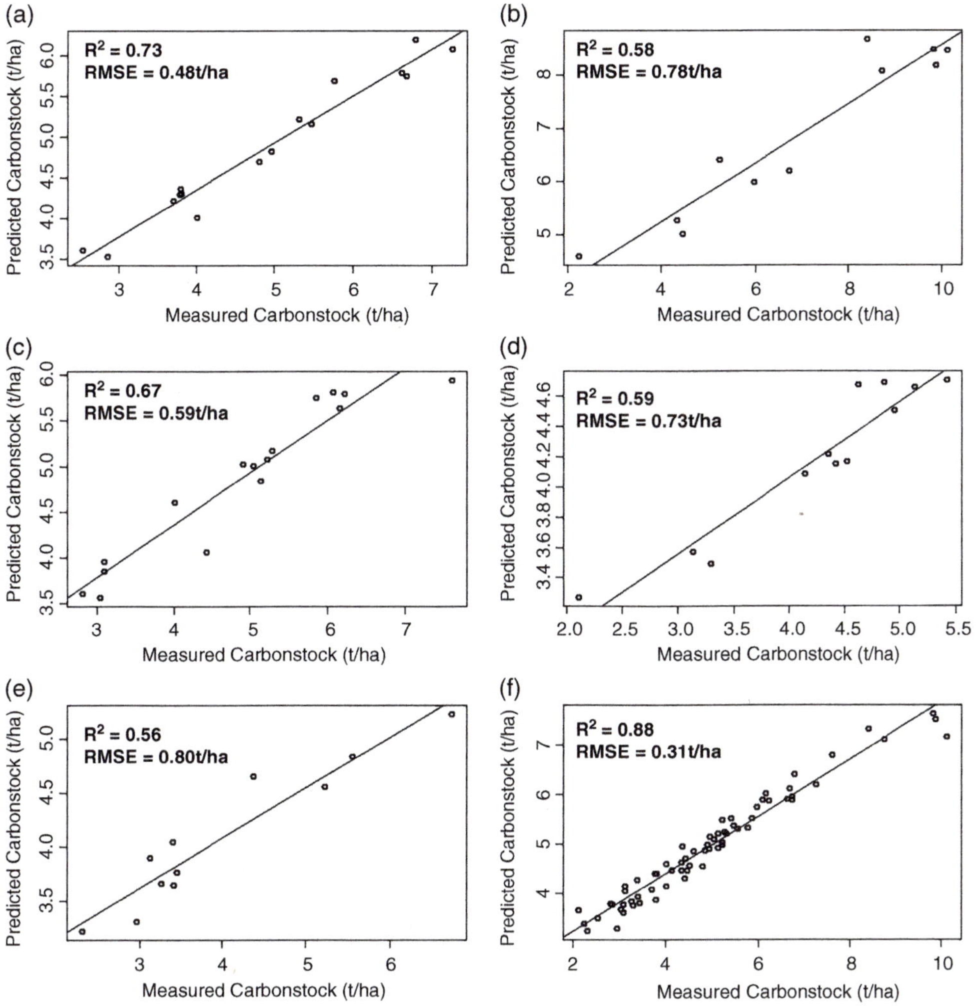

Figure 7.3 Relationship between predicted versus measured carbon stock of (a) *Acacia robusta*, (b) *Acacia caffra*, (c) *Bridelia micrantha*, (d) *Syzygium cordatum*, (e) *Erythrina caffra*, and (f) combined dataset for all species.

between reforested tree species is significantly different ($\alpha \leq 0.05$). The spatial distribution of above ground carbon stock estimates of different reforested species (i.e. *A. caffra*, *A. robusta*, *B. micrancha*, *S. cordatum*, and *E. cafffra*) is shown in Figure 7.5a–f.

7.4 Discussion

Reliable carbon inventory requires critical understanding of carbon stock variability among different tree species. Nonetheless, carbon stock variability for reforested trees has remained unknown, despite the need to understand the contribution and value of reforestation

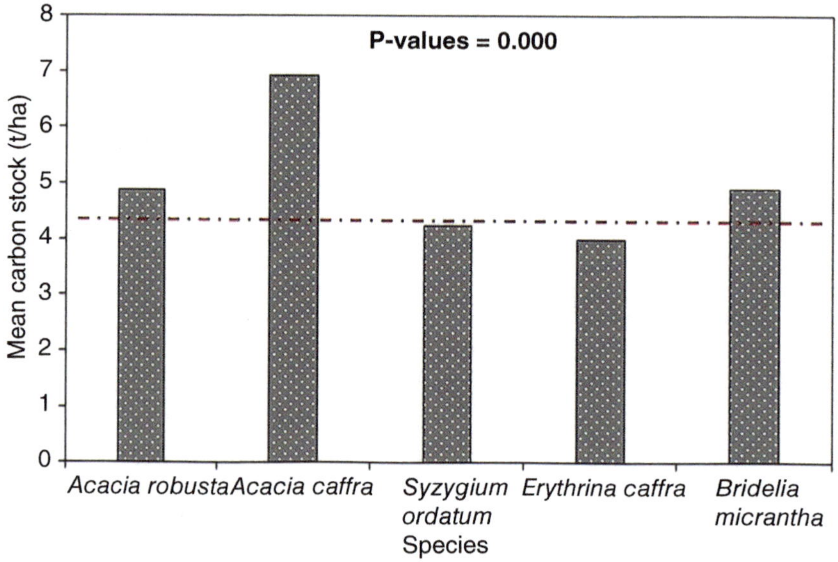

Figure 7.4 Total mean carbon stock variability between different reforested tree species. The red line separates higher and lower mean carbon stock derived from reforested urban landscape.

programs in the global carbon cycle and climate change mitigation. Thus, this study sought to quantify above ground carbon stock variability across different reforested tree species within an urban landscape using remote sensed imagery texture measures.

This study showed that texture measures derived from Sentinel-2 MSI could be used to explicitly determine above ground carbon stock variability among different species of reforested trees within an urban landscape. Based on the results, the consolidation of optimal texture measures (i.e. mean, variance, homogeneity, contrast, dissimilarity, angular second moment, and correlation) selected across different species using cross-validation and backward elimination techniques yielded R^2 ranging from 0.56 to 0.88 and RMSE from 0.31 to 0.80 t/ha in predicting carbon stock allocation. Most of these texture measures were computed from the red, red-edge, and NIR bands, which are highly sensitive to vegetation. The importance of these wavebands in detecting vegetation health and productivity is widely documented in literature (Gara et al. 2016; Sibanda et al. 2016). The results of this study are consistent with Pandit et al. (2020) and Dube and Mutanga (2015) who established that remote sensing derived texture measures are important for improving the estimation accuracy of biomass and carbon stock. The strength of remote sensing derived texture measures in estimating forest carbon stock variability is attributed to the ability to capture crucial information related to biophysical properties of vegetation such as canopy structure, leaf area index, biomass, and tree age (Pandit et al. 2020; Kupidura 2019). According to Lottering and Mutanga (2012), texture measures enhance the pixels' relationship, which increases estimation potential of forest structural attributes, including biomass and carbon stock. Furthermore, advanced texture image processing technique appropriately reduced atmospheric effects, sensor's view angle, and sun-glint (Fan et al. 2014; Mousivand et al. 2014), hence providing pure texture values, which are valuable for concise estimation of forest

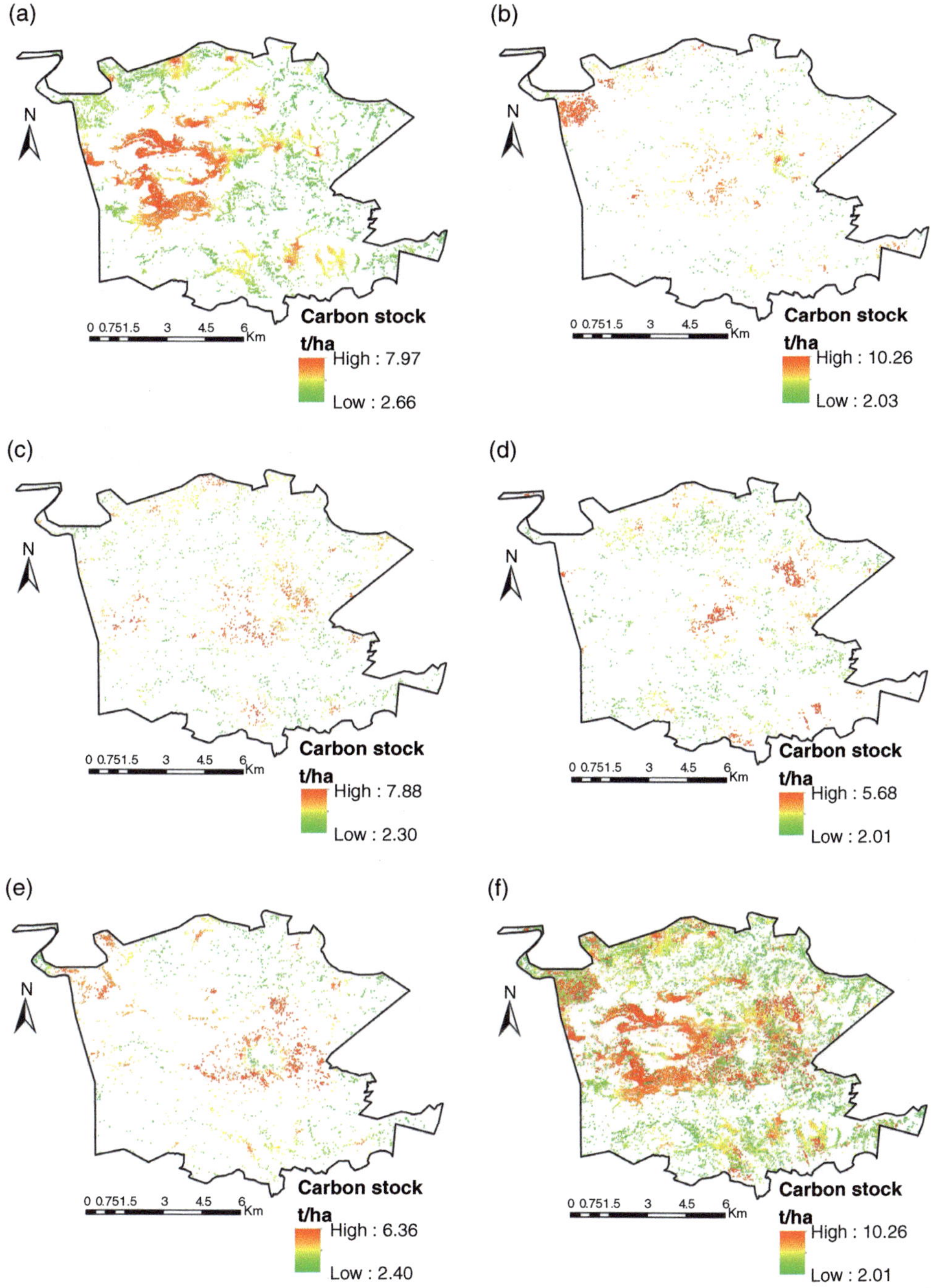

Figure 7.5 Spatial distribution of aboveground carbon stock of (a) *Acacia robusta*, (b) *Acacia caffra*, (c) *Bridelia micrantha*, (d) *Syzygium cordatum*, (e) *Erythrina caffra*, and (f) combined species.

carbon stock. In addition, the sensor's improvement in its push-broom scanning properties such as small signal-to-noise ratio and high radiometric resolution (12 bits) boosted the performance of texture measures derived from Sentinel-2 MSI for carbon stock estimation. Moreover, the adoption of gray level co-occurrence matric allowed for homogenization of 20 m bands to 10 m resolution through resampling, which increased the spatial information of Sentinel-2 MSI for invaluable estimation of carbon stock. Overall, the study presents a cost effective and useful option of remote sensing data processing and acquisition using texture measures technique, especially from Sentinel-2 MSI for monitoring and managing the spatial and temporal dynamics of carbon stock variability in reforested urban environment.

7.4.1 Carbon Stock Variability Between Reforested Tree Species

The results of this study demonstrated a significant variation ($\alpha \leq 0.05$) in aboveground carbon stock among reforested tree species, with *A. robusta*, *B. micrantha*, and *A. caffra* containing higher carbon stock (4.89–6.96 t/ha) than *S. cordatum* and *E. caffra* (3.97 and 4.26 t/ha). Similar results were reported by Dube et al. (2014) who established a significant variation in carbon stock between different tree species (i.e. *Eucalyptus grandis* and *dunii*, and *P. taeda*), while Chen et al. (2015) established a substantial variation in carbon stock between *Acacia crassicarpa*, *E. urophylla*, and *Castanopsis hystrix* species. Such variation in carbon stock may be explained by differences in biochemical (i.e. lignin, carotenoids) and biophysical (i.e. canopy structure, leaf stomata, and area) properties between the taxon's, which significantly control the vegetation's net photosynthetic process and carbon uptake. This is supported by Waring et al. (1998), who noted that unequal carbon uptake due to differing photosynthetic absorption parameters such as pigments and leaf optical properties result in uneven distribution of carbon stock among tree species of different genera. According to Chen et al. (2015) and Dube et al. (2014), variations in carbon stock across different forest species can be attributed to plant traits such as maximum net photosynthetic rate (P_{max}) per unit forest land area (P_{max} multiplied by leaf area index) within different forest species. These studies show that variation in biophysical properties among tree species facilitate differences in carbon stock. Moreover, this study demonstrates that reforestation of different indigenous tree species with great carbon sequestration capacity and storage reserve could promote climate resilient urban landscape and contribute to meeting the requirement of Kyoto Protocol and REDD+ to reduce atmospheric carbon emissions and mitigate climate change. In addition, the study provides basis for determining future optimal reforestation species mix to achieve both carbon assimilation imperatives and provision of other ecosystem goods and services.

7.5 Conclusion

The essence of this study was to quantify above ground carbon stock variability across different reforested tree species using remote sensing derived texture metrics dataset. The results in this study have shown that carbon stock varies significantly among reforested tree species groups. Across species, *A. robusta*, *B. micrantha*, and *A. caffra* have dominant carbon stock, whereas *E. caffra* and *S. cordatum* produced lower carbon stock. The adoption of texture measures derived from freely and readily available Sentinel-2 MSI proved instrumental in estimating carbon stock variability within reforested urban landscape. Results of this study provide knowledge on the contribution of reforestation initiative in the global carbon budget and climate change mitigation potential. Moreover, the study provides necessary information that could benefit forest managers, decision-makers, and policy-makers to establish well-informed management policies and plans for further improvement in sequestration capacity of reforestation programs through large-scale projects, especially in urban and peri-urban landscapes.

Acknowledgments

This study was funded by the DST-NRF SARChI Chair in Land Use Planning and Management at UKZN (Grant No. 84157). A special thanks to eThekwini municipality through the Durban Research Action Partnership (DRAP) for permitting this research project to be conducted on their protected reforestation site.

References

Adamu, B., Rasul, A., Whanda, S.J. et al. (2021). Evaluating the accuracy of spectral indices from Sentinel-2 data for estimating forest biomass in urban areas of the tropical savanna. *Remote Sensing Applications: Society and Environment* 22: 100484.

Baraldi, A. and Panniggiani, F. (1995). An investigation of the textural characteristics associated with gray level cooccurrence matrix statistical parameters. *IEEE Transactions on Geoscience and Remote Sensing* 33: 293–304.

Chen, Y., Liu, Z., Rao, X. et al. (2015). Carbon storage and allocation pattern in plant biomass among different forest plantation stands in Guangdong, China. *Forests* 6: 794–808.

Clar,k A. III, Saucier, J.R., and McNab, W.H. (1986). *Total-tree weight, stem weight, and volume tables for hardwood species in the southeast*. Georgia Forest Research Paper.

Curiel-Esparza, J., Gonzalez-Utrillas, N., Canto-Perello, J., and Martin-Utrillas, M. (2015). Integrating climate change criteria in reforestation projects using a hybrid decision-support system. *Environmental Research Letters* 10: 094022.

Delphin, S., Escobedo, F., Abd-Elrahman, A., and Cropper, W. (2016). Urbanization as a land use change driver of forest ecosystem services. *Land Use Policy* 54: 188–199.

Deo, R.K., Russell, M.B., Domke, G.M. et al. (2017). Evaluating site-specific and generic spatial models of aboveground forest biomass based on Landsat time-series and LiDAR strip samples in the eastern USA. *Remote Sensing* 9: 598.

Dube, T. and Mutanga, O. (2015). Quantifying the variability and allocation patterns of aboveground carbon stocks across plantation forest types, structural attributes and age in sub-tropical coastal region of KwaZulu Natal, South Africa using remote sensing. *Applied Geography* 64: 55–65.

Dube, T., Mutanga, O., Elhadi, A., and Ismail, R. (2014). Intra-and-inter species biomass prediction in a plantation forest: testing the utility of high spatial resolution spaceborne multispectral rapideye sensor and advanced machine learning algorithms. *Sensors* 14: 15348–15370.

Esch, T., Heldens, W., Hirner, A. et al. (2017). Breaking new ground in mapping human settlements from space–the global urban footprint. *ISPRS Journal of Photogrammetry and Remote Sensing* 134: 30–42.

Fan, Y., Koukal, T., and Weisberg, P.J. (2014). A sun–crown–sensor model and adapted C-correction logic for topographic correction of high resolution forest imagery. *ISPRS Journal of Photogrammetry and Remote Sensing* 96: 94–105.

Foody, G.M., Boyd, D.S., and Cutler, M.E. (2003). Predictive relations of tropical forest biomass from Landsat TM data and their transferability between regions. *Remote Sensing of Environment* 85: 463–474.

Gara, T.W., Murwira, A., and Ndaimani, H. (2016). Predicting forest carbon stocks from high resolution satellite data in dry forests of Zimbabwe: exploring the effect of the red-edge band in forest carbon stocks estimation. *Geocarto International* 31: 176–192.

Giardina, C.P. and Ryan, M.G. (2002). Total belowground carbon allocation in a fast-growing Eucalyptus plantation estimated using a carbon balance approach. *Ecosystems* 5: 487–499.

Guo, Z., Fang, J., Pan, Y., and Birdsey, R. (2010). Inventory-based estimates of forest biomass carbon stocks in China: a comparison of three methods. *Forest Ecology and Management* 259: 1225–1231.

Hickey, S., Callow, N., Phinn, S. et al. (2018). Spatial complexities in aboveground carbon stocks of a semi-arid mangrove community: a remote sensing height-biomass-carbon approach. *Estuarine, Coastal and Shelf Science* 200: 194–201.

Hu, Y., Su, Z., Li, W. et al. (2015). Influence of tree species composition and community structure on carbon density in a subtropical forest. *PLoS One* 10: e0136984.

Kayitakire, F., Hamel, C., and Defourny, P. (2006). Retrieving forest structure variables based on image texture analysis and IKONOS-2 imagery. *Remote Sensing of Environment* 102: 390–401.

Keenan, R.J., Reams, G.A., Achard, F. et al. (2015). Dynamics of global forest area: results from the FAO global forest resources assessment 2015. *Forest Ecology and Management* 352: 9–20.

Khan, K., Iqbal, J., Ali, A., and Khan, S. (2020). Assessment of Sentinel-2-derived vegetation indices for the estimation of above-ground biomass/carbon stock, temporal deforestation and carbon emissions estimation in the moist temperate forests of Pakistan. *Applied Ecology and Environmental Research* 18: 783–815.

Kupidura, P. (2019). The comparison of different methods of texture analysis for their efficacy for land use classification in satellite imagery. *Remote Sensing* 11: 1233.

Livesley, S., McPherson, E.G., and Calfapietra, C. (2016). The urban forest and ecosystem services: impacts on urban water, heat, and pollution cycles at the tree, street, and city scale. *Journal of Environmental Quality* 45: 119–124.

Lottering, R. and Mutanga, O. (2012). Estimating the road edge effect on adjacent *Eucalyptus grandis* forests in KwaZulu-Natal, South Africa, using texture measures and an artificial neural network. *Journal of Spatial Science* 57: 153–173.

Lu, D. (2006). The potential and challenge of remote sensing-based biomass estimation. *International Journal of Remote Sensing* 27: 1297–1328.

Mansourian, S. and Vallauri, D. (2005). *Forest Restoration in Landscapes: Beyond Planting Trees*. Springer Science & Business Media.

Materka, A. and Strzelecki, M. (1998). *Texture analysis methods—a review*. Technical University of Łodz, Institute of Electronics, COST B11 Report, Brussels, 10, 4968.

Mngadi, M., Odindi, J., Mutanga, O., and Sibanda, M. (2021). Estimating aboveground net primary productivity of reforested trees in an urban landscape using biophysical variables and remotely sensed data. *Science of the Total Environment* 802: 149958.

Mousivand, A., Menenti, M., Gorte, B., and Verhoef, W. (2014). Global sensitivity analysis of the spectral radiance of a soil–vegetation system. *Remote Sensing of Environment* 145: 131–144.

Odindi, J. and Mhangara, P. (2012). Green spaces trends in the city of Port Elizabeth from 1990 to 2000. In: *Using Remote Sensing*, 653–662.

Pandit, S., Tsuyuki, S., and Dube, T. (2020). Exploring the inclusion of Sentinel-2 MSI texture metrics in above-ground biomass estimation in the community forest of Nepal. *Geocarto International* 35: 1832–1849.

Payn, T., Carnus, J.-M., Freer-Smith, P. et al. (2015). Changes in planted forests and future global implications. *Forest Ecology and Management* 352: 57–67.

Puzicha, J., Buhmann, J.M., Rubner, Y., and Tomasi, C. (1999). *Empirical evaluation of dissimilarity measures for color and texture*. In: *Proceedings of the Seventh IEEE International Conference on Computer Vision*, 1165–1172. IEEE.

Raich, J.W., Clark, D.A., Schwendenmann, L., and Wood, T.E. (2014). Aboveground tree growth varies with belowground carbon allocation in a tropical rainforest environment. *PloS One* 9: e100275.

Rao, P.N., Sai, M.S., Sreenivas, K. et al. (2002). Textural analysis of IRS-1D panchromatic data for land cover classification. *International Journal of Remote Sensing* 23: 3327–3345.

Ribeiro, H.V., Rybski, D., and Kropp, J.P. (2019). Effects of changing population or density on urban carbon dioxide emissions. *Nature Communications* 10: 1–9.

Sarker, L.R. and Nichol, J.E. (2011). Improved forest biomass estimates using ALOS AVNIR-2 texture indices. *Remote Sensing of Environment* 115: 968–977.

Sibanda, M., Mutanga, O., and Rouget, M. (2016). Comparing the spectral settings of the new generation broad and narrow band sensors in estimating biomass of native grasses grown under different management practices. *GI Science & Remote Sensing* 53: 614–633.

Sithole, K., Odindi, J., and Mutanga, O. (2018). Assessing the utility of topographic variables in predicting structural complexity of tree stands in a reforested urban landscape. *Urban Forestry & Urban Greening* 31: 120–129.

Steininger, M. (2000). Satellite estimation of tropical secondary forest above-ground biomass: data from Brazil and Bolivia. *International Journal of Remote Sensing* 21: 1139–1157.

Tang, Y., Chen, A., and Zhao, S. (2016). Carbon storage and sequestration of urban street trees in Beijing, China. *Frontiers in Ecology and Evolution* 4: 53.

Toochi, E. (2018). Carbon sequestration: how much can forestry sequester CO_2. *Forestry Research and Engineering: International Journal* 2: 148–150.

Tuttle, E.M., Jensen, R.R., Formica, V.A., and Gonser, R.A. (2006). Using remote sensing image texture to study habitat use patterns: a case study using the polymorphic white-throated sparrow (*Zonotrichia albicollis*). *Global Ecology and Biogeography* 15: 349–357.

Van der Werf, A. and Nagel, O.W. (1996). Carbon allocation to shoots and roots in relation to nitrogen supply is mediated by cytokinins and sucrose: opinion. *Plant and Soil* 185: 21–32.

Waring, R., Landsberg, J., and Williams, M. (1998). Net primary production of forests: a constant fraction of gross primary production? *Tree Physiology* 18: 129–134.

Wei, Y., Li, M., Chen, H. et al. (2013). Variation in carbon storage and its distribution by stand age and forest type in boreal and temperate forests in northeastern China. *PloS One* 8: e72201.

Xu, G., Jiao, L., Zhao, S. et al. (2016). Examining the impacts of land use on air quality from a spatio-temporal perspective in Wuhan, China. *Atmosphere* 7: 62.

Yuan, X., King, D., and Vlcek, J. (1991). Sugar maple decline assessment based on spectral and textural analysis of multispectral aerial videography. *Remote Sensing of Environment* 37: 47–54.

Zhang, H., Guan, D., and Song, M. (2012). Biomass and carbon storage of eucalyptus and *Acacia* plantations in the pearl river delta, South China. *Forest Ecology and Management* 277: 90–97.

Zhu, C. and Yang, X. (1998). Study of remote sensing image texture analysis and classification using wavelet. *International Journal of Remote Sensing* 19 (16): 3197–3203.

8

Mapping Oil Palm Plantations in the Fringe of Sebangau National Park, Central Kalimantan, Indonesia

Bambang H. Trisasongko[*,1,2], *Dyah R. Panuju*[1,2], *Zulyan Afif*[3], *Yephi Haptadi*[3], *and Abdul R.M. Shariff*[4,5,6]

[1] *Geospatial Information and Technologies for the Integrative and Intelligent Agriculture (GITIIA), Center for Regional Systems Analysis, Planning and Development (CrestPent), Bogor Agricultural University, Jalan Pajajaran, Bogor, 16124, Indonesia*
[2] *Department of Soil Science and Land Resources, Bogor Agricultural University, Jalan Meranti, Dramaga, Bogor, 16680, Indonesia*
[3] *Regional Hazard Mitigation Agency (BPBD), Kotawaringin Timur, Jalan Jenderal Sudirman, Sampit, 74312, Indonesia*
[4] *Faculty of Engineering, Universiti Putra Malaysia, 43400 UPM, Serdang, Selangor Darul Ehsan, Malaysia*
[5] *Smart Farming Technology Research Centre, Faculty of Engineering, Universiti Putra Malaysia, Serdang, 43400, Malaysia*
[6] *Institute of Plantation Studies (IKP), Universiti Putra Malaysia, Serdang, 43400, Malaysia*

8.1 Introduction

Oil palm plantations have attracted controversies for decades. Economies of Southeast Asian countries, in particular Indonesia and Malaysia, are dependent on this sector. Similar conditions occur in African and Latin American countries, notably Nigeria and Colombia. In Indonesia, oil palm has been cultivated by small-scale farmers alongside with other short-term crops, while companies manage large-scale monoculture estates. Greater demand for palm oil further pushes parties to convert forests or other types of plantations (Panuju et al. 2019); therefore, they often ignite debates in environmental issues. For the sake of a better environment and for favorable land use planning, the extent of oil palm plantations needs to be frequently updated.

Considering their large coverage, the role of earth observing data in monitoring plantations is unrivaled. Satellite-borne systems provide rapid data access in observing remote areas, making the image data available within days. Identification of oil palm plantations using multispectral data has long been studied, benefitting from an extended history of optical datasets. Complexity of features, as a result of variations in plant vigor, stand age, nutrient, etc. has been worth mentioning in plantation mapping (Khorramnia et al. 2014; Trisasongko and Paull 2020). Miettinen et al. (2019) mapped oil palm plantations in Borneo, a world hot spot of oil palm plantations. They used visual interpretation as a benchmark to work with a decision tree method utilizing 14 variables derived from medium resolution, multi-source data. While the outcome was promising, a big discrepancy in the correct

*Corresponding author's email: trisasongko@apps.ipb.ac.id

0000-0002-1354-6542 (Bambang H. Trisasongko), 0000-0001-6078-3471 (Dyah R. Panuju), and 0000-0003-4626-4995 (Abdul R.M. Shariff)

Advances in Remote Sensing for Forest Monitoring, First Edition. Edited by Prem C. Pandey and Paul Arellano.

identification of the detected oil palm plantations warrants further research. In addition, spectral confusion may be rooted from inactive plantations. This issue was the core research employing SPOT imageries to identify abandoned oil palm plantations in Malaysia (Yusoff et al. 2017).

While research on utilizing multispectral sensors progresses, cloud coverage may limit large-scale implementation. Atmospheric attenuation is often observed in areas including Kalimantan (Borneo), Papua, Latin America, and Tropical African regions, which restricts information retrieval. While mapping forests and plantations could be done using multi-temporal approaches to minimize cloud effects, success rate may not be sufficient. As such, exploiting Synthetic Aperture Radar (SAR) has often been adopted (Trisasongko et al. 2006; Trisasongko 2009) or alternatively, exploiting data fusion (Razali et al. 2014; Najib et al. 2020). The latter, however, requires larger data feeds and consequently more powerful computing environments.

To some extent, SAR has been used to map plantations in the tropics. With the nature of abundant C-band data, sensors like Radarsat or European Envisat and Sentinel-1 have been the central data providers for mapping purposes. L-band dual polarization, although available for decades, is yet to be thoroughly investigated. Panuju et al. (2019) developed a method to characterize oil palm plantations from a change detection perspective in order to investigate changing plantation types in Indonesia. Multi-frequency SAR data fusion was also investigated to identify oil palm plantations in Malaysia (Oon et al. 2019). Previous efforts have been made by exploiting backscatter coefficient of various polarization states. Survey of the literature suggests that a limited amount of research has been focusing on the use of SLC data, especially on L-band SAR system. This SAR frequency has special distinctiveness, including a greater canopy penetration that might be useful to characterize canopy structure among woody vegetation. Additional SAR features, other than backscatter coefficients, could also be retrieved from SLC data, which could be useful for the identification of oil palm plantations.

This research is aimed to evaluate phase-preserved dual polarimetric SAR data for identifying oil palm plantations through conventional backscatter coefficients and polarimetric decomposition features. In addition, modern machine learners were assessed to better understand its capabilities and constraints in mapping oil palm estates.

8.2 Methodology

8.2.1 Test Site and Datasets

Research site was situated in the western part of the Sebangau National Park (SNP), which is one of the largest Indonesian national parks related to peat and wetlands dominated landscape. The altitude is < 50 m above mean sea level, with flat to undulating terrain. Administratively, the research area is under Katingan and Kotawaringin Timur regencies (Figure 8.1).

Primary dataset in this research was Phased Array-type L-band Synthetic Aperture Radar (PALSAR-2) sensor onboard Advanced Land Observing Satellite-2 (ALOS-2) imagery. It was delivered by Japan Aerospace Exploration Agency (JAXA) in SLC format; hence, it

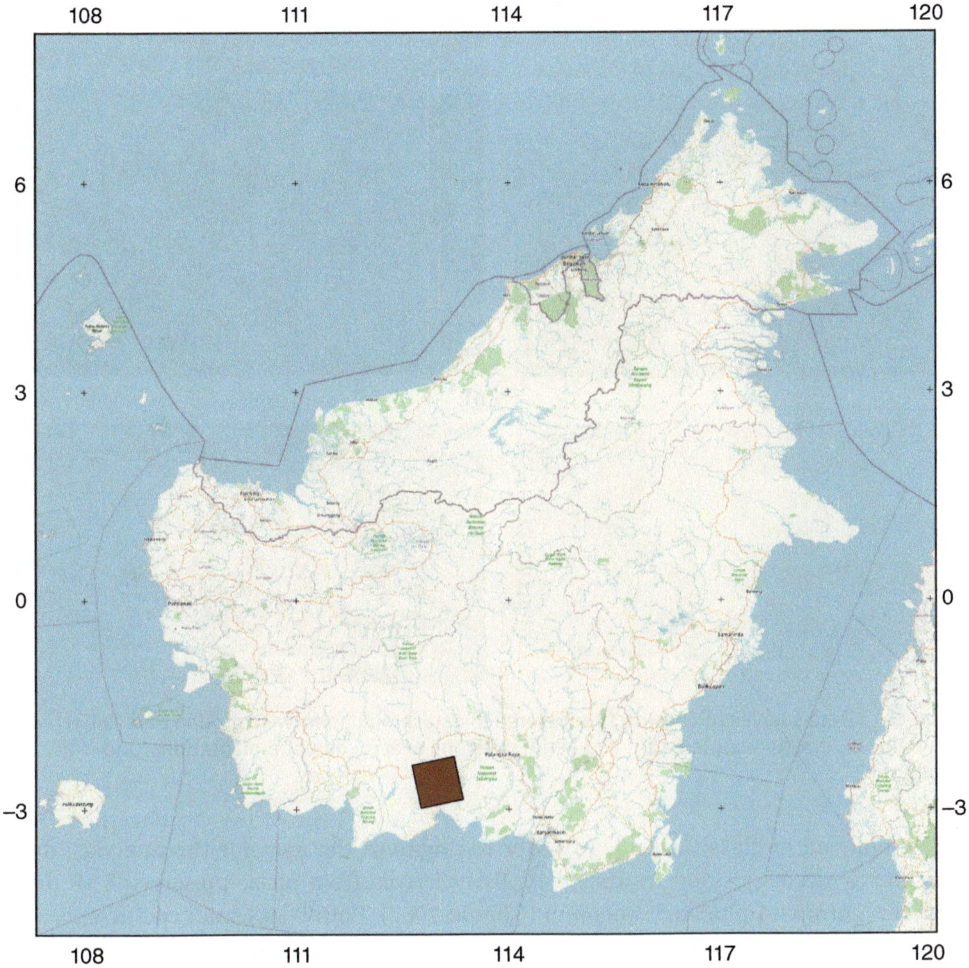

Figure 8.1 Site map and the coverage of PALSAR-2 data. Sampit city, the capital of Kotawaringin Timur regency, is roughly at the center of the image coverage.

allows derivation of dual-polarized Cloude-Pottier decomposition features. In this research, HH and HV backscatter coefficients served as the benchmark for the analysis, in comparison to a combined backscatter and decomposition features. Reference data were constructed employing ground information provided by BPBD field staffs, UAV imaging, and Google Earth data (Figure 8.2).

8.2.2 Data Processing and Analysis

Entire data pre-processing steps were completed with SNAP software, including calibrating backscatter coefficients and deriving polarimetric decomposition features. Backscatter coefficient is conventional consideration for image analyses related to SAR data. In this research, with the nature of dual polarization, only HH and HV backscatter coefficients

Figure 8.2 Data sources to construct the reference dataset: (a) point-wise ground data collected from field surveys; (b) Google Earth; and aerial missions over (c) oil palm plantations; and (d) suburbs.

were employed in classification. In order to augment the data for the analysis, dual polarimetric decomposition features were then derived from phase-preserved SAR data, using the Entropy-Alpha angle theorem (Cloude 2007). Both backscatter coefficients and polarimetric decomposition features were terrain corrected to compensate terrain-related errors. We employed Range-Doppler terrain correction algorithm, with the aid of Shuttle Radar Topography Mission (SRTM) 1 arc-second dataset. In order to simplify the analysis, both data were merged into a single dataset.

Commonly used R statistical software was employed in image analysis. Interface between SNAP data format and R was written in R code using "rgdal" and "raster" packages. Data analysis was mainly performed using packages controlled by "caret" package. All samples were divided into two sets, namely training and validation data with the ratio of 70:30. In order to minimize the bias, a 10-fold cross-validation strategy with three repeats was performed. In this research, two popular machine-learning algorithms, i.e. random forests (RF) and support vector machines (SVM), served as the benchmarks. For the latter, we employed radial basis function (RBF) kernel as it has been shown robust when compared with various applications (Trisasongko et al. 2017). In addition, we also investigated two recent, least explored machine learners. Regularized random forest (RRF), a variant of RF, was recently proposed (Deng and Runger 2013) and therefore has fewer coverage in the literature. Similar situation applies to the extreme gradient boosting (XGB) as introduced by Chen and Guestrin (2016). The performance of these four classifiers was investigated in either the overall accuracy or class-based individual accuracy to examine their ability of separating oil palm plantations among other land cover types.

8.3 Results and Discussion

8.3.1 Identifying Oil Palm

Figure 8.3 shows distribution of targeted classes according to polarization. Single bounce scattering mechanism has contributed to the detection of water, making separation between water and land straightforward using either HH or HV polarization. Due to its distinctive scattering mechanism, i.e. double-bounce scattering, urban landscape was well separated in HH polarization. This research found that overlapping

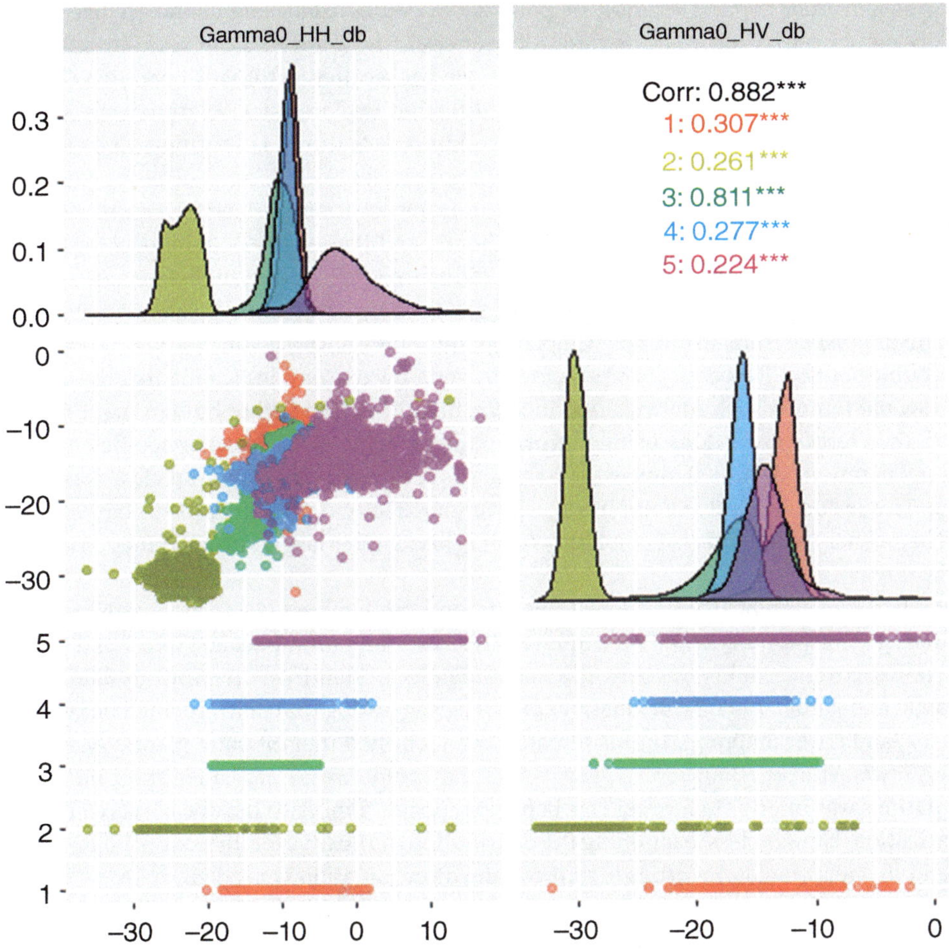

Figure 8.3 Histogram of backscatter coefficients grouped by land cover classes. Code 1: Forest; 2: Waterbodies; 3: Agriculture crops (non-oil plantations); 4: Oil palm plantations; 5: Built-up.

backscatter values, especially in HH polarization, were observed in discriminating natural objects. A previous report similarly discovered difficulties in separating oil palm plantations from either herbaceous or woody vegetation (Ballester-Berman and Rastoll-Gimenez 2021). Similar to the finding of this research, Najib et al. (2020) reported that L-band HH polarization was unable to fully distinguish oil palm plantations from natural forests.

While HH polarization of L-band SAR was less expedient in separating different vegetative covers, an option to use HV polarization would be advantageous. Cross polarization state reflects more interaction with tree structure; hence, HV backscatter coefficient hosts much information about volume scattering, the key scattering mechanism in vegetation. This research indicated that oil palm plantations could be distinguished from natural forests, although Figure 8.3 also suggests that backscatter distribution of plantation was severely overlapped with the one of the agricultural crops (non-oil palm plantations).

Mapping complex landscapes is usually beyond the capability of dual polarimetric SAR data, due to their limited amount of polarization states. Data augmentation has been widely employed to add more features. These could be sourced from textural analysis (Najib et al. 2020); Panuju et al. (2019), or simple backscatter arithmetic operation (Najib et al. 2020). When using SLC data, one could further explore dual polarimetric entropy alpha theorem, which is able to produce three additional features, i.e. entropy, alpha angle, and anisotropy.

Distribution of entropy, alpha angle, and anisotropy according to land cover types is presented in Figure 8.4. Since its introduction (Cloude and Pottier 1997), the model has been used to understand wave-object interactions in diverse applications, including agriculture (Li et al. 2011) and plantations (Trisasongko et al. 2017). Manufactured objects are usually well distinguished from an entropy perspective, since they have a distinctive wave property, i.e. double bounce. This entity is represented by very low entropy, indicating that the object has low randomness in scattering. Identification of built-ups and other human-made structures, therefore, can be done using entropy, or, in the case of backscatter coefficient, the HH polarization. We noted, however, that the range of entropy in built-up class was very large. This is due to fact that Sampit is a relatively small city, with numerous woody vegetation in greenery areas and observable city roads. This justifies large entropy values observed in the city.

Figure 8.4 also indicates that forest has a strong randomness, associated with volume scattering. This property could be employed as a discrimination between forest and other vegetative covers. Ability to help the separation was identified in previous literature (Cloude and Pottier 1997; Trisasongko et al. 2017), suggesting that entropy could be beneficial for land cover mapping. In this research, we identified that oil palm plantations have much lower entropy than forests, suggesting that randomness of scatterers in oil palm plantations was inferior to forests. This behavior appears similar to the one observed in HV polarization. Nonetheless, examining the distribution of those vegetative covers, one could defend that additional properties to improve the discrimination is needed.

Alpha angle represents dominant scattering mechanisms. Single bounce scattering mechanisms have low alpha angle values, while high values embody double bounce scattering. Objects having volume scattering mechanism, like forest in this case, are indicated in mid values of alpha angle. Figure 8.4 depicts that separation among

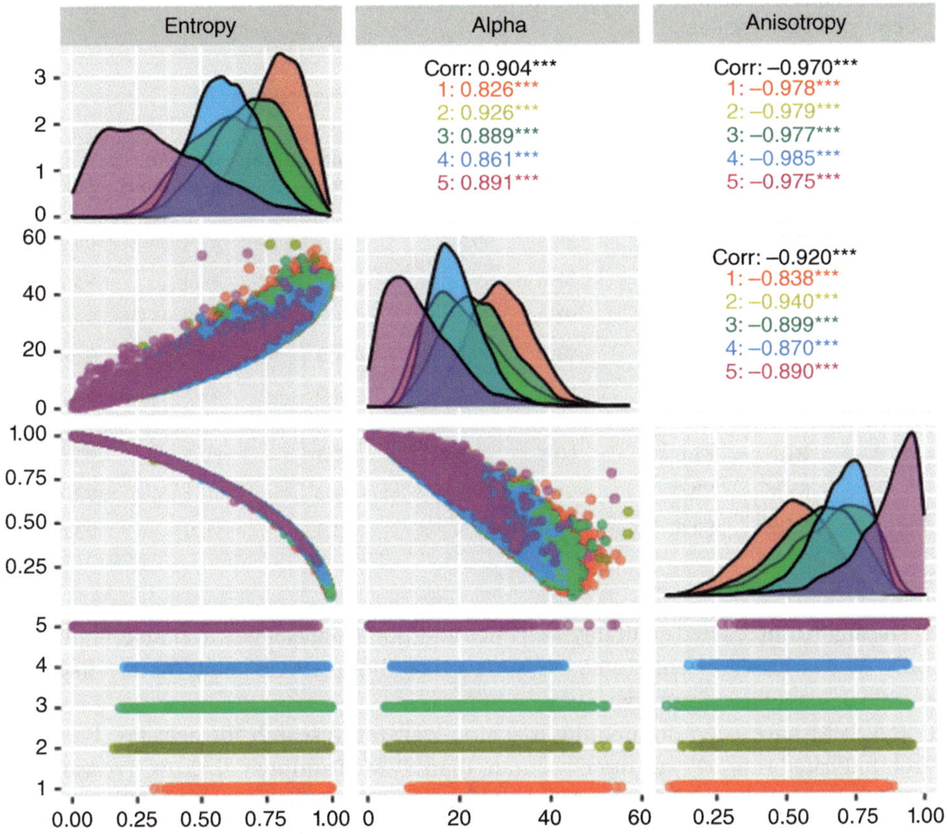

Figure 8.4 Distribution of entropy-alpha angle dual polarimetric decomposition features. Codes 1: Forest; 2: Waterbodies; 3: Agricultural crops (non-oil palm plantations); 4: Oil palm plantations; 5: Built-up.

vegetative covers were partially successful, suggesting that alpha angle was a good candidate for data augmentation of backscatter coefficients.

8.3.2 Classification Accuracies

Summary of the overall accuracy measured from independent samples is presented in Table 8.1. In general, augmented backscatter data were preferable to improve the accuracy. This was consistent with a previous work (Panuju et al. 2019), suggesting that limited predictors would bind the implementation of SAR-based mapping. Nonetheless, Table 8.1 suggests that improvement was trivial and should carefully be assessed. Additional data generated more impact when RF or its variation was applied, which was consistent to preceding findings (Trisasongko et al. 2017; Panuju et al. 2019). In this research, data augmentation had no influence in SVM with RBF, indicating its robustness in separation if small predictors are available. We found that extreme gradient boosting behaved differently. When fed by a limited amount of predictors, it produced overall accuracy equivalent to

Table 8.1 Accuracies obtained from conventional backscatter coefficients and their augmentation using decomposition features.

			Class-wise accuracy (%)				
Data	**Method**	**Overall accuracy (%)**	**Forest**	**Waterbodies**	**Agricultural crops**	**Oil palm plantations**	**Built-up**
Backscatter	RF	79.8	88.5	99.8	62.7	87.9	93.2
	RRF	79.9	88.6	99.8	62.7	88.0	93.2
	SVM	82.5	91.4	99.8	61.5	90.0	93.1
	XGB	82.3	91.4	99.8	61.9	89.8	93.4
Backscatter + Features	RF	82.0	91.1	99.8	62.6	89.5	93.3
	RRF	81.8	90.9	99.8	62.5	89.3	93.4
	SVM	82.5	91.6	99.8	60.2	90.2	92.7
	XGB	82.6	91.6	99.8	62.3	90.0	93.6

SVM. The algorithm reacted equitably well once additional dataset was available, similar to RF, only with a lesser extent. Table 8.1 shows that these general patterns are likewise observable in class-wise accuracy.

RF and SVM have been dominating machine-learning research for decades. Previous research indicated that RF was more accurate than SVM (Panuju et al. 2019; Trisasongko et al. 2020). Nonetheless, in many cases, SVM surpassed the outcome of RF (Trisasongko et al. 2017). In this research, SVM consistently outweighed RF in terms of overall accuracy using either backscatter or its combination with polarimetric decomposition. Varying results suggest that access to both models, or possibly to more advanced models, would be invaluable to seek the best performing approach.

This research discovered that agricultural crop (non-oil palm plantation) was the class with lowest accuracy. This class was largely confused with oil palm plantation class, possibly due to existing oil palm plantations dominated by young oil palm trees. In the early phase of plantation development, canopy cover of young oil palm trees is yet to be fully developed. Hence, the majority of radar returns would be controlled by surface cover, which is similar to the condition in agricultural fields. Confusion between agricultural crops and oil palm was also reported in Sarzynski et al. (2020), which yielded class-wise classification error about 15%.

While the outcome of this research was promising, challenges in accurate mapping of oil palm plantations remain open for future improvement. A report of Darmawan et al. (2020) concluded that ALOS PALSAR produced an oil palm distribution map with overall accuracy about 91% with four distinctive targets; i.e. forest, oil palm plantations, water bodies, and urban structure. Accuracy obtained by the current study was in between the outcome of aforementioned report and Sarzynski et al. (2020). The latter reported an overall accuracy of about 74% in separating oil palm plantations, forest, forest-shrub mixture, mixed bare soil and crops, built-up, and water bodies using dual-pol PALSAR data, augmented with texture features. Further problems appear when employing a shorter SAR wavelength. Santos and Messina (2008), for instance, obtained only 52% accuracy when using Radarsat C-band data. They indicated that enhancement using texture analysis remained unsuccessful to distinguish palms, diseased palms, grass, and secondary forest.

Figure 8.5 depicts predicted map from the XGB fed by combined backscatter and polarimetric features. Sampit city, the capital of Kotawaringin Timur regency, is clearly shown in the center of the figure. In general, plantations in this regency have been more developed than Katingan in the east. Nonetheless, the figure shows extensive activities along the Katingan River in the east, which borders the national park. This suggests that combining conventional backscatter coefficients and polarimetric features could serve as the data input for monitoring oil palm plantations. The outcome of the monitoring could be improved using suitable machine learning models.

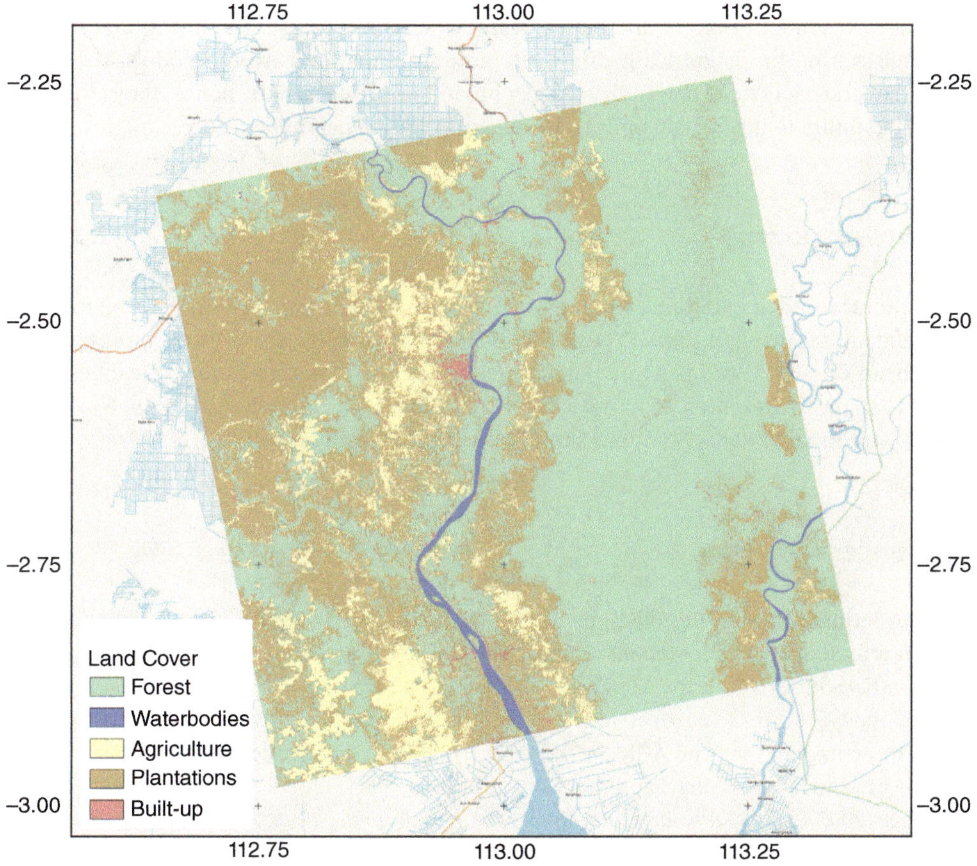

Figure 8.5 Classified map, overlaid on top of Open Street Map (OSM) data. Mapped plantations situated in the western part of the region suit to dense road networks of OSM, which are usually associated with large-scale oil palm estates.

8.4 Conclusion

Due to its utmost importance, forest fringe should be regularly monitored to avoid damaging impacts to natural conservation. In Kalimantan and many tropical regions, challenges have been related to the expansion of plantations, including oil palm. Problems of information retrievals over persistent cloud covered regions have partially been resolved using multi-temporal, multispectral-based data. With the increasing issues of forest fringe, plantations maps, including oil palm, need to be frequently updated.

This research explored the use of polarimetric decomposition as an aid to improve the detection of oil palm plantations. While backscatter coefficients have long been proven useful, additional interpretation capabilities from polarimetric decomposition were indispensable. Entropy and alpha angle, resulting from Cloude dual-polarimetric decomposition theorem helped to distinguish oil palm plantations from forest cover type. It was shown that polarimetric decomposition features could also be beneficial for improving the accuracy of land cover classification, especially in the identification of oil palm plantations. The study also found that modern machine learning models such as extreme gradient boosting provided a marginally better framework to increase the overall accuracy. Nonetheless, due to insignificant difference among machine learning models, it is suggested that end users need to have access to a broad machine learners, hence, they allow for greater flexibility to investigate options in selecting the robust model.

Acknowledgments

The authors would like to thank the Japan Aerospace Exploration Agency (JAXA) for providing data grant through RA6-3004 and RA6-3040 projects. We also express our gratitude to the head of BPBD for his support and encouragement. We thank our colleagues who helped to shape the research, to name a few, Diar Shiddiq, La Ode Syamsul Iman, Nur Etika Karyati, and Rizqi I'anatus Sholihah from Bogor Agricultural University.

References

Ballester-Berman, J.D. and Rastoll-Gimenez, M. (2021). Sensitivity analysis of Sentinel-1 backscatter to oil palm plantations at pluriannual scale: a case study in Gabon, Africa. *Remote Sensing* 13 (11): https://doi.org/10.3390/rs13112075.

Chen, T. and Guestrin, C. (2016). *XGBoost: A scalable tree boosting system*. Paper presented at the Proceedings of the 22nd ACM SIGKDD International Conference on Knowledge Discovery and Data Mining, San Francisco, California, USA.

Cloude, S. (2007). *The dual polarisation Entropy/Alpha decomposition: A PALSAR case study*. Paper presented at the Polinsar, Frascati, Italy.

Cloude, S.R. and Pottier, E. (1997). An entropy based classification scheme for land applications of polarimetric SAR. *IEEE Transactions on Geoscience and Remote Sensing* 35 (1): 68–78. https://doi.org/10.1109/36.551935.

Darmawan, S., Carolita, I., and Ananta, E. (2020). *Identification of Oil Palm Plantation on Multiscatter and Resolution of SAR Data Using Variety of Classifications Algorithm (Case Study: Asahan District, North Sumatera Province)*. Paper presented at the 5th International Conferences of Indonesian Society for Remote Sensing, ICOIRS 2019 and and Indonesian Society for Remote Sensing Congress.

Deng, H. and Runger, G. (2013). Gene selection with guided regularized random forest. *Pattern Recognition* 46 (12): 3483–3489. https://doi.org/10.1016/j.patcog.2013.05.018.

Khorramnia, K., Khot, L.R., Shariff, A.R.B.M. et al. (2014). Oil palm leaf nutrient estimation by optical sensing techniques. *Transactions of the ASABE* 57 (4): 1267–1277. https://doi.org/10.13031/trans.57.10142.

Li, K., Zhang, F., Shao, Y. et al. (2011). Polarization signature analysis of paddy rice in southern China. *Canadian Journal of Remote Sensing* 37 (1): 122–135. https://doi.org/10.5589/m11-018.

Miettinen, J., Gaveau, D.L.A., and Liew, S.C. (2019). Comparison of visual and automated oil palm mapping in Borneo. *International Journal of Remote Sensing* 40 (21): 8174–8185. https://doi.org/10.1080/01431161.2018.1479799.

Najib, M., Ezzaty, N., Kanniah, K.D. et al. (2020). Synergy of active and passive remote sensing data for effective mapping of oil palm plantation in Malaysia. *Forests* 11 (8): 858. https://doi.org/10.3390/f11080858.

Oon, A., Ngo, K.D., Azhar, R. et al. (2019). Assessment of ALOS-2 PALSAR-2L-band and Sentinel-1 C-band SAR backscatter for discriminating between large-scale oil palm plantations and smallholdings on tropical peatlands. *Remote Sensing Applications: Society and Environment* 13: 183–190. https://doi.org/10.1016/j.rsase.2018.11.002.

Panuju, D.R., Paull, D.J., and Trisasongko, B.H. (2019). Combining binary and post-classification change analysis of augmented ALOS backscatter for identifying subtle land cover changes. *Remote Sensing* 11 (1): https://doi.org/10.3390/rs11010100.

Razali, S.M., Marin, A., Nuruddin, A.A. et al. (2014). Capability of integrated MODIS imagery and ALOS for oil palm, rubber and forest areas mapping in tropical forest regions. *Sensors (Switzerland)* 14 (5): 8259–8282. https://doi.org/10.3390/s140508259.

Santos, C. and Messina, J.P. (2008). Multi-sensor data fusion for modeling African palm in the Ecuadorian Amazon. *Photogrammetric Engineering and Remote Sensing* 74 (6): 711–723. https://doi.org/10.14358/PERS.74.6.711.

Sarzynski, T., Giam, X., Carrasco, L., and Huay Lee, J.S. (2020). Combining radar and optical imagery to map oil palm plantations in Sumatra, Indonesia, using the Google Earth engine. *Remote Sensing* 12 (7): https://doi.org/10.3390/rs12071220.

Trisasongko, B.H. (2009). Tropical mangrove mapping using fully-polarimetric radar data. *ITB Journal of Science* 41 A (2): 98–109.

Trisasongko, B. and Paull, D. (2020). A review of remote sensing applications in tropical forestry with a particular emphasis in the plantation sector. *Geocarto International* 35 (3): 317–339. https://doi.org/10.1080/10106049.2018.1516245.

Trisasongko, B., Lees, B., and Paull, D. (2006). Polarimetric classification in a tailings deposition area at the Timika mine site, Indonesia. *Mine Water and the Environment* 25 (4): 246–250. https://doi.org/10.1007/s10230-006-0140-4.

Trisasongko, B.H., Panuju, D.R., Paull, D.J. et al. (2017). Comparing six pixel-wise classifiers for tropical rural land cover mapping using four forms of fully polarimetric SAR data.

International Journal of Remote Sensing 38 (11): 3274–3293. https://doi.org/10.1080/01431161.2017.1292072.

Trisasongko, B.H., Panuju, D.R., and Iman, L.O.S. (2020). Estimating tree age of rubber stands using spaceborne L-band synthetic aperture radar. *IOP Conference Series: Earth and Environmental Science* 418: 012079. https://doi.org/10.1088/1755-1315/418/1/012079.

Yusoff, N.M., Muharam, F.M., and Khairunniza-Bejo, S. (2017). Towards the use of remote-sensing data for monitoring of abandoned oil palm lands in Malaysia: a semi-automatic approach. *International Journal of Remote Sensing* 38 (2): 432–449. https://doi.org/10.1080/01431161.2016.1266111.

Section III

Remote Sensing Technology for Forest Fire Monitoring

9

Forest Fire Susceptibility Mapping by Integrating Remote Sensing and Machine Learning Algorithms

*Shahfahad, Swapan Talukdar, Tanmoy Das, Mohd Waseem Naikoo, Mohd Rihan, and Atiqur Rahman**

Department of Geography, Faculty of Natural Sciences, Jamia Millia Islamia, New Delhi, 110025, India

9.1 Introduction

In India, fire has indeed been recognized as one of the largest threats driving forest loss. The number of forest fires is comparatively higher in the northern Himalayan regions of India, particularly in Himachal Pradesh, Uttarakhand, and northeastern states such as Sikkim, Assam, Nagaland, Arunachal Pradesh, etc. (Joseph et al. 2009; Sannigrahi et al. 2020). Each year, about 20 000 forest fires occur in India, 90% of which occur in only in the Himalayan states (Sahana and Ganaie 2017). Although human activities have been recognized as the primary cause of forest fires in these states, natural phenomena such as severe dryness and lightning activities leading up to the monsoon season also contribute to forest fires (Bahuguna and Singh 2002; Joseph et al. 2009). Among human activities, the slash and burn farming (*Jhum cultivation*) in northeastern states, burning of forest for the growth of grass for pasture lands, use of fire for the collection of forest products such as gum, honey, and other products, and careless throwing of matchsticks or cigarettes by tourists and locals are some of the main causes of forest fires in India (Satendra 2014). The forest fires put several adverse impacts on the ecology, society, economy, and climate of these states.

Uttarakhand and Himachal Pradesh are among the most vulnerable states to forest fires in India. According to an estimate by the State Forest Department of India, each year thousands of hectares of forest are burned by the fires in these states (Babu et al. 2016). Forest fires are common in Uttarakhand and Himachal Pradesh from March to June, resulting in the loss of rich wood resources, endangered species, and substantial harm to the forest ecosystem and environment of the region (Negi 2019; Fulé et al. 2021). Thus, proper forest fire management is required to reduce forest fires in these states. To accomplish this, the researchers have used geospatial approaches to examine various aspects of forest fires, including their origins and effects on the society, environment, economy, and its zonation

*Corresponding author's email: arahman2@jmi.ac.in

0000-0001-5868-1062 (Shahfahad), 0000-0001-6680-9791 (Swapan Talukdar) 0000-0003-3160-6649 (Tanmoy Das), 0000-0001-5779-9325 (Mohd Waseem Naikoo), 0000-0002-2533-8433 (Mohd Rihan), and 0000-0002-9001-5059 (Atiqur Rahman)

Advances in Remote Sensing for Forest Monitoring, First Edition. Edited by Prem C. Pandey and Paul Arellano.

and management (Singh et al. 2016; Sahana and Ganaie 2017; Fulé et al. 2021). More recently, the researchers have applied artificial intelligence and machine learning algorithm (MLA) for the mapping and monitoring of forest fire susceptibility in different parts of the world (Coughlan et al. 2021; Mohajane et al. 2021; Tuyen et al. 2021). The MLAs have been extensively utilized for the study of urban environments and their management (Ali et al. 2020), air quality and aerosols (Liang et al. 2020), land use land cover (LULC) mapping (Talukdar et al. 2020a), wetland inundation mapping (Talukdar et al. 2021), ecosystem services (Willcock et al. 2018), and flood and landslide susceptibility mapping (Talukdar et al. 2020b; Mallick et al. 2021). In their study, Xie and Peng (2019) applied ensemble-learning models to forecast forest fire occurrences in Portugal. Furthermore, Wang et al. (2021) applied game theory algorithm coupled with MLAs for the identification of drivers of forest fires in United States (Wang et al. 2021). However, no systematic research has been conducted in any area of India for forest fire mapping and foresting using MLA.

In Uttarakhand and Himachal Pradesh, the forest fires lead to the large-scale degradation of natural forest cover (Sharma et al. 2012; Kanga et al. 2013; Kanga et al. 2014). More than 50% of forest cover in Uttarakhand and Himachal Pradesh is prone to frequent forest fires mostly due to the anthropogenic activities. Despite large-scale destruction of the forest due to fires each year, not much research has been done regarding forest fire mapping and its susceptibility modeling in these states (Babu et al. 2016; Sahana and Ganaie 2017; Fulé et al. 2021; Tomar et al. 2021), most of which were only studied at the district or regional level. The studies on forest fire susceptibility, factors, and consequences on society and the environment are required for the Uttarakhand and Himachal Pradesh at various scale, for both forest fire mitigation and effective management of forest and forest resources. Therefore, this study is intended analyze the forest fire susceptibility in the Uttarakhand and Himachal Pradesh states of India by coupling geospatial techniques and MLA.

9.2 Study Area

The Indian states of Himachal Pradesh and Uttarakhand (refer to Figure 9.1) are located between latitudes of 28°42'N to 33°15'N and longitudes of 75°35'E to 81°01'E. The regions cover an area of 109 156 km^2, bordered to the north and south by Jammu, Kashmir, and Uttar Pradesh, to the west by Punjab, and southwest by Haryana. The study region has a diverse terrain and is situated on the southern slope of the Himalayas, with climate and vegetation ranging from glaciers at the highest altitudes to subtropical woods at lower elevations. The highest altitudes are ice/rock coated. Grasslands and scrublands are found between 3000 and 5000 m below them. Just below the tree line, temperate coniferous woods develop. The transition to temperate broadleaf forests between 3000- and 2600-m elevation covers a belt from 2600 to 1500 m (Singh and Singh 1987). The Himalayan subtropical pine forests blend with wet deciduous woods in the flat lowland regions, including Uttar Pradesh, Haryana, and Punjab below 1500 m' elevation. The summer season begins in March/April and lasts until May/June, whereas the monsoon season begins in June and lasts until August (Sahana and Ganaie 2017).

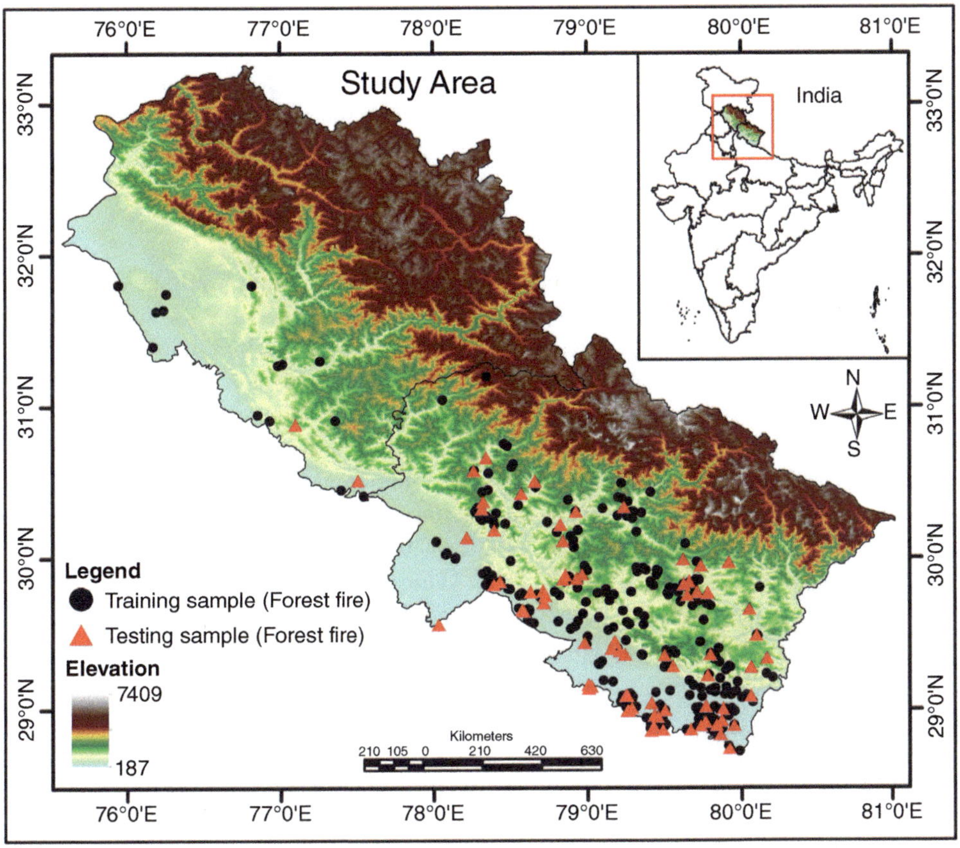

Figure 9.1 Locational and forest fire inventory map of the study area.

9.3 Materials and Methods

9.3.1 Materials

In this study, numerous satellite image-based characteristics were been employed to estimate forest fire susceptibility. First, we used Earth DATA, NASA's Global Fire Atlas with Individual Fire Characteristics, to obtain forest fire locations from 2003 to 2016 (https://firms.modaps.eosdis.nasa.gov/map). Then, using all of the data from 2003 to 2016, we generated forest fire areas in the research area, and constructed training and testing datasets based on the information we gathered. After that, we gathered multiple forest conditioning characteristics from various sources in order to predict forest fire susceptibility. The Indian Meteorological Department provided meteorological data, and Giovanni (a NASA application that allows you to visualize selected geophysical parameters) provided the wind speed and evapotranspiration data (https://giovanni.gsfc.nasa.gov/giovanni). Earth DATA NASA provided the LULC data (https://earthdata.nasa.gov/learn/user-resources/lcluc-information). The International Soil Reference and Information Centre (ISRIC) provided the aridity data.

The elevation data was taken from the Diva GIS (https://www.diva-gis.org) and the slope, aspect, and curvature maps were prepared from this information. The spatial resolution and projection of all of the characteristics were different, so in order to produce consistent spatial resolution (100 m) and projection, we first rectified the projection and then used the resampling approach.

9.3.2 Forest Fire Inventory

Any predictive model requires a large inventory; otherwise, modeling would be impossible. As a result, we used MLA to create a forest fire inventory for estimating susceptibility of forest fire in this work. To create a forest fire inventory, we used yearly fire data from Earth DATA NASA from 2003 to 2016. After that, we analyzed the data and created a forest fire map for the study area. Because using data from a certain period to produce inventory may result in biased findings, experts propose that forest fire inventories be created using temporal data, so we followed their advice. The temporal datasets were then used for preparing the training as well as testing datasets, which include binary data, such as forest and non-forest fires. We gathered non-forest fire data from regions where forest fires had never occurred, just as we did with the forest fire data. Then, using the training data, we retrieved data from factors that would be utilized to estimate forest fire vulnerability. Additionally, the testing data was utilized to validate the susceptibility models.

9.3.3 Ignition Factors for Forest Fire Modeling

The quality of factors used for prediction of fires significantly affects the forecast quality. Because wildfires have a wide range of causes, the ignition parameters used to forecast forest fires also have a wide range. The wildfire size is determined by its origins and the circumstances that contributed to its spread and escalation. Elevation, slope, aspect, plan curvature, LULC, wind speed, yearly temperature, annual rainfall, aridity index, and potential evapotranspiration were among the 10 ignition parameters used in this study to forecast wildfires (Table 9.1 and Figure 9.2).

From SRTM DEM images, four ignition variables were calculated; slope, aspect, elevation, and plan curvature. The slope of forest fire-prone terrain is vital in wildfire advancement because fire spreads faster up the slope than down the slope, and fire rises on steeper slopes owing to closeness to ground surfaces (Bar et al. 2020). Elevation directly influences temperature, rainfall, moisture, and wind, as well as having an indirect impact on vegetation and fuel moisture; hence, it was included in the research (Tiwari et al. 2021). The aspect was included in the analysis because of its link to the amount of solar radiation that a particular region receives (Sachdeva et al. 2018). South-facing surfaces receive more sunshine resulting in greater temperatures, lower humidity levels, and stronger winds. The curvature symbolizes topographic morphology, and was taken into account in the research because of its function in fire propagation (Gupta et al. 2018). Thus, the data of annual rainfall, average temperature, aridity, wind speeds, and evapotranspiration are included in this analysis. The activities done on different LULC types also influences forest fire (Fulé et al. 2021), and hence, the LULC data has been added as the ignition factor.

Table 9.1 Details of the ignition factors used for forest fire modeling.

S. No.	Data layer	Data format	Description	Source
1	Elevation	Raster	SRTM DEM	https://www.diva-gis.org/gdata
2	Slope	Raster		
3	Aspect	Raster		
4	Curvature	Raster		
5	LULC	Raster	Landsat (OLI)	https://earthdata.nasa.gov/learn/user-resources/lcluc-information
6	Temperature	Tabular	Air temperature	India Meteorological Department (IMD)
7	Wind Speed	Tabular	Average Wind Speed	https://giovanni.gsfc.nasa.gov/giovanni
8	Rainfall	Tabular	Average rainfall	India Meteorological Department (IMD)
9	Evapotranspiration	Tabular	Average evapotranspiration	https://giovanni.gsfc.nasa.gov/giovanni
10	Aridity	Raster	Aridity	https://giovanni.gsfc.nasa.gov/giovanni

9.3.4 Method for the Multicollinearity Analysis

When variables in a regression analysis are related (either moderately or highly), the consequence is an inaccurate analysis, which is known as 'multicollinearity.' To quantify the multicollinearity, the variance inflation factors (VIF) and tolerances, variance decomposition proportion, and conditional index, are used (Yoo and Cho 2019). The VIF and tolerances are the most often utilized approach in natural science applications. The VIF is the variance ratio in a multi-attribute model to the variance in a single-attribute model (James et al. 2013). The ratio specifies how the variance of an estimated regression coefficient increases with an increase in collinearity; the more significant the variance, the higher the multicollinearity (Allison 1999). In general, if VIF > 5 or tolerance $= 0.2$, this circumstance indicates that the independent variable in the regression analysis is multi-collinear (Chen et al. 2017).

9.3.5 Methods for Forest Fire Susceptibility Modeling

9.3.5.1 Random Forest (RF)

A significant number of decision trees were created by trial and error to carry out this model into action. Then, all the trees were concatenated (Nachappa et al. 2020). The random forest package was implemented in R studio to develop the RF model. RF starts with forming a decision tree once predictor factors and target variables have been defined. This tree does not utilize all the existing data for the tree, instead of relying on a bootstrap sample that only comprises 66% of the real data, a method known as 'bagging' (Breiman 2001).

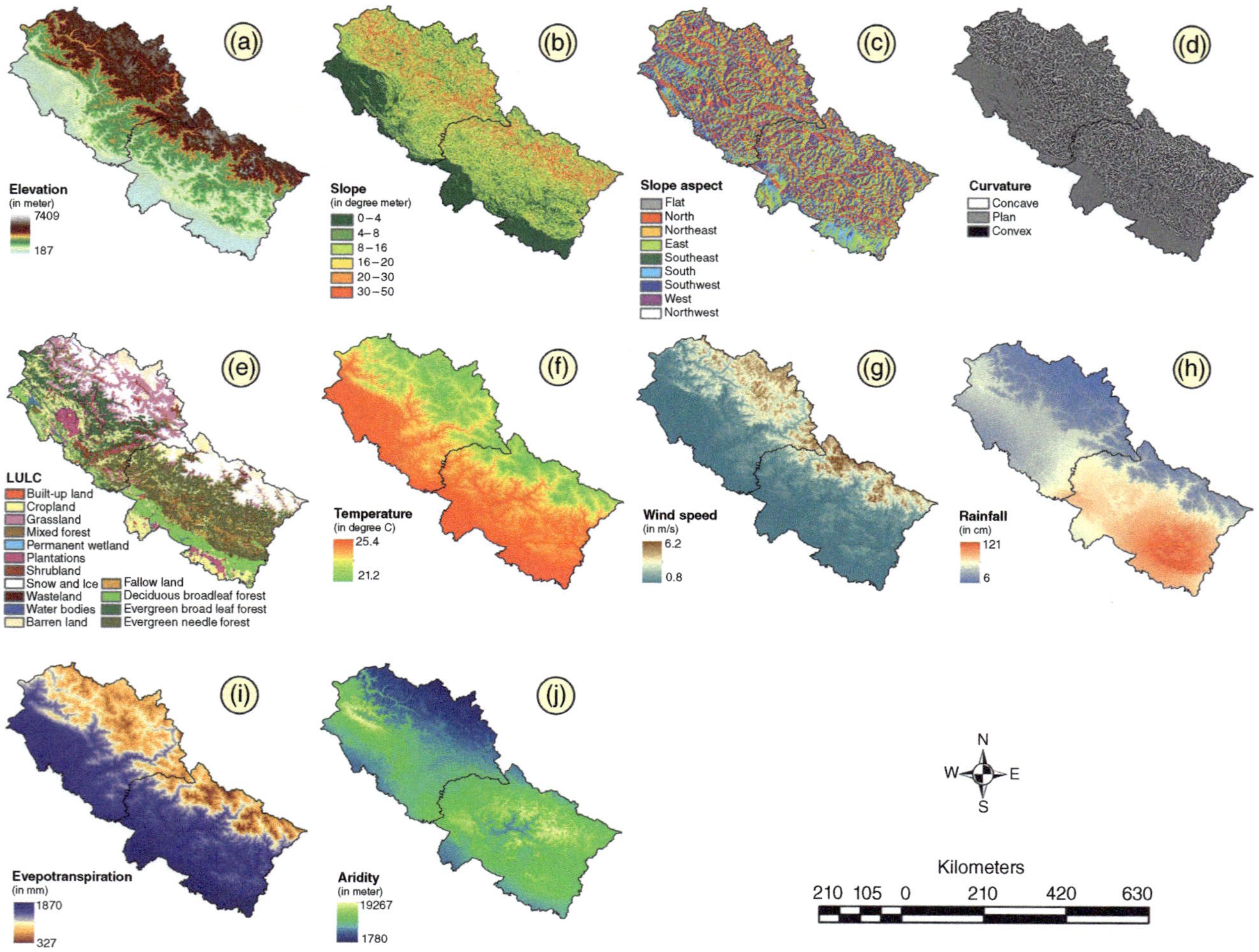

Figure 9.2 Forest fire ignition parameters of the study area; (a) elevation, (b) slope, (c) aspect, (d) curvature, (LULC), (f) temperature, (g) wind speed, (h) rainfall, (i) evapotranspiration, and (j) aridity.

9.3.5.2 Support Vector Machine (SVM)

The SVM is a general MLA for estimating the function of the problem that uses a collection of linear indicator functions (Vapnik 2013). The kernel mathematical function is used in SVM algorithm for the analysis of data. SVM converts the original data into a high-dimensional feature space and creates a hyperplane from it (Kavzoglu et al. 2014). The optimal linear hyperplane separates the original input space, and the kernel function is used to transform data into two classes; non-fire susceptibility, and fire susceptibility 0 and 1. Suitable kernel functions such as polynomial kernel (PL), sigmoid kernel (SIG), radial basis function (RBF), and linear kernel (LN) are required for the SVM's capabilities. According to multiple research outcomes, RBF outperforms other kernels in a forest fire prediction setting (Bui et al. 2012; Hong et al. 2018). As a result, the RBF function was employed in this research.

9.3.5.3 Logistic Regression (LR)

As a multivariate statistical analysis approach, LR has been widely used in natural hazards research (Ali et al. 2020; Chowdhuri et al. 2020). For most statistical procedures, particular assumptions must be defined; however, LR does not need pre-analysis assumptions (Arabameri et al. 2019). Furthermore, LR analysis supports all types of conditioning factors (continuous, nominal, categorical, etc.). LR assesses a dependent variable of a single event (forest) and detects the relationship between that event and other independent variables (conditioning factors) that may influence the event's likelihood. A regression connection between the forest fire inventory and forest fire conditioning variables was investigated using LR binary analysis (Sevgen et al. 2019). In other words, the forest fire inventory is a binary component that indicates whether or not a forest fire exists. The outcome of this approach is the forest fire probability index, which is displayed on an S-shaped curve in a range of 0–1.

9.3.6 Validation of the Models

Testing datasets covering 20% of forest fire and non-forest fire sites were employed in this study for model validation and performance assessment, respectively. The accuracy of the forest fire susceptibility map was then assessed using the receiver operating characteristics (ROC) curve (Hong et al. 2018). The area represents the projected value of the system under the ROC curve, which describes the system's capacity to correctly estimate occurrences (forest fire incidence) and absence of the event (absence of forest fire). The model's accuracy was calculated quantitatively using the ROC curve. The model with the most significant area under the curve (AUC) is optimal, and AUC values range from 1 to 0.5. If the model cannot forecast a forest fire occurrence better than the likely perspective—if the AUC is less than 0.5—the model is not usable for prediction. The AUC and estimate classifications are as follows; 0.5–0.6 poor, 0.6–0.7 moderate, 0.7–0.8 good, 0.8–0.9 very good, and 0.9–1 exceptional (Sevgen et al. 2019).

9.4 Results

9.4.1 Multicollinearity Analysis

The VIF and tolerances were utilized to examine the multicollinearity amid independent variables in this study. A VIF score of > 10, and a tolerance value of 0.2 showed a multicollinearity problem. In the present study, 10 parameters have been generated for fire susceptibility modeling. After applying the collinearity technique, it has been found that all parameters have a VIF score < 5 (refer to Table 9.2). Results show that wind scored the maximum VIF values of 2.75, followed by evapotranspiration, while the aspect and curvature have the lowest VIF values of 1.03 and 1.094, respectively. Consequently, multicollinearity among independent variables is not a worry. Hence, in the current study, 10 ignition-conditioning variables were considered while creating the forest fire susceptibility mapping.

9.4.2 Forest Fire Susceptibility Modeling

The forest fire susceptibility predicted by LR, RF, SVM, and all models' ensembled are shown using four maps (Figure 9.3). Using the natural breaks categorization approach, we divided the risk of fire occurrence into five susceptibility categories for each map, ranging from extremely low to very high. A quantitative study of the maps indicated that the LR model's very high susceptibility level covered 11927.53 km^2 of the land area, followed by high (22084 km^2) and moderate (15089.91 km^2), while the low and very low fire susceptibility levels covered 36538.63 km^2 and 23197.47 km^2 of the land area, respectively (Table 9.3). The RF model predicted the study area into 21316 km^2, 15830.91 km^2, 15092 km^2, 29153 km^2, and 27423.07 km^2 for very high, high, moderate, low, and very low

Table 9.2 Computation of multicollinearity analysis among the ignition conditioning variables.

		Collinearity statistics	
S. No.	**Parameters**	**Tolerance**	**VIF**
1	Aridity	0.646	1.547
2	Aspect	0.971	1.030
3	Curvature	0.914	1.094
4	Elevation	0.521	1.921
5	Evapotranspiration	0.416	2.273
6	LULC	0.766	1.306
7	Rainfall	0.365	2.742
8	Slope	0.591	1.691
9	Temp	0.759	1.241
10	Wind	0.362	2.765

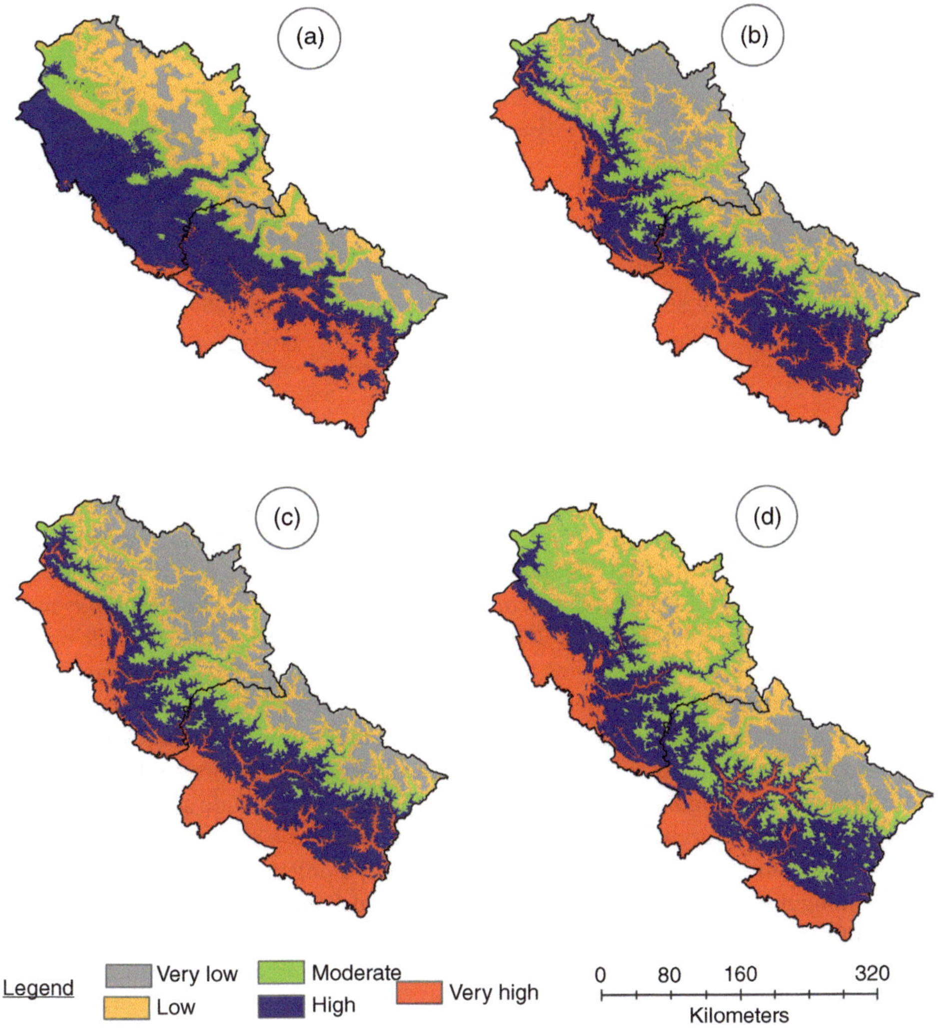

Figure 9.3 Forest fire susceptibility mapping using (a) LR, (b) RF, (c) SVM, and (d) ensemble models.

forest fire susceptibility zones, respectively. Additionally, we found 21 296.9 km^2 and 15 774.9 km^2 areas to be very high and high forest fire susceptibility zones, followed by 14 877.6 km^2 as moderate level of susceptibility, while 28 232.12 km^2 and 28 056.2 km^2 areas were predicted as very low and low level of forest fire susceptibility. The last ensemble model predicted 10 188 km^2, 23 532 km^2, 24 847 km^2, 30 531 km^2, and 19 740 km^2 for very high, high, moderate, low, and very low forest fire susceptibility zones, respectively. We may infer that those anthropogenic pressures turned 28.21% of the study area into a sensitive zone to future fires based on the cantered moderate and high susceptibility classes surrounding regions where human stakes are abundant (i.e. roadways and residential areas).

Table 9.3 Calculated area under the forest fire susceptible zones.

Fire susceptibility zones	Area (in km^2)			
	LR	RF	SVM	Ensemble
Very high	11927.53	21316.57	21296.9	10188
High	22084.15	15830.91	15774.9	23532
Moderate	15089.91	15092.29	14877.6	24847
Low	36538.63	29153.86	28232.12	30531
Very low	23197.47	27423.07	28056.2	19740

9.4.3 Validation Analysis of the Models

The observed and binormal ROC AUC validated the produced forest fire susceptibility maps utilizing the collected GPS points. The resulting AUC under the respective ROC (experimental and binormal) is 0.85 and 0.92 for LR, 0.887 and 0.96 for RF, 0.89 and 0.97 for SVM, and 0.923 and 0.989 for ensemble model (Figure 9.4). The ensemble model appeared as the best model based on ROC curves, followed by SVM, RF, and LR models. However, all models had AUC values above 0.85 and were found to be satisfactory and robust models. Since the AUC values of all the models were greater than 0.85, it can be considered very good results rather than many other study's findings. Therefore, the proposed models can produce robust results, which can be applied in other studies.

Despite these encouraging findings, susceptibility maps may include several unknowns. The edge effect, which occurs when fires that started outside the research area moved into the study region and changed the degree of susceptibility along the borders, is one potential cause of ambiguity. We could not examine the possible edge impact in our research because of a lack of data on flames spreading from other locations to both states. Researchers may use edge detection techniques to guarantee that edge-affected areas are detected and eliminated when the necessary information is available.

9.5 Discussion

Uncontrolled forest fires threaten the forest's overall ecosystem and biodiversity as well as the forest's riches, human life, and regional climate. It is critical to identify forest fire susceptibility zones, where urbanization is increasing close to forest areas, to understand the causes and implications of forest fires. This study used one statistical (LR) and three MLA i.e. RF, SVM, and ensemble forest-fire susceptibility mapping techniques with various spatial and non-spatial causative criteria in Uttarakhand and Himachal Pradesh, two of the most forest-rich and bio-diversified states in India. Using LR, RF, SVM, and ensemble approaches to construct a forest fire susceptibility model and analyze the causative factors helps understand the causes and implications of dramatically rising forest fire events in the area.

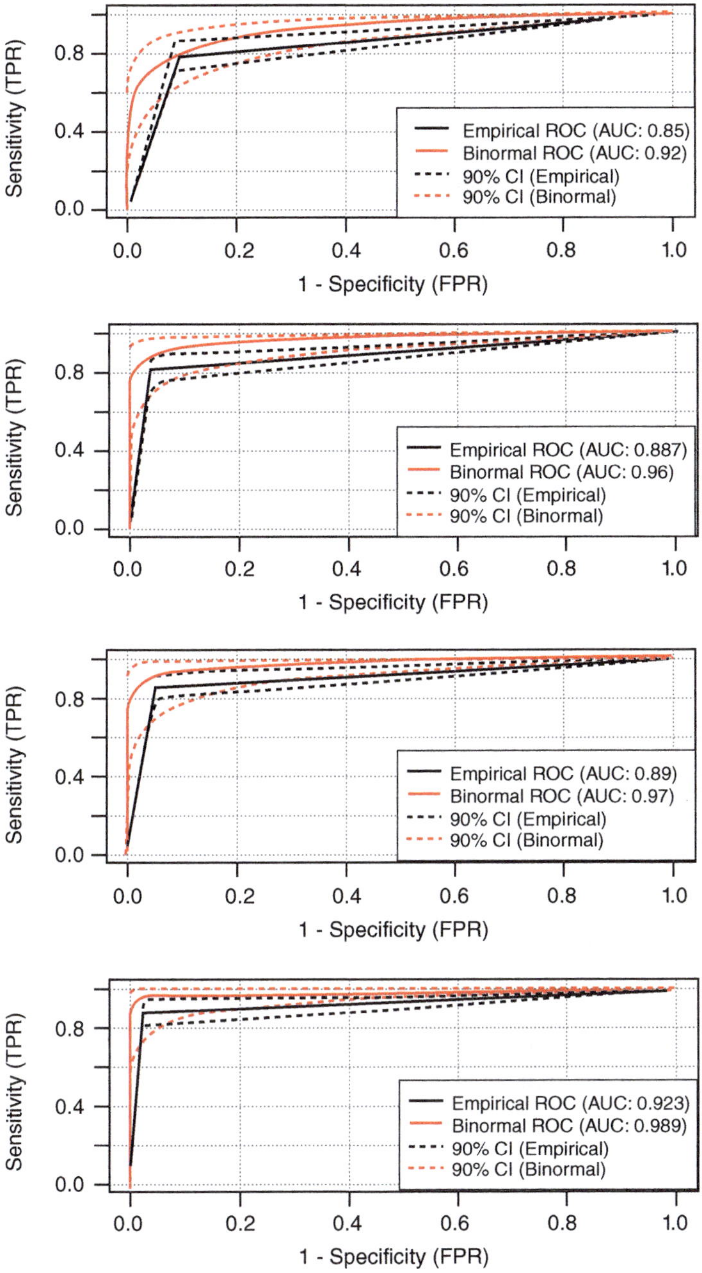

Figure 9.4 Validation of the forest fire susceptibility models; (a) LR, (b) RF, (c) SVM, and (d) ensemble models.

The analysis of results shows that the majority of degraded and deciduous forest cover from both protected (Jim Corbett and Rajaji National Park) and open forest zones (Kot, Pokhara, Rikhnikhal, Yamkeshwar block from Garhwal and Lansdowne Forest region and others) fell into very high and high fire susceptibility zones. The Rajaji National Park is classified as very high susceptibility, while a large portion of Jim Corbett National Park is classified as high susceptibility. Roy and Porwal (2004) and Tiwari et al. (2021), respectively, have reported on the severity of forest fire events in Rajaji and Jim Corbett National Parks, and these studies of forest fire susceptibility groups are compatible with the findings of the current investigation. Conversely, Dwarichal, Garhwal, and other blocks in the Garhwal and Lansdowne Forest divisions have wet forest ranges and are classified as moderate and medium fire susceptibility. These findings are comparable to those of other researchers who performed various locational investigations in Pauri Garhwal (Kumari et al. 2017; Pandey and Ghosh 2018; Tiwari et al. 2021).

The findings were validated by showing that, among the three mapping strategies used in this work, the ensemble outperformed the LR, RF, and SVM models in detecting possible forest fire susceptibility classes. However, after examining their AUC values, the performance of RF and SVM was outstanding, while the performance of LR was reasonable for this research. Compared to the other three models, the ensemble methodology exhibited better prediction accuracy.

9.6 Conclusion

With the changing climatic condition and global warming, the frequency of forest fires has increased around the globe. In India, the maximum increase in forest fire events has been quite noticeable in the Himalayan states. Therefore, this study was intended to analyze the forest fire susceptibility of two Himalayan states of India, i.e. Himachal Pradesh and Uttarakhand, using MLA and remote sensing techniques. For this, three MLAs, i.e. RF, SVM, LR, and an ensemble model were applied for the forest fire susceptibility modeling using 10 ignition factors. The results showed that although the area under forest fire susceptibility zones varies in the models used, overall more than 50% area of the Uttarakhand and Himachal Pradesh lies under the very high, high, and moderate fire susceptibility zones in all the models. Notably, the very high and high forest fire susceptibility zones lie in the lower and foothill regions of the both states (i.e. southern parts) while most of the high hill regions comes under the low and very low forest fire susceptible zones. This is because the high hill regions have either very low or no forest cover, while the low hilly regions have very dense forested land. Further, the anthropogenic influence on forests is very high in the low hill regions. Large-scale tourist movement due to the presence of several hill stations, as well as dense population in this part of Himalayas, also has significant impact on the forest fires. Further, the models applied showed satisfactory results (AUC $>$ 0.85) but the ensemble model showed the best result in this study. Despite encouraging findings, the study had some limitations, such as validation using edge detection and a less comprehensive field survey due to lack of time and resources. Therefore, for the further researchers, it is suggested to employ edge detection techniques to guarantee that edge-affected areas are detected and eliminated when the necessary information is available. Further, it is suggested for future researchers to incorporate some additional parameters such as vegetation health, distance from settlement, and transport network into the forest fire modeling to obtain a more precise result. The findings of this can be beneficial for the state forest departments of Uttarakhand and Himachal Pradesh for the fire mitigation and management of forest ecosystem and biodiversity.

Acknowledgements

The authors are thankful to the EarthData NASA, Giovanni, International Soil Reference and Information Centre (ISRIC), and India Meteorological Department (IMD) for providing necessary data for the study. The authors are also thankful to the Department of Geography, Jamia Millia Islamia, New Delhi for providing the required equipment and laboratory facility for this research.

References

Ali, S.A., Parvin, F., Pham, Q.B. et al. (2020). GIS-based comparative assessment of flood susceptibility mapping using hybrid multi-criteria decision-making approach, naïve Bayes tree, bivariate statistics and logistic regression: a case of Topl'a basin, Slovakia. *Ecological Indicators* 117: 106620.

Allison, P.D. (1999). *Multiple Regression: A Primer*. Newbury Park: Pine Forge Press.

Arabameri, A., Pradhan, B., Rezaei, K. et al. (2019). GIS-based landslide susceptibility mapping using numerical risk factor bivariate model and its ensemble with linear multivariate regression and boosted regression tree algorithms. *Journal of Mountain Science* 16 (3): 595–618.

Babu, S, K.V., Roy, A., and Prasad, P.R. (2016). Forest fire risk modeling in Uttarakhand Himalaya using TERRA satellite datasets. *European Journal of Remote Sensing* 49 (1): 381–395.

Bahuguna, V.K. and Singh, S. (2002). Fire situation in India. *International Forest Fire News* 26: 23–27.

Bar, S., Parida, B.R., and Pandey, A.C. (2020). Landsat-8 and Sentinel-2 based forest fire burn area mapping using machine learning algorithms on GEE cloud platform over Uttarakhand, Western Himalaya. *Remote Sensing Applications: Society and Environment* 18: 100324.

Breiman, L. (2001). Random forests. *Machine Learning* 45 (1): 5–32.

Bui, D.T., Pradhan, B., Lofman, O., and Revhaug, I. (2012). Landslide susceptibility assessment in Vietnam using support vector machines, decision tree, and naïve Bayes models. *Mathematical Problems in Engineering* 2012: 974638.

Chen, W., Shirzadi, A., Shahabi, H. et al. (2017). A novel hybrid artifcial intelligence approach based on the rotation forest ensemble and naïve Bayes tree classifers for a landslide susceptibility assessment in Langao County, China. *Geomatics, Natural Hazards and Risk* 8 (2): 1955–1977.

Chowdhuri, I., Pal, S.C., and Chakrabortty, R. (2020). Flood susceptibility mapping by ensemble evidential belief function and binomial logistic regression model on river basin of eastern India. *Advances in Space Research* 65 (5): 1466–1489.

Coughlan, R., Di Giuseppe, F., Vitolo, C. et al. (2021). Using machine learning to predict fire-ignition occurrences from lightning forecasts. *Meteorological Applications* 28 (1): e1973.

Fulé, P.Z., Garkoti, S.C., and Semwal, R.L. (2021). Frequent burning in chir pine forests, Uttarakhand, India. *Fire Ecology* 17 (1): 1–11.

Gupta, S., Roy, A., Bhavsar, D. et al. (2018). Forest fire burnt area assessment in the biodiversity rich regions using geospatial technology: Uttarakhand Forest fire event 2016. *Journal of the Indian Society of Remote Sensing* 46 (6): 945–955.

Hong, H., Tsangaratos, P., Ilia, I. et al. (2018). Application of fuzzy weight of evidence and data mining techniques in construction of flood susceptibility map of Poyang County, China. *Science of the Total Environment* 625: 575–588.

James, G., Witten, D., Hastie, T., and Tibshirani, R. (2013). *An Introduction to Statistical Learning*, vol. 112. New York: Springer.

Joseph, S., Anitha, K., and Murthy, M.S.R. (2009). Forest fire in India: a review of the knowledge base. *Journal of Forest Research* 14 (3): 127–134.

Kanga, S., Sharma, L.K., Pandey, P.C. et al. (2013). Forest fire modeling to evaluate potential hazard to tourism sites using geospatial approach. *Journal of Geomatics* 7 (1): 93–99.

Kanga, S., Sharma, L.K., Pandey, P.C., and Nathawat, M.S. (2014). GIS modelling approach for forest fire risk assessment and management. *International Journal of Advancement in Remote Sensing, GIS and Geography* 2 (1): 30–44.

Kavzoglu, T., Sahin, E.K., and Colkesen, I. (2014). Landslide susceptibility mapping using GIS-based multi-criteria decision analysis, support vector machines, and logistic regression. *Landslides* 11 (3): 425–439.

Kumari, S., Mehta, J.P., Shafi, S., and Dhiman, P. (2017). Phytosociological analysis of woody vegetation under burnt and unburnt oak dominated forest at Pauri, Garhwal Himalaya, India. *Environment Conservation Journal* 18 (3): 99–106.

Liang, Y.C., Maimury, Y., Chen, A.H.L., and Juarez, J.R.C. (2020). Machine learning-based prediction of air quality. *Applied Sciences* 10 (24): 9151.

Mallick, J., Alqadhi, S., Talukdar, S. et al. (2021). Risk assessment of resources exposed to rainfall induced landslide with the development of GIS and RS based ensemble metaheuristic machine learning algorithms. *Sustainability* 13 (2): 457.

Mohajane, M., Costache, R., Karimi, F., and Pham, Q.B. (2021). Application of remote sensing and machine learning algorithms for forest fire mapping in a Mediterranean area. *Ecological Indicators* 129: 107869.

Nachappa, T.G., Piralilou, S.T., Gholamnia, K. et al. (2020). Flood susceptibility mapping with machine learning, multi-criteria decision analysis and ensemble using Dempster Shafer theory. *Journal of Hydrology* 125275.

Negi, G.C.S. (2019). Forest fire in Uttarakhand: causes, consequences and remedial measures. *International Journal of Ecology and Environmental Sciences* 45 (1): 31–37.

Pandey, K. and Ghosh, S.K. (2018). Modelling of parameters for forest fire risk zone mapping. *ISPRS-International Archives of the Photogrammetry, Remote Sensing and Spatial Information Sciences* XLII-5: 299–304.

Roy, N. and Porwal, M.C. (2004). Forest fire risk zonation using geo-spatial modeling in part of Rajaji National Park, India. *Asian Journal of Geoinformatics* 5 (2): 40–46.

Sachdeva, S., Bhatia, T., and Verma, A.K. (2018). GIS-based evolutionary optimized gradient boosted decision trees for forest fire susceptibility mapping. *Natural Hazards* 92 (3): 1399–1418.

Sahana, M. and Ganaie, T.A. (2017). GIS-based landscape vulnerability assessment to forest fire susceptibility of Rudraprayag district, Uttarakhand, India. *Environmental Earth Sciences* 76 (20): 1–18.

Sannigrahi, S., Pilla, F., Basu, B. et al. (2020). Examining the effects of forest fire on terrestrial carbon emission and ecosystem production in India using remote sensing approaches. *Science of the Total Environment* 725: 138331.

Satendra, K.A. (2014). *Forest Fire Disaster Management.* New Delhi: National Institute of Disaster Management, Ministry of Home Affairs.

Sevgen, E., Kocaman, S., Nefeslioglu, H.A., and Gokceoglu, C. (2019). A novel performance assessment approach using photogrammetric techniques for landslide susceptibility mapping with logistic regression, ANN and random forest. *Sensors* 19 (18): 3940.

Sharma, L.K., Kanga, S., Nathawat, M.S. et al. (2012). Fuzzy AHP for forest fire risk modeling. *Disaster Prevention and Management: An International Journal* 21 (2): 160–171.

Singh, J.S. and Singh, S.P. (1987). Forest vegetation of the Himalaya. *The Botanical Review* 53 (1): 80–192.

Singh, R.D., Gumber, S., Tewari, P., and Singh, S.P. (2016). Nature of forest fires in Uttarakhand: frequency, size and seasonal patterns in relation to pre-monsoonal environment. *Current Science* 398–403.

Talukdar, S., Singha, P., Mahato, S. et al. (2020a). Land-use land-cover classification by machine learning classifiers for satellite observations—a review. *Remote Sensing* 12 (7): 1135.

Talukdar, S., Ghose, B., Shahfahad et al. (2020b). Flood susceptibility modeling in Teesta River basin, Bangladesh using novel ensembles of bagging algorithms. *Stochastic Environmental Research and Risk Assessment* 34 (12): 2277–2300.

Talukdar, S., Mankotia, S., Shamimuzzaman, M., Shahfahad et al. (2021). Wetland-Inundated Area Modeling and Monitoring Using Supervised and Machine Learning Classifiers. In Pandey, P.C and Sharma L.K. (eds.), *Advances in Remote Sensing for Natural Resource Monitoring*. John Wiley & Sons, Inc., Oxford, UK, pp. 346–365.

Tiwari, A., Shoab, M., and Dixit, A. (2021). GIS-based forest fire susceptibility modeling in Pauri Garhwal, India: a comparative assessment of frequency ratio, analytic hierarchy process and fuzzy modeling techniques. *Natural Hazards* 105 (2): 1189–1230.

Tomar, J.S., Kranjčić, N., Đurin, B. et al. (2021). Forest fire hazards vulnerability and risk assessment in Sirmaur District Forest of Himachal Pradesh (India): a geospatial approach. *ISPRS International Journal of Geo-Information* 10 (7): 447.

Tuyen, T.T., Jaafari, A., Yen, H.P.H. et al. (2021). Mapping forest fire susceptibility using spatially explicit ensemble models based on the locally weighted learning algorithm. *Ecological Informatics* 63: 101292.

Vapnik, V.N. (2013). *The Nature of Statistical Learning Theory.* [Sl], –27. N.Y.: Springer Science + Business Media.

Wang, S.S.C., Qian, Y., Leung, L.R., and Zhang, Y. (2021). Identifying key drivers of wildfires in the contiguous US using machine learning and game theory interpretation. *Earth's Future* e2020EF001910.

Willcock, S., Martínez-López, J., Hooftman, D.A. et al. (2018). Machine learning for ecosystem services. *Ecosystem Services* 33: 165–174.

Xie, Y. and Peng, M. (2019). Forest fire forecasting using ensemble learning approaches. *Neural Computing and Applications* 31 (9): 4541–4550.

Yoo, C. and Cho, E. (2019). Effect of multicollinearity on the bivariate frequency analysis of annual maximum rainfall events. *Water* 11 (5): 905.

10

Leveraging Google Earth Engine (GEE) and Landsat Images to Assess Bushfire Severity and Postfire Short-Term Vegetation Recovery

A Case Study of Victoria, Australia

Saroj K. Sharma[*,1,2,3], *Jagannath Aryal*[1,2,3], *and Abbas Rajabifard*[1,2,3]

[1] *Department of Infrastructure Engineering, The University of Melbourne, Melbourne, Parkville, Victoria, Australia*
[2] *Centre for Spatial Data Infrastructures and Land Administration (CSDILA), The University of Melbourne, Parkville, Victoria, Australia*
[3] *Centre for Disaster Management and Public Safety (CDMPS), The University of Melbourne, Parkville, Victoria, Australia*

10.1 Introduction

Bushfires are frequently occurring natural phenomena and have been playing key roles in shaping natural ecosystem, ecological processes, and landscape dynamics across various parts of the world including Australia, the European Mediterranean region, and the United States (Bowman et al. 2019; Burke et al. 2021; Sharples et al. 2016; Turco et al. 2018). In Australia, growing scientific evidence suggests that climate change is causing an increment in the scale, frequency, and severity of bushfires, posing a catastrophic threat to lives, properties, and the environment (Bowman et al. 2009; Dutta et al. 2013, 2016; Halgamuge et al. 2020; Linnenluecke et al. 2011; Prior and Eriksen 2013). Bushfires can have detrimental impacts on biodiversity and human life (Bowman et al. 2009; Prior and Eriksen 2013); however, nature's ability to regenerate can progressively heal the effect in a long run. Additionally, bushfire events are expected to rise with the increment in warmer and drier conditions due to climate change effects in Australia. These events thus will make post-fire vegetation recovery of concern for a sustainable ecosystem.

Understanding vegetation recovery dynamics is important for successful post-fire restoration treatments (Liu 2016; Viana-Soto et al. 2017). A proper understanding of the degree of burn severity as well as influencing drivers determining rate of post-fire vegetation regeneration is necessary for successful implementation of post-fire recovery plan (Bright et al. 2019). Vegetation regeneration depends on several factors, hence a thorough identification and quantification of the drivers is difficult (Viana-Soto et al. 2017). However, different literature has shown that the severity of fires, vegetation species, environmental factors, topographic conditions, and post-fire climatic variability governs the recovery process (Harvey et al. 2016; Shryock et al. 2015). Similarly, temperature, precipitation, solar

*Corresponding author's email: sarojkumars@student.unimelb.edu.au

0000-0002-3333-6008 (Saroj K. Sharma), 0000-0002-4875-2127 (Jagannath Aryal), and 0000-0002-4256-3173 (Abbas Rajabifard)

Advances in Remote Sensing for Forest Monitoring, First Edition. Edited by Prem C. Pandey and Paul Arellano.

radiation, and water availability are identified as climatic drivers influencing post-fire regenerations (Liu 2016; Meng et al. 2015).

Among major variables defining fire regime, burn severity has the greatest ecological importance in defining the post-fire effects and recovery process (Wang and Kemball 2005). Burn severity also describes effects of fires on vegetation and soil organic matters above ground (Keeley 2009). Often forest management activities are focused on the areas to be recovered, hence, land managers are interested in the areas with poor regeneration capabilities for potential interventions to facilitate recovery (Collins et al. 2007). These areas are often attributed to high severity fires (Cocke et al. 2005; Quintano et al. 2018). Hence, burn severity information can be particularly important while formulating post-fire responses and recovery strategies for land managers.

Remote sensing data and methods have been widely used in analyzing ecological dynamics of fires including assessment of post-fire vegetation recovery (Hammill and Bradstock 2006; Pelletier et al. 2021; Quintano et al. 2018; Sunderman and Weisberg 2011). Medium resolution satellite images including Landsat products have proven effective in assessing the vegetation change dynamics due to fire disturbances (Epting et al. 2005; Long et al. 2019; Röder et al. 2008; Storey et al. 2021).

To date, many studies are utilizing spectral indices and remote sensing data and methods to quantify bushfire severity and post-fire vegetation recovery assessment (Cai and Wang 2020; Konkathi and Shetty 2021; Miller and Thode 2007; Pelletier et al. 2021; Sharma et al. 2022; Stambaugh et al. 2015; Storey et al. 2021). Among different spectral indices, dNBR, and RdNBR proposed by Miller and Thode (2007) has been most used in characterizing bushfire severity (Cai and Wang 2020; Soverel et al. 2010). Similarly, NDVI index, proposed by Rouse et al. (1974) is widely used as a proxy to green biomass as it is sensitive to canopy cover and photosynthetic activities. It is reported that after bushfire disturbances, spontaneous drop in vegetation indices was observed due to loss of green vegetation and an increase in green biomass was observed with the recovery of fire (Malak and Pausas 2006). Hence, NDVI can be a good indicator to estimate post-fire vegetation recovery.

GEE is a freely available cloud-based geospatial analysis platform hosting a range of petabyte-scale satellite data for analysis and visualization (Gorelick et al. 2017; Kumar and Mutanga 2018). Since its launch in 2010, it has demonstrated its capability to analyze environmental problems in both local and global scales (Zhao et al. 2021). GEE can be considered a high-performance computing platform and software, hence, any researcher from any part of the world has the access to a planetary-scale data and analysis platform. This has now opened opportunities for remote sensing and geospatial data analysis to many users ranging from remote sensing scientists to environmental analysts. However, despite GEE being used as a cloud-based platform in environmental analytics, the capability of this platform has not been fully utilized in quantifying bushfire severity and post-fire recovery.

In this study, we examined the performance of two widely used burn severity indices, dNBR and RdNBR, in heterogeneous environment. We compared the accuracy of these indices using field-based validation data. Next, we performed time series analysis of NDVI to assess post-fire recovery of different severity classes. Further, we explored and revealed potential influence of climatic and topographic variables on post-fire recovery. We utilized freely available GEE platform and Landsat data to perform bushfire severity and post-fire recovery assessment. More specifically, this research attempted to answer the following questions.

- How the performance of burn severity classification varies among indices, dNBR and RdNBR, in a heterogeneous landscape?
- How variation in burn severity related to post-fire vegetation recovery?
- How climatic and topographic variables influence post-fire recovery process?

The main contributions of this study are as follows:

- A comprehensive study from bushfire severity analysis to recovery assessment utilizing GEE.
- A performance examination of widely used burn severity indices, RdNBR and dNBR, in classifying bushfire severity in a heterogeneous landscape.
- A demonstration of the influence of climatic and topographical variables on the post-fire recovery process in a heterogeneous landscape.

The rest of the chapter is organized as follows: in Section 10.2, we describe the materials and methods including a brief description of the study area and data. Section 10.3 presents the results of severity classification including accuracy assessment. Section 10.4, presents the discussion. Finally, we conclude the paper in Section 10.5 with the recommendations for future research.

10.2 Materials and Methods

10.2.1 Study Area

The study area, Kilmore East region, is in the northern part of Victoria (144°59′44″W, 146°06′07″E, −37°11′32″N, and −37°41′35″S), southeastern Australia (Figure 10.1). The region experienced a severe bushfire on February 7, 2009, also known as the Black Saturday Bushfire, that burned approximately 100 000 ha and destroyed nearly 2200 buildings in less than 12 hours (Cameron et al. 2009). This Kilmore East region fire was merged with another fire of Murrindindi region and continued for three weeks, which burned approximately 400 000 ha when combined (Cruz et al. 2012).

The study area consisted of a wide range of vegetation including plantation forest, grassland, and open woodland in lower elevations, with dry, wet, and mixed types of sclerophyll eucalyptus forest in upper elevations, and dense non-eucalyptus rainforest mostly in deeper gullies (Gill 1981). The industrial plantations include pines (*Pinus radiata*) and eucalypt (*Eucalyptus globulus*) of variable ages, but mostly under 10 years. Similarly, dry sclerophyll eucalyptus consisted of medium to tall over- and understory trees between 20 and 25 m. The major dominant species included peppermints (*Eucalyptus dives, Eucalyptus radiata*), gums (*Eucalyptus. mannifera* sp.) and stringybark (*Eucalyptus obliqua, Eucalyptus macrorhyncha*) (Attiwill and Adams 2013). Wet sclerophyll forest consisted of tall trees up to 40 m dominated by ash (*Eucalyptus regnans*). The fertile and riparian areas were dominated by

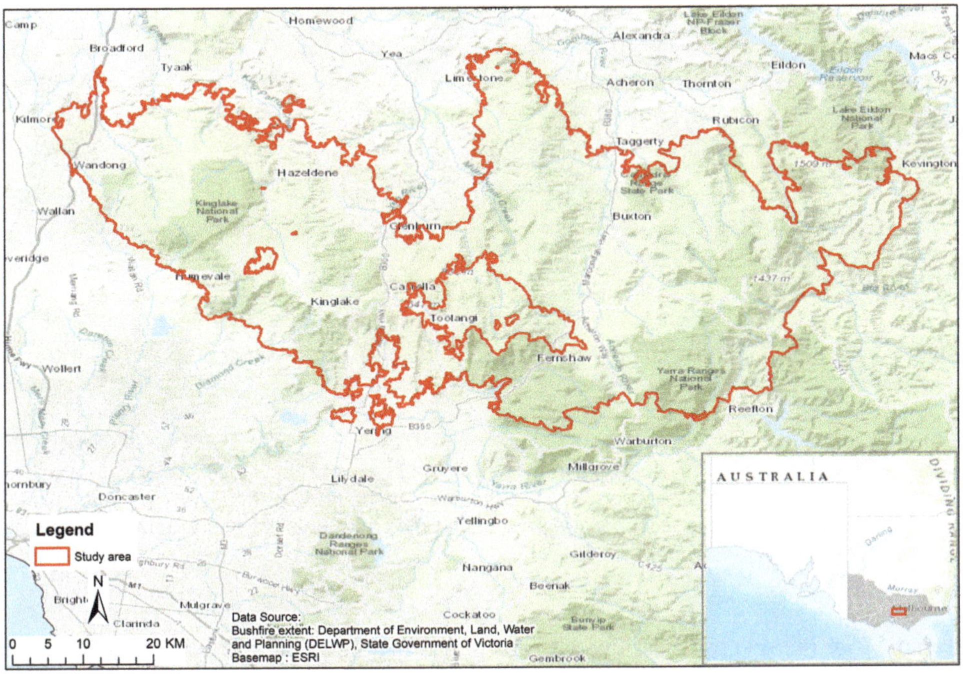

Figure 10.1 Map of the study area, Kilmore East – Murrindindi region of Victoria, Australia. The region experienced a severe bushfire on February 7, 2009 also known as the Black Saturday Bushfires.

tall messmate stringybark (*Eucalyptus obliqua*) and manna gum (*Eucalyptus Viminalis*). Similarly, the rainforest area in deeper gullies had dense multistory non-eucalypt forest with dominant tree species myrtle beech (*Nothofagus cunninghamii*) and southern sassafras (*Atherosperma moschatum*) (Cruz et al. 2012). A comprehensive description of physiography, vegetation, and topography of the Hume region of central Victoria including the study area is provided by Ashton et al. (2000).

The study area had a cool and temperate climate with a long-term (1994–2021) mean annual rainfall of 680 mm and total summer rainfall averaging 142 mm, and mean maximum annual temperature 16.9 °C and the average maximum summer temperature of 24.1 °C (Bureau of Meteorology 2021). However, the region experienced some microclimatic variations due to both diverse topography and vegetation (Ashton 2000). The details of climatic variations and microclimatic conditions during the 2009 Black Saturday bushfire period is explained by Cruz et al. (2012). The altitude of the study area ranged from 60 to 1500 m with higher elevations in the eastern part and the western part.

10.2.2 Conceptual Workflow and Vegetation Recovery Predictors

The overall workflow of this research is categorized into two major stages (Figure 10.2). In the first stage, we generated bushfire severity maps using dNBR and RdNBR (Prichard and Kennedy 2014) indices and classified them into different severity classes. The accuracy

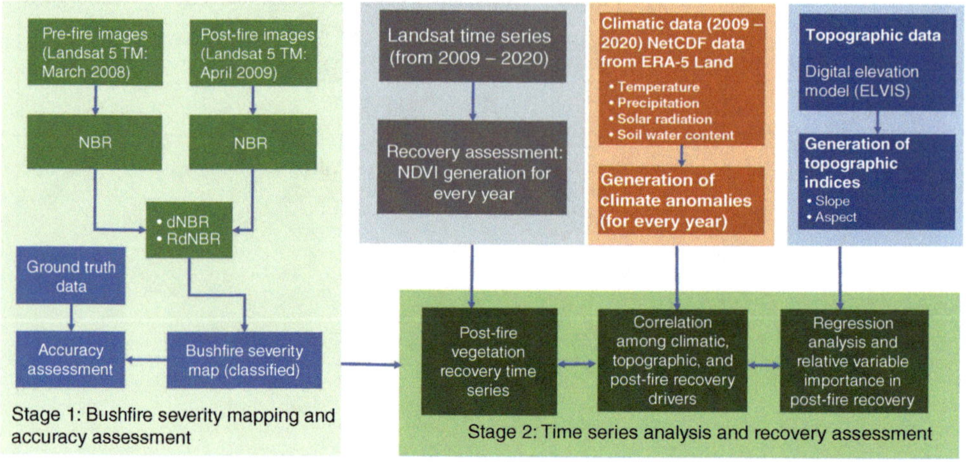

Figure 10.2 Overall workflow of the study. The first stage presents bushfire severity mapping and accuracy assessment whereas second stage presents a time series analysis and vegetation recovery assessment workflow.

assessment of the severity classification was performed based on the validation data collected by the Victorian Government, Department of Environment, Land, Water and Planning (DELWP).

In the second stage, we generated post-fire vegetation recovery time-series and explored how the recovery process was influenced by climatic and topographic variables. Hence, we performed post-fire vegetation recovery assessment by generating NDVI for each year from 2009 to 2020. Further, we assessed correlations among the utilized climatic, topographic, and post-fire recovery drivers. We used random forest regression analysis to identify key climatic and topographic drivers and their relative influence in post-fire recovery.

10.2.3 Dataset

10.2.3.1 Remote Sensing Data

We used multispectral Landsat 5 TM from 2008–2011 and Landsat 8 OLI imageries from 2013–2020 (Table 10.1) for bushfire severity mapping and vegetation recovery assessment. The Landsat data has 30-m spatial resolutions. The assessment area comprised of a part of Landsat scene Worldwide Reference System-2 (WRS-2), Path 92; Row 86. All the Landsat images used in this research were level-1 corrected images available on GEE as image collections. Level-1 Precision Terrain (L1TP) corrected products are radiometrically calibrated, orthorectified, geometrically corrected, and processed through Landsat Surface Reflectance Code (LaSRC). Hence, they are high quality, ready-to-use images suitable for time series analysis (USGS 2021).

The research utilized Landsat 5 TM images from March 2008 (pre-fire) and April 2009 (post-fire) for bushfire severity analysis. We visually confirmed that all the images in the study had excellent quality observations unaffected by clouds, cloud shadows, and terrain

Table 10.1 Specifications of Landsat 5 TM and Landsat 8 OLI images used in this study.

Sensor	Acquisition date	Occurrence	Landsat scene identifier
Landsat 5 TM	March 1, 2008	Pre-fire	LT50920862008061ASA00
Landsat 5 TM	April 21, 2009	Post-fire	LT50920862009111ASA00
Landsat 5 TM	October 1, 2010	Post-fire	LT50920862010274ASA00
Landsat 5 TM	January 5, 2011	Post-fire	LT50920862011005ASA00
Landsat 5 TM	September 2, 2011	Post-fire	LT50920862011245ASA00
Landsat 8 OLI	November 27, 2013	Post-fire	LC80920862013331LGN01
Landsat 8 OLI	March 3, 2014	Post-fire	LC80920862014062LGN01
Landsat 8 OLI	March 22, 2015	Post-fire	LC80920862015081LGN01
Landsat 8 OLI	April 25, 2016	Post-fire	LC80920862016116LGN01
Landsat 8 OLI	January 22, 2017	Post-fire	LC80920862017022LGN01
Landsat 8 OLI	May 1, 2018	Post-fire	LC80920862018121LGN00
Landsat 8 OLI	March 17, 2019	Post-fire	LC80920862019076LGN00
Landsat 8 OLI	September 11, 2020	Post-fire	LC80920862020255LGN00

Table 10.2 Earth Engine Data Catalog for used data in Google Earth Engine.

Data	Source
Landsat 5 TM	https://code.earthengine.google.com/dataset/LANDSAT/LC05/C01/T1_SR
Landsat 8 OLI	https://code.earthengine.google.com/dataset/LANDSAT/LC08/C01/T1_SR

shadows. The time-series analysis based on vegetation indices can be affected by the plant phenological cycle. Hence, we tried to minimize this effect by using images with similar seasonality, mostly of the same month, where possible. Considering the failure of the Scan Line Corrector (SLC) in the ETM+ instrument of the Landsat-7 satellite, we only utilized USGS Landsat-5 and Landsat-8 Surface Reflectance collections. However, the image for 2012 was not available in both Landsat 5 and Landsat 8 datasets, hence we analyzed the two images from 2011, one from January and another from September, instead. The image processing was done in GEE using the atmospherically corrected surface reflectance image dataset (Table 10.2). The detailed specifications of the images and GEE datasets are presented in Tables 10.1 and 10.2.

10.2.3.2 Climate Data

The climatic data were downloaded from ERA-5 Land monthly averaged data. This research utilized four different types of climatic data; temperature, precipitation, solar radiation, and volume of water in the soil layer (0–28 cm) (Table 10.3). The ERA-5 Land is a gridded reanalysis dataset providing climatic measurements of land variables available from

Table 10.3 Description of climatic and topographic variables used in this study.

Variables	Description	Unit
DEM	Elevation data of 10 m spatial resolution	Meter
Slope	Slope map generated from DEM	Degree
Aspect	Aspect map generated from DEM	—
Solar radiation	Solar radiations reaching the surface of the Earth (direct and diffuse)	Joules per square meter (J/m^2)
Temperature	Temperature of air at 2 m above surface level for each year 2009–2020	°Celsius
Precipitation	Accumulated precipitation on Earth surface for each year 2009–2020	Meter
Volumetric soil water	The volume of water in soil layer 1 (0–7 cm) and layer 2 (7–28 cm) of the ECMWF Integrated forecasting system	Cubic meter per cubic meter (m^3/m^3)
RdNBR	Relative difference normalized burn ratio computed by using pre-fire (2008) and post-fire (2009) Landsat TM images	—
dNBR	Difference normalized burn ratio computed by using pre-fire (2008) and post-fire (2009) Landsat TM images	—
NDVI	Normalized difference vegetation index computed for every year from 2008 to 2020 from Landsat TM and ETM+ images	—

Copernicus program of the European Space Agency (ESA). Each climatic grid has a spatial resolution of 9 km (ECMRWF 2021).

We downloaded monthly climatic data in Network Common Data Form (NetCDF) file format. NetCDF data format is getting popular for storing multidimensional scientific data variables (Rew and Davis 1990). The downloaded NetCDF data was processed in Python 3.7 utilizing Xarray library for generating climatic anomalies. The climatic anomalies were calculated for each year (2009–2020) at pixel levels. Anomaly values show a better picture of climatic variations and their impact on vegetation recovery on a local scale (Liu 2016). It is assumed that the variation in temperature, precipitation, solar radiation, and soil water content may influence vegetation recovery process (Liu 2016; Meng et al. 2015; Viana-Soto et al. 2017). For each climatic variable, we computed climatic anomalies from 12 years (2009–2020) average and standardized those variables by computing mean and standard deviation as shown in Eq. (10.1).

$$Z = \frac{X - \mu}{\sigma} \tag{10.1}$$

where Z is the computed anomalies value, X is the climate data pixel value, μ is the mean of climatic variable readings between 2009 and 2020, and σ is the standard deviation.

10.2.3.3 Other Data

Topographic Data The topographic analysis was based on the Digital Elevation Model (DEM) from Elevation and Depth – Foundation Spatial Data (ELVIS) accessed through https://elevation.fsdf.org.au, and have a spatial resolution of 10 m, horizontal accuracy of 12.5 m, and vertical accuracy of 5 m. The utilized DEM data was hydrologically enforced for correctly defining surface drainage and hydrological flow (Vicmap 2020). Before further processing, we identified sinks in DEM and filled them using hydrology tool in ArcGIS. Thus corrected filled DEM was used for generating slope and aspect maps in ArcGIS spatial analyst environment.

Bushfire Severity Validation Data This study utilized subset of burn severity data collected by the Victorian Government, DELWP for the 2009 bushfires. A total of 590 different validation data points, located within our study area were used for validation purpose. The validation data had severity information categorized in five different classes; (i) crown burn, (ii) crown scorch, (iii) moderate crown scorch, (iv) light or no crown scorch, and (v) no crown scorch (DELWP 2021). As there were no significant number of class iv (no crown scorch) data points within the extent of our study area, we did not utilize them during the validation process.

10.2.4 Processing in GEE

GEE provides effective web-based tools, and advanced algorithms for analyzing geospatial big data (Bonney et al. 2020; Zhao et al. 2021). We utilized GEE cloud computing platform for severity mapping and time-series analysis. GEE repository has various multi-temporal remote sensing data including Landsat TM and OLI which were used in this study (Bonney et al. 2020; Zhao et al. 2021)

10.2.5 Fire Severity Characterization

Traditionally, burn severity was estimated by field assessment based on postfire crown scorch (De Santis and Chuvieco 2007). However, with the advancement in remote sensing methods, burn severity indices based on multispectral optical satellite imagery have been developed and used in various literature (Miller et al. 2009; Pelletier et al. 2021). Some of the indices include Composite Burn Index (CBI), dNBR, RdNBR, and Differenced Normalized Difference Vegetation Index (dNDVI) (Klinger et al. 2019; Miller et al. 2009; Soverel et al. 2010). Among them, dNBR and RdNBR are the widely adopted burn indices (Miller and Thode 2007; Prichard and Kennedy 2014). The dNBR index measures absolute change between the spectral indices of pre-fire and post-fire images, whereas RdNBR index calculates the relative change caused by fire based on pre-fire reflectance. Both variations of burn severity indices have demonstrated their effectiveness among other indices (Klinger et al. 2019; Konkathi and Shetty 2021; Miller et al. 2009; Pelletier et al. 2021; Sparks et al. 2014).

Both dNBR and RdNBR indices were calculated by using NIR and SWIR band from Landsat TM images from March 1, 2008 (pre-fire) and April 21, 2009 (post-fire). The NBR and dNBR indices were calculated as shown in Eqs. (10.2) and (10.3):

$$\text{NBR} = \frac{(\text{NIR} - \text{SWIR})}{(\text{NIR} + \text{SWIR})} \tag{10.2}$$

$$\text{dNBR} = (\text{NBR pre-fire} - \text{NBR post-fire}) \times 1000) \tag{10.3}$$

In Landsat TM, the NIR and SWIR are represented by band 4 (0.77–0.90 μm) and band 7 (2.09–2.35 μm) respectively.

Similarly, the RdNBR was calculated as shown in Eq. (10.4) (Miller and Thode 2007):

$$\text{RdNBR} = \frac{\text{NBR pre-fire} - \text{NBR post-fire}}{\sqrt{|\text{NBR prefire}|}} \tag{10.4}$$

In this research, we calculated both dNBR and RdNBR indices and compared their performances against field-collected validation data. Before the validation, we generalized burn severity maps by using a 3×3 pixels smoothed filter to remove the small polygons. Those generalized maps were used in assessing relationships between bushfire severity and post-fire vegetation recovery.

10.2.6 Post-Fire Recovery Indices Calculation

Post-fire recovery assessment was done by calculating inter-annual mean NDVI proposed by Rouse et al. (1974) from one year before the fire (2008) up until 2020. NDVI was calculated by utilizing near-infrared (NIR) and red (R) bands as shown in Eq. (10.5):

$$\text{NDVI} = \frac{(\text{NIR} - \text{RED})}{(\text{NIR} + \text{RED})} \tag{10.5}$$

In Landsat TM band 3 (0.63–0.69 μm) is the red band and band 4 (0.77–0.90 μm) is the NIR band.

10.2.7 Bushfire Severity Accuracy Assessment

The accuracy assessment was performed for each severity class for both dNBR and RdNBR based classifications by generating confusion matrices against validation data. The utilized classification threshold is presented in Table 10.4. Error matrices were generated to assess the fire severity classification error. We generated overall accuracy, kappa coefficient (kappa), user's accuracy and producer's accuracy for each classification scheme.

Table 10.4 Severity classification threshold for dNBR and RdNBR classification schemes.

Severity classification	Class boundary (RdNBR)	Class boundary (dNBR)
Low severity/no crown scorch	< 0.1	< 100
Moderate severity (understory burn/light crown scorch)	0.1–0.3	100–250
High severity (crown scorch)	0.3–0.9	250–700
Very high severity (crown burn)	≥ 0.9	≥ 700

Note: The severity value of dNBR is scaled to 1000.

10.3 Results

10.3.1 Bushfire Severity Assessment

Bushfire severity assessment was performed by calculating dNBR and RdNBR indices in GEE platform. The classification results showed that both dNBR and RdNBR produced close results for all severity classes; low, medium (or moderate), high, and very high (Figure 10.3). The threshold defined for classifying bushfire severity map is presented in Table 10.4. In both schemes, more than half (about 55%) of the area experienced high or very high severity fires resulting in canopy burn or canopy scorch, whereas 14% of the area experienced moderate fires with understory burn. This shows a good correspondence between dNBR and RdNBR classes for each classification group.

Similarly, the regression analysis of severity classes between dNBR and RdNBR methods supported our findings. The results demonstrated that there existed highly significant agreement (p values < 0.001, R^2 0.7128) between these classification schemes, indicating a very close trend between severity classes of dNBR and RdNBR.

Further, the results showed that the dNBR values ranged from −925 to +1600 ($\mu = 362$ and $\sigma = 339$) with the lower quartile 50 and upper quartile 750. Similarly, the RdNBR values range from −10 to 8 ($\mu = 0.445$ and $\sigma = 0.48$) with 0.08–0.90 as first and third quartiles. The dNBR values were computed by using Eq. (10.3), where it was multiplied by 1000 to transform data into integer format, whereas RdNBR was computed by using Eq. (10.4), as originally proposed by Miller and Thode (2007). Histogram diagram of dNBR and RdNBR indices showing frequency distribution of burn pixels is presented in Figure 10.4.

10.3.2 Bushfire Severity Accuracy Assessment Results

The accuracy assessment results against validation data showed that the overall accuracy of RdBNR and dNBR were 90 and 86%, respectively. Similarly, the kappa value, which also reflects the overall accuracy, was 0.86 for RdNBR and 0.81 for dNBR (Figure 10.5).

Similarly, the confusion matrix (Table 10.5) showed that for each class, both producer's accuracy and user's accuracy was higher in RdNBR than in dNBR. Very high classification accuracy (87–95%) was observed in discriminating very high, high, and low severity classes

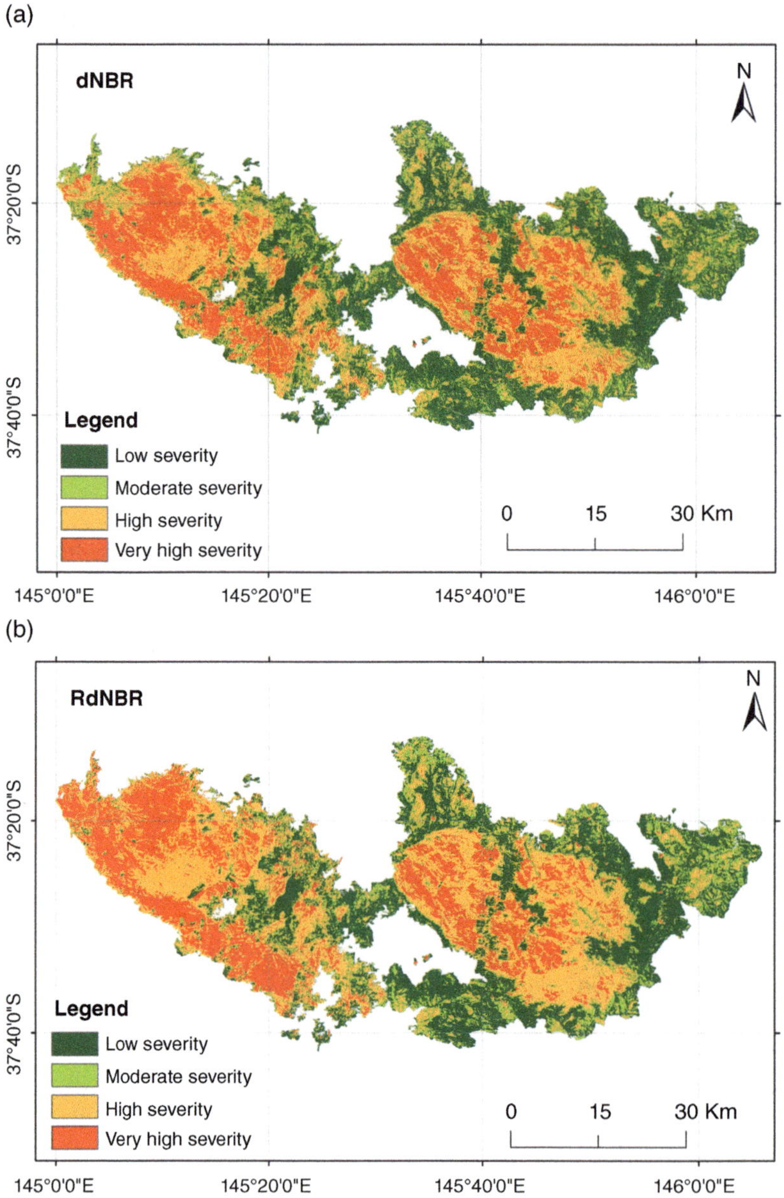

Figure 10.3 Bushfire severity map showing low, moderate, high, and very high severity classes. (a) Classified severity map calculated by using dNBR index, and (b) classified severity map calculated by using RdNBR index.

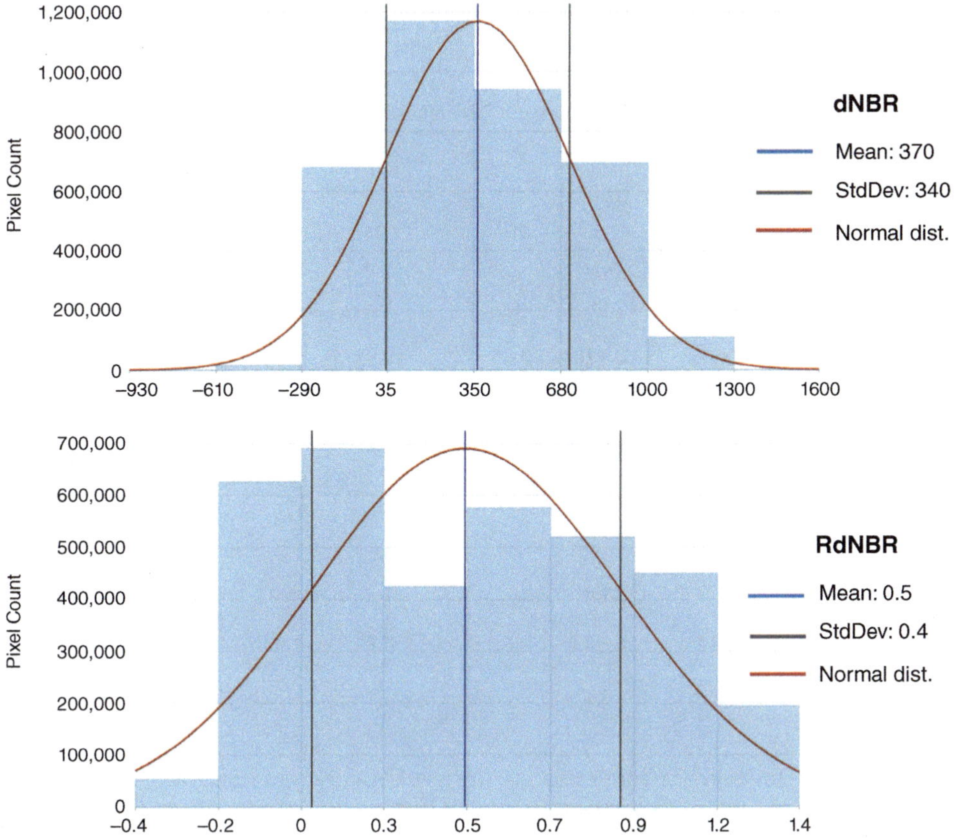

Figure 10.4 Histogram map of dNBR and RdNBR indices showing the frequency distribution.

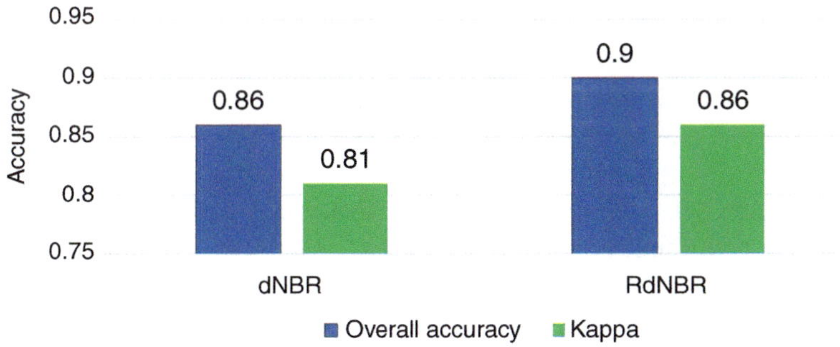

Figure 10.5 Accuracy assessment of bushfire severity classification using dNBR and RdNBR indices.

Table 10.5 Error matrices of bushfire severity classification map.

RdNBR Class	Very high	High	Moderate	Low	Total	User accuracy	Kappa
Very high	133	17	0	0	150	**0.89**	
High	13	211	0	0	224	**0.94**	
Moderate	0	1	41	21	63	**0.65**	
Low	0	1	7	145	153	**0.95**	
Total	146	230	48	166	590	**0.00**	
Producer accuracy	**0.91**	**0.92**	**0.85**	**0.87**	**0.00**	**0.90**	
Kappa							**0.86**
Very high	122	25	0	0	147	**0.83**	
High	24	203	0	0	227	**0.89**	
Moderate	0	1	39	21	61	**0.64**	
Low	0	1	9	145	155	**0.94**	
Total	146	230	48	166	590	**0.00**	
Producer accuracy	**0.84**	**0.88**	**0.81**	**0.87**	**0.00**	**0.86**	
Kappa							**0.81**

in both methods. However, for both methods, the user's accuracy of moderate severity class was relatively low, i.e. 65 and 64% for RdNBR and dNBR respectively. Similarly, the corresponding producer's accuracy was 85 and 81% for RdNBR and dNBR respectively. This means that even though 85% of the moderate severity class in RdNBR (81% for dNBR) was accurately identified as "moderate severity", only 65% of the areas identified as "moderate severity" (64% in dNBR) in the classification were actually of moderate severity. This indicated that both RdNBR and dNBR methods had misclassification problems in distinguishing moderate severity burn pixels including understory burn.

The performance of spectral indices to distinguish between different bushfire severity classes varied, depending on the forest types and assessment methods (Miller et al. 2009; Soverel et al. 2010). In our case, the overall accuracy and Kappa coefficient of RdNBR was higher than dNBR. Further, RdNBR had better classification accuracy (both users and producers) for all severity classes. This demonstrated that RdNBR index had better severity mapping performance than dNBR index heterogeneous forest of varying elevations.

We used RdNBR derived classification map as a base severity map for assessing the post-fire vegetation recovery in the next stage.

10.3.3 Post-Fire Recovery Assessment

The computed NDVI time-series of before and after fires showed how the NDVI value (represented by greenness in Figure 10.6) was changed spatially over time following the fire of 2009 up until 2020. Although some fluctuations due to seasonality can be noticed, a clear

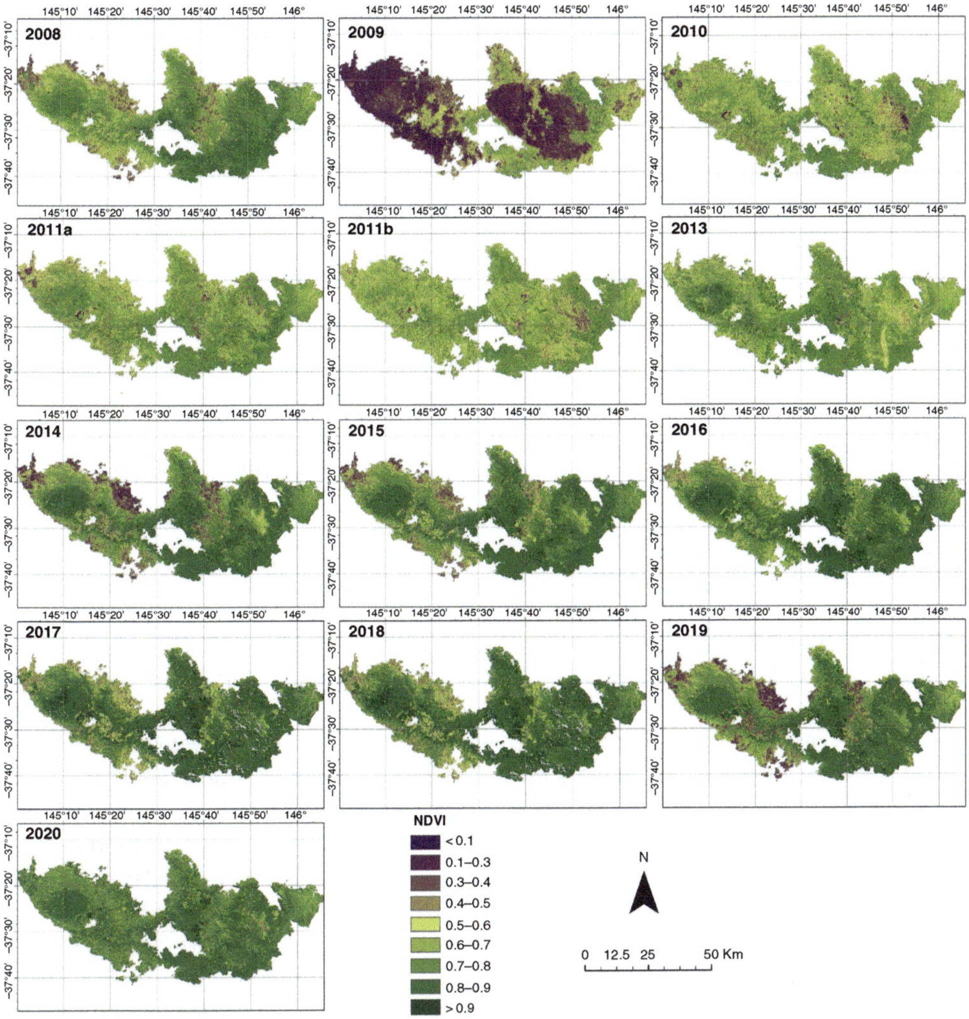

Figure 10.6 Normalized Difference Vegetation Index (NDVI) maps of the study area from 2008 (1 year before the fire), 2009 (bushfire year), and post-fire years (2010–2020).

increment in the surface greenness following the fire (2009) was observed in overall. The interpretation of NDVI time series revealed that the area experienced a rapid recovery of NDVI in the subsequent three years, which was consistently improved and almost reached that of pre-fire NDVI in 2014, five years following the fire. Visual analysis of the time series NDVI showed that highly severe areas required more time for recovery. This trend was supported by correlation analysis between the severity and NDVI changes (proxy of recovery). The results showed that fire severity had strong, negative correlation (−0.79) with vegetation recovery for the first three years.

A dramatic drop in the mean NDVI value was observed after the fire disturbance (Figure 10.7). An approximate 70% drop in the mean NDVI value, i.e. from 0.68 to 0.21, was observed during the fire in those areas with very high severity fires experiencing canopy

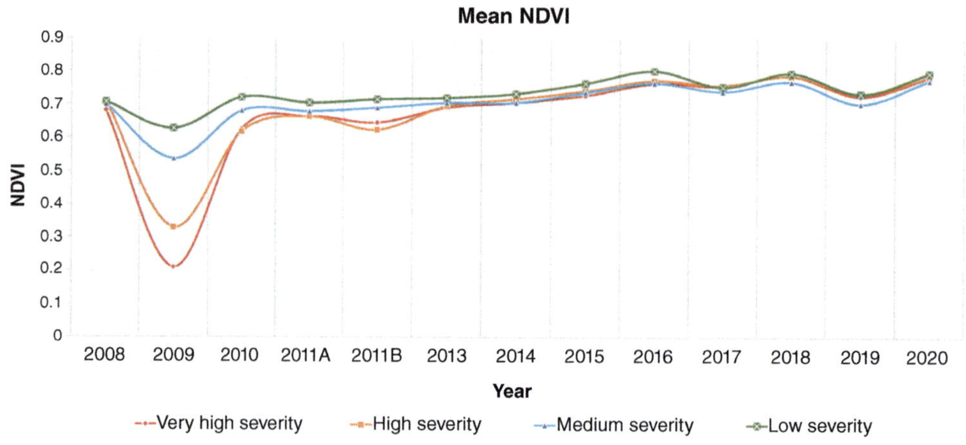

Figure 10.7 Mean NDVI time-series line graph showing how the mean NDVI values changed over time for different burn severity classes.

burn. Similarly, for those areas with canopy scorch, a drop of 53% of NDVI was observed. A smaller drop (23%) in NDVI was observed in those areas of moderate understory burn with low severity fires.

Those areas with less severe fires observed quick recovery than those areas with more severe fires. The low severity fire areas that experienced 11% drop in NDVI were recovered to pre-fire NDVI within a year. Those areas with medium or moderate severity fires regained 95% of their pre-fire NDVI within a year and remained almost stable for a few years. However, the areas with high severity fires that experienced canopy scorch recovered 87% of pre-fire NDVI within three years, following a full phase recovery within four years after the fire. Similarly, the areas with very high severity fires that experienced canopy burn reclaimed their pre-fire NDVI within five years. However, the recovery in the areas with very high severity fires was very rapid for the first year following fires, and gradually slowed down for another four years.

The NDVI value showed a growing trend for up to five years, and after that no specific trend was observed (Figure 10.7). A similar trend was observed by Liu (Liu 2016) in their study. Hence, we assessed the climatic effect on vegetation recovery for five years following the fires.

10.3.4 Correlation Among Climatic, Topographic, and Post-fire Recovery Variables

The climatic anomalies computed from a long-term average (2009–2020) showed that the area received more than average solar radiation and temperature and less than average precipitation and soil water volume for a year following the fire disturbances (2009). In 2009, the year experienced two heatwaves and had record temperatures for August and the warmest winter (BOM 2021a). The case was opposite for the next three years, i.e. in 2010, 2011, and 2012, the area received less than average solar radiations and temperature but more than average soil water volume and precipitation. The results showed a fluctuation in the climatic variables of the area. However, it was observed that more the temperature,

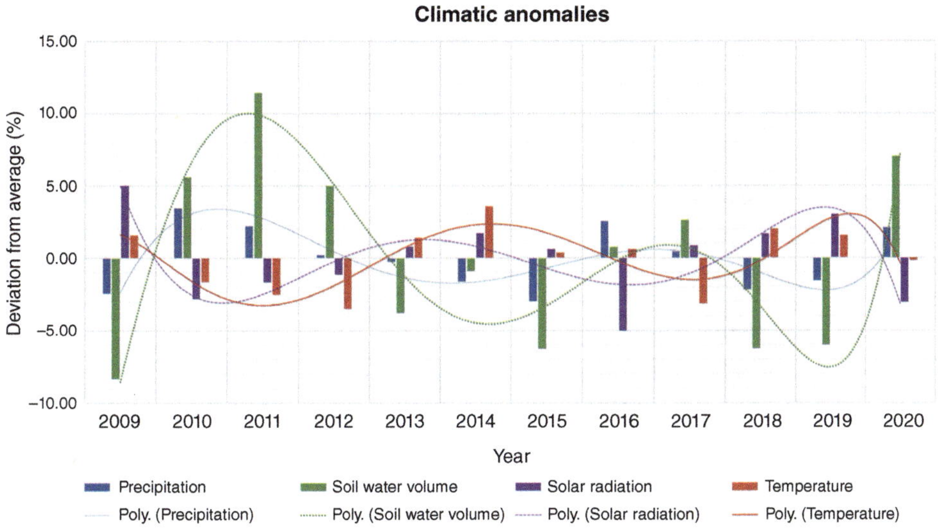

Figure 10.8 Climate anomaly diagram of climatic variables used in this study. The deviation from average in percentage (represented in *y*-axis) is calculated using 12 years climatic average from fire year (2009) to 2020.

more was the solar radiation and less soil water volume and precipitation, and vice versa. The detailed climatic anomalies chart is presented in Figure 10.8.

We performed a correlation analysis between climatic and topographic variables and a change in NDVI as a proxy of vegetation recovery. The results showed a strong correlation in vegetation recovery when assessed against the yearly change in NDVI value for three years after the fire. A significant recovery in NDVI value was observed throughout the area.

The results revealed a high positive correlation between the NDVI change and anomalies of solar radiation ($r = 0.69$) and temperature ($r = 0.72$), calculated against a long-term average (2009–2020). Similarly, the correlation analysis showed that there exists a very low negative correlation ($r = -0.11$) between the elevation and vegetation recovery. Among the predictor variables, temperature anomaly had a strong positive correlation with solar radiation ($r = 0.92$) whereas it had a strong and negative correlation with precipitation ($r = -0.89$) and soil water volume ($r = -0.99$).

10.3.5 Relative Variable Importance in Post-Fire Recovery

The analysis of variable importance for climatic and topographic variables for the short-term vegetation recovery was performed by using random forest regression in python using scikit-sklearn library. The analysis was done by taking a sample size of 2000 with the constraints of features = 9, informative = 5, and random state = 1.

The results showed that temperature had the highest influence (> 55%) in the recovery process, followed by the topography, soil water volume, solar radiation, and precipitation. Among the topographical variables, elevation had the highest influence (~12%) among all.

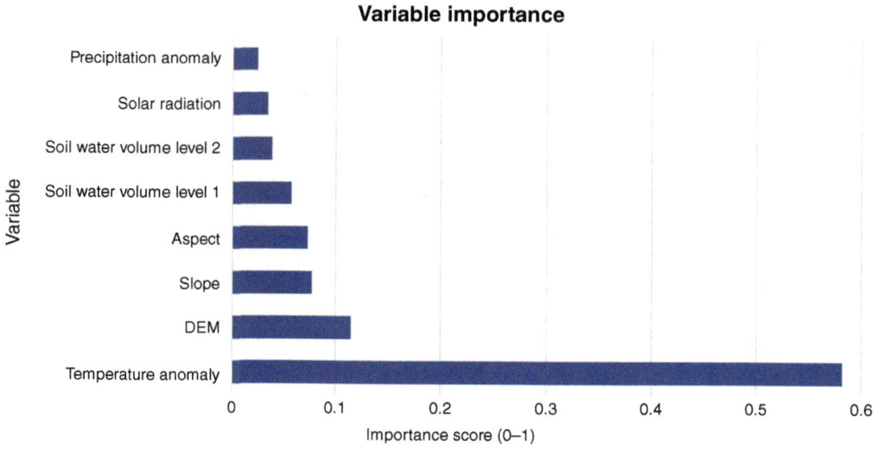

Figure 10.9 Relative importance of climatic and topographical variables in short-term bushfire recovery using random forest regression analysis.

Slope and aspect showed similar importance (7%) for the short-term vegetation recovery (Figure 10.9). Note that the statistical analysis including correlation and variable importance analysis had been calculated by using climatic and NDVI change data for five years.

10.4 Discussion

Landsat TM and OLI imageries have been utilized by several studies to perform bushfire severity and post-fire vegetation recovery assessment. GEE, however, is a relatively new planetary-scale platform, becoming increasingly popular for satellite image analysis. The availability of freely accessible GEE platform with petabytes of archived satellite imageries and their utilization supports for a better understanding of bushfires and post-fire vegetation recovery. Hence, in this study we confirmed from field validated data that the severity of bushfires impact may be reliably estimated by utilizing Landsat TM and OLI imageries in the GEE platform. Further, we also explored how the rate of recovery depends on the severity of impacts and how topographic and climatic variability influence the recovery process.

10.4.1 Bushfire Severity Assessment

This research utilized RdNBR and dNBR indices for bushfire severity classification, which has been widely utilized (Konkathi and Shetty 2021; Miller and Thode 2007; Pelletier et al. 2021; Sharma et al. 2022; Sparks et al. 2014). The accuracy assessment of most of these studies often lacks field-based validation. However, our study utilized field-collected data for validation and generated the accuracy of the performance for both dNBR and RdNBR methods. The overall accuracy range that we obtained in our study (86–90%) is higher, if not comparable to those reported by other researchers; Miller et al. (2009) (>75%),

Konkathi and Shetty (2021) (61%), Stambaugh et al. (2015) (40–50%), Sharma et al. (2022) (86–90%) Sparks et al. (2014) (92–96%), and 94% as reported by Storey et al. (2021).

The accuracy assessment using field-collected validation data showed that the RdNBR method results in higher accuracy than dNBR method for classifying burn severity classes in the heterogeneous forest of hilly terrain (Table 10.5 and Figure 10.5). This finding is following previous bushfire severity studies (Konkathi and Shetty 2021; Miller and Thode 2007; Pelletier et al. 2021; Sharma et al. 2022; Quintano et al. 2018) who demonstrated that the result in RdNBR is substantially higher than dNBR methods, especially for heterogenous landscape. For example, Konkathi and Shetty (2021) reported that the accuracy of RdNBR was higher than dNBR in both Sentinel-2 and Landsat images. Similarly, Pelletier et al. (2021) performed bushfire severity assessment in 446 different plots across 164 fires in California and demonstrated that the RdNBR had more accurate results than dNBR.

However, some of the research indicated that the performance of dNBR is better than RdNBR (Alonso-Gonzalez and Fernandez-Garcia 2021; Hudak et al. 2007; Sparks et al. 2014). Sparks et al. (2014) performed a fire severity assessment using dNBR, and RdNBR indices in patchy shrublands. Similarly, Alonso-Gonzalez and Fernandez-Garcia (2021) performed a landscape-level assessment of bushfire severity using dNBR, and RdNBR and demonstrated that the performance of dNBR was better than RdNBR. The contrasting results in these studies, however, are likely to be closely related to the landscape, vegetation, and fire attributes, such as the homogeneity of landscapes, severity and intensity of fires, or the diversity of vegetation in the area. As the RdNBR is mostly suited for improved performance in heterogeneous landscapes experiencing severe fire (Miller and Thode 2007), which is the characteristic of our study area, hence, the better performance of RdNBR is well justified.

The burn severity classification showed that the severity was high in the western part (Kilmore region) and the central-eastern part (Murrindindi region) of our study area. The fire severity depends on many factors including vegetation structure, fuel availability, and climatic and topographical variables (Harvey et al. 2019; Shryock et al. 2015). However, when we compared the spatial distribution of the severity of fire with the tree density map of the area prepared by the Australian Government, DELWP, we found that those areas with the dense trees experienced a severe fire. It is reasonable to consider that an increment in tree density increases canopy cover and vegetation fuels, which provides better vertical and horizontal fuel continuity. Such a finding in our study is supported by similar results as reported in earlier research (Cocke et al. 2005).

The discrimination of moderate and low severity areas including understory burn experienced misclassification problems, which was also reported in prior studies (Sharma et al. 2022; Stambaugh et al. 2015; Sunderman and Weisberg 2011). Our study showed 65 and 64% of users' accuracy for RdNBR and dNBR respectively, which is comparable to that reported by previous findings, i.e. Konkathi and Shetty (2021) (61%), and Stambaugh et al. (2015) (40–50%). Although, the poor accuracy in distinguishing less severe features can limit our understanding in several ways, this still may not be a major issue from the forest management perspectives. The key information that needs to be supplied for forest management is the areas to be recovered, which is often characterized by high burn areas (Cocke et al. 2005; Quintano et al. 2018). Hence, the remote sensing methods of burn severity classification, especially the RdNBR method, can be effectively applied. However, we acknowledge that research initiatives on new metrics for quantifying burn severity, for example, the relative burn ratio, and hence recommended to test as a successor of RdNBR (Parks et al. 2014).

10.4.2 Post-Fire Recovery Assessment

The NDVI based recovery assessment showed that the study area recovered its pre-fire NDVI within five years, which coincides with the previously reported findings. For example, Liu (2016) compared different post-fire recovery indices including NDVI and found that the indices monotonically increased for five to six years after fires.

However, unlike spectral indices-based recovery assessment, field-based vegetation recovery assessment reported a longer recovery time. For example, Bright et al. (2019) reported around 10+ years was required for complete post-fire recovery (Bright et al. 2019). This is mostly because the spectral indices-based methods can overestimate NDVI as they are largely influenced by the greenness of the area, hence, affected by the presence of weeds, grasses, and understory plant communities in the area. Whereas, field-based methods can only consider vegetation of interests during the assessment. Hence, regaining the NDVI to pre-fire level in five years does not necessarily mean that all the fire impacted vegetations are completely recovered within that time. Rather, vegetation recovery is a long process and may take multiple years as reported in the previous findings (Bright et al. 2019). Hence, remote sensing based recovery assessment are more suitable for quick estimation, whereas, field-based assessment is recommended for better accuracy.

The visual interpretation of the NDVI time series and the yearly statistics showed that although there was a rapid increase in NDVI for three years for all severity classes, a complete recovery of the burn severity had a strong influence on the recovery time. The rapid change in NDVI value within one year, especially in severe burn areas is probably due to the resprouting of weeds and grasses and the regeneration of understory plant communities. This does not necessarily mean that there is a rapid post-fire recovery including understory succession of vegetation communities within the first year to the extent stipulated by NDVI index. This is because the understory regeneration is also influenced by the severity of fires. This is supported by similar findings in previous studies, which reported that regeneration of understory plants is largely controlled by fire severity and species regeneration strategy (Wang and Kemball 2005). Hence, those areas experiencing highly severe fires require a longer time in recovery. This is well justified because highly severe fire burns the crown, which may require several years in the recovery process. In line with our finding, Maia et al. (2012) also reported that fire severity is the key factor governing bushfire recovery.

10.4.3 Climatic and Topographic Influence of Bushfire Recovery Assessment

A significant correlation between the climatic and topographical variables and vegetation recovery was observed in our study. This indicates that climatic and landscape variability may influence the post-fire recovery process. Previous findings show that the post-fire vegetation recovery depends mainly on the climatic condition of one year (next wet season) following a fire including post-fire drought, temperature, and precipitation (Liu 2016; Meng et al. 2015). Post-fire drought is also unfavorable to forest regeneration. Our study area experienced unfavorable climatic conditions for one year, i.e. a heatwave during the fire month (February), Australia's warmest August on record, Victoria's warmest winter on record, and the second heatwave in November. Similarly, the rainfall anomalies showed below average rainfall in 2009 (BOM 2021a). However, the situation was significantly favorable to post-fire recovery in 2010, which received the highest on record rainfall (BOM 2021b). The first change in NDVI of our study area was calculated by using the imageries of October 2010 (over 1.5 years) and it showed a significant increase in NDVI value for all severity classes. This suggested that the improvement in NDVI could have been primarily contributed by the next favorable wet season (summer of 2010) which is what was reported in the previous findings (Meng et al. 2015).

Similarly, the variable importance analysis in our study area showed that temperature was the most important factor, followed by topography. The correlation analysis also showed high correlations of temperature and solar radiation with recovery proxy (change in NDVI). However, although topography (elevation, slope, and aspect) had a weak correlation, it was identified as having a higher score of importance for post-fire recovery. This is justifiable because variable importance in Random Forest (RF) is calculated by estimating how much error in prediction increases when out-of-bag data for that variable is permuted, while other variables are left unchanged (Liaw and Wiener 2002). Similar results were observed in previous studies; Meng et al. (2015) showed that year 0+ temperature anomaly was the most important among climatic variables. Similarly, topography has been identified as an important variable in a previous study by Liu (2016), which is in line with our findings. However, unlike previous studies, our study did not rank precipitation among the most important factors, recognizing that other important factors might have more influence than our results suggested.

10.4.4 Limitations of this Study

The climatic data with 9 km × 9 km resolution could have limited our understanding due to insufficient information on the spatial variability of climatic conditions at a fine granular level. Further, we computed the climatic anomalies being based on an average of 12 years (2009–2020), which could have been calculated from a long-term average. Additionally, the vegetation recovery depends on a range of climatic and topographic factors, more than what we have considered in this study. Further, the remote sensing images utilized in this study had some seasonal variations due to the unavailability of quality images of the study area, which could have introduced phenological error, which is another limitation of this study.

10.5 Conclusions

Putting the identified limitations into perspective, this study compared the performance of widely used severity quantification indices, dNBR and RdNBR, and demonstrated that RdNBR performed better in the heterogeneous landscape environment. In addition, this study highlighted potential use of NDVI as an indicator of vegetation recovery assessment. This study further demonstrated that vegetation recovery is strongly influenced by climatic variability including temperature, solar radiation, and soil water volume and due to topography; slope, aspect, and elevation. We recommend future studies, including comprehensive climatic and topographic variables with higher resolutions as an assessment of recovery governing predictors. Further, we recommend incorporating granularity while considering climatic predictors, such as generating season-specific climatic averages and anomalies in future research. We recognized that optical remote sensing imageries, when collected at the time of fire, are covered with smoke making them difficult to analyze. Hence, as future work, we recommend investigating the capability of weather independent radar remote sensing in bushfire and post-fire recovery characterization.

Acknowledgements

The authors would like to acknowledge The University of Melbourne and the Australian Commonwealth Government for providing the research platform and environment, including Research Training Program Scholarships to undertake doctoral studies of Saroj K Sharma. We thank the Victorian Government, DELWP, for 2009 bushfire severity validation point data, the U.S. Geological Survey for Landsat images, the European Centre of Medium-Range Weather Forecast for climate data variables, Geoscience Australia for Elevation and Land Foundation Data (ELVIS), Digital Elevation Model, and Google for Google Earth Engine platform.

References

Alonso-Gonzalez, E. and Fernandez-Garcia, V. (2021). MOSEV: a global burn severity database from MODIS (2000–2020). *Earth System Science Data* 13 (5): 1925–1938. https://doi.org/10.5194/essd-13-1925-2021.

Ashton, D.H. (2000). The environment and plant ecology of the Hume range, central Victoria. *Proceedings. Royal Society of Victoria* 112 (2): 185–272.

Attiwill, P.M. and Adams, M.A. (2013). Mega-fires, inquiries and politics in the eucalypt forests of Victoria, southeastern Australia. *Forest Ecology and Management* 294: https://doi.org/10.1016/j.foreco.2012.09.015.

BOM (2021a). *Victoria in 2009*. Bureau of Meteorology www.bom.gov.au/climate/current/annual/vic/archive/2009.summary.shtml.

BOM (2021b). *Victoria in 2010*. Bureau of Meteorology www.bom.gov.au/climate/current/annual/vic/archive/2010.summary.shtml.

Bonney, M.T., He, Y., and Myint, S.W. (2020). Contextualizing the 2019–2020 Kangaroo Island bushfires: quantifying landscape-level influences on past severity and recovery with Landsat and Google Earth Engine. *Remote Sensing* 12 (23): https://doi.org/10.3390/rs12233942.

Bowman, D.M.J.S., Balch, J.K., Artaxo, P. et al. (2009). *Fire in the Earth System* 324 (April): 481–485.

Bowman, D.M.J.S., Williamson, G.J., Gibson, R.K. et al. (2019). The severity and extent of the Australia 2019–20 eucalyptus forest fires are not the legacy of forest management. *Nature Ecology & Evolution.* https://doi.org/10.1038/s41559-021-01464-6.

Bright, B.C., Hudak, A.T., Kennedy, R.E. et al. (2019). Examining post-fire vegetation recovery with Landsat time series analysis in three western North American forest types. *Fire Ecology* 15 (1): https://doi.org/10.1186/s42408-018-0021-9.

Bureau of Meteorology (2021). *Climate Statistics for Australian Locations*. Bureau of Meteorology www.bom.gov.au/climate/averages/tables/cw_088162.shtml.

Burke, M., Driscoll, A., Heft-Neal, S. et al. (2021). The changing risk and burden of wildfire in the United States. *Proceedings of the National academy of Sciences of the United States of America* 118 (2): 1–6. https://doi.org/10.1073/PNAS.2011048118.

Cai, L. and Wang, M. (2020). Is the RdNBR a better estimator of wildfire burn severity than the dNBR? A discussion and case study in southeast China. *Geocarto International* 0 (0): 1–15. https://doi.org/10.1080/10106049.2020.1737973.

Cameron, P.A., Mitra, B., Fitzgerald, M. et al. (2009). Black Saturday: the immediate impact of the February 2009 bushfires in Victoria, Australia. *Medical Journal of Australia* 191 (1): https://doi.org/10.5694/j.1326-5377.2009.tb02666.x.

Chuvieco, E., Ventura, G., Martín, M.P., and Gómez, I. (2005). Assessment of multitemporal compositing techniques of MODIS and AVHRR images for burned land mapping. *Remote Sensing of Environment* 94 (4): 450–462. https://doi.org/10.1016/j.rse.2004.11.006.

Cocke, A.E., Fulé, P.Z., and Crouse, J.E. (2005). Comparison of burn severity assessments using differenced normalized burn ratio and ground data. *International Journal of Wildland Fire* 14 (2): 189–198. https://doi.org/10.1071/WF04010.

Collins, B.M., Kelly, Æ.M., Van Wagtendonk, J.W., and Stephens, Æ.S.L. (2007). Spatial patterns of large natural fires in Sierra Nevada wilderness areas. *Landscape Ecology* 545–557: https://doi.org/10.1007/s10980-006-9047-5.

Cruz, M.G., Sullivan, A.L., Gould, J.S. et al. (2012). Anatomy of a catastrophic wildfire: the Black Saturday Kilmore East fire in Victoria, Australia. *Forest Ecology and Management* 284: 269–285. https://doi.org/10.1016/j.foreco.2012.02.035.

De Santis, A. and Chuvieco, E. (2007). Burn severity estimation from remotely sensed data: performance of simulation versus empirical models. *Remote Sensing of Environment* 108 (4): 422–435. https://doi.org/10.1016/j.rse.2006.11.022.

Dutta, R., Aryal, J., Das, A., and Kirkpatrick, J.B. (2013). Deep cognitive imaging systems enable estimation of continental-scale fire incidence from climate data. *Scientific Reports* 3: 3–6. https://doi.org/10.1038/srep03188.

Dutta, R., Das, A., and Aryal, J. (2016). Big data integration shows Australian bush-fire frequency is increasing significantly. *Royal Society Open Science* 3 (2): https://doi.org/10.1098/rsos.150241.

DELWP. (2021). *Fire Severity 2009 Ground Control Plot Locations*. Data Vic. https://discover.data.vic.gov.au/dataset/fire-severity-2009-ground-control-plot-locations

ECMRWF. (2021). *ERA5-Land*. European Centre for Medium-Range Weather Forecast. https://www.ecmwf.int/en/era5-land

Epting, J., Verbyla, D., and Sorbel, B. (2005). Evaluation of remotely sensed indices for assessing burn severity in interior Alaska using Landsat TM and ETM+. *Remote Sensing of Environment* 96 (3–4): 328–339. https://doi.org/10.1016/j.rse.2005.03.002.

Gill, A.M. (1981). Patterns and processes in open-forests of eucalyptus in southern Australia vegetation. In: *Australian Vegetation* (ed. R.H. Groves), 152–176. Cambridge University Press.

Gorelick, N., Hancher, M., Dixon, M. et al. (2017). Google Earth Engine: planetary-scale geospatial analysis for everyone. *Remote Sensing of Environment* 202: 18–27. https://doi.org/10.1016/j.rse.2017.06.031.

Halgamuge, M.N., Daminda, E., and Nirmalathas, A. (2020). Best optimizer selection for predicting bushfire occurrences using deep learning. *Natural Hazards* 103 (1): 845–860. https://doi.org/10.1007/s11069-020-04015-7.

Hammill, K.A. and Bradstock, R.A. (2006). Remote sensing of fire severity in the Blue Mountains: influence of vegetation type and inferring fire intensity. *International Journal of Wildland Fire* 15 (2): 213–226. https://doi.org/10.1071/WF05051.

Harvey, B.J., Donato, D.C., and Turner, M.G. (2016). High and dry: post-fire tree seedling establishment in subalpine forests decreases with post-fire drought and large stand-replacing burn patches. *Global Ecology and Biogeography* 25 (6): 655–669. https://doi.org/10.1111/geb.12443.

Harvey, B.J., Andrus, R.A., and Anderson, S.C. (2019). Incorporating biophysical gradients and uncertainty into burn severity maps in a temperate fire-prone forested region. *Ecosphere* 10 (2): https://doi.org/10.1002/ecs2.2600.

Hudak, A.T., Morgan, P., Bobbitt, M.J. et al. (2007). The relationship of multispectral satellite imagery. *Fire Ecology* 3 (1): 64–90.

Keeley, J.E. (2009). Fire intensity, fire severity and burn severity: a brief review and suggested usage. *International Journal of Wildland Fire* 18 (1): 116–126. https://doi.org/10.1071/WF07049.

Klinger, R., McKinley, R., and Brooks, M. (2019). An evaluation of remotely sensed indices for quantifying burn severity in arid ecoregions. *International Journal of Wildland Fire* 28 (12): https://doi.org/10.1071/WF19025.

Konkathi, P. and Shetty, A. (2021). Inter comparison of post-fire burn severity indices of Landsat-8 and Sentinel-2 imagery using Google Earth Engine. *Earth Science Informatics* 14 (2): 645–653. https://doi.org/10.1007/s12145-020-00566-2.

Kumar, L. and Mutanga, O. (2018). Google Earth Engine applications since inception: usage, trends, and potential. *Remote Sensing* 10 (10): 1–15. https://doi.org/10.3390/rs10101509.

Liaw, A. and Wiener, M. (2002). Classification and regression by random forest. *R News* 2 (3): 18–22.

Linnenluecke, M.K., Stathakis, A., and Griffiths, A. (2011). Firm relocation as adaptive response to climate change and weather extremes. *Global Environmental Change* 21 (1): 123–133. https://doi.org/10.1016/j.gloenvcha.2010.09.010.

Liu, Z. (2016). Effects of climate and fire on short-term vegetation recovery in the boreal larch forests of northeastern China. *Scientific Reports* 6 (June): 1–14. https://doi.org/10.1038/srep37572.

Long, T., Zhang, Z., He, G. et al. (2019). 30m resolution global annual burned area mapping based on landsat images and Google Earth Engine. *Remote Sensing* 11 (5): 1–30. https://doi.org/10.3390/rs11050489.

Maia, P., Pausas, J.G., Vasques, A., and Keizer, J.J. (2012). Fire severity as a key factor in post-fire regeneration of *Pinus pinaster* (Ait.) in central Portugal. *Annals of Forest Science* 69 (4): 489–498. https://doi.org/10.1007/s13595-012-0203-6.

Malak, D.A. and Pausas, J.G. (2006). Fire regime and post-fire normalized difference vegetation index changes in the eastern Iberian peninsula (Mediterranean basin). *International Journal of Wildland Fire* 15 (3): 407–413. https://doi.org/10.1071/WF05052.

Meng, R., Dennison, P.E., Huang, C. et al. (2015). Effects of fire severity and post-fire climate on short-term vegetation recovery of mixed-conifer and red fir forests in the Sierra Nevada Mountains of California. *Remote Sensing of Environment* 171: 311–325. https://doi.org/10.1016/j.rse.2015.10.024.

Miller, J.D. and Thode, A.E. (2007). Quantifying burn severity in a heterogeneous landscape with a relative version of the delta Normalized Burn Ratio (dNBR). *Remote Sensing of Environment* 109 (1): 66–80. https://doi.org/10.1016/j.rse.2006.12.006.

Miller, J.D., Knapp, E.E., Key, C.H. et al. (2009). Calibration and validation of the relative differenced Normalized Burn Ratio (RdNBR) to three measures of fire severity in the Sierra Nevada and Klamath Mountains, California, USA. *Remote Sensing of Environment* 113 (3): 645–656. https://doi.org/10.1016/j.rse.2008.11.009.

Parks, S.A., Dillon, G.K., and Miller, C. (2014). A new metric for quantifying burn severity: the relativized burn ratio. *Remote Sensing* 6 (3): 1827–1844. https://doi.org/10.3390/rs6031827.

Pelletier, F., Eskelson, B.N.I., Monleon, V.J., and Tseng, Y.C. (2021). Using Landsat imagery to assess burn severity of national forest inventory plots. *Remote Sensing* 13 (10): 1–20. https://doi.org/10.3390/rs13101935.

Prichard, S.J. and Kennedy, M.C. (2014). Fuel treatments and landform modify landscape patterns of burn severity in an extreme fire event. *Ecological Applications* 24 (3): 571–590. https://doi.org/10.1890/13-0343.1.

Prior, T. and Eriksen, C. (2013). Wildfire preparedness, community cohesion and social-ecological systems. *Global Environmental Change* 23 (6): 1575–1586. https://doi.org/10.1016/j.gloenvcha.2013.09.016.

Quintano, C., Fernández-Manso, A., and Fernández-Manso, O. (2018). Combination of Landsat and Sentinel-2 MSI data for initial assessing of burn severity. *International Journal of Applied Earth Observation and Geoinformation* 64 (July 2017): 221–225. https://doi.org/10.1016/j.jag.2017.09.014.

Rew, R. and Davis, G. (1990). NetCDF: an interface for scientific data access. *IEEE Computer Graphics and Applications* 10 (4): 76–82.

Röder, A., Hill, J., Duguy, B. et al. (2008). Using long time series of Landsat data to monitor fire events and post-fire dynamics and identify driving factors. A case study in the Ayora region (eastern Spain). *Remote Sensing of Environment* 112 (1): 259–273. https://doi.org/10.1016/j.rse.2007.05.001.

Rouse, J.W., Haas, R.H., Schell, J.A., and Deering, D.W. (1974). Monitoring vegetation systems in the Great Plains with ERTS. *NASA Special Publication* 351: 309.

Sharma, S.K., Aryal, J., and Rajabifard, A. (2022). Remote sensing and meteorological data fusion in predicting bushfire severity: a case study from Victoria. *Australia. Remote Sensing* 14: 1645. https://doi.org/10.3390/rs14071645.

Sharples, J.J., Cary, G.J., Fox-hughes, P. et al. (2016). Natural hazards in Australia: extreme bushfire. *Climatic Change* 85–99: https://doi.org/10.1007/s10584-016-1811-1.

Shryock, D.F., Esque, T.C., and Chen, F.C. (2015). Topography and climate are more important drivers of long-term, post-fire vegetation assembly than time-since-fire in the Sonoran Desert, US. *Journal of Vegetation Science* 26 (6): 1134–1147. https://doi.org/10.1111/jvs.12324.

Soverel, N.O., Perrakis, D.D.B., and Coops, N.C. (2010). Estimating burn severity from Landsat dNBR and RdNBR indices across western Canada. *Remote Sensing of Environment* 114 (9): https://doi.org/10.1016/j.rse.2010.03.013.

Sparks, A.M., Boschetti, L., Tinkham, W.T. et al. (2014). An accuracy assessment of the MTBS burned area product for shrub–steppe fires in the northern Great Basin, United States. *International Journal of Wildland Fire* 70–78: https://doi.org/10.1071/wf13206.

Stambaugh, M.C., Hammer, L.D., and Godfrey, R. (2015). Performance of burn-severity metrics and classification in oak woodlands and grasslands. *Remote Sensing* 7 (8): 10501–10522. https://doi.org/10.3390/rs70810501.

Storey, E.A., Lee West, K.R., and Stow, D.A. (2021). Utility and optimization of LANDSAT-derived burned area maps for southern California. *International Journal of Remote Sensing* 42 (2): 486–505. https://doi.org/10.1080/01431161.2020.1809741.

Sunderman, S.O. and Weisberg, P.J. (2011). Remote sensing approaches for reconstructing fire perimeters and burn severity mosaics in desert spring ecosystems. *Remote Sensing of Environment* 115 (9): 2384–2389. https://doi.org/10.1016/j.rse.2011.05.001.

Turco, M., Rosa-Cánovas, J.J., Bedia, J. et al. (2018). Exacerbated fires in Mediterranean Europe due to anthropogenic warming projected with non-stationary climate-fire models. *Nature Communications* 9 (1): 1–9. https://doi.org/10.1038/s41467-018-06358-z.

USGS (2021). *Landsat Mission*. USGS https://www.usgs.gov/core-science-systems/nli/landsat/landsat-level-1-processing-details.

Viana-Soto, A., Aguado, I., and Martínez, S. (2017). Assessment of post-fire vegetation recovery using fire severity and geographical data in the Mediterranean region (Spain). *Environments – MDPI* 4 (4): 1–17. https://doi.org/10.3390/environments4040090.

Vicmap. (2020). *Product Data Specification Vicmap Elevation DEM 10m*.

Wang, G.G. and Kemball, K.J. (2005). Effects of fire severity on early development of understory vegetation. *Canadian Journal of Forest Research* 35 (2): 254–262. https://doi.org/10.1139/x04-177.

Zhao, Q., Yu, L., Li, X. et al. (2021). Progress and trends in the application of Google Earth and Google Earth Engine. *Remote Sensing* 13 (18): 1–21. https://doi.org/10.3390/rs13183778.

Section IV

Advancement in RS-Drones and Multi-Sensors Multi-Source for Forest Monitoring

11

Recent Advancement and Role of Drones in Forest Monitoring: Research and Practices

Neeraj K. Maurya[1], *Amit K. Tripathi*[1], *Aashri Chauhan*[1], *Prem C. Pandey*[*,1], *and Salim Lamine*[2,3]

[1] *Center for Environmental Sciences & Engineering, School of Natural Sciences, Shiv Nadar Institution of Eminence (Deemed to be University), Greater Noida, Uttar Pradesh, 201314, India*
[2] *Faculty of Biological Sciences, University of Sciences and Technology Houari Boumediene (USTHB), PO Box 32 Bab Ezzouar, 16111, Algiers, Algeria*
[3] *Department of Geography and Earth Sciences, University of Aberystwyth, SY23 3DB, Ceredigion, Wales, UK*

11.1 Introduction

Forests are home to 80% of the world's terrestrial biodiversity and play a vital role in sustaining ecological balance. The world's forests are approximately 4 billion ha, i.e. 31% of the total land area (Lugten 2008). Russia, the US, Brazil, Canada, and China are the five most forest-rich countries, accounting for more than 50% of the total forest area (FAO and UNEP, 2020). Forests provide habitat to numerous species of flora and fauna, and millions of people worldwide depend on forests. India is one of the four mega-diversity countries of Asia, harboring huge floral and faunal diversity in different forest ecosystems. It has four important biodiversity hotspots; Indo-Burma, the Himalayas, and the Western Ghats and Sri Lanka. According to a recent report, the total forest cover of India is approximately 712,249 km^2, which is 21.67% of the country's total geographical area (FSI, 2019). Qualitative types of forest monitoring mainly refer to the seasonal or annual changes of different biophysical areas (e.g. LAI, tree height, biomass, *f*PAR, *f*COVER) and biochemical areas (chlorophyll, water content, protein, cellulose), which are estimated using EO datasets. Traditional methods involve intensive human resources, which are time-consuming and very expensive. However, the derived output is point-based data, but when it comes to the large spatial extent, spaceborne, and airborne platforms, both types of techniques show more potential than conventional methods.

Remote sensing coupled with geographic information systems (GIS) is an effective tool, as it is cost-effective and known for fast data acquisition, processing, and analysis. For quantification of tropical deforestation and habitat fragmentation, multitemporal satellite data were used (Skole and Tucker, 1993). With this technique, the accurate information of forest

*Corresponding author's email: prem26bit@gmail.com

0000-0003-4370-4086 (Neeraj K. Maurya), 0000-0001-9220-9001 (Amit K. Tripathi), 0000-0002-0361-8549 (Aashri Chauhan), 0000-0002-0049-1415 (Prem C. Pandey), and 0000-0002-1354-6542 (Salim Lamine)

ecosystems can be extracted from remotely sensed imagery, which could be helpful for sustainable forest management. Remote sensing (RS) has become an important data source in forestry applications, with different sensors (such as multispectral, hyperspectral, and synthetic aperture radar) and platforms (like unmanned aerial vehicles and satellites) available, which are used to map the distribution of forest vegetation and describe the three-dimensional forest structure (Lechner et al., 2020), species diversity (Malhi et al. 2020; Kumar et al. 2014), and several other applications (Pandey et al. 2020b; Pandey and Sharma 2021). In India, uses of remote sensing has been seen in diverse forestry applications such as in the study of biodiversity conservation (Menon and Bawa, 1997), forest fragmentation (Roy and Tomar, 2000; Jha et al., 2005), mapping and forest vegetation characterization (Behera et al., 2001), assessment and monitoring of tropical forest resources (Roy et al., 2002), deforestation (Kumar et al. 2010; Rai et al. 2017), forest cover change detection (Kumar et al. 2010), and mapping and classifying vegetation/forest types (Reddy et al., 2015; Roy et al., 2015).

The aim of this chapter is to deliver the importance of EO datasets and elucidate the role of drones in forest monitoring. It also illustrates how the research and practices in forestry are currently mostly dependent upon EO datasets, and thus highlighting capability in estimating biophysical and biochemical properties with recent advancements in drone-based forest monitoring.

11.2 Field Sampling Methods in Forest Application: Traditional to Present

The population structure of the forest ecosystems is generally characterized by the number or percentage of individuals in different girth/diameter classes and height classes by different researchers (Pandey and Shukla, 2003; Tripathi and Shankar, 2014; Shankar and Tripathi, 2017). In general, for tree sampling in forest ecosystems, the girth at breast height (GBH) or diameter at breast height (DBH) are frequently measured at 1.3 m above the ground level. Usually, 10 cm DBH or 30 cm GBH is used for tree size (Lieberman et al., 1996; Parthasarathy, 2001; Huang et al., 2003). Girth or size class distribution is commonly used to understand the forest's stand structure and regeneration status. The structure of the forests generally simplifies the stratum of the canopy layer, such as top canopy, sub-canopy, and understory. The girth class distribution analysis provides information about forest stand structure, while height class distribution tells about forest canopy structure; together, they provide complete scenarios of the forest structure. The past disturbance in the forest area is usually determined by the population structure of tree species (Tripathi and Shankar, 2014).

Globally, over the past four decades, there have been several sampling methods used in forest inventory, in which plot methods were widespread. The plots of different sizes and shapes have been in trend, for example, square plots (Parthasarathy, 2001; Lü et al., 2010), rectangular plots (Huang et al., 2003; Banda et al., 2006; Pandey et al 2019), long belt transects (Tripathi and Shankar, 2014), and large-scale permanent plots (Sukumar et al., 1992; Lieberman et al., 1996, Condit et al., 1996). In continuation, global positioning system (GPS) technology has been used in various applications recently (Pandey et al., 2021). It is used for taking ground control points (GCP) as it offers precise location correction for spatial remote sensing images.

Remote sensing as a tool provides a synoptic view of forests and their condition in real-time. Field surveys, literature reviews, and map interpretation are traditional methods and are time-consuming, data logged, and often too expensive; in this scenario, the importance of remote sensing is increasing. When used over traditional methods of monitoring forests, remote sensing technology can survey a large geographical area, including inaccessible areas, in a short span of time. The last few decades have witnessed explosive growth in the application of remote sensing and GIS to study vegetation's biophysical and biochemical properties, and address the problem of deforestation, landscape-level biodiversity conservation, land use, and land cover changes. The collected data from forests by remote sensors are used to analyze multiple applications such as monitoring the risk of wildfires to study forests' biophysical and biochemical properties, and for forest fire modeling (Sharma et al. 2012). The remotely sensed data of a particular place is frequently available in a range of spatial and temporal scales. Since the 1970s, Landsat satellites have provided imagery for the entire planet (Lechner et al. 2020). In addition, the technique uses different software (Arc-GIS, Erdas Imagine, and Google Earth) for georeferencing, digitization, vector analysis, image processing, classification, and other raster analysis calculations. Vegetative characteristics derived from remotely sensed data during forest monitoring and assessment depend mainly on the spatial and spectral resolution of the imagery used (Roy et al., 2002).

Aerial photography is considered the oldest remote sensing practice, and at first, it provided the necessary information in many forestry applications. The canopy density accuracy of forests depended on the scale of aerial photographs; with higher resolution of aerial imagery conveying more accuracy and due to high cost and difficulty in procurement, aerial photographs canopy density estimation could not become prevalent (Roy et al., 2002).

During development, remote sensing uses a series of satellite imageries such as IRS-1B, IRS-1C, Landsat TM (Thematic Mapper), and ETM+ (Enhanced Thematic Mapper Plus), Landsat MSS (Multispectral Scanner), IRS-1C LISS III, IRS-1D LISS III, IRS-1B LISS-I, IRS-P6, and Advanced Wide Field Sensor (AWiFS). Landsat TM and ETM+ are essential tools for studying tropical forests (Trigg et al., 2006). These images have been used to prepare a current vegetation map or a particular region/place. Supervised classification, a digital classification technique, has been widely used for forest cover mapping. For vegetation type classification, this algorithm was frequently used in image classification. In supervised classification, the pixels of a known area were used to train the model to classify an unknown area of landscape. Analysts closely control training sites. The spectral behaviour of training sites gives information on vegetation cover classes. Normalized difference vegetation index (NDVI) is the most generalized index for vegetation health monitoring and has been used to interpret remote sensing images to observe changes in forest health over time. During the past few decades, Land Use and Land Cover (LULC) is one of the most common remote sensing applications.

In remote sensing, a sensor has been a virtual device that captures data about an object, including forest vegetation. Each object is characterized by its distinctive spectral features, i.e. reflectance. Based on their unique spectral signature, the object can be identified from remote sensing imagery (Xie et al., 2008). During forest cover mapping, spectral reflectance in the red and near-infrared regions is mainly considered. Optical sensors acquire images in many spectral bands. Multispectral sensors acquire images in more than two broad spectral bands; conversely, hyperspectral sensors are known for a large number of much narrower bands (Lechner et al., 2020). The most widely used sensor is optical sensors, where the functions and design are similar to a normal digital camera, and it can record data in and beyond visible wavelengths, i.e. infrared and thermal wavelengths.

The multispectral sensors are the most commonly used in forest-related studies, as they measure reflectance in visible, near-, middle-, and far-infrared regions of the electromagnetic spectrum. In addition, hyperspectral remote sensing provides a wide range of applications in forest management, precision agriculture, water resources, soil, and geology (Pandey et al., 2020a). Hyperspectral remote sensing with the help of devices such as spectroradiometers or some spaceborne sensors, have higher spectral range with very narrow bandwidth, and deal with the spectral recording information of ground objects (Pandey et al., 2020b; Roy et al., 2015). As a result, minute changes may be mapped and high spectral and spatial resolution are used to classify the tree stand in the forest (Roy et al., 2002; Pandey et al., 2020b; Srivastava et al. 2020a). Furthermore, Sentinel-1A satellites, launched in 2014, provide high-resolution SAR data, which can be effectively used for forest change detection during consecutive years since it carries a C-band SAR, giving images in all light and weather conditions.

The rapid advancement in remote sensing technology leads us to acquire more Radio Detection and Ranging (RADAR) and Light Detection and Ranging data (LiDAR). The first use of Radar technology was seen during World War II to locate aircraft and ships. Imaging radar has been used in forestry applications such as land cover mapping, estimation of forest attributes including average tree height, timber volume, basal area, and biomass, and regrowth monitoring since 1960 (Roy et al., 2002). In recent times, LiDAR is considered worthy for characterizing vegetation structure, which is equivalent to Radar but uses LiDAR light. LiDAR is an important development in remote sensing technology, as it can directly measure the vertical structure of the forest and measure tree height by constructing a 3D structure of the forest. Using high-resolution unmanned aerial vehicles (UAVs) and digital aerial photography are also essential, as these can provide millimeter resolution multispectral data and 2.5D representations of forest structure, which can be used to identify individual tree species (Lechner et al., 2020). A list of the multispectral sensors compatible with drones and the bands they acquired are mentioned in Table 11.1. LiDAR sensors precisely describe canopy heights, stand volume, basal area, and above ground biomass (AGB) in different forest types. LiDAR is used to create high-resolution models of ground elevation with a vertical accuracy of 4 inches. It is usually mounted on a small aircraft with a laser scanner, a GPS, and an Inertial Navigation System (INS). With remote sensing technology, point cloud data collected by LiDAR can be quickly and easily analyzed with point cloud software.

UAVs, commonly known as drones, are operated remotely with a ground-based controller and mounted with a digital camera or sensors. The UAVs platforms' are cost-effective, highly flexible (positioned rapidly), and provide high spatial resolution imagery since they fly at low altitudes (Lechner et al., 2020). Currently, the multi-temporal images acquired by UAVs as low altitude remote sensing platforms are extensively used to estimate the biophysical and biochemical properties of the forests. For monitoring the forest with the help of remote sensing, the first step is the collection of forest ground data, which are biophysical and biochemical parameters that play an essential role in monitoring. The following sections will cover the biophysical, biochemical, and other applications in forestry by drones and illustrate their significant role in monitoring the forest through remote sensing.

Table 11.1 Common multispectral sensors available for UAV in market.

Model	Resolution	Frame rate	Weight (gm)	Spectral bands (nm)
Tetracam ADC Lite	2048 × 1536 pixel	—	200	Green (520–600) Red (630–690) NIR (760–900)
Mica Sense				
Red Edge-MX	1.280 × 960 pixel	1 capture per second (all bands)	231.9	Red (668) Green (560) Blue (475) Red edge (717) NIR (842)
Sentera	CMOS 12.3 Mp	30 fps	80	RGB, NIR
Ricoh GR2	16.2 MP	4 fps	221	RGB
Hasselblad				Visible
H6D-50C	50 Mp, CMOS	30 fps	2115	
H6D-100c	100 Mp, CMOS	30 fps	2130	
Sentera NDVI and red edge sensor	1.2 Mp cmos	7 fps	30	Red, red edge, and NIR
GXRA16	16.2 Mp	3 fps	550	Visible
Phase One iXU 1000	100 MP	0.7	1700	RGB
Mapir Survey 3	12 MP (RGB)	—	76	RG, NIR
Sentek GEMS	CMOS	1.4 (0.7 seconds between images)	170	RGB, NIR
GoPro Hero 4 Silver	12 MP	15	81.6	Visible

11.3 Biophysical Parameters Assessment Using Remote Sensing

Biophysical vegetation parameters play a significant role in dynamic environmental changes such as water, carbon, and understanding the nitrogen cycle (Gitelson, 2004). Numerous remote sensing devices are used to derive various sorts of biophysical data based on reflectance values or backscatter energy using various methods. These methods can be divided into (i) simple regression, (ii) vegetation index, and (iii) modeling approach. The simple regression model is nothing but the derivation of the correlation between the ground data and one of the bands; for example, we take LAI field data increasing with an escalation in the near-infrared (NIR) values but a decrease in the red region values. Large amounts of ground data is needed because that is the limitation of correlation. The vegetation index (VI) technique, also known as band rationing, is a spectral enhancement extensively used to estimate LAI, biomass, *f*PAR, *f*COVER, and other parameters, with semi-empirical and empirical models being used to create the relationship between VI and ground data (Kumar et al. 2013; Pandey et al. 2019a; Pandey et al. 2019b). One of the most widely used band ratios to estimate the above parameters is the NDVI. Other rationing approaches for this strategy include the enhanced vegetation index (EVI), normalized difference red-edge index (NDREI), and plant sense reflectance index (PSRI), among many others. Techniques such as these are considered simple modeling approaches, where models use radiative transfer through empirical models. PROSAIL, an empirical model-based approach, was introduced as a combination of (PROSPECT) and scattering by arbitrary inclined leaves (SAIL). PROSPECT is a radiative transfer model based of Allen's generalized "plate model" that represents the optical properties of plant leaves from 400 nm to 2500 nm. PROSPECT utilises two classes of input variables; the leaf structure parameter and the leaf biochemical content. Several biochemical and biophysical inputs are required for this model such as leaf chlorophyll content, LAI, water thickness, leaf mesophyll structure, mean leaf inclination angle, hotspot size parameter, and soil reflectance value, along with some external factors such as zenith, azimuth, solar angle, and horizontal visibility (Jacquemoud et al., 1995).

Multispectral remote sensing requires red bands; primarily red-edge and near-infrared bands. The green vegetation shows maximum and minimum reflectance values in this band spectrum, highlighting the particular characteristics (Gitelson, 2004). Many spaceborne multispectral sensors have high to moderate spatial resolution used for this type of study, for example, Sentinel-2, Landsat-8, and SPOT-5 (Guerini Filho et al., 2020; Shelestov et al., 2017; Shelestov et al., 2015). Hyperspectral sensors acquire data in large numbers of narrow bands; they derive species-level information for mapping (Pandey et al., 2014), absorb in blue and red wavelength regions, and reflect in green and NIR wavelength regions (Asner, 1998). Looking at the availability of hyperspectral sensor data, we may find that most of them are commercial; however, very few open source data are available. The drone-based hyperspectral sensors show their potential and are mentioned with the sensor name, manufacturer, spectral range, and number of bands acquired by them are mentioned in Table 11.2.

Microwave remote sensing techniques are also used to analyze biophysical properties (Ulaby et al., 2014). The semi-empirical (Water Cloud) model and machine-learning algorithm (Random Forests [RF] and Support Vector Machine [SVM]) are calculated based on backscatter energy. Ground data are mainly used to calculate crop phenology, LAI, Plant Area Index (PAI), crop geometry, and biomass using microwave data from the L & C bands. Full-polarised (HH-HV-VV) data can be used to calculate these parameters precisely. The dual-channel (VV-VH) method, on the other hand, provides high accuracy for this type of study (Mandal et al., 2021).

LiDAR, an active remote sensing activity, is also suitable for this type of study. It opens new doors for vegetation monitoring as it can gather more information about vegetation

Table 11.2 Hyperspectral sensors and their specifications.

Manufacturer and model	Resolution	Frame rate (Hz)	Weight (gm)	Spectral range (wavelength)
Thermotenix system	Silicom ampohorous	—	105	800–1200 nm
MS M350	640 × 512 pixel	4980	—	1.5–5.4 μm
Telops MSM2K	320 × 256 pixel	90 000	—	1.5–5.5 μm
MS M100k	640 × 512 pixel	12 0000	—	3–4.9 μm
Headwall-Photonics	InGaAs	—	1025	400–1000 nm

Table 11.3 Thermal sensors and their specifications.

Manufacturer and model	Resolution	Frame rate	Weight (gm)	Spectral range (μm)
FLIR Vue Pro	1920 × 1080 pixel	30 Hz	—	7.5–13.5
FLIR Duo Pro R	336 × 256 pixel	7.5 Hz	—	7.5–13.5
Workswell WIRIS 640	640 × 512 pixel	—	390	7.5–13.5
Zenmuse X7	336 × 256 pixel	30 Hz	449	7.5–13.5
YUNEEC E10T Thermal	320 × 256	30 Hz	364	8–14

than optical and microwave remote sensing. It can measure tree distribution vertically, produce three-dimensional mapping canopy structure patterns, and calculate biomass, tree height, LAI, and other valuable information (Hmida et al., 2017; Lefsky et al., 1999).

Thermal remote sensing can also monitor vegetation of any area based on the relationship between land surface temperature (LST) (Sinha et al. 2014) and several vegetation indices, such as LAI, canopy height, above ground biomass (AGB), vegetation optical depth, and continuous vegetation field (Kumar et al., 2020). Long-term phenomena could be monitored by implying these techniques, such as change in the pattern of vegetation over certain periods and climate change (see Table 11.3).

Currently, the remote sensing technologies discussed are being employed in research with a new approach by using multi-source, multi-sensor, multi-temporal data in estimating biophysical parameters for better accuracy, such as SAR (Sentinel-1), optical (WorldView-2), or integration/fusion of datasets at signal, pixel, features, or decision levels. By integrating two or more datasets, researchers are generating better products than individual images or data; examples of these types of studies are provided by Nuthammachot et al. (2022). Digital elevation model (DEM) data was used to calculate LAI of grassland (Lu and He, 2019) above ground biomass (ABG) using multi-sensors datasets such as like Sentinel-1 (microwave) and Sentinel-2 (MX) imagery (Nuthammachot et al. 2022).

EO datasets, especially hyperspectral images, are able to provide a large number of broadband and narrowband VIs for predicting forest parameters, biophysical parameters through NDVI, EVI, Atmospherically Resistant Vegetation Index (ARVI), Red Edge Normalized Difference Vegetation Index, Modified Red Edge Simple Ratio Index, Modified Red Edge Normalized Difference Vegetation Index, Structure Insensitive Pigment Index (SIPI), and Red Green Ratio (Tuominen et al., 2009). The model-based approach may also be used in hyperspectral data (Zhang et al., 2021) for biophysical parameter estimation. A huge number of published literature can be seen which utilizes multi-sensor-multi-source datasets for biophysical parameter estimation across the globe.

11.3.1 Above Ground Biomass (AGB)

Worldwide, due to carbon emission in the atmosphere from different sources, global warming is becoming a massive problem for society. The forest ecosystem is one of the crucial agents for carbon storage in soil, oceans, and atmosphere. The forest is the largest terrestrial ecosystem, where its biomass is an essential variable for estimating the ecosystems' carbon sequestration and carbon storage capacity. The precise forest biomass estimation is essential for acquiring an idea about the carbon cycle of the large forest ecosystem (Lu et al., 2005). Globally, the tropical forests account for a significant percentage of the global forests and play a crucial role in the global C-cycle (Carbon cycle) in terms of C-flux and the C-balance. Tropical forests are known for their higher net productivity, making them more efficient in carbon sequestration when compared with other forests.

Remote sensing techniques offer an advantage over traditional field methods of forest biomass estimation. The AGB of the forest ecosystem was evaluated by traditional field measurements, which was not feasible for large forest stands (Lu et al., 2020). It was measured directly with destructive harvesting for wood density, DBH or GBH, and height measurement, which was expensive and time-consuming. Furthermore, gathering all forest biomass data within a satellite pixel for the biomass-estimating model is not scientifically reasonable. It results in mismatching of spatial scale between ground-based observation and satellite data due to a difference in extent caused by pixel size. Moreover, accommodating the two or more quadrats to a single pixel and taking an average of all field measured values would determine the exact pixel values of field sites, and this was demonstrated as a valid and accurate approach by Dusseux et al. (2015) and Barrachina et al. (2015). It is therefore imperative to have more sampling sites to cover additional parts of the study site to overcome the mismatch issue and increase the accuracy of AGB estimation, to effectively measure and monitor the forest AGB of small to large forest stands. Many researchers in different parts of the world have used remote sensing datasets to estimate forest AGB using Landsat 7 ETM+ data, active remote sensing radar backscatter data, multi-temporal Landsat and PALSAR data, Landsat 8, and Sentinel-1A data (Zheng et al., 2004; Mitchard et al., 2009; Shen et al., 2016). Because of its easy accessibility and affordability (Basuki et al., 2013; Kajisa et al., 2009; Lu et al., 2012; Giannetti et al., 2018), researchers across the globe have used optical remote sensing for forest AGB estimation. Optical remote sensing forms a two-dimensional view of forests and other earth surfaces based on natural radiation from the sun (Timothy et al., 2016). In this respect, Landsat TM, AVHRR, and MODIS sensors provide globally consistent spatial data at local to global scales (Timothy et al., 2016).

The new emerging technology using UAVs has made it possible to solve both problems mentioned. It has its merits in estimating terrestrial biomass (Bendig et al., 2014), and its flight mode is more adjustable compared to the satellite remote sensing data. According to the requirements, it can be flown at an altitude of a few meters to hundreds of meters. UAVs are equipped with various sensors that invert vegetation information. Since the AGB cannot be calculated directly, it is calculated with the help of allometric equations (Ota et al., 2017; Giannetti et al., 2018; Alonzo et al., 2018) and regression (Cao et al., 2019; Alonzo et al., 2018). Therefore, the data should be high resolution for an accurate ABG estimation.

For example, UAV-Lidar systems have been used to estimate the biomass of forests (Wallace et al., 2012), crops (Tilly et al., 2014), and grassland (Cooper et al., 2017). The long-wavelength bands are considered more beneficial for biomass estimation. The estimation of AGB using airborne LiDAR remote sensing is precise compared with radar or passive optical data (Zolkos et al., 2013). Radar is an active type of remote sensing and has the ability to penetrate through forest canopy and interact with the tree trunks and branches, which contributes to the main portion of AGB (Liu et al. 2020). However, active sensors, like Radar and LiDAR, have some limitations as they are known for complex preprocessing and the high cost of data acquisition. Zhang et al. (2018) used UAVs to estimate grassland canopy height and AGB in China. With high-resolution UAV acquired data, they developed canopy height model (CHM) and biomass model, and reported the performance, i.e. the field height measurement and biomass model shows high correlation. Authors extolled the role of RS datasets acquired from UAVs to estimate biomass and height estimation at the quadrant scale, as demonstrated in Figure 11.1 in a study conducted by Zhang et al. (2018).

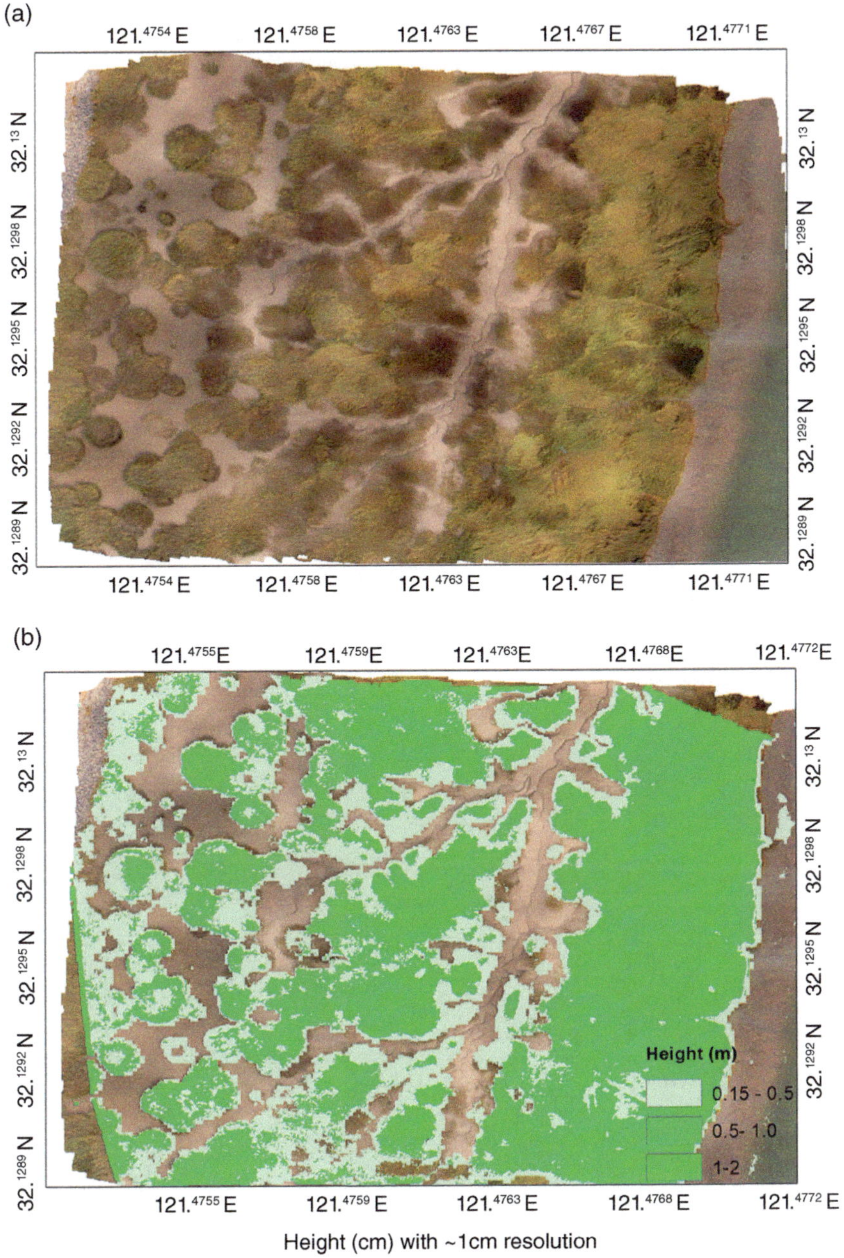

Figure 11.1 (a) The UAV orthophotoquad image of the spatial distribution of *Spartina alterniflora* in the Jiangsu study site; (b) height distribution map with ~1 cm resolution;

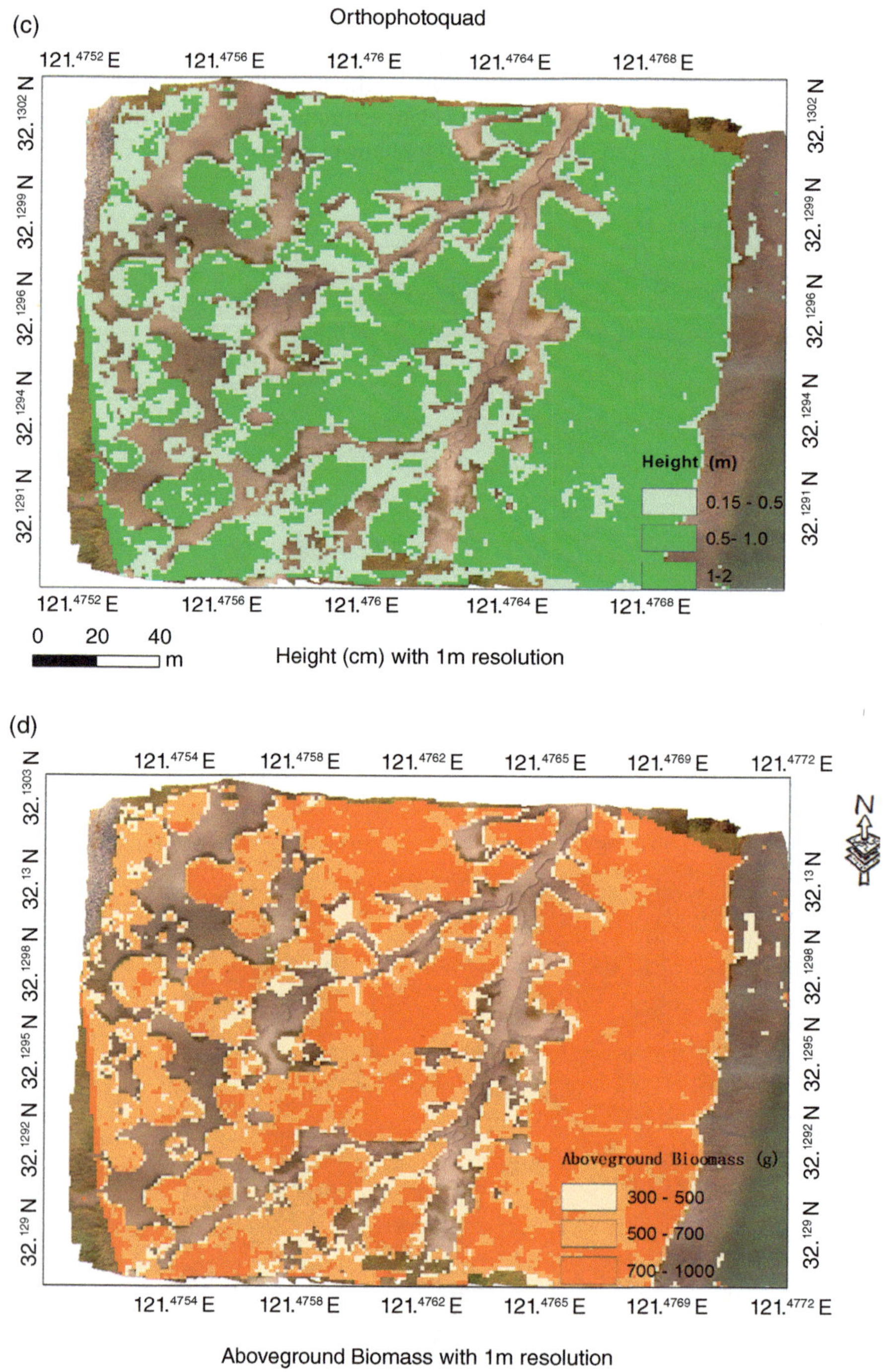

Figure 11.1 (Continued) (c) resampling height distribution map with 1 m resolution; and (d) inverted aboveground biomass distribution map. Source: Adopted from Zhang H et al. (2018), MDPI, CC BY 4.0.

Santini et al. (2019) employed UAVs to acquire RGB, MX, and thermal images for phenotyping of forest genetic trails. Authors utilized MX/RGB to estimate vegetation indices such as NDVI, EVI, Renormalized Difference Vegetation Index (RDVI), Optimized-Soil-Adjusted Vegetation Index (OSAVI), MCARI, TCARI, ARI2, CRI2, WBI, and a combination of ratios. At the same time, thermal images were used to assess the canopy temperature. Environmental conditions were also taken into account to measure VPD (vapor pressure deficit) for the completion of the study. The authors discussed the potential of UAVs for assessing morphophysiological differences among tree populations compared to usually costly and time-consuming traditional techniques (see Figure 11.2).

11.3.2 Tree Height and Diameter at Breast Height (DBH)

Tree height and diameter at breast height (DBH) are essential for estimating a forest stand's economic and ecological significance and quantifying timber resources. Both play an

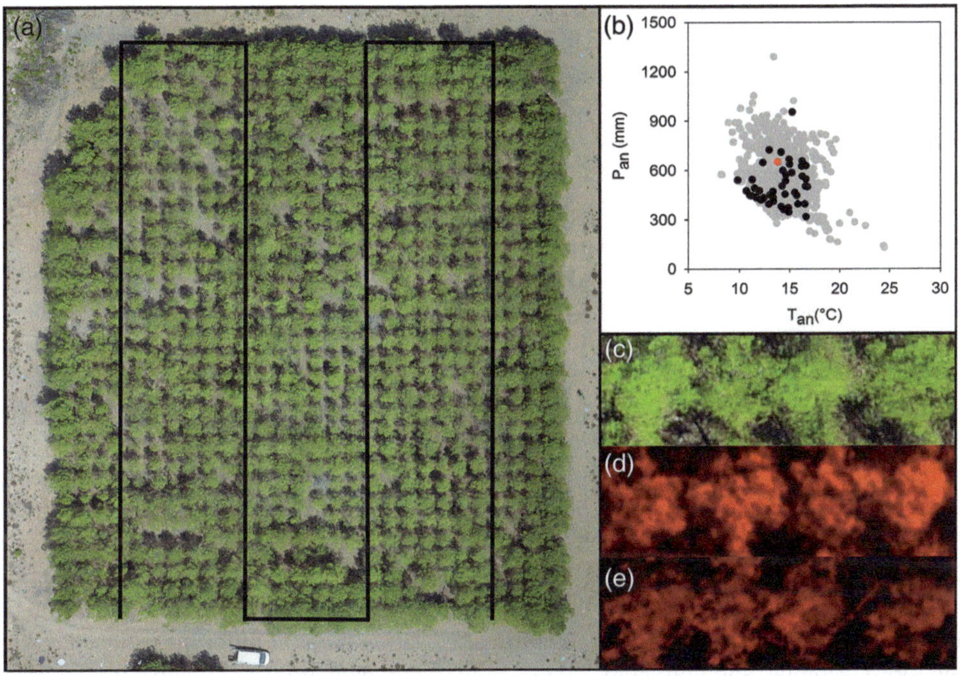

Figure 11.2 Aerial images of the genetic trial of *Pinus halepensis* considered for this study and the main characteristics of the 56 populations tested. (a) Aerial image of the complete trial; the black line represents the approximate trajectory followed by the unmanned aerial vehicle. (b) Mean nnual precipitation (Pan) and temperature (Tan) for the distribution range-Distribution map European Forest Genetic Resource Program [EUFORGEN] (http://www.euforgen.org/species/pinus-halepensis/) of *P. halepensis* (grey dots) calculated in 10′ resolution grids from the WorldClim database (period 1960–1990). The site's temperature and precipitation where the 56 populations were located at the origin (black dots) and the trial site (red dot) is shown. (c) Aerial image of one experimental unit in RGB. (d) Aerial image of one experimental unit [as in (c)] in infrared (false color). (e) Aerial image of one experimental unit [as in (c)] in infrared, but cropped for pixels containing vegetation (NDVI > 0.5). Source: Santini et al. (2019), with permission of Elsevier.

essential role in calculating individual and total stand volumes (Husch et al., 2002), assessing a site's overall productive capacity (West and West, 2009), and identifying the social status of a single tree's ability to access resources (Thenkabail, 2015). Furthermore, both the parameters are essential in estimating forest biomass and carbon stocks (Andersen et al., 2006; Anand et al. 2020), which are factors that are becoming increasingly important for climate change mitigation through forest activities (Bolte et al., 2009; Spathelf et al., 2018; Krause et al., 2019). The traditional methodology for measuring tree height in the field is direct or indirect. The direct method, i.e. destructive, is in which a tree is harvested and measured using meter tape from base to tip (Van Laar and Akça, 2007). On the other hand, the indirect method, which is a non-destructive method, engages trigonometric methods (West and West, 2009) using the hypsometer (Apostol et al., 2016), laser (Falkowski et al., 2006), or total station (Anderson and Gaston, 2013). The number and arrangement of trees, shrubs, and herbs provide forest canopy structure at a regional scale in the particular forest.

Moreover, for DBH, tree trunks are measured at the breast height using the meter tape, and the calliper is the direct measurement. These methods are limited to certain areas only. For DBH measurements, additional indirect methods such as relascopes and dendrometers have been developed and used. Under normal forest conditions, Kalliovirta et al. (2005) evaluated a laser-relascope, a mix of a relascope and a dendrometer. They calculated a standard error of 8.2 mm by comparing the readings to the values obtained using a steel calliper. The accuracy is dependent on the distance between the observer and the tree. Measuring tree heights and DBH by these instruments is labour-intensive and time-consuming. Alternatively, copious research has been recently performed globally to determine both using remote sensing techniques, which are cost-effective and have the advantages of fast data acquisition, processing, and analysis.

LiDAR is one of the latest among EO datasets for estimating forest stand height, 3D structure, and canopy cover. Most widely used are airborne and terrestrial LiDAR scanners. LiDAR integrated with optical data is utilized by many researchers in forest survey and assessment. Over the years, researchers have developed numerous algorithms for tree height estimation from different types of LiDAR data. Nilsson (1996) estimated tree height and standing volume with an airborne LiDAR system with accurate outcomes. Dubayah et al. (2010) studied tropical forest tree height using LiDAR remote sensing in Costa Rica. Furthermore, Shendryk et al. (2014) combined LiDAR data and SPOT-5 multispectral images to evaluate forest canopy height. Figure 11.3 illustrates the utilization of UAV LiDAR point cloud data for generation of elevation maps at different regions including forest, ecotone, and grassland. LiDAR point cloud data are very effective in detecting individual trees and their stems, and extract the DBH accurately as illustrated in Figure 11.4.

11.3.3 Leaf Area Index (LAI)

The total one-sided green leaf area per unit of ground surface known as LAI provides information about the amount of light that can be intercepted by leaves. Mathematically, LAI can be demonstrated as leaf area (m^2) per ground area (m^{-2}) and is a unit-less

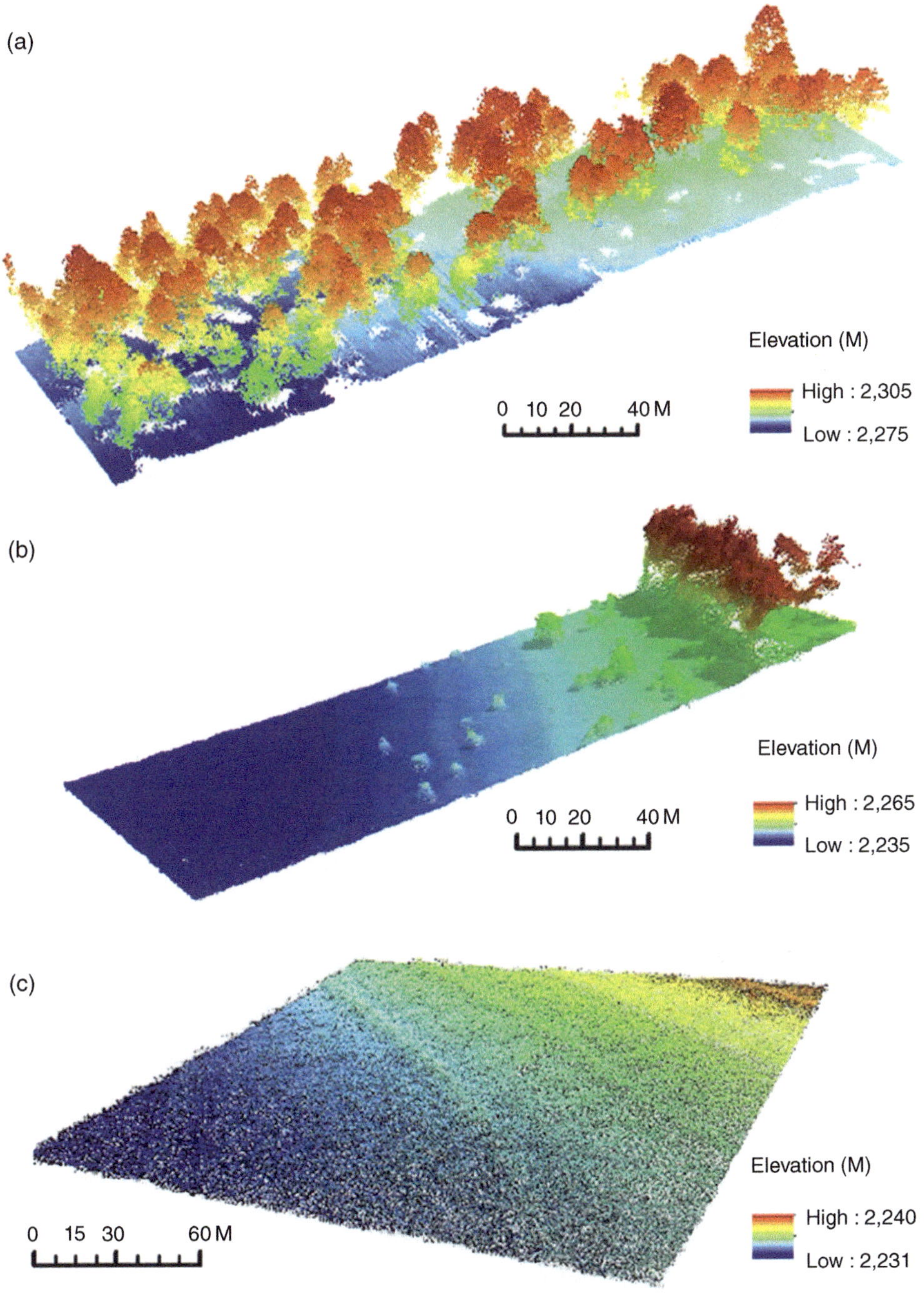

Figure 11.3 Examples of (a) the octocopter UAV LiDAR point cloud data in a forested site; (b) fixed-wing UAV-derived structure from motion (SfM) point cloud data at an ecotone site; and (c) fixed-wing UAV-derived SfM point cloud from a grassland environment with micro-topographic variability created by small man-made furrows. (Adopted from Sankey et al., 2017; with permission of Elsevier.)

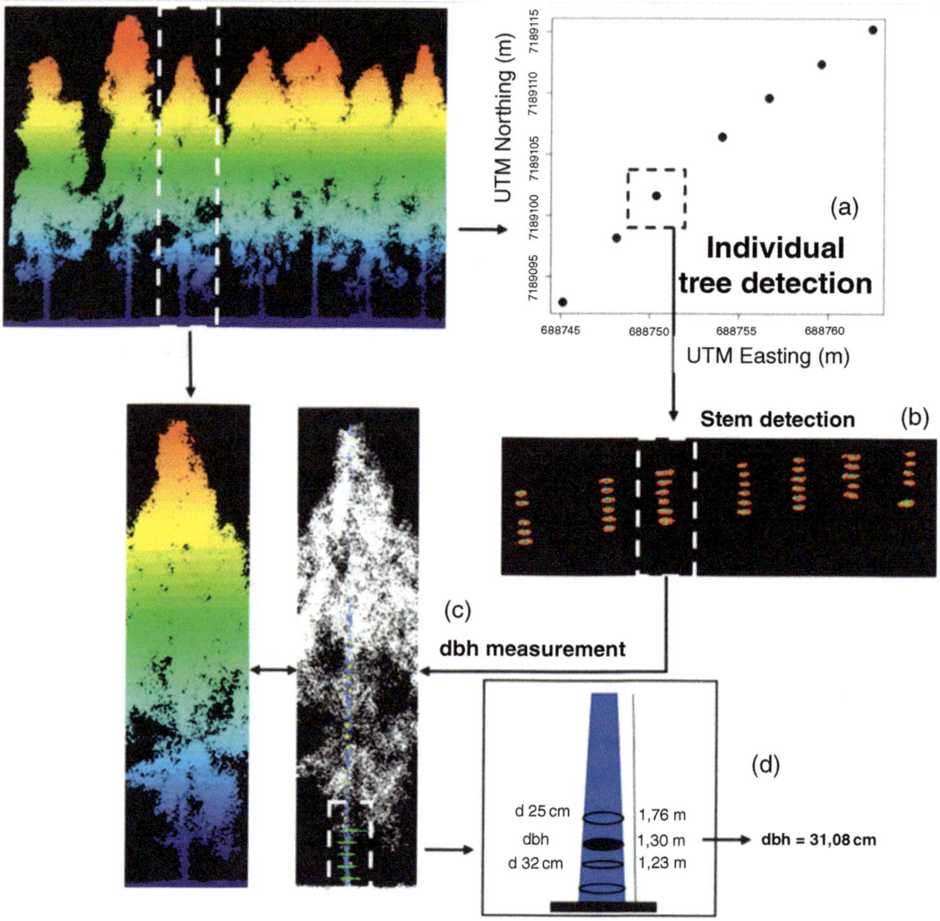

Figure 11.4 Automatic isolation of stems of isolated trees and taper extraction using TreeLS R package. (a) Extract tree map (positioning); (b) classification of stem points (classify stem points); (c) extraction of diameter measurements (extract measure); (d) interpolation algorithm, which locates the two closest diameters of the DBH to obtain the value at the height of 1.30 m. Where: DBH is diameter at breast height; D is two closest diameters of the DBH. [(c) Source: Adapted from Dalla Corte et al. (2020), MDPI, CC BY 4.0.]

notation. Therefore, it is an essential structural feature of vegetation to gather information about the forest density as well as other functioning of the plants (Chen and Cihlar, 1996). The plant functions governed by LAI directly and indirectly are canopy interception, evapotranspiration, and gross photosynthesis; these are essential activities closely related to the LAI because leaf surfaces are the primary energy and mass exchange sites. The long-term assessment of LAI provides an understanding of dynamic changes in productivity and climate impacts on forest ecosystems (Zheng and Moskal, 2009). LAI is used as a critical indicator of stress in forests; it is a measure of photosynthetically active forest canopy structure. In this way, LAI estimation is essential to understand net primary production, carbon balance, biogeochemical cycles, the mass exchange between the atmosphere and the forest ecosystem.

In traditional methodologies, estimation of LAI was not easily estimated for large spatial extents showing its work-intensive and time-consuming nature; however, emerging remote sensing technologies, especially optical and active remote sensing techniques, upgraded these issues significantly. There have been two broad methods for estimation of LAI, the direct (point quadrat sampling) and indirect methods (Jonckheere et al., 2004). The direct measure or point quadrat sampling (early method) involves destructive sampling and litter fall collection. However, estimating in the heterogeneous forest is complicated because it depends on extrapolating the allometric equations (Chen and Cihlar, 1995). In contrast, indirect methods are based on optical instruments such as LI-COR-LAI-200, TRAC, or digital hemispherical photography (DHP) are used widely for LAI estimation (Jonckheere et al., 2004; Chen and Cihlar, 1995; Schleppi et al., 2007). These instruments and models are less time-consuming and can be applied to a broader sampling field location (Leblanc et al., 2005), as the theories behind them are well developed. The remotely sensed, fine-resolution spatial and spectral (hyperspectral) data can be used to retrieve LAI (Lee et al., 2004; Colombo et al., 2008). Currently, significantly less work has been done using Hyperspectral sensors on UAVs.

LiDAR is an indirect method for LAI estimation. LiDAR is used for many forest monitoring studies related to LAI (Solberg et al., 2009; Tang et al., 2012) and estimates LAI based on the correlation with the canopy gap fraction, derived by various laser penetration matrices (LPMs) (Zhao and Popescu, 2009; Richardson et al., 2009). Most of the time, allometric equations were employed to estimate LAI using LiDAR derived canopy height and foliage density (Luo et al., 2015; Pope and Treitz, 2013). Table 11.4 illustrates the widely used LiDAR sensors across the globe with their specifications and model/make, which mostly depend on their categories. Based on the return information, LiDAR

Table 11.4 LiDAR sensors and their specifications.

LiDAR Sensors
RedTail LiDAR systems' RTL-450
Velodyne - HDL-32E
RIEGL - VUX-1UAV
Leica ALS80
FARO Focus 3D × 130
LeddarTech - Vu8
Routescene - UAV LidarPod
YellowScan Surveyor UAV
Geodetics - Geo-MMS SAASM
Phoenix Scout-16
Snoopy A-Series

Table 11.5 List of widely used drone type along with their pros, cons, and use.

Types of drones	Pros	Cons	Typical uses
Multi-rotor	Accessibility	Short flight duration	Aerial photography and video aerial inspection
	Ease of use	Small payload capacity	
	VTOL and hover flight	—	
	Good camera control	—	
	Can operate in a confined area	—	
Fixed-wing	Long endurance	Launch and recovery needs large space	Aerial mapping, pipeline and power line inspection
	Large area coverage	no VTOL/hover	
	Fast flight speed	Harder to fly, more training needed	
	—	Expensive	
Single-rotor	VTOL and hover flight	More dangerous	Aerial LiDAR laser scanning
	Long endurance (with gas power)	Harder to fly, more training needed	
	Heavier payload capability	Expensive	
Fixed-wing hybrid	VTOL and long-endurance flight	Not perfect at either hovering or forward flight	Drone delivery
		Still in development	

is divided into three categories; (a) discrete return, (b) full waveform, and (c) photon counting based on the detection methods (Wulder et al., 2012). As per the convenience to mount LiDAR sensors on different platforms, the LiDAR data can be categorized into three groups; (a) terrestrial laser scanning (TLS), (b) airborne laser scanning (ALS), and (c) spaceborne laser scanning (SLS). Table 11.5 compiles the list of widely used drone types, their use and utilization in research along with their pros and cons.

ALS is used on UAVs for estimating multiscale LAI (Vincent et al., 2017). The advantage of employing ALS is that it covers a relatively large area, especially in a contiguous manner. The ALS point cloud data is collected using a small laser pulse footprint ranging from 0.1 m to 3 m at altitudes between 500 and 3000 m. ALS point cloud data is the most prevalent data type used in forest inventory applications, several system developers, and an increasing number of commercial client support data acquisition and analysis (Hyyppä et al., 2008) from 10 m to 30 m. A standard ALS full-waveform data system is the airborne Land, Vegetation, and Ice Sensor (LVIS) (Blair et al., 1999). Figure 11.5 shows the point cloud data of a forest plot (colored by height), with an average point density of 4.3 points/m^2 obtained by an ALS (a), and the LVIS waveform data with a footprint size of 20 m (b).

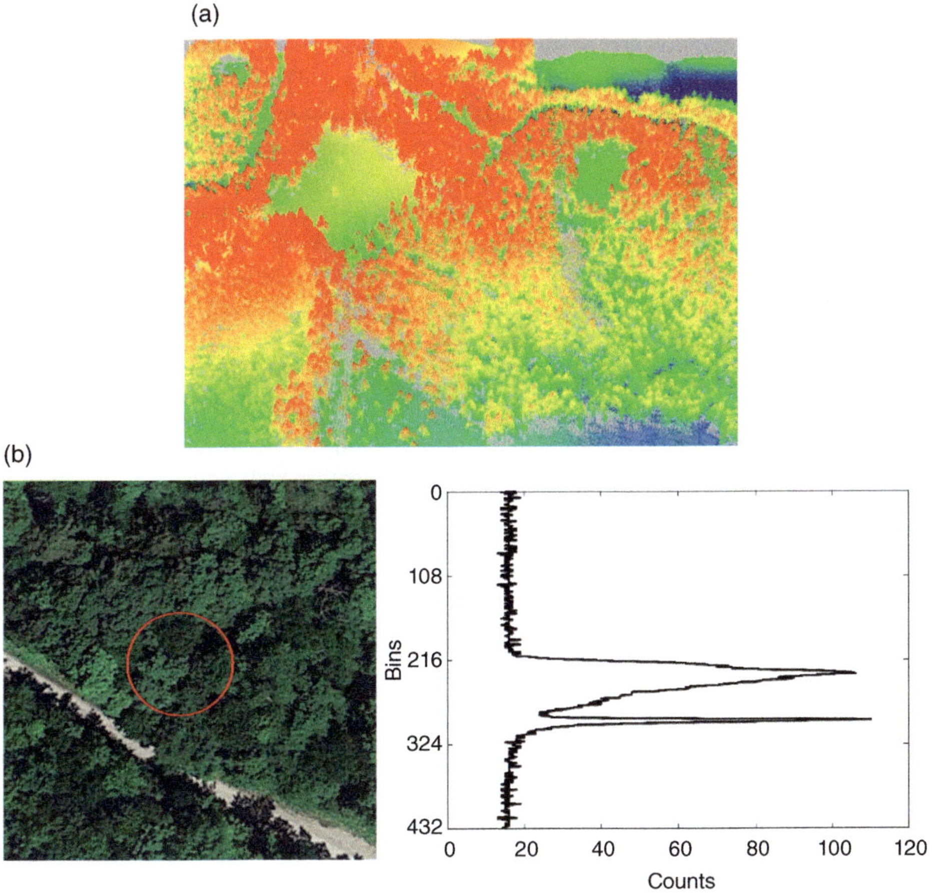

Figure 11.5 Airborne laser scanning (ALS) data examples. (a) Airborne point cloud data colored by height (4.3 points/m^2); (b) airborne large-footprint full-waveform data (LVIS) ~20 m in diameter (44°2′59″N, 71°17′18″W), 2009. The image in the left panel is from Google Earth®, and the right panel is a waveform return where the upper and lower peaks come from the forest canopy and the ground surface, respectively. Source: Adapted from Wang and Fang (2020), MDPI, CC BY 4.0.

11.4 Biochemical Parameter Assessment Using Remote Sensing

Biochemical parameters such as lignin, protein, cellulose, hemicellulose, sugar, starch, chlorophyll, and water also play crucial roles in monitoring the vegetation. These parameters incorporate the photosynthetic process and maintain ecosystem sustainability (Shengyan et al., 2002; Du et al., 2020). These parameters are estimated by three methods, (i) index method, (ii) stepwise multiple linear regression, or proposing a statistical analysis for the generated regression equation between high-resolution remote sensing data and laboratory-based chemical information acquired from semi-empirical or empirical models, and (iii) stepwise multiple linear regression based on a model of the forest crown. For correction, the previous method 3-D models are merged (e.g. DART model merged with PROSPECT model, RSADU merged with LIBERTY leaf model) (Peterson et al., 1988; Shengyan et al., 2002). For plant monitoring, various sensors are available, including multispectral, hyperspectral, and thermal. When these sensors are mounted on UAVs, a considerable amount of comprehensive knowledge mandates the creation and use of powerful analysis techniques capable of extracting information about the structural and biochemical composition of plants, allowing for a better knowledge of relevant plant features (Hassler and Baysal-Gurel, 2019). Hyperspectral data are more suitable for estimating biophysical information than traditional multispectral data (Darvishzadeh et al., 2010). However, using the empirical model, two and three-band based rationing named 2BDA (two-band algorithm), 3BDA (three-band algorithm), and NDCI (Normalized Difference Chlorophyll Index) on Sentinel-2 data for chlorophyll calculation show higher accuracy (Xu et al., 2019). Along with the three pioneer methods, vegetation classification techniques like maximum likelihood classifier (MLC) and biochemical content classifier (BCC) on Hyperion (EO-1) sensors give accurate value (Jia et al., 2011).

11.4.1 Canopy Chlorophyll Content (CCC)

Chlorophylls are miraculous plant pigments that use light energy for photosynthesis. Their amount in the leaves determines the nutrient status in the vegetation. In other words, chlorophyll content (Ca + b) is a vital sign of plant photosynthetic status related to productivity. Canopy chlorophyll content (CCC) is defined as "the total amount of chlorophyll a and b pigments in a contiguous group of plants per unit ground area" (Gitelson et al., 2005). Often expressed in g/m^2, CCC is a product of leaf chlorophyll content per unit area and LAI (Darvishzadeh et al., 2010). This way, it describes the total photosynthetically active radiation absorbed by the canopy (Gitelson et al., 2015). The estimation of CCC provides valuable information related to biotic stress factors and vegetation physiological status, as it could be due to plant disease, nutritional stress, and environmental stresses, so, the estimation of CCC is used as an indicator of forest health (Ali et al., 2020).

Remote sensing is used to retrieve chlorophyll content over large areas with an accurate, low-cost, and real-time approach by establishing empirical relationships between different vegetation indices (VIs) and chlorophyll content or through physical-based approaches relying on the inversion of canopy reflectance models. CCC's remote sensing is mainly based on optical remote sensing, covering the visible to NIR spectral region and the short wave infrared region (SWIR). Apart from CCC (Ca + b), the other parameters such as LAI, fractional cover, biomass and absorbed photosynthetic active radiation, and leaf fraction exposed to light are essential to understand plant functioning (Clevers and Gitelson, 2013).

With the advent of Sentinel-2A and Sentinel-2B, it has become much easier to estimate Chl-a and Chl-b and other biophysical parameters in forest research (Clevers and Gitelson, 2013). The Sentinel-2 multispectral imager (MSI) is characterized by three spectral bands corresponding to band numbers 5, 6, and 7 at 705 nm, 740nm, and 783 nm wavelengths, respectively, that exploit the red-edge information suitable for quantifying CCC (Ali et al., 2020) and vegetation health stress. It is now possible with Sentinel-2 to produce large-scale CCC maps with a spatial and temporal resolution, opposite than those used for monitoring ecological processes like vegetative decline or stress (Zarco-Tejada et al., 2019). In addition, hyperspectral remote sensing imagery has great potential for accurate retrieval of total leaf chlorophyll content in complex forest canopies (Zhang et al., 2008). Zarco-Tejada et al. (2019) conducted chlorophyll (C_{a+b}) content estimation in a sparse pine forest dominated by *Pinus pinaster* with Sentinel-2A and hyperspectral imagery in the context of forest decline in Spain. Gitelson (2004) investigated the spectral reflectance of several plant leaves such as maple, chestnut, wild vine, and beech in a wide range of pigment content. Zhang et al. (2008) developed a hyperspectral remote sensing algorithm to retrieve chlorophyll content in open spruce and closed forests in Ontario. Ali et al., (2020) implemented several algorithms to explore the potential of sentinel-2 data in CCC retrieval in a heterogeneous mixed mountain forest in Germany. It has been concluded that reflectance measurement is quite helpful in quick and accurate measurement of Chl-content with non-destructive approach.

11.4.2 Canopy Water Content (CWC)

Canopy water content (CWC) is the ratio of mass of water content to unit ground area (Clevers et al., 2010) It comprises water, which makes up 40–80 percent of the volume of plant leaves, and is the key factor in monitoring biochemical (Shen et al., 2005) and biophysical parameters in agricultural and forestry applications. Its determination leads to assessing a variety of applications in drought monitoring (Tucker 1980), forest fire susceptibility (Chuvieco et al., 2002; Yebra et al., 2013), soil moisture retrieval (Yilmaz et al., 2008), and other applications. Employing a high-resolution sensor may allow us to detect the spatial changes of CWC within a field, which can be used for forest monitoring. In the recent past, several studies have been conducted and demonstrated the link between the CWC and canopy reflectance. CWC estimation using remote sensing relies on the water absorbing bands of wavelength 970 nm, 1450 mn, and 1950 nm (Clevers et al., 2010).

Different vegetation indices developed at different wavelengths have shown to have the best relationships with canopy reflectance to leaf water content in several studies (Wang and Fang, 2020). According to Danson and Bowyer (2004), the optimal wavelength at which the most significant link with leaf water content is found depends on the quantity and water content ranges in the leaf sample under examination. Each of the wavelengths chosen is thought to be suitable for determining leaf water content in a certain plant species.

11.5 UAV-Based Remote Sensing

One of the specific applications where remote sensing has been actively employed is monitoring forest ecosystems, particularly national forest inventory (NFI) programs to improve estimates (Barrett et al., 2016). Forest ecosystems are highly dynamic, making accurate and up-to-date data extremely crucial (Goodbody et al., 2019). Regular data collection could be costly depending on which remote sensing platform is employed (Colomina and Molina, 2014). The geographical and temporal resolutions given by traditional satellite-based data are typically unsuitable for achieving regional or local objectives. Airborne platforms, on

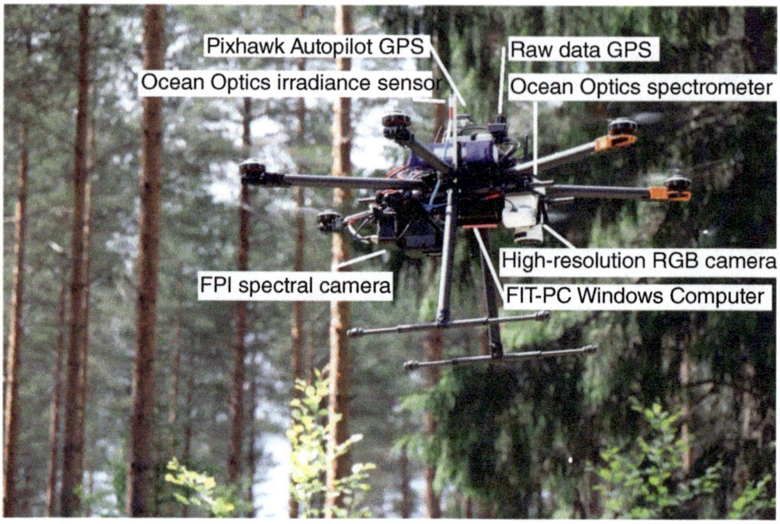

Figure 11.6 Unmanned aerial vehicle (UAV) remote sensing system of the Finnish Geospatial Research Institute (FGI) is based on the Tarot 960 hexacopter. (Source: Adapted from Nevalainen et al. (2017), MDPI, CC BY 4.0.)

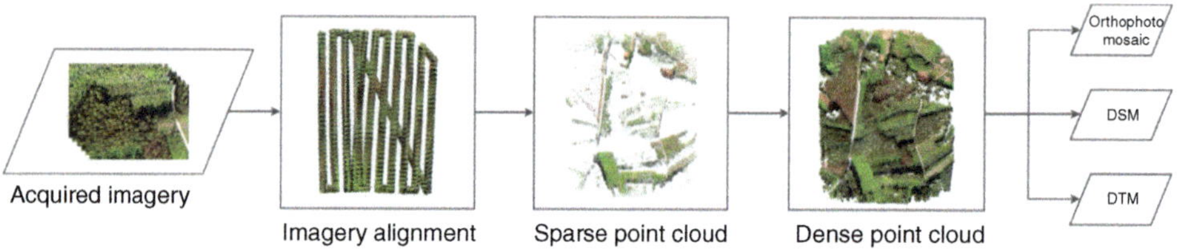

Figure 11.7 Typical photogrammetric processing pipeline. The acquired UAVs imagery goes through an imagery alignment process, resulting in a sparse point cloud, which in turn, is densified into a dense point cloud. By interpolating it, several raster outcomes can result; digital surface model (DSM), digital terrain model (DTM), and orthophoto mosaic, among others (depending on the sensor type). (Source: Adapted from Guimarães et al. (2020), MDPI, CC BY 4.0.)

the other hand, could be utilized to gather more scale-appropriate data (see Figure 11.6). Nevertheless, they are costly when time-series monitoring is required regularly. Compared to a UAV, it is evident that traditional remote sensing platforms are unsuitable for real-time applications, as UAVs combine high spatial resolution, quick turnaround times, and cheaper operational costs (Chianucci et al., 2016). These are the main factors for the popularity of UAVs over space and airborne data (see Figure 11.7).

UAVs are the complete system that include the necessary equipment, network, and technicians to control the unmanned aircraft (Gupta et al., 2013). These platforms could be utilized for various purposes including environmental, emergency/rescue, surveillance agroforestry (Santamaria et al., 2012), and precision agriculture such as crop health assessment (Srivastava et al. 2019), route planning, and optimization (Srivastava et al. 2020b). UAVs (the platform itself) do not require a human pilot on board, unlike airborne platforms. A pilot controls it via a ground station or remote control, which helps to reduce the cost of these operations (Pádua et al., 2017).

Several characteristics make UAVs one of the most appropriate solutions to consider in the management of forests and agricultural resources. First, because of low material and operational costs and high-intensity data gathering (Wargo et al., 2014); second, because UAVs can host a wide range of sensors that can be customized to be task-oriented (Shakhatreh et al., 2019); third, because UAV missions can be planned in a flexible manner, avoiding inclement weather and information on the spatial on-demand (Dunford et al., 2009); and finally, because UAVs can be employed in real-time activities, such as employing thermal sensors to detect forest fire, and find the point of outburst to stop it. Furthermore, because of UAVs flexibility and operability, the higher cost of high-resolution satellite data and challenges linked to frequent cloud cover encourage UAV usage. Local conservation workers and researchers in the developing tropics can save time, workers, and money by using UAVs (Koh and Wich, 2012). There are many types of UAVs used for vegetation monitoring. UAVs can be classified into four widely used classes based on their mechanism. Every type has merits and cons with different zones of use. The widely used drone is summarized in the Table 11.4.

As mentioned earlier, drones can carry every type of sensor for biophysical and biochemical parameters estimations like capturing vegetation spectrum in different wavelengths (visible, near-infrared, and thermal). They can create the DSM using the three-dimensional measurement of trees. Traditional methods such as linear regression, vegetation indices, and pixel-based image analysis could be done using UAV data to create a powerful impact with extreme accuracy. However, it also opens the path for object-based image analysis and the use of machine learning algorithms (de Castro et al., 2021). Tables 11.1 through 11.4 show different sensor names widely used for drone-based monitoring.

11.6 Other Important Forest Research Applications and Practices

The previous discussion demonstrates the role of drones in estimation of biophysical and biochemical parameters. Apart from these applications, drones may be utilized by forest management for other applications such as wildlife monitoring, faunal assessment, throwing seeds for propagation, burn area check, fire point confirmation, keeping a track of tourists in wildlife sanctuary/national park/reserve forests, land use/land cover, or monitoring local climate over forest areas. Drones mounted with high-resolution RGB cameras may be deployed to monitor and record behavior of animals and birds for their habitat assessment, to record their movement, calls/sounds, migration patterns, and interaction. Drones are currently being used to transmit real time data to processing research stations or store them in google drive or cloud services for real time assessment (such as behaviour of animals, sounds, movement patterns) which is facilitated with a data storage and data transmission facility. Drones are also being incorporated with weather monitoring sensors for recording, monitoring, surveillance, and transferring to data centers for a few parameters such as humidity, temperature, rainfall, wind speed, and direction (installed with anemometers, temperature sensors, and humidity sensors). Due to their robust nature and access, forest authorities exploit drones for several other purposes as well, such as guiding tourists inside forest (tourism), illegal tree cutting, checking for poachers, unauthorized entry in the core zones of forests, detecting forest fires (equipped with smoke detector sensors), and so on. For small fires or if any fire point is detected, drones equipped with carbon dioxide powder can be sent to extinguish the fire by spraying powder at the point of fire in the forest. Surveillance technology using drones can prevent the theft of precious wood and medicinal plants, such as sandalwood from the forest. Because of these practices, major preventive measures could be taken by governments for illegal poaching, hunting, entry, unauthorized tree cutting, or fire intensity assessment for better planning and management. The only major limitation with drones is that power backup and battery life become exhausted very quickly. To counteract this problem, scientists came up with an idea of automatic charging stations and wireless charging of drones while in flight.

11.7 Conclusions

Due to less spatial resolution, satellite or airborne data may derive coarse resolution in spatial and temporal resolution, and are influenced by cloud cover. However, UAVs allow adjusting flight altitudes according to the study's needs and help to acquire high-resolution data. The availability of various UAV-based sensors plays a crucial part in monitoring different vegetation and other ecosystem landscapes. Different UAV sensors such as multispectral, hyperspectral, LiDAR, and thermal derive reflectance values can estimate diverse information levels such as the forest's biophysical (AGB, height, DBH, and LAI) and biochemical (chlorophyll content, canopy water content) levels. These types of long-term data may even predict natural calamities such as drought by studying photochemical factors (e.g. changes in the xanthophyll's cycle) (Jiao et al., 2021). The methods (models and algorithms) of deriving any biophysical or biochemical information are almost the same; for example, LAI and AGB can estimate from hyperspectral UAV data using vegetation indices and red-edge parameters, and stepwise regression (SWR) and partial least squares regression (PLSR) methods such as spaceborne hyperspectral data, but with better accuracy (Zhao et al., 2020). Nowadays, drone-based analyses create new paths for forest monitoring by creating 3-D point cloud models (Sun et al., 2019). This type of research is usually performed based on LiDAR sensors. Some 3-D modeling research was done using optical bands like NIR for tree-classifications based on different photogrammetric techniques (Gini et al., 2012). Integration of GIS and photogrammetry also create accurate 3-D models by using RGB drone data (Budiharto et al., 2021). UAVs show their potential in monitoring the forests biophysical and biochemical parameters more accurately compared with images acquired through spaceborne platforms. For biophysical parameters, LiDAR sensors are good, whereas hyperspectral are good for biochemical parameters. In India, the government is relaxing restrictions on drone deployment and releasing drone policy 2021 for efficient image acquisition and research practices. The Drones Rules 2021 was published on August 25, 2021 and thereafter few changes were incorporated and released as Drone (Amendment) Rules 2022, dated February 11, 2022. This encompasses several conditions and points, a few of them are listed here. Aerial coverage of drones has been increased from 10 minutes to > 30 minutes with increased payload capacity, including drone taxis. and manufacturing drones purely for exports are exempted from type certification and unique identification numbers, drone corridors will be developed for cargo deliveries and many more important changes, such as prior permission, will be required to fly only in the red and yellow zones. As discussed, the role of drones is increasing in forest research and practices, due to their robust capabilities, flexible operational nature, and as per users' needs and requirements. The next era will soon be called the drone era because of their increasing use in several research themes including agriculture, urban, forest, disasters, and hydrology.

References

Ali, A.M., Darvishzadeh, R., Skidmore, A. et al. (2020). Comparing methods for mapping canopy chlorophyll content in a mixed mountain forest using Sentinel-2 data. *International Journal of Applied Earth Observation and Geoinformation* 87: 102037.

Alonzo, M., Andersen, H.E., Morton, D.C., and Cook, B.D. (2018). Quantifying boreal forest structure and composition using UAV structure from motion. *Forests* 9 (3): 119.

Anand, A., Pandey, P.C., Petropoulos, G.P. et al. (2020). Use of Hyperion for mangrove forest carbon stock assessment in Bhitarkanika forest reserve: a contribution towards blue carbon initiative. *Remote Sensing* 12 (4): 597.

Andersen, H.E., Reutebuch, S.E., and McGaughey, R.J. (2006). A rigorous assessment of tree height measurements obtained using airborne LiDAR and conventional field methods. *Canadian Journal of Remote Sensing* 32 (5): 355–366.

Anderson, K. and Gaston, K.J. (2013). Lightweight unmanned aerial vehicles will revolutionize spatial ecology. *Frontiers in Ecology and the Environment* 11 (3): 138–146.

Apostol, B., Lorent, A., Petrila, M. et al. (2016). Height extraction and stand volume estimation based on fusion airborne LiDAR data and terrestrial measurements for a Norway spruce [*Picea abies (L.) Karst.*] test site in Romania. *Notulae Botanicae Horti Agrobotanici Cluj-Napoca* 44 (1): 313–323.

Asner, G.P. (1998). Biophysical and biochemical sources of variability in canopy reflectance. *Remote Sensing of Environment* 64 (3): 234–253.

Banda, T., Schwartz, M.W., and Caro, T. (2006). Woody vegetation structure and composition along a protection gradient in a miombo ecosystem of western Tanzania. *Forest Ecology and Management* 230 (1–3): 179–185.

Barrachina, M., Cristóbal, J., and Tulla, A.F. (2015). Estimating above-ground biomass on mountain meadows and pastures through remote sensing. *International Journal of Applied Earth Observation and Geoinformation* 38: 184–192.

Barrett, F., McRoberts, R.E., Tomppo, E. et al. (2016). A questionnaire-based review of the operational use of remotely sensed data by national forest inventories. *Remote Sensing of Environment* 174: 279–289.

Basuki, T.M., Skidmore, A.K., Hussin, Y.A., and Van Duren, I. (2013). Estimating tropical forest biomass more accurately by integrating ALOS PALSAR and Landsat-7 ETM+ data. *International Journal of Remote Sensing* 34 (13): 4871–4888.

Behera, M.D., Kushwaha, S.P.S., and Roy, P.S. (2001). Forest vegetation characterization and mapping using IRS-1C satellite images in eastern Himalayan region. *Geocarto International* 16 (3): 53–62.

Bendig, J., Bolten, A., Bennertz, S. et al. (2014). Estimating biomass of barley using crop surface models (CSMs) derived from UAV-based RGB imaging. *Remote Sensing* 6 (11): 10395–10412.

Blair, J.B., Rabine, D.L., and Hofton, M.A. (1999). The laser vegetation imaging sensor: a medium-altitude, digitisation-only, airborne laser altimeter for mapping vegetation and topography. *ISPRS Journal of Photogrammetry and Remote Sensing* 54 (2–3): 115–122.

Bolte, A., Ammer, C., Löf, M. et al. (2009). Adaptive forest management in central Europe: climate change impacts, strategies and integrative concept. *Scandinavian Journal of Forest Research* 24 (6): 473–482.

Cao, L., Liu, H., Fu, X. et al. (2019). Comparison of UAV LiDAR and digital aerial photogrammetry point clouds for estimating forest structural attributes in subtropical planted forests. *Forests* 10 (2): 145.

Chen, J.M. and Cihlar, J. (1996). Retrieving leaf area index of boreal conifer forests using Landsat TM images. *Remote Sensing of Environment* 55 (2): 153–162.

Chen, J.M. and Cihlar, J. (1995). Quantifying the effect of canopy architecture on optical measurements of leaf area index using two gap size analysis methods. *IEEE Transactions on Geoscience Remote Sensing* 33 (3): 777–787.

Chianucci, F., Disperati, L., Guzzi, D. et al. (2016). Estimation of canopy attributes in beech forests using true colour digital images from a small fixed-wing UAV. *International Journal of Applied Earth Observation and Geoinformation* 47: 60–68.

Chuvieco, E., Riaño, D., Aguado, I., and Cocero, D. (2002). Estimation of fuel moisture content from multitemporal analysis of Landsat Thematic Mapper reflectance data: applications in fire danger assessment. *International Journal of Remote Sensing* 23 (11): 2145–2162.

Clevers, J.G., Kooistra, L., and Schaepman, M.E. (2010). Estimating canopy water content using hyperspectral remote sensing data. *International Journal of Applied Earth Observation and Geoinformation* 12 (2): 119–125.

Clevers, J.G. and Gitelson, A.A. (2013). Remote estimation of crop and grass chlorophyll and nitrogen content using red-edge bands on Sentinel-2 and -3. *International Journal of Applied Earth Observation and Geoinformation* 23: 344–351.

Colombo, R., Meroni, M., Marchesi, A. et al. (2008). Estimation of leaf and canopy water content in poplar plantations by means of hyperspectral indices and inverse modeling. *Remote Sensing of Environment* 112 (4): 1820–1834.

Colomina, I. and Molina, P. (2014). Unmanned aerial systems for photogrammetry and remote sensing: a review. *ISPRS Journal of Photogrammetry and Remote Sensing* 92: 79–97.

Condit, R., Hubbell, S.P., Lafrankie, J.V. et al. (1996). Species-area and species-individual relationships for tropical trees: a comparison of three 50-ha plots. *Journal of Ecology* 549–562.

Cooper, S.D., Roy, D.P., Schaaf, C.B., and Paynter, I. (2017). Examination of the potential of terrestrial laser scanning and structure-from-motion photogrammetry for rapid nondestructive field measurement of grass biomass. *Remote Sensing* 9 (6): 531.

Dalla Corte, A.P., Rex, F.E., Almeida, D.R.A.D. et al. (2020). Measuring individual tree diameter and height using GatorEye High-Density UAV-LiDAR in an integrated crop-livestock-forest system. *Remote Sensing* 12 (5): 863.

Danson, F.M. and Bowyer, P. (2004). Estimating live fuel moisture content from remotely sensed reflectance. *Remote Sensing of Environment* 92 (3): 309–321.

Darvishzadeh, R., Atzberger, C., Skidmore, A., and Schlerf, M. (2010). Retrieval of vegetation biochemicals using a radiative transfer model and hyperspectral data. In: *ISPRS TC VII Symposium – 100 Years ISPRS*. Vienna, Austria, July 5–7, 2010, IAPRS, vol. XXXVIII (ed. W. Wagner and B. Székely), Part 7B.

Du, L., Yang, J., Sun, J. et al. (2020). Leaf biochemistry parameters estimation of vegetation using the appropriate inversion strategy. *Frontiers in Plant Science* 11 (1–13): 533.

Dubayah, R.O., Sheldon, S.L., Clark, D.B. et al. (2010). Estimation of tropical forest height and biomass dynamics using LiDAR remote sensing at La Selva, Costa Rica. *Journal of Geophysical Research: Biogeosciences* 115 (G2).

Dunford, R., Michel, K., Gagnage, M. et al. (2009). Potential and constraints of Unmanned Aerial Vehicle technology for the characterization of Mediterranean riparian forest. *International Journal of Remote Sensing* 30 (19): 4915–4935.

Dusseux, P., Hubert-Moy, L., Corpetti, T., and Vertès, F. (2015). Evaluation of SPOT imagery for the estimation of grassland biomass. *International Journal of Applied Earth Observation and Geoinformation* 38: 72–77.

Falkowski, M.J., Smith, A.M., Hudak, A.T. et al. (2006). Automated estimation of individual conifer tree height and crown diameter via two-dimensional spatial wavelet analysis of LiDAR data. *Canadian Journal of Remote Sensing* 32 (2): 153–161.

FSI. (2019). *India State of Forest Report 2019*. Forest survey of India, Ministry of Environment, Forests & Climate Change, Dehradun, India.

Giannetti, F., Chirici, G., Gobakken, T. et al. (2018). A new approach with DTM-independent metrics for forest growing stock prediction using UAV photogrammetric data. *Remote Sensing of Environment* 213: 195–205.

Gini, F., De Maio, A., and Patton, L. (ed.) (2012). *Waveform design and diversity for advanced radar systems*, 20–24. London, UK: Institution of engineering and technology.

Gitelson, A.A. (2004). Wide dynamic range vegetation index for remote quantification of biophysical characteristics of vegetation. *Journal of Plant Physiology* 161 (2): 165–173.

Gitelson, A.A., Viña, A., Ciganda, V. et al. (2005). Remote estimation of canopy chlorophyll content in crops. *Geophysical Research Letters* 32 (8).

Gitelson, A.A., Peng, Y., Arkebauer, T.J., and Suyker, A.E. (2015). Productivity, absorbed photosynthetically active radiation, and light use efficiency in crops: implications for remote sensing of crop primary production. *Journal of Plant Physiology* 177: 100–109.

Goodbody, T.R., Coops, N.C., and White, J.C. (2019). Digital aerial photogrammetry for updating area-based forest inventories: a review of opportunities, challenges, and future directions. *Current Forestry Reports* 5 (2): 55–75.

Guerini Filho, M., Kuplich, T.M., and Quadros, F.L.D. (2020). Estimating natural grassland biomass by vegetation indices using Sentinel-2 remote sensing data. *International Journal of Remote Sensing* 41 (8): 2861–2876.

Guimarães, N., Pádua, L., Marques, P. et al. (2020). Forestry remote sensing from unmanned aerial vehicles: a review focusing on the data, processing and potentialities. *Remote Sensing* 12 (6): 1046.

Gupta, S.G., Ghonge, D., and Jawandhiya, P.M. (2013). Review of unmanned aircraft system (UAS). *International Journal of Advanced Research in Computer Engineering & Technology (IJARCET)* 2.

Hassler, S.C. and Baysal-Gurel, F. (2019). Unmanned aircraft system (UAS) technology and applications in agriculture. *Agronomy* 9 (10): 618.

Hmida, S.B., Kallel, A., Gastellu-Etchegorry, J.P., and Roujean, J.L. (2017). Crop biophysical properties estimation based on LiDAR full-waveform inversion using the DART RTM. *IEEE Journal of Selected Topics in Applied Earth Observations and Remote Sensing* 10 (11): 4853–4868.

Huang, W., Pohjonen, V., Johansson, S. et al. (2003). Species diversity, forest structure and species composition in Tanzanian tropical forests. *Forest Ecology and Management* 173 (1–3): 11–24.

Husch, B., Beers, T.W., and Kershaw, J.A. Jr. (2002). *Forest Mensuration*. Wiley.

Hyyppä, J., Hyyppä, H., Leckie, D. et al. (2008). Review of methods of small-footprint airborne laser scanning for extracting forest inventory data in boreal forests. *International Journal of Remote Sensing* 29 (5): 1339–1366.

Jacquemoud, S., Baret, F., Andrieu, B. et al. (1995). Extraction of vegetation biophysical parameters by inversion of the PROSPECT+ SAIL models on sugar beet canopy reflectance data. Application to TM and AVIRIS sensors. *Remote Sensing of Environment* 52 (3): 163–172.

Jha, C.S., Goparaju, L., Tripathi, A. et al. (2005). Forest fragmentation and its impact on species diversity: an analysis using remote sensing and GIS. *Biodiversity & Conservation* 14 (7): 1681–1698.

Jia, K., Wu, B., Tian, Y. et al. (2011). Vegetation classification method with biochemical composition estimated from remote sensing data. *International Journal of Remote Sensing* 32 (24): 9307–9325.

Jiao, W., Wang, L., Smith, W.K. et al. (2021). Observed increasing water constraint on vegetation growth over the last three decades. *Nature Communications* 12 (1): 1–9.

Jonckheere, I., Fleck, S., Nackaerts, K. et al. (2004). Review of methods for in situ leaf area index determination: Part I. Theories, sensors and hemispherical photography. *Agricultural and Forest Meteorology* 121 (1–2): 19–35.

Kajisa, T., Murakami, T., Mizoue, N. et al. (2009). Object-based forest biomass estimation using Landsat ETM+ in Kampong Thom Province. Cambodia. *Journal of Forest Research* 14 (4): 203–211.

Koh, L.P. and Wich, S.A. (2012). Dawn of drone ecology: low-cost autonomous aerial vehicles for conservation. *Tropical Conservation Science* 5 (2): 121–132.

Krause, S., Sanders, T.G., Mund, J.P., and Greve, K. (2019). UAV-based photogrammetric tree height measurement for intensive forest monitoring. *Remote Sensing* 11 (7): 758.

Kumar, P., Pandey, P.C., Kumar, V. et al. (2014). Efficient recognition of forest species biodiversity by inventory-based geospatial approach using LISS IV sensor. *IEEE Sensors Journal* 15 (3): 1884–1891.

Kumar, P., Rani, M., Pandey, P.C. et al. (2010). Monitoring of deforestation and forest degradation using remote sensing and GIS: A case study of Ranchi in Jharkhand (India). *Report and Opinion* 2 (4): 14–20.

Kumar, P., Sharma, L.K., Pandey, P.C. et al. (2013). Geospatial strategy for tropical forest-wildlife reserve biomass estimation. *IEEE Journal of Selected Topics in Applied Earth Observations and Remote Sensing* 6 (2): 917–923.

Kumar, S.S., Prihodko, L., Lind, B.M. et al. (2020). Remotely sensed thermal decay rate: an index for vegetation monitoring. *Scientific Reports* 10 (1): 1–11.

Leblanc, S.G., Chen, J.M., Fernandes, R. et al. (2005). Methodology comparison for canopy structure parameters extraction from digital hemispherical photography in boreal forests. *Agricultural and Forest Meteorology* 129 (3–4): 187–207.

Lechner, A.M., Foody, G.M., and Boyd, D.S. (2020). Applications in remote sensing to forest ecology and management. *One Earth* 2 (5): 405–412.

Lee, K.S., Cohen, W.B., Kennedy, R.E. et al. (2004). Hyperspectral versus multispectral data for estimating leaf area index in four different biomes. *Remote Sensing of Environment* 91 (3–4): 508–520.

Lefsky, M.A., Cohen, W.B., Acker, S.A. et al. (1999). Lidar remote sensing of the canopy structure and biophysical properties of Douglas-fir western hemlock forests. *Remote Sensing of Environment* 70 (3): 339–361.

Lieberman, D., Lieberman, M., Peralta, R., and Hartshorn, G.S. (1996). Tropical forest structure and composition on a large-scale altitudinal gradient in Costa Rica. *Journal of Ecology* 137–152.

Lu, D., Batistella, M., and Moran, E. (2005). Satellite estimation of aboveground biomass and impacts of forest stand structure. *Photogrammetric Engineering & Remote Sensing* 71 (8): 967–974.

Lu, D., Chen, Q., Wang, G. et al. (2012, 2012). Aboveground forest biomass estimation with Landsat and LiDAR data and uncertainty analysis of the estimates. *International Journal of Forestry Research* 1–17, article ID 436537.

Lü, X.T., Yin, J.X., and Tang, J.W. (2010). Structure, tree species diversity and composition of tropical seasonal rainforests in Xishuangbanna, southwest China. *Journal of Tropical Forest Science* 260–270.

Lugten, G. (2008). Food and agriculture organization. *International Journal of Marine and Coastal Law* 23: 761.

Luo, S., Wang, C., Pan, F. et al. (2015). Estimation of wetland vegetation height and leaf area index using airborne laser scanning data. *Ecological Indicators* 48: 550–559.

Malhi, R.K.M., Anand, A., Mudaliar, A.N. et al. (2020). Synergetic use of in-situ and hyperspectral data for mapping species diversity and above ground biomass in Shoolpaneshwar Wildlife Sanctuary, Gujarat. *Tropical Ecology* 61 (1): 106–115.

Mandal, D., Bhattacharya, A., and Rao, Y.S. (2021). *Radar Remote Sensing for Crop Biophysical Parameter Estimation*. Springer.

Menon, S. and Bawa, K.S. (1997). Applications of geographic information systems, remote-sensing, and a landscape ecology approach to biodiversity conservation in the western Ghats. *Current Science* 134–145.

Mitchard, E.T., Saatchi, S.S., Woodhouse, I.H. et al. (2009). Using satellite radar backscatter to predict above-ground woody biomass: A consistent relationship across four different African landscapes. *Geophysical Research Letters* 36 (23).

Nevalainen, O., Honkavaara, E., Tuominen, S. et al. (2017). Individual tree detection and classification with UAV-based photogrammetric point clouds and hyperspectral imaging. *Remote Sensing* 9 (3): 185.

Nilsson, M. (1996). Estimation of tree heights and stand volume using an airborne LiDAR system. *Remote Sensing of Environment* 56 (1): 1–7.

Nuthammachot, N., Askar, A., Stratoulias, D., and Wicaksono, P. (2022). Combined use of Sentinel-1 and Sentinel-2 data for improving above-ground biomass estimation. *Geocarto International* 37 (2): 366–376.

Ota, T., Ogawa, M., Mizoue, N. et al. (2017). Forest structure estimation from a UAV-based photogrammetric point cloud in managed temperate coniferous forests. *Forests* 8 (9): 343.

Pádua, L., Vanko, J., Hruška, J. et al. (2017). UAS, sensors, and data processing in agroforestry: A review towards practical applications. *International Journal of Remote Sensing* 38 (8–10): 2349–2391.

Pandey, P.C., Anand, A., and Srivastava, P.K. (2019b). Spatial distribution of mangrove forest species and biomass assessment using field inventory and earth observation hyperspectral data. *Biodiversity and Conservation* 28 (8): 2143–2162.

Pandey, P.C., Tate, N.J., and Balzter, H. (2014). Mapping tree species in coastal Portugal using statistically segmented principal component analysis and other methods. *IEEE Sensors Journal* 14 (12): 4434–4441.

Pandey, P.C., Balzter, H., Srivastava, P.K. et al. (ed.) (2020a). Future Perspectives and Challenges in Hyperspectral Remote Sensing. In: *Hyperspectral Remote Sensing*, 429–439. Elsevier.

Pandey, P.C., Srivastava, P.K., Chetri, T. et al. (2019a). Forest biomass estimation using remote sensing and field inventory: a case study of Tripura, India. *Environmental Monitoring and Assessment* 191 (9): 1–15.

Pandey, P.C., Srivastava, P.K., Balzter, H. et al. (ed.) (2020b). *Hyperspectral Remote Sensing: Theory and Applications*. Elsevier.

Pandey, P.C. and Sharma, L.K. (ed.) (2021). *Advances in Remote Sensing for Natural Resource Monitoring*. Wiley.

Pandey, P.C., Amit, K.T., and Jyoti, K.S. (2021). An Evaluation of GPS Opportunity in Market for Precision Agriculture. In: *GPS and GNSS Technology in Geosciences*, 337–349. Elsevier.

Pandey, S.K. and Shukla, R.P. (2003). Plant diversity in managed sal (*Shorea robusta Gaertn.*) forests of Gorakhpur, India: species composition, regeneration and conservation. *Biodiversity & Conservation* 12 (11): 2295–2319.

Parthasarathy, N. (2001). Changes in forest composition and structure in three sites of tropical evergreen forest around Sengaltheri, western Ghats. *Current Science* 389–393.

Peterson, D.L., Aber, J.D., Matson, P.A. et al. (1988). Remote sensing of forest canopy and leaf biochemical contents. *Remote Sensing of Environment* 24 (1): 85–108.

Pope, G. and Treitz, P. (2013). Leaf area index (LAI) estimation in boreal mixed wood forest of Ontario, Canada using light detection and ranging (LiDAR) and WorldView-2 imagery. *Remote Sensing* 5 (10): 5040–5063.

Rai, A.T., Pandey, P.C., Kumar, P., and Srivastava, P.K. (2017). Assessment of forest deforestation of Tripura State, India using earth observation data. *Bulletin of Environmental and Scientific Research* 6 (1).

Reddy, C.S., Jha, C.S., Diwakar, P.G., and Dadhwal, V.K. (2015). Nationwide classification of forest types of India using remote sensing and GIS. *Environmental Monitoring and Assessment* 187 (12): 1–30.

Richardson, J.J., Moskal, L.M., and Kim, S.H. (2009). Modeling approaches to estimate effective leaf area index from aerial discrete-return LIDAR. *Agricultural and Forest Meteorology* 149 (6–7): 1152–1160.

Roy, P.S. and Tomar, S. (2000). Biodiversity characterization at landscape level using geospatial modelling technique. *Biological Conservation* 95 (1): 95–109.

Roy, P.S., Behera, M.D., Murthy, M.S.R. et al. (2015). New vegetation type map of India prepared using satellite remote sensing: comparison with global vegetation maps and utilities. *International Journal of Applied Earth Observation and Geoinformation* 39: 142–159.

Roy, P.S., Dutt, C.B.S., and Joshi, P.K. (2002). Tropical forest resource assessment and monitoring. *Tropical Ecology* 43 (1): 21–37.

Sankey, T., Donager, J., McVay, J., and Sankey, J.B. (2017). UAV LiDAR and hyperspectral fusion for forest monitoring in the southwestern USA. *Remote Sensing of Environment* 195: 30–43.

Santamaria, E., Barrado, C., Pastor, E. et al. (2012). Reconfigurable automated behavior for UAS applications. *Aerospace Science and Technology* 23 (1): 372–386.

Santini, F., Kefauver, S.C., Resco de Dios, V. et al. (2019). Using unmanned aerial vehicle-based multispectral, RGB and thermal imagery for phenotyping of forest genetic trials: A case study in *Pinus halepensis*. *Annals of Applied Biology* 174 (2): 262–276.

Schleppi, P., Conedera, M., Sedivy, I., and Thimonier, A. (2007). Correcting non-linearity and slope effects in the estimation of the leaf area index of forests from hemispherical photographs. *Agricultural and Forest Meteorology* 144 (3–4): 236–242.

Shakhatreh, H., Sawalmeh, A.H., Al-Fuqaha, A. et al. (2019). Unmanned aerial vehicles (UAVs): a survey on civil applications and key research challenges. *IEEE Access* 7: 48572–48634.

Shankar, U. and Tripathi, A.K. (2017). Rainforests north of the Tropic of Cancer: Physiognomy, floristics and diversity in 'lowland rainforests' of Meghalaya. *India. Plant Diversity* 39 (1): 20–36.

Sharma, L.K., Kanga, S., Nathawat, M.S. et al. (2012). Fuzzy AHP for forest fire risk modeling. *Disaster Prevention and Management: An International Journal* 21 (2): 160–171.

Shelestov, A., Kolotii, A., Camacho, F. et al. (2015). Mapping of biophysical parameters based on high-resolution EO imagery for JECAM test site in Ukraine. In: *2015 IEEE International Geoscience and Remote Sensing Symposium (IGARSS)*, 1733–1736. IEEE.

Shelestov, A., Kolotii, A., Skakun, S. et al. (2017). Biophysical parameters mapping within the SPOT-5 Take 5 initiative. *European Journal of Remote Sensing* 50 (1): 300–309.

Shen, W., Li, M., Huang, C., and Wei, A. (2016). Quantifying live aboveground biomass and forest disturbance of mountainous natural and plantation forests in Northern Guangdong, China, based on multi-temporal Landsat, PALSAR and field plot data. *Remote Sensing* 8 (7): 595.

Shen, Y., Niu, Z., and Yan, C.Y. (2005). Estimation models for vegetation water content at both leaf and canopy levels. *The Journal of Applied Ecology* 16 (7): 1218–1223.

Shendryk, I., Hellström, M., Klemedtsson, L., and Kljun, N. (2014). Low-density LiDAR and optical imagery for biomass estimation over boreal forest in Sweden. *Forests* 5 (5): 992–1010.

Shengyan, D., Jidong, G., and Lexiang, Q. (2002). Assessment of biochemical concentrations of vegetation using remote sensing technology. *Journal of Geographical Sciences* 12 (3): 321–330.

Sinha, S., Pandey, P.C., Sharma, L.K. et al. (2014). Remote estimation of land surface temperature for different LULC features of a moist deciduous tropical forest region. In: *Remote Sensing Applications in Environmental Research*, 57–68. Cham: Springer.

Skole, D. and Tucker, C. (1993). Tropical deforestation and habitat fragmentation in the Amazon: satellite data from 1978 to 1988. *Science* 260 (5116): 1905–1910.

Solberg, S., Brunner, A., Hanssen, K.H. et al. (2009). Mapping LAI in a Norway spruce forest using airborne laser scanning. *Remote Sensing of Environment* 113 (11): 2317–2327.

Spathelf, P., Stanturf, J., Kleine, M. et al. (2018). Adaptive measures: integrating adaptive forest management and forest landscape restoration. *Annals of Forest Science* 75 (2): 1–6.

Srivastava, K., Bhutoria, A.J., Sharma, J.K. et al. (2019). UAVs technology for the development of GUI based application for precision agriculture and environmental research. *Remote Sensing Applications: Society and Environment* 16: 100258.

Srivastava, K., Pandey, P.C., and Sharma, J.K. (2020b). An approach for route optimization in applications of precision agriculture using UAVs. *Drones* 4 (3): 58.

Srivastava, P.K., Malhi, R.K.M., Pandey, P.C. et al. (2020a). Revisiting hyperspectral remote sensing: origin, processing, applications and way forward. In: *Hyperspectral Remote Sensing: Theory and Applications* (ed. P.C. Pandey, P.K. Srivastava, H. Balzter, et al.), 3–21. Elsevier.

Sukumar, R., Dattaraja, H.S., Suresh, H.S. et al. (1992). Long-term monitoring of vegetation in a tropical deciduous forest in Mudumalai, southern India. *Current Science Association* 62 (9): 608–616.

Sun, P., Zhao, X., Xu, Z. et al. (2019). A 3D LiDAR data-based dedicated road boundary detection algorithm for autonomous vehicles. *IEEE Access* 7: 29623–29638.

Tang, H., Dubayah, R., Swatantran, A. et al. (2012). Retrieval of vertical LAI profiles over tropical rain forests using waveform LiDAR at La Selva, Costa Rica. *Remote Sensing of Environment* 124: 242–250.

Thenkabail, P.S. (2015). *Land Resources Monitoring, Modeling, and Mapping with Remote Sensing*. Boca Raton, FL: CRC Press.

Tilly, N., Hoffmeister, D., Cao, Q. et al. (2014). Multitemporal crop surface models: accurate plant height measurement and biomass estimation with terrestrial laser scanning in paddy rice. *Journal of Applied Remote Sensing* 8 (1): 083671.

Timothy, D., Onisimo, M., and Riyad, I. (2016). Quantifying aboveground biomass in African environments: a review of the trade-offs between sensor estimation accuracy and costs. *Tropical Ecology* 57 (3): 393–405.

Trigg, S.N., Curran, L.M., and McDonald, A.K. (2006). Utility of Landsat 7 satellite data for continued monitoring of forest cover change in protected areas in Southeast Asia. *Singapore Journal of Tropical Geography* 27 (1): 49–66.

Tripathi, A.K. and Shankar, U. (2014). Patterns of species dominance, diversity and dispersion in 'Khasi hill sal' forest ecosystem in northeast India. *Forest Ecosystems* 1 (1): 1–20.

Tucker, C.J. (1980). Remote sensing of leaf water content in the near infrared. *Remote Sensing of Environment* 10 (1): 23–32.

Tuominen, J., Lipping, T., Kuosmanen, V., and Haapanen, R. (2009). Remote sensing of forest health. *Geoscience and Remote Sensing* 161–166.

Ulaby, F.T., Long, D.G., Blackwell, W.J. et al. (2014). *Miacrowave Radar and Radiometric Remote Sensing*. Ann Arbor: The University of Michigan Press.

Van Laar, A. and Akça, A. (2007). *Forest Mensuration*, vol. 13. Springer Science & Business Media.

Vincent, G., Antin, C., Laurans, M. et al. (2017). Mapping plant area index of tropical evergreen forest by airborne laser scanning. A cross-validation study using LAI2200 optical sensor. *Remote Sensing of Environment* 198: 254–266.

Wallace, L., Lucieer, A., Watson, C., and Turner, D. (2012). Development of a UAV-LiDAR system with application to forest inventory. *Remote Sensing* 4 (6): 1519–1543.

Wang, Y. and Fang, H. (2020). Estimation of LAI with the LiDAR technology: a review. *Remote Sensing* 12 (20): 3457.

Wargo, C.A., Church, G.C., Glaneueski, J., and Strout, M. (2014). *Unmanned Aircraft Systems (UAS) research and future analysis*. In: *2014 IEEE Aerospace Conference*, 1–16. IEEE.

West, P.W. and West, P.W. (2009). *Tree and forest measurement*, vol. 20. Berlin: Springer.

Wulder, M.A., White, J.C., Nelson, R.F. et al. (2012). LiDAR sampling for large-area forest characterization: a review. *Remote Sensing of Environment* 121: 196–209.

Xie, Z., Roberts, C., and Johnson, B. (2008). Object-based target search using remotely sensed data: a case study in detecting invasive exotic Australian pine in south Florida. *ISPRS Journal of Photogrammetry and Remote Sensing* 63 (6): 647–660.

Xu, M., Liu, H., Beck, R. et al. (2019). A spectral space partition guided ensemble method for retrieving chlorophyll-a concentration in inland waters from Sentinel-2A satellite imagery. *Journal of Great Lakes Research* 45 (3): 454–465.

Yebra, M., Dennison, P.E., Chuvieco, E. et al. (2013). A global review of remote sensing of live fuel moisture content for fire danger assessment: moving towards operational products. *Remote Sensing of Environment* 136: 455–468.

Yilmaz, M.T., Hunt, E.R. Jr., and Jackson, T.J. (2008). Remote sensing of vegetation water content from equivalent water thickness using satellite imagery. *Remote Sensing of Environment* 112 (5): 2514–2522.

Zarco-Tejada, P.J., Hornero, A., Beck, P.S.A. et al. (2019). Chlorophyll content estimation in an open-canopy conifer forest with Sentinel-2A and hyperspectral imagery in the context of forest decline. *Remote Sensing of Environment* 223: 320–335.

Zhang, H., Sun, Y., Chang, L. et al. (2018). Estimation of grassland canopy height and aboveground biomass at the quadrat scale using unmanned aerial vehicle. *Remote Sensing* 10 (6): 851.

Zhang, Y., Chen, J.M., Miller, J.R., and Noland, T.L. (2008). Leaf chlorophyll content retrieval from airborne hyperspectral remote sensing imagery. *Remote Sensing of Environment* 112 (7): 3234–3247.

Zhang, Y., Yang, J., and Du, L. (2021). Analyzing the effects of hyperspectral ZhuHai-1 band combinations on LAI estimation based on the PROSAIL model. *Sensors* 21 (5): 1869.

Zhao, K. and Popescu, S. (2009). LiDAR-based mapping of leaf area index and its use for validating GLOBCARBON satellite LAI product in a temperate forest of the southern USA. *Remote Sensing of Environment* 113 (8): 1628–1645.

Zhao, X., Tao, R., Li, W. et al. (2020). Joint classification of hyperspectral and LiDAR data using hierarchical random walk and deep CNN architecture. *IEEE Transactions on Geoscience and Remote Sensing* 58 (10): 7355–7370.

Zheng, D., Rademacher, J., Chen, J. et al. (2004). Estimating aboveground biomass using Landsat 7 ETM+ data across a managed landscape in northern Wisconsin, USA. *Remote Sensing of Environment* 93 (3): 402–411.

Zheng, G. and Moskal, L.M. (2009). Retrieving leaf area index (LAI) using remote sensing: theories, methods and sensors. *Sensors* 9 (4): 2719–2745.

Zolkos, S.G., Goetz, S.J., and Dubayah, R. (2013). A meta-analysis of terrestrial aboveground biomass estimation using LiDAR remote sensing. *Remote Sensing of Environment* 128: 289–298.

12

Applications of Multi-Source and Multi-Sensor Data Fusion of Remote Sensing for Forest Species Mapping

Pegah Mohammadpour[*,1,2] and *Carlos Viegas*[1]

[1] *University of Coimbra, ADAI, Department of Mechanical Engineering, Rua Luís Reis Santos, Pólo II, 3030-788 Coimbra, Portugal*
[2] *Universidad de Alcalá, Department of Geology, Geography and Environment Science, C/Colegios, 2, 28801 Alcalá de Henares, Spain*

12.1 Introduction

The forest's role in the environment and society goes far beyond their ecological value as global carbon sinks, sustainers of vast biodiversity and ecosystems, and regulators of hydrological and atmospheric cycles. They also provide a wide range of valuable ecosystem services, with the forest sector contributing to roughly 1% of Europe's Gross Domestic Product (GDP).

Vegetation mapping and classification play an essential role in managing the forest's natural, biological, ecological, and economic resources, helping to monitor deforestation, natural and human hazards, and climate change effects (Mayaux et al. 1999). However, vegetation mapping and classification techniques face some limitations, such as a lack of up-to-date, reliable, and accurate vegetation data and a shortage of historical data to process its evolution (Mayaux et al. 1999). Remote sensing can solve some of these problems, though it requires detailed knowledge of the vegetation type of the region and the unique vegetation spectral signature in different bands (Guan et al. 2019; Johansen 2004; Mayaux et al. 1999).

Remote sensing provides a different source of imagery with variable spectral, spatial, radioactive, and temporal characteristics that may be applied for the various objectives of vegetation mapping (Xie et al. 2008; Zeng et al. 2020). Satellite imagery data can be used for mapping vegetation and land cover, land cover change over time, seasonal changes, inventory of natural resources, and damage assessment after natural hazards such as wildfires or hurricanes (Langley et al. 2001). Each vegetation cover has unique and definable spectral characteristics, which are highly dependent on the variation of vegetation structure and map scale (Jarman 1985). Conventional mapping methods include field surveys, map interpretation, collateral, and ancillary data analysis. The main problems of the conventional methods are the use of out-of-date data, gaps in data, inaccessibility of regions, and time- and cost-consuming surveys, which may pose significant constraints in large-scale areas or

* Corresponding author's email: Pegish2014@gmail.com

0000-0001-8628-3563 (Pegah Mohammadpour) and 0000-0002-9242-3195 (Carlos Viegas)

Advances in Remote Sensing for Forest Monitoring, First Edition. Edited by Prem C. Pandey and Paul Arellano.

for increased temporal resolution mapping (Barrett et al. 2016; Xie et al. 2008). Currently, satellite imagery is the main data source for Earth observation, particularly for large-scale mapping (Xie et al. 2008). Using satellite imagery and computer analysis for large-scale mapping has several merits in comparison with traditional methods. Some of the main advantages include increasing temporal frequency as well as cost- and time-effectiveness, providing historical data and high-quality data availability, consistency between locations especially in borders and boundaries, converting satellite image formats automatically to Geographic Information Systems (GIS), and ease of obtaining computer-based maps for forest management purpose such as floods and fires (Barrett et al. 2016; Jarman 1985; Sheykhmousa et al. 2020; Weih and Riggan 2010; Woodcock et al. 1994; Xie et al. 2008; Zeng et al. 2020). Mayaux et al. (1999) provided a vegetation map of central Africa by using coarse resolution satellite imagery based on the spectral response of the vegetation cover. This investigation performed under the name of the TREES project. Its main objective was to establish the forest monitoring system and vegetation map quickly and conveniently, to be modified at the regional or continental scale. They stated that the TREES vegetation map has greater spatial details and consistency than the conventional map, although vegetation mapping by using satellite imagery requires detailed processing., They also suggested the synchronizing of conventional and remote sensing-derived maps to use the advantage of both. Another result of their investigation was using the regional map as a baseline layer of mapping stratification to improve the details (Mayaux et al. 1999).

Vegetation mapping was mainly separated into two general phases; image pre-processing and image classification. However, some research defined various procedures for vegetation mapping (Jarman 1985; Johansen 2004; Xie et al. 2008). In Jarman's (1985) paper, computer-aided vegetation mapping of the fynbos biome using Landsat multispectral scanning was divided into three main sections; data collection, interpretation, and product generation. Jarman also broke down the digital image processing into three main categories; image restoration process, image enhancement process, and information extraction process. Jarman (1985) applied six main steps to map the vegetation in the fynbos biome as follows; (i) identification and extraction of the study area, (ii) pre-processing of the raw digital data and test area selection, (iii) classification of test areas and selection of spectral classes, (iv) selection of map classes, (v) final map classification, and (vi) final map production. In 2004, Johansen (2004) divided satellite image processing for vegetation mapping into the primary processes of pre- and post-classification. Pre-classification includes classifying the satellite image by visual interpretation, spectral classification, and segmentation resulting in the first-order map product. The interpretation and correction process should be performed using different layers of ancillary data and aerial analysis, called post-classification. Finally, the vegetation map may be produced by gathering the map database, vegetation database, and other geodata.

In this study, a review of vegetation mapping by using satellite imagery and fusion of different satellite sources and Unmanned Aerial Vehicle (UAV) based on previous research was conducted. In Section 12.2, the main steps of vegetation mapping based on gathering all the information of earlier studies are demonstrated step by step, from image acquisition to classification and accuracy assessment. Furthermore, choosing appropriate sensors, image quality, selection criteria, and the effect of selecting suitable indices on image classification are illustrated. In Section 12.3, the importance of satellite data in comparison with traditional databases is discussed. The merits of the fusion of various satellite imagery sources and their combination with UAV data over a single source are explained. The findings are discussed in Section 12.4. Finally, in Section 12.5, the main conclusion of this review literature and potential future work are presented.

12.2 Forest Mapping Process

Accurate and timely land cover mapping affects the precision of all subsequent applications (Phan et al. 2020). Approximately 31% percent of the Earth's land surface is covered by forests and grasslands, which indicates the key role of vegetation management in the terrestrial ecosystem, biodiversity and sustainability, and human communities in rural, urban, and Wild–Urban Interfaces (WUI) (FAO and UNEP 2020; Phan et al. 2020).

The most widely used fuel type classification was developed by the Northern Forest Fire Laboratory (NFFL) in the US and the Canadian Fire Behavior Prediction (FBP) System (Domingo et al. 2020). Some local adaptations to the NFFL have been proposed, such as the Prometheus project, which adapted the NFFL system for Mediterranean basin conditions. Fuel mapping and fuel type survey and classification are generally performed by field survey. Although the traditional methods have advantages, including limited error and reliable sources for reference datasets, their disadvantages overcome the advantages; time and cost consuming methods, not suitable for operational fuel characterization, the limited quantity of ground-based reference data, area access, sampling and observer bias, difficulty in a direct transferability to the remote sensing perspective and scale, and finally not explicitly matching the continuous representation of earth observation data and its spatial scales due to measuring field data as discrete point observations or plots (Kattenborn et al. 2019).

Different vegetation species and their particular characteristics may be identified by satellite sensors, mostly in the red or near-infrared spectrum, by their unique spectral features either in reflectance or emission regions. Raw satellite data goes through an automated process of image pre-processing and image classification, leading to vegetation mapping. In this work, five steps of the vegetation mapping process are considered; (i) image acquisition, (ii) image pre-processing, (iii) image enhancement, (iv) image classification, and (v) accuracy assessments.

12.2.1 Image Acquisition

Finding appropriate remote-sensing sensors for the study purpose could be a challenge mainly because there are several factors to consider. Remote sensing imaging sensors are split into two types of sensors based on how they measure the energy emitted or reflected from the target on Earth; (i) passive sensors depend on an external source such as the sun, and (ii) active sensors that transmit ElectroMagnetic Radiation (EMR) signal to the target and measure the reflected signals. RADAR and LiDAR are examples of active remote sensing, while multispectral and hyperspectral sensors are passive sensors (Brown et al. 2005). Xie et al. (2008) believed that a suitable sensor for vegetation mapping should be selected by considering different spatial, temporal, spectral, and radiometric characteristics of the sensors, and data fusion of sensors or different sources, including spaceborne or airborne (Guan et al. 2019; Xie et al. 2008). Defining four important criteria can ease the selection of adequate sensors and image quality; (i) mapping objective, (ii) cost of images, (iii) climate and atmospheric condition, and (iv) technical issues for image interpretation. Mapping objectives are directly related to the required level of spatial resolution and classification accuracy. For example, detailed vegetation classification needs relatively higher resolution than mapping high-level vegetation classification. The cost of images may vary from very expensive imagery with high resolution, to purchased or free data available with medium to low resolution with respect to the mapping scale (Xie et al. 2008). Atmospheric and climate effects usually reduce the apparent resolution of remote sensing imagery and cause noise and errors in the classification of cover surface (Kaufman 1984). To overcome this issue, cloud-free images may be obtained through multi-sensors or different time periods, in addition to atmospheric correction. Possible technical issues regarding the quality of images, image preprocessing, and interpretation should be figured out regarding identifying required characteristics and features of images such as different spectral features, texture, size, and tone. (Xie et al. 2008).

Choosing a suitable spatial resolution based on the project requirements is of critical importance. For example, high spatial resolution often introduces more challenges in terms of maximum accessible coverage and financial constraints to acquire or purchase the data (Barrett et al. 2016). Navulur (2006) defined spatial resolution of remote sensing data in four categories based on pixel size with Ground Sampling Distance (GSD); (i) low or coarse resolution with GSD of 30 m or greater, (ii) medium resolution with GSD 2–30 m, (ii) high resolution with GSD 0.5–2 m, and (iv) very high resolution with GSD lower than 0.5 m. Some studies mentioned that medium spatial resolution data between 10 and 20 m may be sufficient for most research instead of detailed imagery (Barrett et al. 2016; Feilhauer et al. 2014). Barrett et al.'s (2016) result demonstrated that medium spatial resolution optical and radar satellite data, such as Advanced Land Observation Satellite (ALOS) and Phased Array-type L-band Synthetic Aperture Radar (PALSAR), together with ancillary soil and topographic data, allow for accurate upland vegetation mapping and monitoring (Barrett et al. 2016). Likewise, a comparison between the mapping accuracy using 4 m spatial resolution of IKONOS imagery and 30 m spatial resolution of Landsat showed that the former was not comprehensively superior to the latter. Although the higher spatial resolution generally impacts the characterization of the objects and overall spatial accuracy, spatial resolution is not the only determinant of classification accuracy, and other factors such as temporal resolution are also effective (Holland and Aplin 2013).

The time of image acquisition by satellite has a substantial impact on the classification accuracy and discrimination of different vegetation types. Each vegetation has a unique spectral signature that evolves during the plant life cycle and affects reflectance such as pigmentation, water content, and physiological structure. So multitemporal data may account for the seasonal variation of vegetation and improve the classification accuracy (Gillanders et al. 2008; Guan et al. 2019; Holland and Aplin 2013; Langley et al. 2001).

High spatiotemporal resolution and historical data are essential for any support decision-making applications such as land or forest monitoring (Phan et al. 2020). Spatiotemporal data fusion is a methodology that aims to produce data with a high spatial and temporal resolution by combining images from two various sensors. There are two types of imagery in spatiotemporal data fusion; (i) fine imagery with high spatial resolution and low temporal resolution, and (ii) coarse imagery with low spatial resolution and high temporal resolution. Most spatiotemporal approaches are extremely dependent on the correctness of their similar pixel selection, since incorrect similar pixels might result in inaccuracies in the final spatiotemporal fusion outcomes. The Flexible Spatiotemporal Data-Fusion model (FSDAF), for example, employs auxiliary land-cover categorization information to assist in identifying comparable pixels by making sure they have the same land cover type as the target pixel. The similar pixel selection methods based on spectral similarity approaches now meet the criteria for applications such as land-cover mapping, where the spectral variations between various land cover types are visible. However, these approaches may not work effectively in vegetation mapping applications. Because spectral reflectance differences between vegetation types are substantially lower than between land cover categories, spectral similarity-based approaches may result in vegetation mapping errors (Guan et al. 2019). Guan et al. (2019) investigated new object-based image analysis methods to develop spatiotemporal data fusion imagery by studying the combination of Sentinel 2 and Landsat 8 images, Landsat 8 and Moderate-resolution Imaging Spectroradiometer (MODIS) images, and Sentinel 2 and MODIS images for the vegetation mapping application. Their findings indicate that the proposed framework greatly enhanced spatiotemporal data fusion techniques by more precisely identifying vegetation borders and reducing the "pepper–salt" effect, especially in integrating Sentinel 2 and Landsat 8 images.

Landsat Tm/ETM+, SPOT, MODIS, NOAA-AVHRR, IKONOS, and QuickBird are the most used sensors from the previous research until 2008 (Xie et al. 2008). Furthermore, the Sentinel-1 and Sentinel-2 data have enabled a variety of remote sensing applications, especially for land cover and vegetation mapping, by presenting the idea of the combination of optical and radar sensors to improve the accuracy of classification and map products (Barrett et al. 2016; Chakhar et al. 2021; Lastovicka et al. 2020; Sun et al. 2020; Zeng et al. 2020). The combination of Sentinel-1 with multitemporal SAR and Sentinel-2 with optical imagery sensors enables researchers to detect different spectral and physical characteristics of vegetation (Amani et al. 2020).

The next step is controlling the quality of image products delivered by the image distributors, which usually includes levels 0, 1A, 1B, 2A, 2B, 3A, and 3B with image quality gradually increased (Baillarin et al. 2012; Xie et al. 2008). Image quality may be measured by parameters such as radiometric resolution, geometric resolution and distortion, sharpness, noise level, and data errors (Crespi and de Vendictis 2009; Jarman 1985). Baillarin et al. (2012) carried out a study exclusively on Sentinel-2 Level-1 products. Based on their research, the Level-1 product system of Sentinel-2 breaks down into three sub-sections: (i) Level-0 and Level-1A that correspond to raw compressed and uncompressed data, respectively; (ii) Level-1B that provides products with radiometric and geometric correction; and (iii) Level-1C which provides ortho-rectified top of atmosphere reflectance with a sub-pixel, multi-spectral, and multi-date registration and cloud mask (Baillarin et al. 2012). Image processing includes preprocessing, enhancement, and classification. Image quality issues may be partially improved by applying image preprocessing and enhancement (Schowengerdt 2012; Xie et al. 2008).

12.2.2 Image Pre-processing

As defined by Xie et al. (2008), image pre-processing is mainly devoted to preparation phases to amend the quality of the original images, such as removing noise and increasing the interpretability of image data. Image processing is essential for vegetation extraction and mapping, and depends on the level of image products published by the sources (e.g. Level-1A, -1B, or -1C). The remote sensing data are usually downlinked to the ground station, pre-processed, and released to the users, however, on-board imagery pre-processing was proposed recently to reduce the cost and time delay in downlinking data (often the data size is more than 10 GB), and providing products for rapid decision-making in response to emergencies (Zhang and Zhang 2007).

Consequently, the image pre-processing stage largely depends on the source image quality, characteristics of sensors, and issues. Image pre-processing contains the following operations; (i) bad line replacement and noise removal, (ii) radiometric correction, and (iii) geometric correction (Schowengerdt 2012; Xie et al. 2008). Bad line replacement is the process of determining the overall quality of images by reviewing all the bands and repairing the missing lines or data in each band (Xie et al. 2008). The noise removal process eliminates the undesirable image characteristics produced by sensors and reduces the effects of atmospheric scattering measured by sensors (Schowengerdt 2012). Radiometric and geometric corrections are related to correcting radiometric errors, digital image distortion, and geometric distortion (Xie et al. 2008). Furthermore, Zhang and Zhang (2007) included multi-resolution image fusion as an additional operation for image pre-processing and mentioned that the fusion of images from different sensors, or the same sensor with different spatial resolutions, enhances the land cover interpretation.

After applying the necessary correction, the corrected images can be shared in the image enhancement process or directly to the classification process, or both (Schowengerdt 2012).

12.2.3 Image Enhancement

Image enhancement and masking (e.g. for clouds, water, irrelevant features) is a process aimed at refining favorable or useful image features, such as particular species of vegetation, for visualization purposes (Shyu and Leou 1998; Xie et al. 2008). The choice of image enhancement techniques depends on image content, specific task and objective, observer characteristics, and viewing conditions (Chourasiya and Khare 2019; Singh and Mittal 2014). There are several methods for image enhancement that are generally separated into two categories: (i) traditional methods such as scale conversion, histogram conversion, color composition, color conversion between Red-Green-Blue (RGB), and Hue–Saturation–Intensity transform (HSI), etc.; and (ii) new methods such as genetic algorithm approach (Shyu and Leou 1998), and Artificial Neural Network (ANN) (Jang et al. 2006; Xie et al. 2008). New methods have been developed upon the traditional methods and their gaps eliminated (Chourasiya and Khare 2019; Jang et al. 2006; Shyu and Leou 1998; Singh and Mittal 2014). Shyu and Leou (1998) analyzed some of the traditional methods and mentioned their weaknesses. To begin with, color conversion in RGB is inappropriate for the human visual system. Furthermore, most of the traditional methods are only applicable for one type of degradation (Shyu and Leou 1998). Finally, most color-based methods struggle to fetch objects from the background when there is low contrast between such objects and the background (Singh and Mittal 2014). Another broad category of image enhancement methods is the spatial-based domain that operates directly on image elements and pixels (e.g. logarithmic transforms, power-law transforms, genetic algorithms, and histogram equalization), and frequency-based domain that operates directly on the transform coefficients of the image (e.g. discrete cosine transform, discrete Fourier transform, and Hartley Transform) (Chourasiya and Khare 2019; Singh and Mittal 2014).

12.2.4 Image Classification

The process of extracting different classes (e.g. vegetation species) from raw satellite data is known, in a broad term, as image classification. In other words, the image interpretation and classification process is a translation of spectral classes of the imagery into the vegetation types. Multiple factors have an impact on the efficiency and computational cost of image classification, such as the choice of the classification algorithm, sensor types, training samples, input features, pre- and post-processing techniques, ancillary data, target classes, and accuracy of the final product (Corcoran et al. 2013; Huang and Zhang 2013; Sheykhmousa et al. 2020). Vegetation extraction is defined as the interpretation of satellite images upon image or texture, tone, pattern, and association information to derive vegetation information either by supervised or unsupervised methods (Xie et al. 2008). Image classification methods may be categorized into two types; supervised classification and unsupervised classification. Supervised classification requires previously classified reference samples to train the classifier and then classify unknown data. The first step for this method is to train representative parameters for each class of interest by using available sufficient known pixels. According to the trained parameters, the classifier attaches the labels to all the images (Liu 2003).

Alternatively, unsupervised methods are typically used when reliable geologic information or field observation is unavailable. For unsupervised methods, analysis of images is done without using previously classified reference samples, so the classifier itself determines pixels and assigns them to specific classes. There is an additional type of classification that combines supervised and unsupervised classification and the advantages of both approaches, called hybrid classification (Liu 2003).

Object-based classification works by segmenting images into discrete objects or features and then classifying the objects. This method targets objects instead of pixels. Based on data distribution assumptions, image classification can also be categorized as parametric or non-parametric (Sheykhmousa et al. 2020). One of the mentioned literature reviews divided image classification into two categories: traditional methods (e.g. K-mean and ISODATA for unsupervised classification, the Maximum Likelihood Classification (MLC) for supervised classification); and improved methods (e.g. continuous classification, developed supervised and unsupervised Spectral Angle Classifiers (SAC), ANN, and fuzzy logic, Decision Tree (DT), Random Forests (RF), Support Vector Machines (SVM), SPAtial Reclassification Kernel, Ancillary Data Classification Model (ADCM), and hybrid methods). Each one of these includes supervised and unsupervised approaches (Xie et al. 2008). The reason for such a division is not only that the same vegetation may have different spectral features, but also different vegetation may possess the same spectral signature in remotely sensed images. Therefore, the accuracy of classification may decrease either by applying supervised or unsupervised methods. As a result, improved methods need to be applied to obtain better accuracy assessments (Xie et al. 2008). The most commonly used classification algorithm in the remote sensing community is SVM, RF, ensemble classifiers, MLC, Means Clustering Classifier (MCC), and deep learning algorithms, as shown in Figure 12.1 (Liu 2003; Sheykhmousa et al. 2020).

Compared with other machine learning approaches, decision tree-based models fared better in Above-Ground Biomass (AGB) prediction (Tamiminia et al. 2021). The DT technique provides the groundwork for tree-based models, where it creates a regression model based on the conditional statement's tree structure. DT makes decisions based on characteristics in the dataset to divide the data down into smaller groupings. Nodes are classified as decision nodes or leaf nodes. The former describes how the data will be separated based on decisions, while the latter defines the attributes' values. DT enables simple interpretation and management of non-linear data, although it is susceptible to overfitting, and small noise in the input data may have a significant impact on the predictions (Tamiminia et al. 2021). RF is a non-parametric ensemble approach that combines several decision trees in parallel, and employs a combination of bagging and replacement to train the trees. RF not only employs a bagging technique to ensure tree diversity, which decreases over-fitting, but it is also capable of dealing with noisy datasets (Tamiminia et al. 2021). An alternative to deep neural networks (DNNs) with fewer hyper-parameters and complexity is Deep Forest, a novel decision tree ensemble technique. DFs are quicker and simpler to train than DNNs. Unlike DNNs, which needs a large amount of data to train, DF may function well with small-scale training data. Multi-grained cascade forest is another name for this method (gcForest) (Tamiminia et al. 2021).

12.2.5 Accuracy Assessments

Accuracy assessment is the result evaluation and final step of the vegetation mapping process; it determines the degree of correctness of the vegetation classification. A vegetation map is accurate if it offers a faithful representation of vegetation in the region (Xie et al. 2008). There are two kinds of accuracy assessments; visual assessment and statistical assessment (Amani et al. 2020). Most of the accuracy assessment methods perform their evaluation using a confusion or error matrix. The confusion matrix describes the fitness between the derived classes and the reference data by measuring the Overall Accuracy (OA), Producer Accuracy (PA), User Accuracy (UA), and Kappa coefficient of agreement (K) (Sheykhmousa et al. 2020; Xie et al. 2008). Sheykhmousa et al. (2020) stated that the use of OA and K for accuracy assessment by including ground truth data might be the most common and reliable approach for reporting the accuracy of thematic maps. These accuracy measurements make the classification algorithms comparable when independent training and validation data are incorporated into the classification scheme (Sheykhmousa et al. 2020). Figure 12.2 shows the two

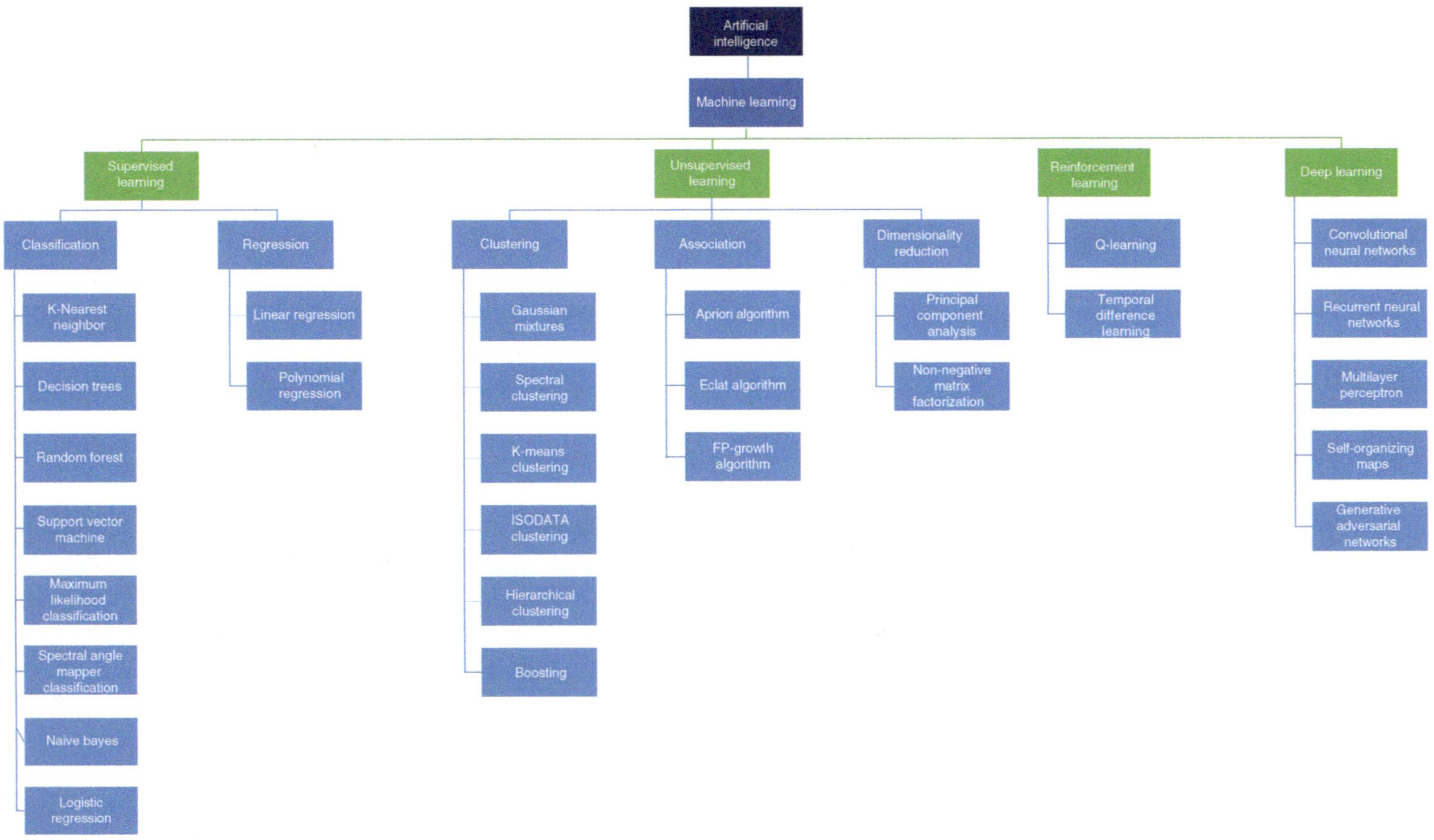

Figure 12.1 Commonly used machine learning methods.

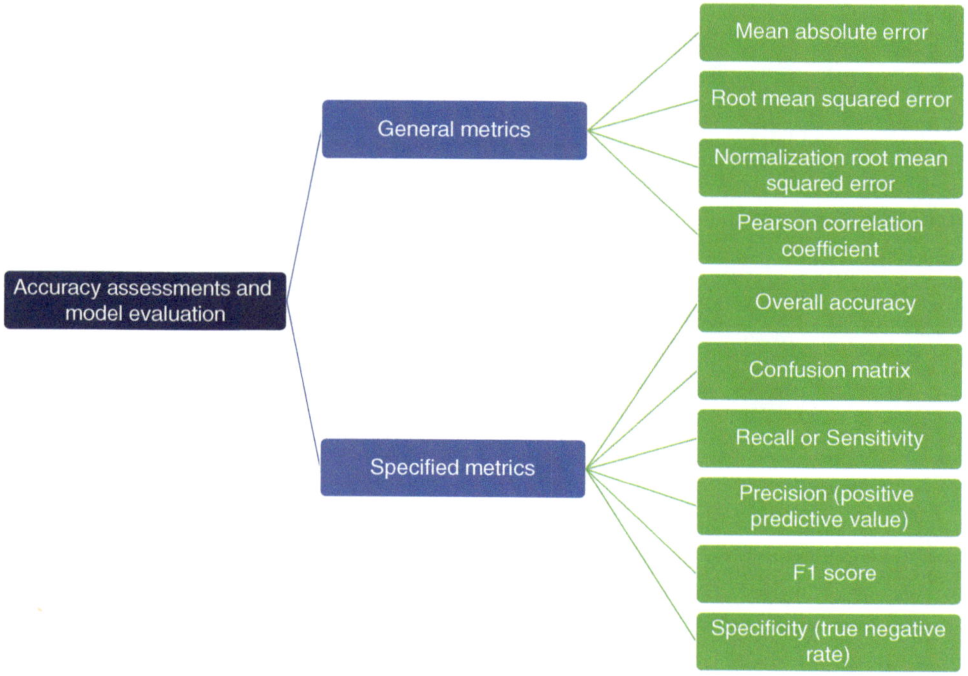

Figure 12.2 Accuracy assessment and model evaluation metrics.

categories for assessments; (i) general metrics and (ii) metrics specifying for classification problems.

12.2.6 Vegetation Indices

A well-characterized vegetation classification should present actual vegetation community composition and exact and adequate classification of vegetation patterns and complexity (Xie et al. 2008). Many sensors provide imagery for producing vegetation indices calculated from visible and near-infrared regions (Xie et al. 2008). Vegetation indices have been used to monitor, analyze, and map vegetation and to remove miscalculate vegetation parameters due to atmospheric conditions or illumination angles (Barrett et al. 2016). Consequently, spectral bands are sensitive to different vegetation features (e.g. structure, pigments, and water content). Therefore, several vegetation indices have been defined concerning different vegetation features and contribute to improving vegetation classification, as presented in Table 12.1. The Normalized Difference Vegetation Index (NDVI) is the most commonly used index for vegetation phenology and is easily calculated in red and near-infrared bands of most optical sensors. Most of the other indices have been designed for special purposes. For example, Water-Adjusted Vegetation Index (WAVI) is intended to identify the aquatic vegetation characteristics and distinguish them from terrestrial vegetation via integrating the spectral response in the shorter visible wavelength range (Sun et al. 2020). Some recent studies declared that using multiple vegetation indices could offer a complete characterization of the vegetation classes and enhance the

Table 12.1 Commonly used vegetation indices available in this Chapter, by reference.

Index	Application	Reference
CTVI (Corrected Transformed Vegetation Index)	Vegetation classification	Perry and Lautenschlager (1984)
DVI (Difference Vegetation Index)	Distinguishing between soil and vegetation and land cover classification	Tucker (1979)
Entropy	Land cover classification	Phan et al. (2020)
EVI (Enhanced Vegetation Index)	Detection of vegetation communities in various seasons (Bhatnagar et al. 2021); Land cover classification (Phan et al. 2020)	Bhatnagar et al. (2021), Phan et al. (2020)
GNDVI (Green Normalized Difference Vegetation Index)	Estimating photosynthetic activity and is a commonly used vegetation index to determine water and nitrogen uptake into the plant canopy	Gitelson and Merzlyak (1994)
LAI (Leaf Area Index)	Controlling physical and biological processes of plant canopies (Chen and Black 1992); Provides an understanding of dynamic changes in productivity and climate impacts on forest ecosystems (Zheng and Moskal 2009)	Chen and Black (1992), Zheng and Moskal (2009)
MNDWI (Modified Normalized Difference Water Index)	Using green and SWIR bands for the enhancement of open water features; It diminishes built-up area features that are often correlated with open water in other indices (Domingo et al. 2020); Land cover classification (Phan et al. 2020)	Domingo et al. (2020), Phan et al. (2020)
MNLI (Modified Nonlinear Index)	An enhancement to the Non-Linear Index (NLI) that incorporates the Soil Adjusted Vegetation Index (SAVI) to account for the soil background	Yang et al. (2008)
MSAVI (Modified Soil Adjusted Vegetation Index)	Determines the density of greenness by reducing the soil background influences	Agarwal et al. (2019a)
MSAVI2 (Modified Secondary Soil Adjusted Vegetation Index)	Land cover classification	Phan et al. (2020)
NBR (Normalized Burn Ratio)	Highlighting burnt areas in large fire zones	Domingo et al. (2020)
NDBI (Normalized Difference Build-Up Index)	Highlighting urban areas where there is typically a higher reflectance in the shortwave-infrared (SWIR) region, compared to the near-infrared (NIR) region; Applications include watershed runoff predictions and land-use planning; Land cover classification	Phan et al. (2020)
NDII (Normalized Difference Infrared Index)	A reflectance measurement that is sensitive to changes in water content of plant canopies; The NDII uses a normalized difference formulation instead of a simple ratio, and the index values increase with increasing water content	Domingo et al. (2020)

(Continued)

Table 12.1 (Continued)

Index	Application	Reference
NDPI (Normalized Difference Phenology Index)	Vegetation phenology and classification	Cao et al. (2018)
NDVI (Normalized Difference Vegetation Index)	The detection of vegetation communities in various seasons (Bhatnagar et al. 2021); determine the density of greenness (Agarwal et al. 2019a)	Agarwal et al. (2019a), Domingo et al. (2020),
NDWBI (Normalized Difference Water Body Index)	Land cover classification	Phan et al. (2020)
NDWI (Normalized Difference Water Index)	Characterizing the wetness of the ground (Bhatnagar et al. 2021); Land cover classification (Phan et al. 2020)	Domingo et al. (2020), Phan et al. (2020)
OSAVI (Optimized Soil Adjusted Vegetation Index)	Providing greater soil variation than SAVI for low vegetation cover, while demonstrating increased sensitivity to vegetation cover greater than 50%	Rondeaux et al. (1996)
PI (Phenology Index)	Developing specifically to monitor vegetation phenology. It combines the merits of NDVI and the NDII, and aims to remove the confounding effects of brightness and wetness from greenness signals	Gonsamo et al. (2012)
RDVI (Renormalized Difference Vegetation Index)	Using the difference between near-infrared and red wavelengths, along with the NDVI, to highlight healthy vegetation	Roujean and Breon (1995)
RVI (Ratio Vegetation Index)	Estimating the green biomass	Agarwal et al. (2019a)
SR (Simple Ratio)	Monitoring vegetation, Land cover classification, this index correlates with foliage intensity; Sensitive to three main external factors, solar and viewing geometry, soil background, and atmospheric effects	Noi Phan et al. (2020)
SAVI (Soil Adjusted Vegetation Index)	Minimizing soil brightness influences using a soil-brightness correction factor; This is often used in arid regions where vegetative cover is low; Land cover classification	Huete (1988), Phan et al. (2020)
TPI (Topographical Position Index)	Measuring topographic slope positions and automating landform classifications	Räsänen and Virtanen (2019)
TTVI (Thiam's Transformed Vegetation Index)	Vegetation classification	Thiam (1997)
TVI (Transformed Vegetation Index)	Vegetation classification	Deering et al. (1975)
TWI (Topographical Wetness Index)	Quantifying topographic controls on hydrological processes; TWI can also be used to characterize biological processes including forest site quality, vegetation patterns, and annual net primary production	Räsänen and Virtanen (2019)

Table 12.1 (Continued)

Index	Application	Reference
WDRVI (Wide Dynamic Range Vegetation Index)	Similarly measuring levels of near-infrared and red light but is at least three times more accurate than NDVI because it can pick up on subtle differences in the crop canopy in moderate-to-high vegetation density, which is especially important for crops with dense canopies and mature crops	Gitelson (2004)

vegetation classification exclusively when using the combination of different spectral bands (Barrett et al. 2016; Lastovicka et al. 2020; Sun et al. 2020).

12.3 Data Fusion

Forest species mapping and classification have been widely explored since they are an important indicator for forest inventory and management (Szantoi et al. 2017). Classification of remote sensing data is a challenging task, due to the complexity of the landscape and vegetation species, the large volume of data, and limited or imbalanced training data (Pouteau and Stoll 2012; Sheykhmousa et al. 2020; Wilschut et al. 2013). Remote sensing for species classification started with human interpretation of aerial photography at large scales, and Heller et al. (1964) reliably recognized tree species on color film, suggesting the feasibility of applying the technique to real inventory concerns. Application of remote sensing for species classification has progressed to multispectral image classification (Vieira et al. 2003; Walsh 1980), multitemporal analysis (Brown De Colstoun et al. 2003; Wolter et al. 1995), hyperspectral analysis (Goodenough et al. 2003; Martin et al. 1998), and multi-sensor data fusion, which typically refers to the combination of (2D and 3D) data from several sensors (Alonzo et al. 2014; Goodenough et al. 2005; Ke et al. 2010). Most of the multispectral and multitemporal development was originated by using data of Landsat (Agarwal et al. 2019b) and Landsat 7 Enhanced Thematic Mapper (ETM) (Agarwal et al. 2018) to distinguish the tree species, Landsat Multi-Spectral Scanner (MSS) and Landsat Thematic Mapper (TM) for forest classification (Wolter et al. 1995), and Landsat 7 Enhanced Thematic Mapper-Plus (ETM+) to classify and map land cover (Brown De Colstoun et al. 2003). Application of hyperspectral remote sensing images have been initiated by the use of Airborne Visible/Infrared Imaging Spectrometer (AVIRIS) (Martin et al. 1998) and Earth Observing-1 (EO-1) satellite with hyperspectral sensor (Goodenough et al. 2003) to classify forest cover and species. Finally, multisensor data fusion has been introduced by fusion of hyperspectral data from Hyperion, ASAR radar data, Advanced Land Imager data from the EO-1 satellite, Quickbird imagery, and airborne AVIRIS for aboveground carbon estimation by Goodenough et al. (2005).

Although the potential of satellite data in agriculture production was discovered in the mid-1970s, the need for accurate mapping has been discussed by the use of freely available satellite data with sufficient resolution and temporal frequency since the launch of Sentinel satellites (Hegarty-Craver et al. 2020). Medium resolution of satellite data such as the 10 m resolution of Sentinel-2 exposes some limitations on distinguishing different vegetation classes in heterogeneous vegetation communities. As a result, a key parameter to have a high accuracy classification depends on the availability and accuracy of the ground truth data such as field survey and UAV data (Bhatnagar et al. 2021; Hegarty-Craver et al. 2020). On the other hand, high-resolution satellite imagery can be costly with low temporal frequency and low area coverage. Due to spectrally mixed pixels and segments, they can cause blur or confusion between land cover types (Gray et al. 2018).

By considering machine learning development and launching high technology satellite constellations and UAVs, integrating new machine learning algorithms or object-based classification, and fusing different remote sensing data sources may contribute to the accurate map of complex land cover with interfaces between classes and mixed classes (Gray et al. 2018; Murugan et al. 2016). Fusion of satellites and drones has several applications, such as wetland vegetation mapping and dynamics (Bhatnagar et al. 2021; Gray et al. 2018; Kaplan 2018), vegetation and crop mapping (Hegarty-Craver et al. 2020; Räsänen and Virtanen 2019), high precision agriculture such as sugarcane production, and within class classification of lands (Agarwal et al. 2018, 2019b; Murugan et al. 2016, 2017), fuel type classification (Domingo et al. 2020; Erdody and Moskal 2010; García et al. 2011), identifying animal habitats by classifying land cover to the forest, reforested areas, bare areas, and oil palm agriculture (Szantoi et al. 2017), flood inundation map (Xia et al. 2017.), and identification of invasive woody species (Kattenborn et al. 2019). UAV and satellite data share the same perspective (top of canopy) and can be matched on a spatial scale by resampling (Kattenborn et al. 2019). UAV data can distinguish water bodies and non-water bodies such as small inundation, grass, and soil patches, which cannot be detected easily by the low to moderate satellite imagery or conventional manned aerial photography (Xia et al. 2017). However, UAV flight is highly dependent on the favorable weather conditions plus the limitation of remote control during the survey due to safety issues and image quality, and these limitations could make the analysis of large-scale projects inefficient and difficult (Milewski and Niepodległości 2016).

For forest mapping, there is a need for spectral and structural indices (Szantoi et al. 2017). Remote sensing can provide horizontal and vertical forest structure components for forest mapping and characterization (Domingo et al. 2020). Different sources (UAV, aerial, and satellite) provide different spectral resolutions, and each extra spectral data can reveal new features hidden from other sensors. Combining optical datasets with topography, vegetation height, and structure elements from different sources can boost classification accuracy (Räsänen and Virtanen 2019). Active remote sensing such as LiDAR and SAR provides vertical forest characterization, while passive remote sensing cannot penetrate forest canopies and indirect relationship with vertical structure. Integration of multispectral and active remote sensing shows the capacity to improve fuel characterization (Domingo et al. 2020).

12.3.1 Fusion of Satellite and UAV/Drone

The drone and satellite data combination application was first examined in precision agriculture. Population growth is placing increasing pressure on the agriculture sector to manage resources more sustainably while simultaneously expanding productivity to fulfill the needs of people. Precision agriculture, which is described as a combination of technologies that combine sensors, information systems, equipment, and informed management to increase productivity by accounting for spatiotemporal variability and uncertainty in agricultural systems, may help solve these challenges (Revill et al. 2020). Murugan et al. (2016) investigated drone images to retrieve satellite data for precision agriculture monitoring and the within-class classification of vegetation. Their methodology was the data fusion of Landsat 8 and the drone imagery to classify vegetation into sparse and dense classes over the Roorkee Tehsil in Uttarakhand, India, covered mainly by agricultural land and sugarcane. After data preprocessing of two sources, they developed their model by classifying with spectral angle mapper (SAM), computing Threshold value, higher Separability Index (SI), and NDVI. They extracted vegetation areas by comparing them with drone images, and then separated them into sparse and dense classes by collecting ground truth points from Landsat images. They also verified the results visually and quantitatively by comparing the estimated area to Landsat data with drone imagery. Their study showed that some classes were classified incorrectly in Landsat image classification due to the presence of mixed pixels. The computed area for dense vegetation by Landsat image is significantly less than the computed area by drone. Although the authors found their methodologies effective for their study, they did not mention the percentage of accuracy, compare the results with the data interpretation of Landsat data alone, and evaluate the effectiveness of their methodology on improving accuracy. Furthermore, the authors only applied the drone data to identify visually, compare, and validate vegetation classes.

In their subsequent investigation, they proposed an adaptive thresholding technique based on a multi-objective optimization technique to maximize the OA and minimize the False Alarm Rate (FAR) to find threshold (Murugan et al. 2017). By obtaining the classification accuracy of 87% for the testing data and 73% for the validation data, they declared the ability of their adaptive method to produce highly consistent classification outcomes. They also found the SWIR-1 band with high SI of 2.5, which provides better separability among other bands. However, they did not explain that this reason is due to SWIR-1 sensitivity to the leaf moisture content or, as previous studies mentioned, infrared and near-infrared bands are most suitable for vegetation classification.

During the same year as Murugan et al.'s (2016) investigations, Milewski and Niepodległości (2016) estimated the production of permanent grassland by use of UAV multicopter with six-rotor, Landsat 8, and ground measurement. They declared that a high level of details such as drone high spatial resolution data is impossible to obtain with satellite data, which is necessary for precise determination and modification of mathematical models used for image analysis and processing.

Agarwal et al. (2018) analyzed the performance of commonly used classification methods, including k-means, Parallelepiped, SVM, Maximum Likelihood, Minimum Distance, Neural Net, Mahalanobis Distance, and Binary Encode to compute the area estimation of three different classes of bare land, dense vegetation, and sparse vegetation by using Sentinel-2 and drone data. They applied image processing on drone data to upscale to the pixel resolution of Sentinel-2 data. They generally concluded the better performance of SVM and Mahalanobis Distance than other classification methods with an OA of 76.488 and 75.595%, respectively. However, they observed that some methods had the highest PA for some classes and low OA. For example, Parallelepiped has 96% accuracy in classifying bare land with a weak classification accuracy of 23.27% for dense vegetation. They also mentioned the usefulness of drones in within-class classification. Similar research was developed on a machine learning-based approach to calibrate the NDVI values obtained from Sentinel-2 with drone data on the same study area in India (Agarwal et al. 2019b). Using up-scaled drone images, they provided the mask of three classes of fallow land, sparse vegetation, and dense vegetation. They developed a neural network based on back-propagation to train and calibrate the NDVI value of Sentinel-2 with drone data. They observed improved classification in Sentinel-2 data using calibrated NDVI, and the calibrated image is more similar to in-situ drone imagery. The true-positive value of calibrated data was compared with the drone imagery and primary classification of Sentinel-2 images for two different datasets. Although this methodology improved the classification accuracy, the true-positive value of calibrated Sentinel-2 has an error of 30% in some classes for both data sets compared with drone imagery.

Precise and reliable data on terrain mapping and change detection are of paramount importance for civilian and defense applications, which may be obtained through the combination of satellite and drone. Agarwal et al. (2019a) proposed a neural network by inputting satellite and drone data to detect the changes and terrain behavior. They used NDVI, Modified Soil Adjusted Vegetation Index (MSAVI), and Ratio Vegetation Index (RVI) to identify the classes. The mask was also generated from up-scaled drone imagery to make it Sentinel-2 equivalent and overlaid on the classified Sentinel-2 data of each date to detect the change in class label. An OA between 70.7 and 89.83% was obtained by implementing their proposed neural network method.

Sentinel-2 can monitor crops worldwide at sub-field spatial resolutions (10–20 m) using observations in the red-edge spectral range. On the other hand, satellite-based Leaf Area Index (LAI) estimations need calibration using ground observations. Calibration is a difficult task because of spatial heterogeneity and scale mismatches between field observations and satellite data. UAVs provide high-resolution (cm-scale) LAI estimations, which Revill et al. (2020) utilize to characterize uncertainty and eliminate spatial scaling differences between Sentinel-2 observations and field surveys. Revill et al. (2020) studied LAI by combining Sentinel-2 with 20 m resolution multispectral observation and a novel UAV multispectral sensor matching Sentinel-2 spectral bands. They tested a two-stage calibration strategy for deriving robust LAI estimations, which may aid crop management decision-making, from Sentinel-2 using the UAV LAI.

Along with LAI ground observations at various field locations and growth phases, they deployed a UAV-mounted multispectral sensor with the same red-edge and near-infrared central wavelengths as that of the first nine bands (i.e. bands 1–8A) of the Sentinel-2 Multispectral Instrument. Figure 12.3a illustrates the single-stage calibration using ground measurements and indirect calibration utilizing high-resolution UAV data for upscaling ground measurements and calibrating the Sentinel-2 retrieval model using UAV LAI data. The bias associated with the two-stage calibration procedure exhibited a narrower and symmetrical distribution, while the single-stage technique gives very imprecise estimates in certain circumstances, as seen by the error distribution's large tails in Figure 12.3b. Both algorithms were validated using averaged UAV data per Sentinel-2 pixel. The authors demonstrated that, even within managed agricultural systems, there is substantial and significant within-field variation in LAI at scales finer than the resolution of existing

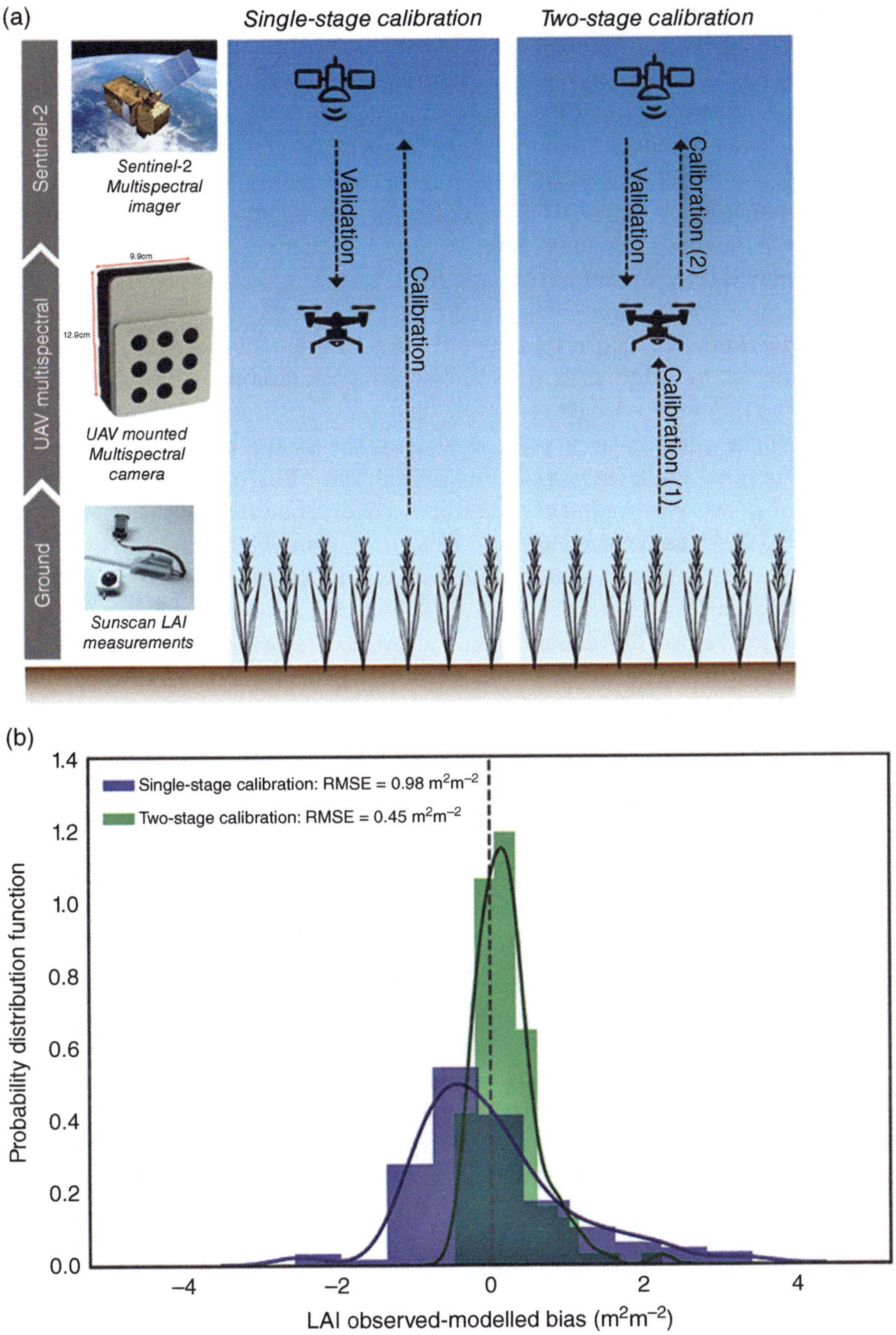

Figure 12.3 (a) Overview of Single-stage and Two-stage approaches for calibrating GPR models used for retrieving LAI from Sentinel, and (b) comparison of observed–modeled LAI bias probability distribution function plots for the single-stage (blue line and shading) and two-stage (green line and shading) calibration approaches. (Source: Adopted from Revill et al. 2020, MDPI CCBY 4.0).

satellite imaging instruments. This paper demonstrates that UAV multispectral observations at the cm scale, taken using a sensor tailored to match the spectral bands of the Sentinel-2 satellite signal, may aid in the interpretation of the satellite's signal. They confirmed that UAV multispectral observations at the centimeter scale, taken with a sensor tuned to Sentinel-2 spectral bands, enhance satellite signal interpretation.

On a large scale, forest fuel classification is important for sustainable forest management, especially one prone to higher wildfire frequency. The quantity and distribution of fuels in forest mapping are crucial parameters to assess fire danger and fir behavior. The Mediterranean forest is a good example of large fire-affected environments to date. Fires change and shape the current cover land composition and species richness in this region. Domingo et al. (2020) provided a fuel type classification of three different Mediterranean forests in Spain affected by wildfires in 1994. They used a combination of Airborne Laser Scanning (ALS) data, Sentinel-2 data, and 136 field plots for this classification. They reached the overall classification accuracy of 0.59 using the SVM method with the radial kernel. However, the other researchers, who combined airborne LiDAR and high-resolution color near-infrared imagery or multispectral airborne data, obtained classification accuracy for Prometheus fuel type between 84% and 88% due to using higher point density and multispectral images with higher resolution and different structural complexity of vegetation (Erdody and Moskal 2010; García et al. 2011). All of these studies found data fusion a practical solution for improving classification accuracy. However, they declared the strength of LiDAR and an active sensor for vertical forest structure characterization compared to optical sensors. However, they mentioned three main problems of active sensors; application of these methodologies on large-scale areas, inaccessible areas blocking the laser pulses due to topography or vegetation, and some confusion in a different fuel type Prometheus. Domingo et al. examined the classification accuracy of only one data source, ALS or Sentinel-2 data, and the value was lower than the OA of fusion of these two sources.

One of the major threats to forest management is the unpredictable growth of invasive plant species, which can cause a negative impact on biodiversity, ecosystem functioning and services, and financial loss (Kattenborn et al. 2019). There is an essential need for high spatial resolution remote sensing data to detect and map invasive species at early stages, especially for training data. Kattenborn et al. (2019) assessed the potential of UAV data for semi-automatic reference data acquisition on mapping invasive species. Moreover, they evaluated the combination of UAVs as upscaling reference data with Sentinel-1 and Sentinel-2 data to map species cover over large spatial extents operationally. 2D texture metrics, 3D structure metrics, RGB orthomosaics, and hyperspectral reflectance were derived to examine the different combinations of predictors on the accuracy of the final map. In general, the highest accuracy was obtained for the combination of hyperspectral, 2D texture, and 3D structure with a very little difference for the combination of RGB, 2D texture, and 3D structure, which means an inexpensive camera could be sufficient for mapping. However, it depends on the homogeneity of the surrounding vegetation. Referring to the usage of a single predictor, the hyperspectral data were the most accurate, and the 3D structural metrics alone had the lowest accuracy for mapping in their study. Nevertheless, the accuracy increased mainly by combining different predictors since they generated more sensitivity to different plant species' biochemical and structural variation. In the case of a combination of Sentinel-1 and Sentinel-2 in their study, they asserted that Sentinel-2 multispectral data has more importance to estimate the cover fraction than Sentinel-1 SAR data. However, Sentinel-1 data has generally increased the accuracy and is helpful for the region with frequent cloud cover, for which Sentinel-2 had some limitations (Kattenborn et al. 2019).

Different structural tree metrics may be used to measure tropical forest health and temporal dynamics such as vegetation height, canopy cover, stem volume, and AGB. AGB is defined as the entire quantity of above ground live organic matter in vegetation, and represented as oven-dry tons per unit area (Brown 1997; Urbazaev et al. 2018). For sustainable forest management, reliable and accurate AGB information is essential because vegetation biomass influences a wide variety of ecological processes, including carbon and water cycles, energy fluxes, and local and regional climate (Urbazaev et al. 2018).

AGB has been estimated using a variety of remote sensors, including both radar and optical sensors (Li et al. 2020; Tamiminia et al. 2021). Radar sensors can penetrate the canopy and interact with the primary biomass components, namely tree trunks, and branches (Li et al. 2020). Saturation at medium-high biomass levels (depending on wavelength) and a substantial dependency on environmental conditions (e.g. rainfall, freezing, different moisture conditions) are two drawbacks of radar data for AGB assessment (Urbazaev et al. 2018). Optical remotely-sensed data offer different vegetation indices and texture images, which have a strong correlation with biomass but saturate at high biomass (Li et al. 2020; Urbazaev et al. 2018). When estimating AGB from optical data, there is a high reliance on environmental, seasonal, and acquisition conditions (e.g. rain, shadow, cloud cover, and solar zenith angle), leading to inaccurate results in the tropics (Urbazaev et al. 2018). Consequently, the most typical problem with optical and radar datasets is saturation (Tamiminia et al. 2021; Urbazaev et al. 2018). Saturation occurs in forests with multilayer canopies or dense biomass when spectral reflectance values of pixels are not responsive to biomass changes, affecting the accuracy of AGB calculation (Tamiminia et al. 2021). It is possible to get detailed 3D information of an object on the earth's surface, including trees and buildings, using LiDAR. Laser pulses transmitted from a LiDAR sensor penetrate forest canopy and provide information about their vertical structure (e.g. height, canopy volume). Accurate AGB estimations may be made using LiDAR data, which does not suffer from signal saturation but is costly and limited to small spatial coverage (Urbazaev et al. 2018). Urbazaev et al. (2018) explained that LiDAR is an essential tool for repeating reference data collections across difficult areas (e.g. restricted or inaccessible areas). It can capture spatial variability, which is useful in heterogeneous tropical forests. However, the LiDAR-calibrated AGB map showed higher uncertainties in comparison to the NFI-calibrated map in Mexico at a national scale (Urbazaev et al. 2018).

Decision tree-based models have performed better in AGB prediction than other machine learning techniques (Tamiminia et al. 2021). The results of the decision tree, RF, and deep forest regression models were examined using a LiDAR aerial laser scanner, Landsat 5 TM, and a combination of them to determine the combined capacity of LiDAR and optical data for accurate AGB estimate in contrast to a single sensor. The most accurate AGB estimation was obtained by combining vertical features generated by LiDAR and spectral information collected from Landsat 5 TM data and using a deep forest regression model, as seen in Figure 12.4. The deep forest was shown to outperform RF and DT models, indicating that it is a highly competitive alternative to DNNs and biomass can be predicted more precisely due to the specific properties of Deep forest. In all three regression models, Tamiminia et al. (2021) have shown the trend of improved performance combining both LiDAR and Landsat data.

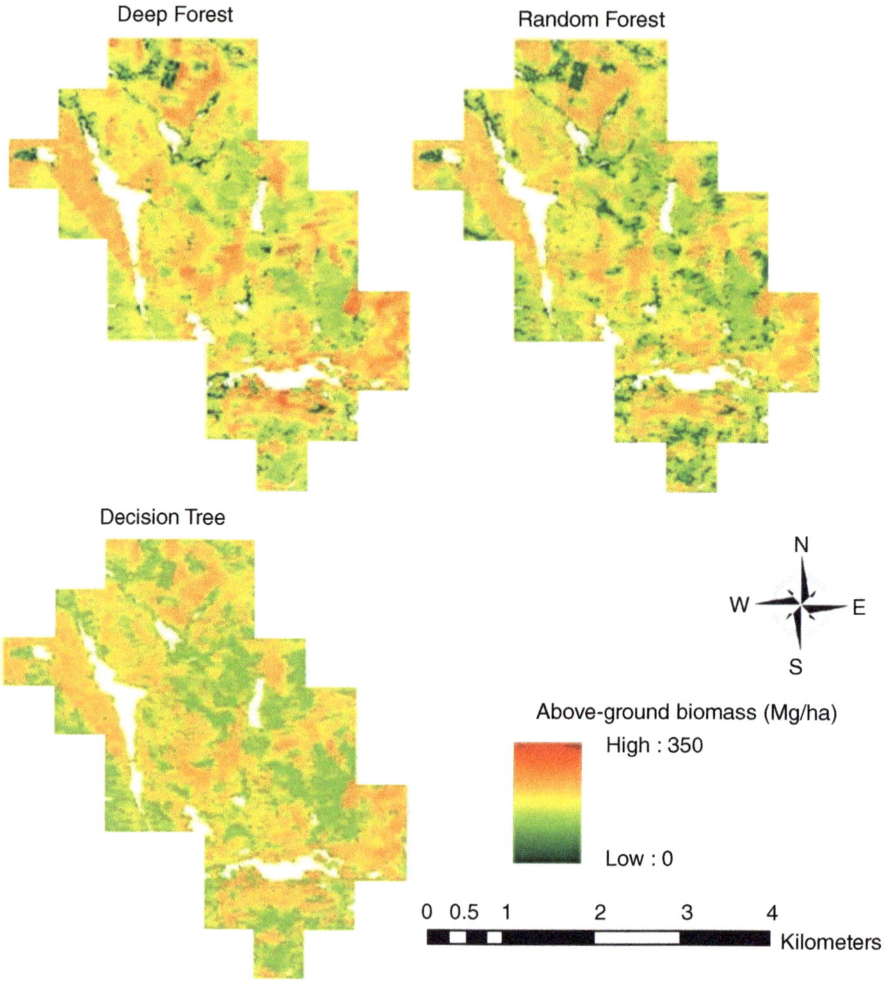

Figure 12.4 Above-ground biomass maps of HWF generated using decision tree-based regression models and remote sensing data. The maps were produced based on the combination of LiDAR and Landsat 5 TM data, retrieved from Tamiminia et al. (2021) (Source: From Tamiminia et al. (2021), to reuse under CCBY 4.0).

The centimeter-level resolution has been argued as a prerequisite to mapping vegetation patterns and features accurately, especially in spatially heterogeneous landscapes (Sheykhmousa et al. 2020). There has always been a trade-off between spatial resolution and area coverage by increasing the dataset resolution due to the required cost and time, the technology of remote sensing sources, and the methodology of using different remote sensing sources and sensors. Object-based classification widely provides an accurate land cover classification for high or medium spatial resolution remote sensing imagery, with dominant images resolution of 0–2 m (Ma et al. 2017; Räsänen and Virtanen 2019). A review study on the supervised object-based classification showed the highest mean OA of 86.33% for the UAV and then SPOT-5, QuickBird, and IKONOS by mean classification accuracy of 85.72, 85.50, and 85.49%, respectively (Ma et al. 2017). Object-based classification mostly used fuzzy methods or supervised methods such as Neural Network (NN), SVM, RF, DT, and MLC (Ma et al. 2017). Ma et al. (2017) found the highest mean classification accuracy of 85.81% for the RF classifier, followed by the SVM classifier (85.19%) and DT classifier (84.15%). NN and MLC had relatively lower mean classification accuracy at 81.58 and 81.5%; however, the authors mentioned that the advantages of each classifier vary under different conditions, e.g. study area, land cover type, and spatial resolution. Object-based classification offers more advantages for agriculture land mapping among other types of land cover (Ma et al. 2017).

For forest species classification utilizing an object-based technique, a synergistic combination of high spatial resolution multispectral images of QuickBird, with 2.4 m resolution, and low-posting-density LIDAR data, with 3 m resolution, was explored by Ke et al. (2010). The combination of spectral and LIDAR data in both image segmentation and object-based classification enhanced forest classification compared to utilizing either data source alone, according to quantitative segmentation quality assessment and classification accuracy findings (Ke et al. 2010). Furthermore, they declared that multispectral imagery with high spatial resolution satellite data such as QuickBird aided in defining forest stand boundaries and offered spectral separation between forest species. On the other hand, high spatial resolution imagery offers difficulties since the spectral response of individual pixels no longer accurately depicts the properties of an object of interest (e.g. forest stand). Solar canopy light and topography influences may create substantial reflectance variations within a single forest stand in the high spatial resolution imagery. Consequently, typical pixel-based classification based purely on spectral data may fail, resulting in salt-and-pepper noise in classification output. Ke et al. (2010) mentioned that the issues associated with high spatial resolution were solved through object-based classification and using LiDAR data which can distinguish between forest species.

Segmentation scale, classification, and segmentation input data from remote sensing are influential factors in the OA of the object-based classification. Räsänen and Virtanen (2019) examined 78 different classifications, 13 segmentations, and six different feature sets for each segmentation using different datasets, including optical imagery, topography, and vegetation height from UAV, aerial, and satellite sources with spatial resolution ranging from 5 cm to 3 m. They applied supervised object-based classification using RF since they used high-resolution datasets. Five cm resolution UAV images with small segment size and all calculated features excluding textural features had the highest OA of 76.7%. Satellite images of WorldView-2 with the resolution of 2 m and Planet Scope with a 3 m had an OA of 69.1% and 64.8%, respectively. They concluded that although the UAV data is not necessary for the classification, UAV or aerial image usage notably increases the overall classification accuracy as a training dataset, especially for the coarser-resolution image or upscaling purposes, and to track changes in vegetation patches and fine-scale phenology. They also presented that implementing multiple remote sensing datasets such as multiple optical, topographical, and vegetation height datasets instead of using only spectral bands and indices increases the classification accuracy. It is essential to combine optical datasets, whether multiple or single datasets, with topography and vegetation structure features to accurately map the vegetation in spatially heterogeneous landscapes (Räsänen and Virtanen 2019). Most of the studies utilized high spatial resolution UAV images over small areas as training datasets and validation data; however, it is also possible to capture lower spatial resolution UAV imagery over a larger area for classification and other mapping purposes (Räsänen and Virtanen 2019).

It should be mentioned that even by using satellite data with a high spatial resolution of 1.24 or 5 m compared to satellite data with a low-to-medium resolution of 10–250 m, the UAS augmented the classification accuracy of satellite data (Gray et al. 2018; Szantoi et al. 2017). Gray et al. (2018) investigated the combination of drone data with two commercial satellites of WorldView-3 (WV) with fine spatial resolution and RapidEye with higher temporal frequency to monitor details changes and map the Estuarine environment. Their methodology was to use drone imagery to assist the training of supervised SVM classifier for classifying four different images products of satellites data; WV with the standard 8 bands, WV 8 bands + NDVI + texture, RE with the standard 5 bands, and RE 5 band + NDVI + texture. They reached the best OA of 93% for the standard 8-band WV-3 image; however, the results from both satellites are promising to map the wetland accurately. Although the overall accuracies of previous studies with the fusion of UAV and free available satellite data were between 76 and 88%, these value differences between commercial and free access satellite data are negligible. There is always a trade-off when considering the cost, spatial resolution, temporal frequency, and area coverage (Gray et al. 2018; Räsänen and Virtanen 2019; Szantoi et al. 2017). It should be mentioned that using satellite data with a high spatial resolution of 1.24 or 5 m, effective sources to map highly heterogeneous areas, or with a low-to-medium resolution of 10–250 m, the UAS augmented the classification accuracy of satellite data (Gray et al. 2018; Szantoi et al. 2017).

Furthermore, they used both drone data and field-based sample points to assess the accuracy of each image classification. They received an overall agreement of 96% between the drone and field-based data. They concluded that Unmanned Aerial System (UAS) has a high capacity to train and validate satellite imagery (Gray et al. 2018).

Most of the data fusion studies aim to optimize the cost and time of producing maps. Bhantnagar et al. (2021) investigated using limited drone images for augmenting satellite imagery to improve the OA of vegetation classification of wetlands. Instead of using hyperspectral or multispectral sensors, they captured the images from drone with an RGB camera, and they applied color correction based on an iterative linear transformation of the Probability Density Function (PDF) of the image and color re-graining to compensate for the absence of infrared band information or multispectral sensors. Then, drone imagery was classified using the RF model and up-sampled to use as labeled training data for semi-supervised Mapping Vegetation Community (MVC) algorithm classification of Sentinel-2 imagery. They repeated this approach for summer, spring, and winter in the study area to create the annual map to monitor any natural changes in the wetland over time. Although their methodology has two main sources of error, such as model accuracy of drone classifier and loss via up-sampling, they reached the OA of 87% for the final optimal-multi-season-labeled vegetation map. They also examined the unsupervised K-means clustering analysis on Sentinel-2 imagery, resulting in an average accuracy of 67%. This result showed that the essential use of drone imagery as ground truth data and training data for accurate vegetation mapping aims to reduce the dependency of mapping based on field survey, which is time- and cost-consuming and requires human resources. For example, observing and identifying animal habitats is an expensive and challenging task if done by conventional methods. A combination of UAV and satellite data can enhance the performance and accuracy of the result and monitor them in every location and time of the year. Szantoi et al. (2017) mapped orangutan habitats mostly found in logged forest and reforested areas to identify conservation priorities and plan actions. They combined Landsat 8 imagery with high resolution UAS to classify the land cover by RF classifier into seven different classes with the OA of 75.82%; (i) water, (ii) logged forest, (iii) reforested, (iv) bare soil, (v) mature oil palm plantation, (vi) riverine forest, and (vii) young oil palm plantation. According to the results of their investigation, they found that UAS images might be utilized instead of ground truth data to produce a more accurate classification. They believe their approach is a viable alternative for the time-consuming and expensive practice of conducting field surveys and ground-truthing of Land Use/Land Cover (LULC) classes in widely dispersed locations and remote areas. Furthermore, they mentioned that both the medium-resolution satellite images and the UAS pseudo ground data must be collected in a similar time frame (or at the very least during the same year period) to avoid potential biases due to seasonality. Additionally, due to the lack of national regulations or changes without prior notice regarding UAS operations, flying a UAS is considered relatively uncertain. Such considerations should be taken into account for a semi-automated LULC monitoring system using UAS.

"Big data" processing and the availability of cloud-free images over a vast area are the two fundamental obstacles when mapping land cover over a large area. "Big data" necessitate a large amount of storage space and access to high-performance computers. It is also challenging to get clear images over a vast area in a short period because of cloud cover. Cloud computing platforms such as Google Earth Engine (GEE) can overcome major issues with large-area land cover mapping. Research into vegetation and forest, followed by land use and land cover, was GEE's most often used application discipline. Researches on satellite data over the past decade reveal that the RF algorithm is the most commonly employed classification method (Phan et al. 2020). Since it can handle datasets with high dimensional and multiple sources, increase processing speed by focusing on important variables, and outperform other common classifiers like SVM, kNN, or MLC in terms of achieving higher accuracy. Various composition approaches (metrics, time series, and the various annual composition strategies), as well as diverse input images and feature selection (spectral and auxiliary variables), were examined by Phan et al. (2020) to create and assess land cover maps, employ Landsat 8 Surface Reflectance (L8SR) and RF classifier with eight different combination strategies. Based on their studies, elevation was consistently the most significant classification variable across all datasets, followed by entropy, B7, and B1. The three most significant factors in the time series data were elevation, entropy, and B5 of the July composite image. Spectral indices were scored relatively low in all datasets, and elevation and entropy are the two most important factors. In terms of big data, the temporal aggregation method has significantly reduced the data volume (e.g. time-series data volume between 13.5 and 20.8 GB was reduced to approximately 4.3 GB median data volume), resulting in a simpler and faster analysis, especially in the GEE platform, filling the gaps due to atmospheric and climate conditions, and almost equal accuracy (88.74%) as time-series datasets (89.80 or 89.70%). Land cover classification using remote sensing has been shown to be more accurate when employing popular spectral indices such as NDVI, EVI, and SAVI. However, these indices consistently performed poorly in comparison to other feature bands of Landsat 8 in the Mongolia study site (Phan et al. 2020). This is due to the research area's predominant cover type being vegetation, which undergoes rapid color shifts (from green to brown). These changes are not linear in time due to the climate gradient in the study area. Another factor that has a critical role in land cover classification is the selection of input images for composition, especially in a snowy, cloudy, and expansive area like Mongolia, which can cause OA ranging between 77.66 and 89.80%. According to Phan et al. (2020), GEE supports both Landsat (TM, ETM+, and OLI) and Sentinel-2 (A and B), and integration of these satellite data can solve the difficulties such as cloud and snow cover, and lack of close time acquisition images of Landsat 8.

Figure 12.5 illustrates the land cover classification process, especially grassland, and uses validation data and image inspection with aerial photography to make a confusion matrix to assess the accuracy. In this figure, Spagnuolo et al. (2020) compared the prediction of RF and validation data to measure the accuracy and refine after classification. They combined multi-date radar (PALSAR-2 and Sentinel-1) and optical (Sentinel-2) imagery with the field data and aerial imagery to classify the land cover of Kenya. They used 20% of their

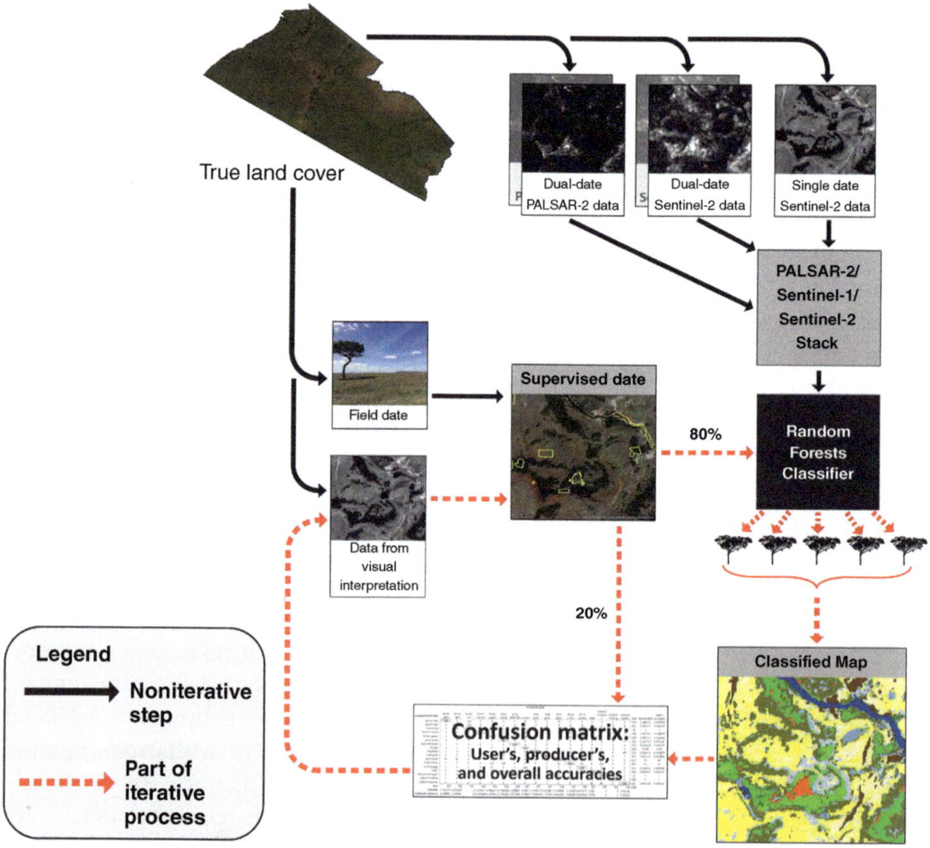

Figure 12.5 An overview of the land cover classification process. (Source: Adopted from Spagnuolo et al. (2020), MDPI, CCBY 2.0).

supervised data for validation and accuracy assessments, and 80% of data and multi-sensor composite imagery were input into RF classifiers. In the end, they achieved fine-resolution and highly accurate land cover maps by the UA and PA ranging from 66 to 100% and OA of 86% (Spagnuolo et al. 2020). Not only did they successfully categorize a diverse range of land cover types, ranging from open water to forests, they were also successful in distinguishing distinct grass heights (short, medium, and tall), with UA rates of 83, 82, and 85%, respectively.

Inundation mapping and dynamics are essential to detect the effect of climate change, human activities, and the reaction of the terrestrial ecosystem. Xia et al. (2017) examined the wetland inundation mapping with three different methods of Linear Spectral Unmixing (LSU), ANN, and Regression Tree (RT) by combining high-resolution UAV data and Landsat-8 Operational Land Imager (OLI) data. Generally, they found the RT method had the best performance for the mapping wetland Subpixel Inundation Percentage (SIP) with the highest R2 value (0.935) and lowest RMSE value (8.54%) in comparison with LSU and ANN. Furthermore, they mentioned the two main sources of uncertainty in their studies; (i) spectral uncertainty of the UAV data due to similar spectral characteristics of land cover and limitation of using only RGB band for the UAV image, and (ii) uncertainty of geolocation matching between the UAV and satellite image. The other goal of their study, which distinguished it from others, was that they labeled the proportion of the wetland inundation for each pixel at a subpixel scale to improve the classification accuracy. In their correlation analysis between Reference SIP and Landsat spectral data, EVI had the best correlation. Band 5, 6, 7, NDWI-2, and MNDWI were also highly correlated with SIP. The correlation analysis can provide good information to choose the most suitable indexes and bands for the aim of the study. Kaplan (2018) implemented the UAV imagery to classify six different classes in the study area and analyzed the correlation between Land Surface Temperature (LST) and NDVI that extracted from Landsat 8 and dual-polarization backscatter values (VH-VV) from the Sentinel-1 at a monthly period over a year based on the UAV classification. The author did not mention the effect of using UAVs instead of conventional field surveys in his research. He remarked the high correlation between NDVI and LST, which is expectable as the LST was calculated based on NDVI value. In the correlation between data extracted from Sentinel-1 and Landsat 8, a strong correlation was observed between VH and NDVI and LST data. At the same time, the VV polarization has no strong correlation with none of the parameters. In addition, he detected there is no strong correlation between these parameters in heterogenic structure, especially for low spatial resolution data since different land cover types can be seen there, such as Sediment Bog.

The challenge is about not only the large-scale mapping and monitoring, but analysis of smallholder farms on a regional scale remains a challenge. Hegarty-Craver et al. (2020) studied a scalable approach by combining optical images from Sentinel 2 and SAR images from the sentinel 1 with drone ground truth data to create a cropland cover map for smallholder agriculture systems. They used 80% of the labeled ground truth dataset and 20% of datasets to train the RF model to classify the map into seven classes (maize, beans, cassava, bananas, other vegetation, trees, and non-vegetation land cover). The OA of their model was 83%, with the three highest accuracies of 91% for the maize, 90% for the trees, and 84% for the beans, with the lowest accuracies at 67% for Cassava. The reasons for some misclassification were mainly related to similar spectral features of some classes, spectral overlaps, the number of data points, quality of ground truth data, intercropping patterns, and their strategies for labeling the cells that dominated by 75% of single class type. They confirmed the potential of using UAVs as ground truth data, especially for their studies. Although Domingo et al. (2020) mentioned the overfitting of the RF model for reduced sample plots or low-density point data, Hegarty-Craver et al. (2020) used only the RF model for the classification due to its ability to control high data dimensionality without overfitting.

12.4 Discussion

The most often used combination among the data fusion techniques was an integration of UAV, airborne, or ALS (either with LiDAR or RGB camera) and satellite data (13 out of 21 relevant cases). Only eight cases specifically belong to the fusion of drone data and satellite data. In addition, most of the LiDAR ALS data were captured previously and were not up-to-date. Current research has focused on employing UAVs to bridge the scaling gap between the ground and Earth Observation (EO) data, including the calibration vegetation indices derived by EO using UAVs. Based on Revill et al.'s (2020) research, one of the shortcomings of these previous investigations is that the UAV sensors have measured spectra that are different from those of the satellite. They believed, for reliable scaling studies, UAV sensors should have the same spectral bands as satellite-based sensors (Revill et al. 2020). According to the reviewed articles, data fusion aims to merge diverse and complementary data to enrich information apparent in the images and increase interpretability. As a result, more accurate and detailed data is produced in segmentation and classification models (Al-Wassai et al. 2014). Based on the review papers, data fusion can occur at pixel-level, feature-level, and decision-level depending on the stage at which fusion takes place (Alonzo et al. 2014; Al-Wassai et al. 2014; Zhang et al. 2020). Each method for image fusion operates at one of these three levels or a combination of these levels (Kayani et al. 2007).

At the pixel level, the object's feature information is extracted after processing and analyzing. The advantage is that the image's fine detail can be preserved, but the processing data is large and time-consuming (Yuan and Zhu 2018). At the feature level, fusion involves feature extraction from the input data such as pixel intensities or edge and texture features (Zhang et al. 2020). The purpose of feature-level image fusion is to work with raw data from sensors with varying spatial and temporal resolutions (e.g. MODIS, SAR, Landsat TM, and so on) to extract feature information and merge images of the same kind. Image fusion at the feature level enables the generation of new images with greater accuracy and the compression of data (Yuan and Zhu 2018). In comparison to pixel-level fusion, feature-level fusion considers a variety of aspects, including the feature information and correlation contained within the image. It enables the extraction of more macro-level feature-level information than is possible with pixel-level fusion (Zhang et al. 2020). At the decision level, the result of the first object detection and classification from the source image is sent into the fusion algorithm as input (Kayani et al. 2007). Although decision-level is a high-level fusion and provides more accurate and real-time outputs, the main disadvantage is the high level of data loss in this sort of fusion (Xiao et al. 2020).

12.5 Conclusion and Future Trends

Extensive studies were conducted concerning vegetation mapping by using satellite imagery and data fusion of satellite data and UAV data in recent decades. Although several types of research were performed, still gaps exist and the current database and methodology need further investigation and improvements. This chapter presented a literature review of previous researches on the vegetation mapping process using remote sensing data, data fusion specifically the integration of satellite and drone data.

Two key themes have emerged in the development of remote sensing sensors in recent years. First, there is a new generation of satellite sensors, including the Sentinel constellations, which can provide free access on a global scale at short time intervals. Second, more comprehensive and detailed data collected by UAVs and micro-satellites are becoming more widely accessible at the operational level. The literature review found that UAVs have become a reliable source of data with very high spatial resolution and highly variable temporal resolution at an affordable price. UAVs data may be used directly as databases to classify tree species on small scales or as a source of additional reference information to supplement spaceborne and airborne datasets. To improve classification algorithms and processing chains, future research on land cover and species classification should often include more than one test location and different types of land cover. Because most of the previous studies focused on precision agriculture, crop monitoring, and wetland, there is a need to study with more details on forest mapping and species. LiDAR data might be used in conjunction with medium resolution passive optical data (Landsat, Sentinel-2) to map forest composition across greater geographic areas. One topic of particular importance when identifying forest species, in terms of classification and validation procedures, is the question of how to categorize uncommon species. One of the possible solutions is the integration of pre-knowledge data such as existing inventory data into an ensemble classifier to avoid strong over-classification of a rare species. In some studies other than multiple remote sensing sources, local experts or forest analysts were hired as consultants to select the most representative land cover areas as study areas and to know the characterization of the land cover and vegetation.

Acknowledgments

This work is financed by national funds through the Portuguese Foundation for Science and Technology – FCT, under grant agreement 2021.05971.BD attributed to the first author. The authors gratefully acknowledge the Portuguese Foundation for Science and Technology – FCT for its support under the framework of the research project PCIF/SSI/0096/2017 – FIREFRONT – Real-Time Forest Fire Mapping and Spread Forecast Using Unmanned Aerial Vehicles, financed by national funds through the Ministry of Science, Technology and Higher Education, Portugal.

References

Agarwal, A., Singh, A.K., Kumar, S., and Singh, D. (2018). *Critical Analysis of Classification Techniques for Precision Agriculture Monitoring Using Satellite and Drone*. In: *2018 IEEE 13th International Conference on Industrial and Information Systems (ICIIS)*, vol. 978, 83–88.

Agarwal, A., Kumar, S., and Singh, D. (2019a). Development of neural network based adaptive change detection technique for land terrain monitoring with satellite and drone images. *Defence Science Journal* 69 (5): 474–480. https://doi.org/10.14429/dsj.69.14954.

Agarwal, A., Kumar, S., and Singh, D. (2019b). Development of machine learning based approach for computing optimal vegetation index with the use of Sentinel-2 and drone data. *The International Geoscience and Remote Sensing Symposium* 5832–5835. https://doi.org/10.1109/IGARSS.2019.8897896.

Alonzo, M., Bookhagen, B., and Roberts, D.A. (2014). Urban tree species mapping using hyperspectral and LiDAR data fusion. *Remote Sensing of Environment* 148: 70–83. https://doi.org/10.1016/j.rse.2014.03.018.

Al-Wassai, F.A., Kalyankar, N.V., and Al-Zaky, A.A. (2014). *Multisensor Images Fusion Based on Feature-Level*. April, 2011, [Online]. Available: http://arxiv.org/abs/1108.4098

Amani, M., Kakooei, M., Moghimi, A. et al. (2020). Application of google earth engine cloud computing platform, sentinel imagery, and neural networks for crop mapping in Canada. *Remote Sensing* 12 (21): 3561. https://doi.org/10.3390/rs12213561.

Baillarin, S.J., Meygret, A., Dechoz, C. et al. (2012). *Sentinel-2 Level 1 Products and Image Processing Performances*. The International Archives of the Photogrammetry, Remote Sensing and Spatial Information Sciences https://doi.org/10.1109/IGARSS.2012.6351959.

Barrett, B., Raab, C., Cawkwell, F., and Green, S. (2016). Upland vegetation mapping using random forests with optical and radar satellite data. *Remote Sensing in Ecology and Conservation* 2 (4): 212–231. https://doi.org/10.1002/rse2.32.

Bhatnagar, S., Gill, L., Regan, S. et al. (2021). A nested drone-satellite approach to monitoring the ecological conditions of wetlands. *ISPRS Journal of Photogrammetry and Remote Sensing* 174: 151–165. https://doi.org/10.1016/j.isprsjprs.2021.01.012.

Brown, S. (1997). *Estimating Biomass and Biomass Change of Tropical Forests: A Primer*. (FAO Forestry Paper 134, January, 1–4), Available at http://www.fao.org/docrep/w4095e/w4095e00.htm

Brown, C.W., Connor, L.N., Lillibridge, J.L. et al. (2005). An introduction to satellite sensors, observations and techniques. In: *Remote Sensing of Coastal Aquatic Environments*. Remote Sensing and Digital Image Processing, vol. 7 (ed. R.L. Miller, C.E. Del Castillo and B.A. McKee). Dordrecht: Springer https://doi.org/10.1007/978-1-4020-3100-7_2.

Brown-De Colstoun, E.C., Story, M.H., Thompson, C. et al. (2003). National Park vegetation mapping using multitemporal Landsat 7 data and a decision tree classifier. *Remote Sensing of Environment* 85 (3)): 316–327. https://doi.org/10.1016/S0034-4257(03)00010-5.

Cao, R., Chen, Y., Shen, M. et al. (2018). A simple method to improve the quality of NDVI time-series data by integrating spatiotemporal information with the Savitzky–Golay filter. *Remote Sensing of Environment* 217: 244–257. https://doi.org/10.1016/j.rse.2018.08.022.

Chakhar, A., Hernández-López, D., Ballesteros, R., and Moreno, M.A. (2021). Improving the accuracy of multiple algorithms for crop classification by integrating sentinel-1 observations with sentinel-2 data. *Remote Sensing* 3 (2): 243. https://doi.org/10.3390/rs13020243.

Chen, J.M. and Black, T.A. (1992). Defining leaf area index for non-flat leaves. *Plant, Cell & Environment* 15 (4): 421–429. https://doi.org/10.1111/j.1365-3040.1992.tb00992.x.

Chourasiya, A. and Khare, N. (2019). A comprehensive review of image enhancement techniques. *International Journal of Innovative Research & Growth* 8 (6): https://doi.org/10.26671/ijirg.2019.6.8.101.

Corcoran, J.M., Knight, J.F., and Gallant, A.L. (2013). Influence of multi-source and multi-temporal remotely sensed and ancillary data on the accuracy of random forest classification of wetlands in northern Minnesota. *Remote Sensing* 5 (7): 3212–3238. https://doi.org/10.3390/rs5073212.

Crespi, M. and de Vendictis, L. (2009). A procedure for high resolution satellite imagery quality assessment. *Sensors* 9 (5): 3289–3313. https://doi.org/10.3390/s90503289.

Deering, D.W., Rouse, J.W., Haas, R.H., and Schell, J.A. (1975). *Measuring Forage Production of Grazing Units from Landsat MSS Data*. In: *10th Symposium on Remote Sensing of Environment (Ann Arbor ERIM)*, 1169–1178.

Domingo, D., de la Riva, J., Lamelas, M.T. et al. (2020). Fuel type classification using airborne laser scanning and Sentinel-2 data in mediterranean forest affected by wildfires. *Remote Sensing* 12 (21)): 1–22. https://doi.org/10.3390/rs12213660.

Erdody, T.L. and Moskal, L.M. (2010). Fusion of LiDAR and imagery for estimating forest canopy fuels. *Remote Sensing of Environment* 114 (4): 725–737. https://doi.org/10.1016/j.rse.2009.11.002.

FAO and UNEP (2020). *The State of the World's Forests 2020*. Rome, Italy: FAO and UNEP.

Feilhauer, H., Dahlke, C., Doktor, D. et al. (2014). Mapping the local variability of Natura 2000 habitats with remote sensing. *Applied Vegetation Science* 17 (4): 675–779. https://doi.org/10.1111/avsc.12115.

García, M., Riaño, D., Chuvieco, E. et al. (2011). Multispectral and LiDAR data fusion for fuel type mapping using support vector machine and decision rules. *Remote Sensing of Environment* 115 (6): 1369–1379. https://doi.org/10.1016/j.rse.2011.01.017.

Gillanders, S.N., Coops, N.C., Wulder, M.A. et al. (2008). Multitemporal remote sensing of landscape dynamics and pattern change: describing natural and anthropogenic trends. *Progress in Physical Geography* 13 (5): 503–528. https://doi.org/10.1177/0309133308098363.

Gitelson, A.A. (2004). Wide dynamic range vegetation index for remote quantification of biophysical characteristics of vegetation. *Journal of Plant Physiology* 161 (2): 165–173. https://doi.org/10.1078/0176-1617-01176.

Gitelson, A. and Merzlyak, M.N. (1994). Quantitative estimation of chlorophyll-a using reflectance spectra: experiments with autumn chestnut and maple leaves. *Journal of Photochemistry and Photobiology B: Biology* 22 (3): 247–252. https://doi.org/10.1016/1011-1344(93)06963-4.

Gonsamo, A., Chen, J.M., David, T.P. et al. (2012). Land surface phenology from optical satellite measurement and CO_2 eddy covariance technique. *Journal of Geophysical Research. Biogeosciences* 117 (G3): G03032. https://doi.org/10.1029/2012JG002070.

Goodenough, D.G., Dyk, A., Niemann, K.O. et al. (2003). Processing Hyperion and ALI for forest classification. *IEEE Transactions on Geoscience and Remote Sensing* 41 (6 PART I): 1321–1331. https://doi.org/10.1109/TGRS.2003.813214.

Goodenough, D.G., Chen, H., Dyk, A. et al. (2005). *Multisensor data fusion for aboveground carbon estimation*. In: *Proc. XXVIIIth Gen. Assem. Int. Union Radio Sci.*, No. 1, 11–14.

Gray, P.C., Ridge, J.T., Poulin, S.K. et al. (2018). Integrating drone imagery into high resolution satellite remote sensing assessments of estuarine environments. *Remote Sensing* 10, 1257 (8): https://doi.org/10.3390/rs10081257.

Guan, H., Su, Y., Hu, T. et al. (2019). An object-based strategy for improving the accuracy of spatiotemporal satellite imagery fusion for vegetation-mapping applications. *Remote Sensing* 11: 2927. https://doi.org/10.3390/rs11242927.

Hegarty-Craver, M., Polly, J., O'Neil, M. et al. (2020). Remote crop mapping at scale: using satellite imagery and UAV-acquired data as ground truth. *Remote Sensing* 12 (12): 1–15. https://doi.org/10.3390/rs12121984.

Heller, R.C., Doverspike, G.E., and Aldrich, R.C. (1964). Identification of tree species on large - scale panchromatic and color aerial photographs. *Agricultural Handbook* 261: 17.

Holland, J. and Aplin, P. (2013). Super-resolution image analysis as a means of monitoring bracken (*Pteridium aquilinum*) distributions. *ISPRS Journal of Photogrammetry and Remote Sensing* 75: 48–63. https://doi.org/10.1016/j.isprsjprs.2012.10.002.

Huang, X. and Zhang, L. (2013). An SVM ensemble approach combining spectral, structural, and semantic features for the classification of high-resolution remotely sensed imagery. *IEEE Transactions on Geoscience and Remote Sensing* 51 (1): 257–272. https://doi.org/10.1109/TGRS.2012.2202912.

Huete, A.R. (1988). A soil-adjusted vegetation index (SAVI). *Remote Sensing of Environment* 25 (3): 295–309. https://doi.org/10.1016/0034-4257(88)90106-X.

Jang, J.D., Viau, A.A., Anctil, F., and Bartholomé, E. (2006). Neural network application for cloud detection in SPOT VEGETATION images. *International Journal of Remote Sensing* 27 (4): 719–736. https://doi.org/10.1080/01431160500106892.

Jarman, M.L. (1985). Computer-aided vegetation mapping. *Southern African Forestry Journal* 132 (1): 40–45. https://doi.org/10.1080/00382167.1985.9629550.

Johansen, B.E. (2004). Mountain Vegetation Mapping in Dovre Area, Norway, Using Landsat TM Data and GIS. In: *Proceedings SPIE 5239, Remote Sensing for Environmental Monitoring, GIS Applications, and Geology III*, (13 February 2004) http://dx.doi.org/10.1117/12.514305.

Kaplan, G. (2018). Monthly analysis of wetlands dynamics using remote sensing data. *ISPRS International Journal of Geo-Information* 7 (10): 411. https://doi.org/10.3390/ijgi7100411.

Kattenborn, T., Lopatin, J., Förster, M. et al. (2019). Remote sensing of environment UAV data as alternative to field sampling to map woody invasive species based on combined Sentinel-1 and Sentinel-2 data. *Remote Sensing of Environment* 227 (January): 61–73. https://doi.org/10.1016/j.rse.2019.03.025.

Kaufman, Y.J. (1984). *Atmospheric Effects On Remote Sensing Of Surface Reflectance*. Proceedings SPIE 0475, Remote Sensing. *Critical Review of Technology* 20–33. https://doi.org/10.1117/12.966238.

Kayani, B.N., Mirza, A.M., Bangash, A., and Iftikhar, H. (2007). Pixel & feature level multi-resolution image fusion based on fuzzy logic. In: *Innovations and Advanced Techniques in Computer and Information Sciences and Engineering* (ed. T. Sobh). Dordrecht: Springer https://doi.org/10.1007/978-1-4020-6268-1_24.

Ke, Y., Quackenbush, L.J., and Im, J. (2010). Synergistic use of QuickBird multispectral imagery and LIDAR data for object-based forest species classification. *Remote Sensing of Environment* 114 (6): 1141–1154. https://doi.org/10.1016/j.rse.2010.01.002.

Langley, S.K., Cheshire, H.M., and Humes, K.S. (2001). A comparison of single date and multitemporal satellite image classifications in a semi-arid grassland. *Journal of Arid Environments* 49 (2): 401–411. https://doi.org/10.1006/jare.2000.0771.

Lastovicka, J., Svec, P., Paluba, D. et al. (2020). Sentinel-2 data in an evaluation of the impact of the disturbances on forest vegetation. *Remote Sensing* 12: 1914. https://doi.org/10.3390/rs12121914.

Li, Y., Li, M., Li, C., and Liu, Z. (2020). Forest aboveground biomass estimation using Landsat 8 and Sentinel-1A data with machine learning algorithms. *Scientific Reports* 10 (1): 1)–12. https://doi.org/10.1038/s41598-020-67024-3.

Liu, X. (2003). *Supervised Classification and Unsupervised Classification*. Available at: http://Cfa.Harvard.Edu

Ma, L., Li, M., Ma, X. et al. (2017). A review of supervised object-based land-cover image classification. *ISPRS Journal of Photogrammetry and Remote Sensing* 130: 277–293. https://doi.org/10.1016/j.isprsjprs.2017.06.001.

Martin, M.E., Newman, S.D., Aber, J.D., and Congalton, R.G. (1998). Determining forest species composition using high spectral resolution remote sensing data. *Remote Sensing of Environment* 65 (3): 249–254. https://doi.org/10.1016/S0034-4257(98)00035-2.

Mayaux, P., Richards, T., and Janodet, E. (1999). A vegetation map of Central Africa derived from satellite imagery. *Journal of Biogeography* 26 (2): 353–366. https://doi.org/10.1046/j.1365-2699.1999.00270.x.

Milewski, T. and Niepodległości, A. (2016). *The use of satellite pictures and data provided by drones for the purposes of identification of crops and assessment of plant production*. October, 1–8. https://doi.org/10.1481/icasVII.2016.f30b

Murugan, D., Garg, A., Ahmed, T., and Singh, D. (2016). *Fusion of Drone and Satellite Data for Precision Agriculture Monitoring*. 11th International Conference on Industrial and Information Systems (ICIIS). https://doi.org/10.1109/ICIINFS.2016.8263068

Murugan, D., Garg, A., and Singh, D. (2017). Development of an adaptive approach for precision agriculture monitoring with drone and satellite data. *IEEE Journal of Selected Topics in Applied Earth Observations and Remote Sensing* 10 (12)): 5322–5328. https://doi.org/10.1109/JSTARS.2017.2746185.

Navulur, K. (2006). *Multispectral image analysis using the object-oriented paradigm*.

Perry, C.R. and Lautenschlager, L.F. (1984). Functional equivalence of spectral vegetation indices. *Remote Sensing of Environment* 14 (1–3): 169–182. https://doi.org/10.1016/0034-4257(84)90013-0.

Phan, T.N., Kuch, V., and Lehnert, L.W. (2020). Land cover classification using google earth engine and random forest classifier: the role of image composition. *Remote Sensing* 12 (15): 1–22. https://doi.org/10.3390/RS12152411.

Pouteau, R. and Stoll, B. (2012). SVM selective fusion (SELF) for multi-source classification of structurally complex tropical rainforest. *IEEE Journal of Selected Topics in Applied Earth Observations and Remote Sensing* 5 (4): 1203–1212. https://doi.org/10.1109/JSTARS.2012.2183857.

Räsänen, A. and Virtanen, T. (2019). Remote sensing of environment data and resolution requirements in mapping vegetation in spatially heterogeneous landscapes. *Remote Sensing of Environment* 230: 111207. https://doi.org/10.1016/j.rse.2019.05.026.

Revill, A., Florence, A., Macarthur, A. et al. (2020). Quantifying uncertainty and bridging the scaling gap in the retrieval of leaf area index by coupling Sentinel-2 and UAV observations. *Remote Sensing* 12: 1843. https://doi.org/10.3390/rs12111843.

Rondeaux, G., Steven, M., and Baret, F. (1996). Optimization of soil-adjusted vegetation indices. *Remote Sensing of Environment* 55 (2): 95–107. https://doi.org/10.1016/0034-4257(95)00186-7.

Roujean, J.L. and Breon, F.M. (1995). Estimating PAR absorbed by vegetation from bidirectional reflectance measurements. *Remote Sensing of Environment* 51 (3): 375–384. https://doi.org/10.1016/0034-4257(94)00114-3.

Schowengerdt, R.A. (2012). *Techniques for Image Processing and Classifications in Remote Sensing*. Academic Press.

Sheykhmousa, M., Mahdianpari, M., Ghanbari, H. et al. (2020). Support vector machine versus random Forest for remote sensing image classification: a meta-analysis and systematic review. *IEEE Journal of Selected Topics in Applied Earth Observations and Remote Sensing* 13: 6308–6325. https://doi.org/10.1109/JSTARS.2020.3026724.

Shyu, M.S. and Leou, J.J. (1998). A genetic algorithm approach to color image enhancement. *Pattern Recognition* 31 (7): 871–880. https://doi.org/10.1016/S0031-3203(97)00073-3.

Singh, G. and Mittal, A. (2014). Various image enhancement techniques – a critical review. *International Journal of Innovation Science and Research* 10 (2): 267–274.

Spagnuolo, O.S.B., Jarvey, J.C., Battaglia, M.J. et al. (2020). Mapping Kenyan Grassland Heights across large spatial scales with combined optical and radar satellite imagery. *Remote Sensing* 12: 1086. https://doi.org/10.3390/rs12071086.

Sun, C., Li, J., Cao, L. et al. (2020). Evaluation of vegetation index-based curve fitting models for accurate classification of salt marsh vegetation using Sentinel-2 time-series. *Sensors (Switzerland)* 20 (19): 5551. https://doi.org/10.3390/s20195551.

Szantoi, Z., Smith, S.E., Strona, G. et al. (2017). Mapping orangutan habitat and agricultural areas using Landsat OLI imagery augmented with unmanned aircraft system aerial photography. *International Journal of Remote Sensing* 38 (8–10)): 2231–2245. https://doi.org/10.1080/01431161.2017.1280638.

Tamiminia, H., Salehi, B., Mahdianpari, M. et al. (2021). A comparison of decision tree-based models for forest above-ground biomass estimation using a combination of airborne lidar and landsat data. *Archives of the Photogrammetry, Remote Sensing and Spatial Information Sciences* 5 (3): 235–241. https://doi.org/10.5194/isprs-annals-V-3-2021-235-2021.

Thiam, A.K. (1997). *Geographic information system and remote sensing methods for assessing and monitoring land degradation in the Shale: The Case of Southern Mauritania*. Darks University.

Tucker, C.J. (1979). Red and photographic infrared linear combinations for monitoring vegetation. *Remote Sensing of Environment* 8 (2): 127–150. https://doi.org/10.1016/0034-4257(79)90013-0.

Urbazaev, M., Thiel, C., Cremer, F. et al. (2018). Estimation of forest aboveground biomass and uncertainties by integration of field measurements, airborne LiDAR, and SAR and optical satellite data in Mexico. *Carbon Balance and Management* 13 (1): https://doi.org/10.1186/s13021-018-0093-5.

Vieira, I.C.G., De Almeida, A.S., Davidson, E.A. et al. (2003). Classifying successional forests using Landsat spectral properties and ecological characteristics in eastern Amazônia. *Remote Sensing of Environment* 87 (4)): 470–481. https://doi.org/10.1016/j.rse.2002.09.002.

Walsh, S.J. (1980). Coniferous tree species mapping using LANDSAT data. *Remote Sensing of Environment* 9 (1): 11–26. https://doi.org/10.1016/0034-4257(80)90044-9.

Weih, R. and Riggan, N. (2010). Object-based classification vs. pixel-based classification: comparative importance of multi-resolution imagery. *The International Archives of the Photogrammetry, Remote Sensing and Spatial Information Sciences*, XXXVIII-4/C7.

Wilschut, L.I., Addink, E.A., Heesterbeek, J.A.P. et al. (2013). Mapping the distribution of the main host for plague in a complex landscape in Kazakhstan: an object-based approach using SPOT-5 XS, Landsat 7 ETM+, SRTM and multiple random forests. *International Journal of Applied Earth Observation and Geoinformation* 23: 81–94. https://doi.org/10.1016/j.jag.2012.11.007.

Wolter, P.T., Mladenoff, D.J., Host, G.E., and Crow, T.R. (1995). Improved forest classification in the northern lake states using multi-temporal Landsat imagery. *Photogrammetric Engineering and Remote Sensing* 61 (9)): 1129–1143.

Woodcock, C.E., Collins, J.B., Gopal, S. et al. (1994). Mapping forest vegetation using Landsat TM imagery and a canopy reflectance model. *Remote Sensing of Environment* 5 (3): 240–254. https://doi.org/10.1016/0034-4257(94)90074-4.

Xia, H., Zhao, W., Li, A. et al. (2017). *Subpixel Inundation Mapping Using Landsat-8 OLI and UAV Data for a Wetland Region on the Zoige*, 1–22. https://doi.org/10.3390/rs9010031.

Xiao, Z.X., Bavirisetti, D.P., Liu, G., and Zhang, X. (2020). Decision-level image fusion. In: *Image Fusion*, 149–170. Singapore: Springer https://doi.org/10.1007/978-981-15-4867-3_4.

Xie, Y., Sha, Z., and Yu, M. (2008, 2021, Article ID 6659831). Remote sensing imagery in vegetation mapping: a review. *Journal of Plant Ecology* 14: https://doi.org/10.1093/jpe/rtm005.

Yang, Z., Willis, P., and Mueller, R. (2008). Impact of Band-Ratio Enhanced Awifs Image To Crop Classification Accuracy. *Pecora 17 – The Future of Land Imaging . . . Going Operational*.

Yuan, L. and Zhu, G. (2018). Research on remote sensing image classification based on feature level fusion. *International Archives of Photogrammetry, Remote Sensing and Spatial Information Sciences – ISPRS Archives* 42 (3): 2185–2189. https://doi.org/10.5194/isprs-archives-XLII-3-2185-2018.

Zeng, L., Wardlow, B.D., Xiang, D. et al. (2020). A review of vegetation phenological metrics extraction using time-series, multispectral satellite data. *Remote Sensing of Environment* 237(February): 111511. https://doi.org/10.1016/j.rse.2019.111511.

Zhang, J. and Zhang, Y. (2007). Remote sensing research issues of the National Land Use Change Program of China. *ISPRS Journal of Photogrammetry and Remote Sensing* 62 (6)): 461–472. https://doi.org/10.1016/j.isprsjprs.2007.07.002.

Zhang, R., Tang, X., You, S. et al. (2020). A novel feature-level fusion framework using optical and SAR remote sensing images for land use/land cover (LULC) classification in cloudy mountainous area. *Applied Sciences* 10 (8): 1–24. https://doi.org/10.3390/APP10082928.

Zheng, G. and Moskal, L.M. (2009). Retrieving leaf area index (LAI) using remote sensing: theories, methods and sensors. *Sensors* 9 (4): 2719–2745. https://doi.org/10.3390/s90402719.

Section V

Opportunities, Challenges, and Future Aspects in Forest Monitoring

13

Challenges and Monitoring Methods of Forest Management Through Geospatial Application: A Review

Soumya Pandey, Neeta Kumari, Soumesh K. Dash, and Shah Al Nawajish*

Department of Civil and Environmental Engineering, Birla Institute of Technology, Mesra, Ranchi, Jharkhand, India

13.1 Introduction

Vegetation cover supports the myriad life on earth. Vegetation can be broadly defined as the accumulation of plant varieties and green cover on land. It is classified into forests, grasslands, tundra, desert, and ice sheet (Nowak et al. 2004; Rustad et al. 2001). The growth of different vegetation species depends on varying topography, soil, water holding capacity of the soil, slope, and weather patterns. According to the Intergovernmental Panel on Climate Change (IPCC) 2013 and 2017, forests occupy approximately one-third of land area and contain about 70% of carbon reserves (IPCC 2013, 2017; Keenan et al. 2015). It is home to 80% of land biodiversity in the world, and supports overall ecological health. Forests consist of not only trees but also all the plants, animals, and microbes that are inhabitants of the forest area (Jenkins and Schaap 2018). The interface between and within these different bio-diversities are overly complex, but their complexity allows organisms to evolve and sustain in a continuously changing environment (Dijkstra et al. 2010; Rachid et al. 2021). Although it is estimated that 4 billion hectares of land are forest areas, they are not evenly distributed across the globe (Sahani and Behera 2001; Yergeau and Kowalchuk 2008) (Table 13.1). Latitude forests are classified based on the temperature (rise above 10 °C in summer), precipitation (above 200 mm), type of soil condition (according to depth, fertility, roots), varying climatic condition, and type of flora–fauna that grow and inhabit the areas. As per the latitude, there are three types of forests, i.e. Tropical, Temperate, and Boreal (Taiga). Furthermore, they are also classified based on density, seasonality, location, types of trees, types of leaves (evergreen, deciduous, thorn, mangrove forests), etc. (Table 13.2). Forests have a multifarious effect on all the components of the biosphere starting from bio-geochemical cycle, water and soil resources, microbes, climate change (Ahiablame et al. 2017; Johnson et al. 2012; Junges and Ducati 2018; Ladd et al. 2013), etc. Loss in the forest cover results in predominance of land degradation processes such as soil erosion,

* Corresponding author's email: neetak@bitmesra.ac.in

0000-0001-5525-6079 (Soumya Pandey), 0000-0002-0830-2084 (Neeta Kumari), 0000-0002-8283-5698 (Soumesh K. Dash), and 0000-0002-9663-6200 (Shah Al Nawajish)

Advances in Remote Sensing for Forest Monitoring, First Edition. Edited by Prem C. Pandey and Paul Arellano.

Table 13.1 Major classification of forest and other vegetation.

Vegetation type	Area	Climate	Soil conditions	Features
Tropical grasslands or savannas	Major areas of Central Africa, Australia, South America, and India	Found in warmer climates; annual rainfall from 508 to 1270 mm	Soil is porous, has high permeability; rich in organic content and nutrients; thin cover	6–8 months of rainfall followed by a long period of droughts; both rainy and dry seasons
Tropical forest	Near the equator between area 23.5°N and 23.5°S latitude	Avg.Temp. 20–25 °C varying by not more than 5 °C; more than 200 mm rainfall is distributed annually	Contains fewer nutrients and is acidic; faster decomposition; prominent soil leaching	Trees are 25–35 m tall; no winter, and only two seasons are present (rainy and dry); daylight is 12 hours
Temperate grasslands	Mostly in South Africa, Hungary, Argentina and Uruguay, Former Soviet Union, Central North America	Hot summers (over 38 °C) and cold winters (−40 °C); moderate rainfall between 508 to 889 mm	Soil is deep and dark, upper fertile layers, nutrient-rich, strongly held by roots	Annual rainfall, seasonal drought and fires are important in maintaining the biomes in grasslands
Temperate forest	North America, Northeast Asia, western and central Europe	Temps vary from −30 to 30 °C; rainfall 750–1500 mm is distributed annually	Fertile, enriched with decaying litter	Well-defined seasons with a distinct winter; moderately dense and allows light to penetrate
Boreal forest (TAIGA)	Between 50° and 60° N lat.; Eurasia and North America	Exceptionally low temp; snow precipitation 40–100 cm annually	Nutrient deficient and acidic	Largest land biome; growing season of 130 days; low light penetration
Artic Tundra	North hemisphere, around the North Pole, branching to taiga forests	Cold, desert-like conditions; avg. temps between −34 and 3–12 °C; rainfall or snowfall between 150 and 250 mm	Slowly formed soil layer; permafrost is present consisting of gravel and finer materials; no deep root systems	Growing seasons range from 50 to 60 days; 1700 types of plants present
Alpine tundra	High altitude region where trees cannot grow	Temperature below freezing at night	Well-drained soil	The growing season is 180 days
Hot and dry desert	North America, Southern Asian Realm, South and Central America, Africa, Australia	Temps range from 43.5–49 °C max to min of 18 °C; extremely low rainfall of 15 mm/yr and evaporation rates	Course-textured, shallow, rocky or gravely, with good drainage and no subsurface water	Extreme heat during the day and cold at night; canopies are rare; leaves of plants are nutrient-rich, have a hard outer cover and conserve water

Table 13.1 (Continued)

Vegetation type	Area	Climate	Soil conditions	Features
Semiarid desert	North America, Greenland, Russia, Europe, and northern Asia	The average temperature between 21 and 27 °C, ranging between 10 and 38 °C; rainfall from 20– 40 mm	The soil ranges from sandy fine-textured to rock fragments; low salt content; no subsurface water	Condensation of dew at night is high; plants have a spine and some have shiny textures to reflect light
Coastal desert	Near Arctic, Neotropical region	Temps range from 35 to −4 °C; avg. rainfall 80–130 mm	Fine-textured soil, porous, good drainage; roots system close to the surface	Cool winters followed by long, warm summers; leaves are thick and fleshy to hold water
Cold desert	Antarctica, Greenland, Near Arctic realm	Temperature between 21 and 26 °C max to −2 to 4 °C min; rainfall averages between 150 and 260 mm	Soil is heavy silty, has high salt content, is porous, and has good drainage properties	Long, cold winters with prominent snowfall, high rainfall, and short warm summers; plants are deciduous with spiny leaves

Source: The World's UC Museum of Paleontology (online resource), Berkeley University of California. (https://ucmp.berkeley.edu/exhibits/biomes/forests.php)

Table 13.2 World distribution of forests.

Country	Area of forest in million hectares	% of the worlds forest reserves
Russian Federation	815	20.1
Brazil	497	12.2
Canada	347	8.5
USA	310	7.6
China	220	5.4
Australia	134	3.3
Democratic Republic of Congo	126	3.1
Indonesia	92	2.3
Peru	72	1.8
India	72	1.8
Rest of the world	1375	33.9

Source: Food and Agriculture Organization of the United Nations. (Data from FAO 2020.)

nutrient leaching, loss of microbial biomes, water stress and pollution, air pollution, an increase of greenhouse gases, etc. Some of the major causes of forest loss are extensive agriculture, grazing, deforestation, urbanization, industrialization, alteration of biogeochemical cycles, mining, and other anthropogenic activities.

According to the Food and Agriculture Organization (FAO) (2002) report, half of all wood cultivated is used as fuels. Developing countries use wood as their primary source of fuel (15%) and those forests produce more than 90% of wood globally. These countries use wood in the form of charcoal obtained from other industries like paper making. Forests act as a sink (2.6 billion tons) as well as a source of carbon (25%) in the atmosphere. It has been found that the forest productivity and carbon sink have increased since the 1980s. The total forest sink in the span of years 1990–2007 was found to be approximately 2.4 pentagrams of carbon per year. Although the carbon emission reported from deforestation and change in land use have been approximately 4.2 pentagrams of carbon per year, the study does not include ocean and atmospheric sinks (Pan et al. 2011). The carbon sink varies in different types of forest. Around 35% of carbon sink and 60% of carbon stock occurs in harvested wood, living biomass, deadwood, litter, and soil of the forests. Currently, the carbon sink in terms of forests occurs mostly in the tropical and boreal forests, but with a difference in the storage. The tropical forest has more sink in the biomass (56%) and less in the soil (32%), whereas the boreal forest has less sink in the biomass (20%) and more in the soil (60%) (IPCC 2000; Pan et al. 2011).

Overall vegetation products and services offer socioeconomic stability to around 1.6 billion people, mostly residing in developing and low-income countries. Although, loss in the vegetation cover is also more frequent in these countries as the population is heavily dependent on the products and services offered by the forests, especially in tropical areas (Rahman et al. 2017). Unfortunately, around 5–10 gigatons of carbon dioxide is emitted because of deforestation and loss of vegetation cover. This causes a massive increase in greenhouse gases and overall climate change of the Earth (Buckland et al., 2019).

The forest is capable of channeling its microclimate, which impacts every component and activity of the region including agriculture, wildlife, human settlement, microbes, etc. Climate change has resulted in increased dry weather in many regions of the world, leading to frequent forest fires. For example, in India, Uttarakhand state suffered from the most drastic climate change and wildfires that occurred in the years 2016 and 2020 resulting in loss of lives and vegetation in and around 4538 and 71 ha, respectively (Pandey and Tyagi 2012). Similarly in Russia, the taiga forests caught on fire in 2021, brought extensive heat and drought, and the smoke generated reached as far as the North Pole. Many such cases have been reported in Greece, Poland, Portugal, South Korea, Syria, Israel, Indonesia, Australia, New Zealand, and many more countries in the northern hemisphere (Calviño-Cancela and Cañizo-Novelle 2018; Halofsky et al. 2020; Sujatha and Sridhar 2018).

Various methods of monitoring and assessment of characteristics of forest stands have evolved from traditional to modern techniques. The conventional methods of studying forest characteristics such as health, cover, extent, biophysical and biochemical features have been a tedious, costly, and complicated task (Apostol et al. 2018). For instance, the Reduce Emissions from Deforestation and Forest Degradation (REDD+) Framework proposed new criteria for forest monitoring and degradation assessment at United Nations Framework Convention on Climate Change (UNFCCC [COP 16]) 2011 (Gascón and Eva 2014). They defined deforestation as a "measurable decrease in tree crown below 10–30% of forest areas with a minimum size of 0.05–1 ha," which is difficult to detect even with a low spatial resolution of 30 m. The techniques of many types of forest monitoring studies, especially field surveys, have been used to study various aspects of the forest. Collection of data from field survey itself cannot be efficient, as many factors are synergistic with growth and deterioration of forest cover and health. The UNFCCC also states that the REDD+ framework to monitor the forest stand needs to be based on the combination of remote sensing and ground-based inventory (Gascón and Eva 2014).

Assessment of forest biodiversity has been based on-site surveys, which classifies diversity into levels such as alpha (locally present), beta (present between dissimilar sites), and a combination of two, which help in analyzing data such as type, frequency, and composition of species (Malhi et al. 2020). Pilot surveys are done by estimating suitable plots and shapes, and sampling size. Based on these factors systematic cluster sampling, stratified sampling, and random sampling is compiled by dividing the area into grids. These are extensive, tedious, and require extreme precision and skilled staff. National and regional forest inventories (NFI) have been providing data regarding all aspects of forest stands an reserves. Usually, they collect the information by repeated sampling, growth estimates, field surveys, and change in the area and climate (Tewari, 2016). NFI provides unbiased information of forest characteristics, columns, stocks, species, carbon sink and emission, ecosystem services, change in area, and biodiversity. In India, the Forest Survey of India (FSI) manages the collection of data and information at the regional and global levels. Some of these assessments are for long term and some are seasonal, and planning of such surveys is always problematic. Manual monitoring of forest inventory is difficult, as the areas are large and illegal logging, deforestation, and fires need to be monitored daily. Some other challenges commonly faced are accessibility, funding, and trained labor, as well as the availability of technology, tools, and equipment (Chen et al. 2015; Chen et al. 2016; Lipinski et al. 2016). Accessibility of road networks, mode of transportation, and ease of traveling is usually not provided. Furthermore, the dense canopy of the forest does not favor visibility in all seasons (Lausch et al. 2017; Deka et al. 2019; Clark et al. 1995; Liu et al. 2017). Lack of funding influences the inventory assessment planning, design, and work plan, as the more intensive studies require more skilled labor and long-term planning. Although modern technologies can be easily used to overcome these issues, manual field techniques require costly equipment, trained staff, and a well-connected network, which is difficult in varying terrain, especially in hilly areas (Shuaibu and Dagba 2013).

In hilly or mountainous regions, major forest degradation occurs due to agricultural expansion, fire, windstorm, logging, land conversion, flooding, landslides, and avalanches. Agriculture has been known to be the key driver of global deforestation. The research claimed that for better development, these two landscapes have to be addressed synergistically (Van de Wiel et al. 2019). Manual inventory of forest data is quite challenging; the uneven terrain, steep slopes, lack of oxygen, temperature moderation, lack of funds, and transport are some common challenges faced by surveyors. Likewise, anthropogenic influences, such as agricultural shift, extensive wood logging, and frequent natural hazards like landslides, fires, etc., cause a massive shift in forest area, yet they are hard to monitor in the uneven landscape (Luo et al. 2001; Kosmas et al. 2016; Ladd et al. 2013; Moudrý et al. 2018). These challenges have been overcome using GIS and remote sensing. The collection of data regarding various species of trees, and other biodiversity, wildlife movement, extent, cover, natural calamities, among other are all performed using data obtained from satellite data.

To monitor the changes in the forest cover, forest health, forest physiochemical and biological aspects, and natural hazards, many technological applications have been used in recent decades (Anand et al. 2020; Pandey et al. 2019; Srivastava et al. 2020). These techniques provide a unique edge over the traditional field monitoring and assessment of the forest data. Advancement in the form of geospatial techniques such as GPS, GIS, and remote sensing, and their components, which constitute sensors like LISS, SENTINEL, LANDSAT TM based on LIDAR, RADAR, and varying platforms help in data monitoring and collection. Some other important tools such as geostatistical analyses, cluster analysis, indices, Normalized Difference Water Index (NDWI), Normalized Difference Vegetation Index (NDVI), Soil Adjusted Vegetation Index (SAVI), Enhanced Vegetation Index (EVI), Perpendicular Vegetation Index (PVI), Multispectral Imaging (MSI), and Land Surface Water Index (LSWI) help in the assessment of forest data. Future change prediction and time series analysis can be performed using machine learning and neural networking models such as Bayesian network, Cellular automata artificial neural networking, fuzzy neural networking, Deep learning, among others. With the help of these techniques and geospatial tools, it becomes easy to monitor the change in forest reserves at a spatiotemporal scale.

Some global agencies like FAO, United Nations (UN), UNFCCC, Environment Programme World Conservation Monitoring (UNEP-WCMC), Intergovernmental Panel on Climate Change (IPCC), National Forest Resources Monitoring and Assessment (NAFORMA), and others produce a report on global forest data monitoring, assessment, inventory, crown cover, biomass estimations, and other biophysical parameters and organize global conferences to discuss the current advances in the ecological status of the world (FAO 2021; Lynch et al. 2020). These reports and conferences help in preparing frameworks like REDD+, which help set the country and global goals for sustainable and judicial environmental management. For example, the IPCC Sixth Assessment Report 2021 on climate change stressed visible sea rise and faster warming of earth surfaces. The impact is mostly seen in oxygen levels, and hydrological cycles of the world and extreme weather such as heat waves, flash floods, landslides and many others. Most of the planet is suffering from water stress and pollution. Moreover, only a decrease or arrest of carbon emission could control the warming (Abdalla et al. 2018; Etminan et al. 2016; Lenka et al. 2015; Liu et al. 2015; Raffel 2016). As forests are found to be the greatest carbon sink, increase in forest cover may improve the climate. This also is subjected to a reduction in other Anthropocene carbon sources like the use of woods, petroleum, and other fuels. REDD+ acts as a framework for the development of national strategies. The assessment report must be generated, especially in developing countries, based on carbon emission per year, the forest reference level of carbon in the country, procedure, application of REDD+ policies applied in the country, and management techniques of forests. The purpose is to provide financial aid, promote public–private partnerships, and form a global network of information for monitoring and management of forest reserves. This review study addresses the importance, challenges, techniques, and measures in the field of forests.

13.2 Importance of Forest Cover

The need for sustainable forest reserves is necessary, as all forms of life are dependent on them for their survival (Figure 13.1). From monitoring biogeochemical cycles, providing habitat to wildlife, humans, and various plants, and microbial species to providing food, water, and fuel security, controlling natural hazards, conserving water and soil, everything

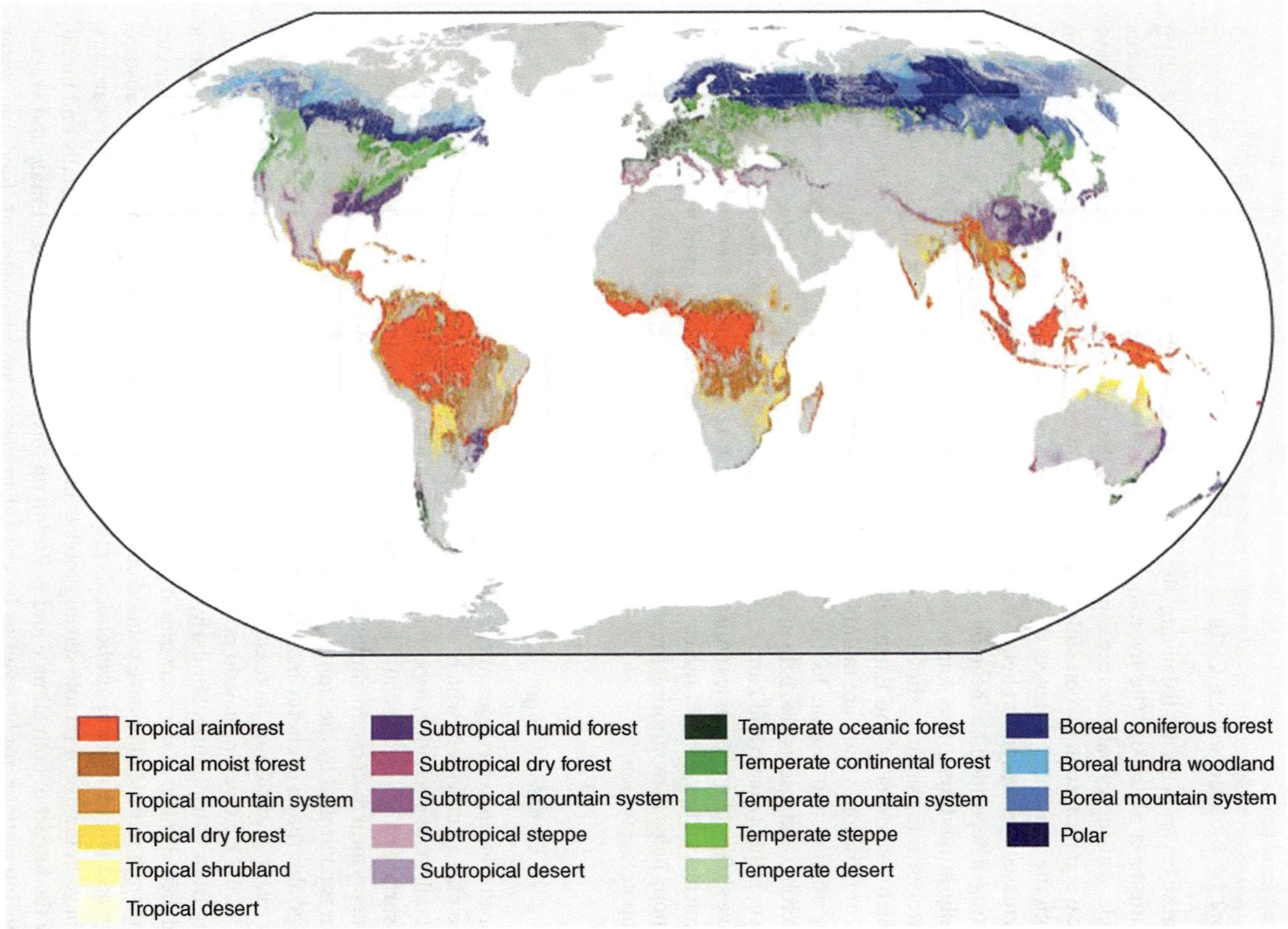

Figure 13.1 The most intact forests by global ecological zone, 2015. Source: The State of the World's Forests 2020.

is dependent on forest cover (FAO, 2012; FAO and UNEP 2020). The multifarious impact derived from forest cover is broadly described further.

13.2.1 Biogeochemical Cycle

Forest cover helps in balancing the biogeochemical cycles (Austin and Zanne 2015; Hungate et al. 2000). The hydrological cycle, carbon cycle, nitrogen cycle, phosphorus cycle, and energy balances are especially dependent on them. Forests and grasslands have great significance in carbon sinks forests and other vegetation cover tend to fixate carbon from the atmosphere. Hence, they help in pollution control and mitigate climate change. Increased carbon content helps the trees and other plants to save water in a dry ecosystem. Nitrogen deposition has been proven to have a positive effect on soil's nitrogen cycle, especially in the forest. This happens mostly due to enhanced decomposition of litters and nitrogen rates sinking, showing prolonged fertilization (Anjali et al. 2021). Arid lands, which cover 40% of the Earth's land surface, have low biotic movement but high litter decomposition, which are mainly due to climate and solar radiation that promotes photo degradation and carbon loss from litter above ground. Other than photodegradation, microbes and insects also help in the process of decomposition, especially when vegetation cover is thin (Austin and Zanne 2015). Forests are huge land mass carbon sink pools as they have slow-cycling processes (Austin and Zanne 2015; Chave et al. 2009). Decaying of litter, branches, plants, etc., all increase the nutrient and carbon content of the soil. The forest canopy helps in monitoring the moisture returns to the atmosphere through the process of evapotranspiration.

13.2.2 Climate Change

Forest cover plays a major role in channeling climate and weather change. Plant processes like photosynthesis in which water vapor is released, and energy emission are drivers for overall weather change, especially cloud formation. Cloud formations hinder the amount of surface radiation and could lead to precipitation (Chen et al. 2019). Likewise, forest ecosystems form their micro-climate as the temperature and moisture surrounding the ecosystem is controlled by the transpiration process. Most forests have low reflecting power and large absorption of energy mainly occur as sunlight, which is necessary for photosynthesis for plant growth and increasing atmospheric oxygen and photo degeneration of litter (Ali 2013). However, they do not increase the temperature, as cooling effects keep occurring from transpiration. Similarly, in the process of photosynthesis, atmospheric carbon is reduced. Hence, many countries have planted fields of forests like China, the UK, etc. Although some studies have pointed out that dark leaves may increase the sunlight absorption and release other toxic chemicals making the earth warmer. Hence, some researchers believe in the use of remote sensing tools to enhance the forest view seasonally and monitor the impact of both climate and pollution on forests. Overall, these climate and vegetation interactions are believed to be critical to vegetation management, food and water security and monitoring extreme weather (Figure 13.2) (Buchhorn et al. 2020; FAO 2012, 2020; Mohammed et al. 2020).

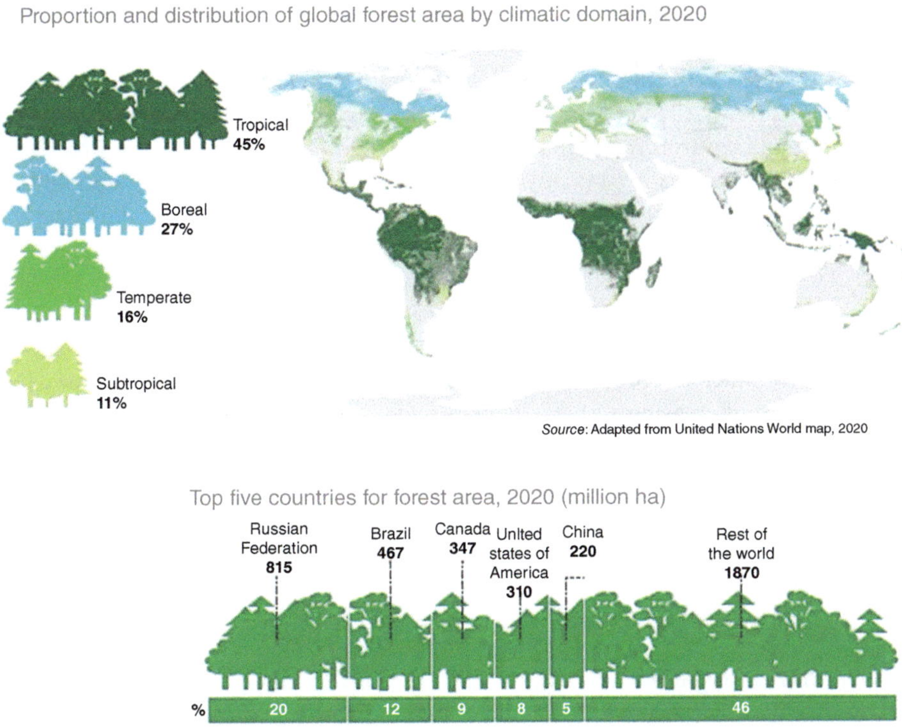

Figure 13.2 Proportion and distribution of global forest area by climatic domain & main countries for forest area. (Source: Adopted from FAO 2020. Prepared by FAO based on FAO global ecological zone map (FAO 2012) and global Copernicus Land Cover map for 2015 (Buchhorn et al. 2020) available under the CC BY-NC-SA 3.0 IGO.)

13.2.3 Soil and Nutrients

Forests help balance the soil environment and physical characteristics such as soil quality and quantity, soil volume, soil texture, soil structure, and soil nutrient pools. In return, healthy soil helps the forest and other forms of vegetation to thrive, and hence this interdependent cyclic process keeps the ecosystem in balance (Austin and Ballare 2014; Cornwell et al. 2008; Dijkstra et al. 2005). In grasslands, regional soils have rich carbon and nutrient pools. Depending upon the soil volume, depth, and water retention, the roots of plants adapt and try to hold onto the soil to prevent weathering and erosion. This also conserves soil and its nutrient content (N, P, K, and C content) (Ahmad et al. 2018; Häger 2012; Steinauer et al. 2015). Conservation of soil may help in better management of water resources too, although this is dependent on the soil structure or soil formation. Vegetation positively affects the soil porosity and permeability (Martin et al. 2010; Wang et al. 2012)

13.2.4 Soil Conservation

A plants species, structure, pattern, and role of vegetation cover have a major influence on the runoff, soil erosion, and sediment yield (Ahmad et al. 2018; Bautista et al. 2007; Bochet et al. 1998; Fu et al. 2009; Martin et al. 2010; Wang et al. 2015; Zhu et al. 2015), vegetation canopies, rainfall patterns, hydrological processes, soil characteristics, plant root and litter all impact the production of runoff and erosion. Vegetation and forest cover intercept rainfall by breaking its flow (Allen et al. 2014) and weakening the intensity of splash, and this limits the runoff and controls inter-rill and rill soil erosion (Ahmad et al. 2018; Crockford and Richardson 2000; Li et al. 2014; Ouyang et al. 2013). It enhances infiltration and soil properties, decreases runoff flow rate, and thus help in soil erosion. Vegetation has more impact on sediment transfer than on detachment of soil (Li et al. 2014; Pannkuk and Robichaud 2003; Smets et al. 2008; Wang and Fu 2013). Conservation of water and soil by litter layers are dependent on its biomass, type, coverage rates, and thickness (Findeling et al. 2003; Li et al. 2014; Liu et al. 2014). Furthermore, the plant roots control soil erosion by holding the soil intact and preventing detaching, increasing infiltration, and improving soil organic content, increasing soil stable aggregates, and forming strong soil structure. Plant roots form a web that binds particles of soil together and enhance cohesion, while their vertical and lateral growth provide shear and tensile strength to soil, further controlling soil detachment and erosion loss (De Baets and Poesen 2010; Gyssels et al. 2006; Mamo and Bubenzer 2001; Pohl et al. 2009; Reubens et al. 2007; Zhang et al. 2018).

13.2.5 Microbes

Along with the other reasons mentioned, sustainability of vegetation cover and forest cover are important as they serve as wildlife habitat, and are home to various microbial diversities, and algal communities. These microbes, especially algae, help monitor the food chain in various ecosystems and maintain the oxygen supply in the atmosphere, which is basic for the survival of life. Forest cover also has a huge socio-economic and ecological impact on humans. Ecosystem services provided by forests such are biodiversity conservation (Dijkstra et al. 2010), climate regulation (Izaurralde et al. 2011), soil conservation and prevention of land degradation (Boillat et al. 2017; Trisurat et al. 2016), control desertification (Moussa and Omar 2020), watershed services (such as water conservation and regulation, recharging of aquifers, improving aquatic productivity, etc.) (Garrett et al. 2018; Himanshu et al. 2019; Katayama et al. 2019), disaster management (prevents flood and landslides) (Djukem et al. 2020; Jenkins and Schaap 2018), livelihoods, farming lands, timber production, construction material, food and fuel security, medicines, and fresh air; these are all necessities for human sustainable socio-economic development (Jenkins and Schaap 2018).

13.3 Challenges in the Sustainability of Forest Management

Although forest biomes channel and develop all forms of life, this is a most misused and neglected resource. Various natural processes like erosion, flood, desertification (Adio et al. 2010), extreme weather, climate, along with Anthropocene activities like deforestation, overgrazing, urbanization (Pandey et al. 2021), industrialization (Mannan et al. 2019), droughts, and farming have threatened the sustainability of forest cover and health. The impact of which results in the form of global warming, extreme weather, soil and water pollution, deforestation, loss of microbial and aquatic diversities, and increased occurrence of natural hazards.

13.3.1 Challenges Due to Anthropogenic Activities

Forest ecosystems have a dynamic response to human activities. These mainly include land-use changes, global warming, greenhouse effect, increased erosion, and pollution (Table 13.3) (Beckline 2017; Daniel et al. 2017; Foley et al. 2005; Mannan et al. 2019; Pandey et al. 2021). Human lives need various forms of vegetation cover for their survival, including forest and grasslands. From gaining ecosystem services, food and water security, extraction of fuel and energy resources, creating a sustainable livelihood, and overall socio-economic development humans are dependent on vegetation cover. Through ancient times, the main occupations for humans were agricultural, which has caused the forest cover to be actively destroyed. Even for settlement and infrastructure development, forest cover has been destroyed in the form of deforestation, human lit fires, and logging (Beckline et al. 2018; Mannan et al. 2019).

13.3.1.1 Deforestation

According to FAO, the global forest cover was reduced from 2.5 million ha (1990) to 2.1 million ha (2000) to 1.7 million ha (2010) over the two-decade span from 1990 to 2010 (Mannan et al. 2019; Mateo-Sagasta et al. 2017; OECD-FAO 2012; Ridder 2007), and carbon loss of 2.1% per year from forest areas takes place mainly because of deforestation (Saeed et al. 2016). Half of the forests in eastern North America have been deforested for timber and agricultural use from 1600 to 1800. In Nepal, 64% of the rural population relies on the forest for energy, timber production, and livelihood (Poudel 2018). In the Republic of Srpska, 0.5% of the land was lost from the forest and converted to croplands, grassland, shrublands, and sparsely vegetative lands (Hishe et al. 2017). Deforestation is the process of removal of trees for clearing land for other activities like agriculture or grazing, infrastructure construction, or extracting construction material (like timber, palm, etc.), fuel and energy, or mining. Burning forests would clear the land for agriculture and ashes would make the land nutrient-rich with the carbon content present in ash (Beckline 2017; Daniel et al. 2017; Foley et al. 2005).

Table 13.3 Challenges in sustainability of forest.

Deforestation
Agriculture and water stress
Mining
Land degradation
Climate change
Change in land cover land use
Weathering and erosion
Forest fires

Source: Adapted from Literature survey.

Currently, the global status of vegetation cover has reached its critical mark due to Anthropocene activities, especially in Asian countries. In developing countries, along with the loss of forest cover, the farming land has degraded with loss of nutrients and organic matter content (Pokhriyal et al. 2020; Poudel 2018; Rahman et al. 2017). Furthermore, vegetation cover in barren lands or unusable land grows with time, but forests do not regrow naturally, as there is too much competition for tree seedlings from other forms of vegetation in terms of gaining water, nutrients, and sunlight for their survival (McCarthy et al. 2011). According to the United Nations Sustainable Development Group's (UNSDG) 2019 report, deforestation for commodity is 27%, forestry 26%, agricultural shift 24%, wildfires 23%, and urbanization 0.6%.

13.3.1.2 Agriculture and Water Stress

During the period of the late 1940s to the early 1960s global use of fertilizers, pesticides, herbicides, and commercial equipment improved the vegetation state of countries globally. Many countries practiced "controlled vegetation" for growing better plants or required types of vegetation, especially forests, while suppressing the growth of unwanted plants. The main aim of this practice was to reduce the competition for resources during the growth period with other vegetation in the same area (McCarthy et al. 2011). To do so, a greater use of chemicals was adopted in the form of artificial fertilizers, herbicides, and pesticides (Little et al. 2006; McCarthy et al. 2011).

However, seeing the overall degradation in environment and land resources, the use of chemicals in all forms of vegetation were either minimized, controlled, or banned. However, the practice of the use of chemicals to enrich soil quality is still very prominent in poor developing countries. In the long term, these practices not only damage the soil ecosystem but also cause loss of microbial diversities, which changes the functioning of various biogeochemical cycles. Furthermore, these are greatly responsible for nutrition loading (Ma et al. 2016; Zhang et al. 2011) and pollution in both soil and water (Zhang et al. 2020), as the chemicals are washed away with surface runoff during precipitation or over-irrigation from farm lands to local rivers and large water bodies, especially in lowlands or plains, dams, and reservoirs, leading to eutrophication. Eutrophication causes loss of aquatic biomes and over time, the entire water body becomes dead.

These ecological problems are further enhanced by urbanization and industrialization. Leachate from landfills, open dumping of municipal waste, industrial effluents, and mine drains end up depositing many sorts of organic, inorganic, and heavy metals in water bodies and sediments bed (Pandey et al. 2021). Leachates also flow through soil pores and contaminate the soil. One of the main concerns among these pollutants is the presence of heavy metals, as they bio-accumulate as well as bio-magnify through the food chain, resulting in some profoundly serious health hazards for all forms of life. These forms of chemicals significantly decrease the overall crop yield, along with lowering their nutritional value in agricultural farms, or sometimes make the entire area barren within a few years. Moreover, an increase in built-up area at the cost of vegetation, forest cover, and agricultural land has started to threaten food security as well as create many water and soil problems. Water stress is a common problem almost globally; lack of percolation and infiltration area, no mode of holding catchment runoff, an increase of pollutants in water bodies, and extreme demand for clean water has stressed all resources both above and below the surface.

13.3.1.3 Mining

Mining activities also have a prominent impact on the loss of forest cover and health. Forests are burned and are compromised to clear the land for mining. In addition to this, all the mined areas produce a huge quantity of wastes, which are usually left in the open. Rainfall causes runoff from these areas, called "mine drains," to flow and cause pollution of water bodies. Heavy metals present in these mined areas also contaminate soil to the point where soil loses its productivity. Soil and water quality degrades greatly in the nearby areas, which hinders crop production and vegetation growth. This process leaves a large amount of loose soil, which easily gets eroded causing further siltation to occur in river beds and low land areas.

13.3.1.4 Land Degradation

Desertification is a process by which natural or human factors, or both, causes reduction in bio-productivity of drylands consisting of arid and semi-arid lands, grasslands, and scrub-lands (Moussa and Omar 2020; Zalidis et al. 2002). According to the United Nations Convention to Combat Desertification (Moussa and Omar 2020), Desertification includes permanent or temporary degradation of soil quality, vegetation cover, water resources, or wildlife, along with economic productivity of land in drylands where potential-evapotranspiration falls within the range of 0.05 to 0.65 inches per year. Loss of vegetation cover also makes the soil loose and vulnerable, leading to extreme erosion susceptibility. These eroded soil particles along with pollutants settle in riverbeds and reservoirs. Soil erosion is a global challenge as it mainly degrades soil and water quality results in sedimentation, and increases chances of flood. Around 55.6% of the global land area has been degraded by erosion. Land degradation in the form of soil erosion is occurring naturally due to water, wind, waves, etc. particularly in undulated, areas but these are enhanced by excessive plowing, tillage, grazing, and deforestation. Erosion is also responsible for the loss of fertile topsoil, which is rich in organic carbon and nutrients like nitrogen, phosphorus, and potassium. Topsoil loss makes the area difficult for revegetation in the area; to compensate for the nutrient loss, the use of fertilizers becomes necessary, hence a vicious cycle is formed. Overuse of fertilizers also results in increasing the salinization, acidification, hardness of soil, etc. Increased salinity and acidity cause the loss of microbial and fungal diversity, deteriorate the growth of roots, and decrease their nutrient absorption, ultimately worsening the plant growth in the region.

13.3.1.5 Climate Change

Climate change is a major driver for environmental degradation, and is even more susceptible to forest cover loss (IPCC 2017; Sheik et al. 2011). Forest cover areas are the major carbon sink. Loss in forest cover due to deforestation, loss in soil quantity and quality, overgrazing, intensified cultivation, and over-irrigation have been responsible for less carbon absorption. Hence, the increase of carbon dioxide and other harmful greenhouse gases like carbon monoxide, CFCs, and methane due to industrialization, the use of modern technology in urban areas, burning fuel and waste in rural areas, all add to global warming. Global warming in return has resulted in global temperature increase causing droughts, wildfires (Daniel et al. 2017; Sheik et al. 2011; van Bergen et al. 2019), melting of glaciers, and increasing the global sea levels. Increased CO_2 can raise global temperature by 3 °C by the year 2030 (Ali 2013) and can increase sea levels globally. This may drown, flood, or salinize one-third of the world's croplands. Change in climate affects the vegetation cover altering microbial functioning, biogeochemical cycles, and structural and physical forms of plants. Microbial diversities are compromised due to their slower rate of adaptation to changing weather. Reduction in agricultural areas, increase in salinity, floods, and droughts, acid rain (USGS, 2007) due to extreme weather forces the agricultural-dependent population to relocate, and hence further damage to new ecosystems may occur. Change in climate and vegetation cover has a synergistic effect; change in one alters the other.

In summary, all the natural processes and Anthropocene activities have a combined effect on vegetation and forest cover. These processes threaten the sustainability of global greenery posing some serious challenges, but managing and monitoring the overall vegetation cover is not easy.

13.3.2 Application of Geospatial Technology in Monitoring of the Forests

In the past, monitoring and management of global forest reserves was an exceedingly difficult task, when extensive research studies through manual sample collection, and field and laboratory experiments were performed to collect data and analyze it. It was a tedious task as many geographical, topographical, and climatic factors played a major role in defining the feasibility and accessibility of sampling areas (Peterson et al. 2012; Kosmas et al. 2016). Understanding the synergistic effect of all factors was nearly impossible, hence planning and policies were rarely implemented with success. With the development of engineering and technologies, various Geospatial tools were introduced for the overall environment and climate management.

Geospatial tools such as remote sensing (RS), geographical information system (GIS), and geographical positioning system (GPS) are frequently used to monitor forest cover. From resource data collection to utilization, planning, modeling, map production, hazard management, pollution interferences, and climate changes to land-use changes are all activities done using these geospatial tools. In recent decades, many advances of these tools have been developed and used, such as the use of thermal microwave and image sensors to obtain real-time view and spectral differences in biomass, which have become great tools for globally monitoring large plots of the forests. These techniques have been integrated with the use of machine learning and statistical tools such as multivariate statistics, cluster analysis, and regression analysis to make the data more informative and more widely applicable.

Remote sensing helps acquire information about the physical characteristics of objects or phenomena on earth without any physical contact but by measuring reflection and emitted radiation from the surface of objects (USGS, 2020). Through the use of this tool, environmental hazards like wildfires or forest fires, dust storms, volcanoes, and changes in land use, land cover, vegetation cover, etc. can be observed without being in the area. This gives a broader and clearer view of the earth's surface and helps in warning the inhabitants before any calamity occurs, saving lives. GPS is used to navigate through land and water areas, especially through dense forests, wetlands, and oceans, where a direct view is blocked from the naked eye. GPS uses satellite navigation and helps in developing maps, geotagging, mining operations, touring, cartography, etc. Forests are dense and the source of light is usually dependent on the density of canopies, whereas wetlands, marshy lands, vast croplands, deserts, ice cover, and mountains all have undulated surfaces and navigating through these areas could be dangerous. With the help of GPS and satellite imagery, one can easily know what is ahead and use directions to move through these areas without getting lost.

Another crucial and advanced tool is GIS, which is a framework for gathering, managing, and analyzing data. GIS has various features that help us monitor the geographical, topographical, water resources, and climatic changes around the globe. It helps in collecting data of specified locations and has provisions for layering various types of resource data for real-time analysis in both 2D in forms of maps and 3D in forms of models. It also helps in assessing the interdependencies, patterns, relations, and changing situations to make better plans for ecological sustainability and judicial use of resources. Almost all topographical, geographical, and climatic data are collected through RS and GPS, which are then layered through the GIS where changes in land use/land cover, vegetation cover, and climate are mapped and compared with changing space and time. The use of one or more geospatial tools depends upon the range of application, severity, and complexity of the problem at hand. Some of the most common applications are in sampling sites (fixing site location), creating maps (to visualize topographical undulations, water bodies, roads and networking, landscape changes), and determining soil changes horizontally and vertically.

Some other recent advances in geospatial applications in the forest are the use of sensors like Aerial LiDAR, terrestrial LiDAR (Behrenfeld 2017; Hilmers et al. 2018), RADAR (Agapiou 2020; Pal and Chakrabortty 2019), microwave, miniaturized cameras, platforms like drones, microsatellites, computational tools, Geo databases, and Geostatic neural network methods (Meraj et al. 2018). Moreover, many spectral indices such as Normalized Difference Vegetation Index (NDVI) (Kumar et al. 2010), Soil-Adjusted Vegetation Index (SAVI) (Eid et al. 2020), and Normalized Difference Wetland Index (NDWI) (Al Kafy et al. 2021) are also used to monitor the change in land use land cover, vegetation growth, and the impacts of urbanization on the vegetation cover and climate change. Sensors can be active or passive; passive sensors use solar radiation to detect the surface of objects present on earth, whereas active sensors use their source of energy to measure the surface of the objects. Sensors can be classified as thermal, optical, and microwave. For example, LiDAR is an electromagnetic wave sensor, RADAR is a microwave sensor, and they are advanced forms of remote sensing measurement. Additionally, many recent technologies have been developed which can combine the data obtained from any form of sensor with another to provide better resolution of the area.

LiDAR is an example of a non-imaging sensor, which uses light as a pulsed laser to measure the variable distances of objects to the earth. When these data are combined with other aerial measure system data, they provide a 3D image of the land. LiDAR uses datasets collected from airborne platforms like a helicopter, unmanned aerial systems (UAS), unmanned aerial vehicles (UAV), and airplanes, terrestrial laser scanning (TLS) (Xue and Su 2017). Among these applications, the use of LiDAR integrated with TLS offers several features such as high-resolution data, below canopy data, and field vertical and horizontal view, while being cost-effective. Its application has been successfully proved in vegetation and biomass monitoring, fuel bed studies, and forest inventory. The method is sometimes complicated, as the light pulse tends to mingle with the present cloud cover over objects, but this issue may be solved by applying interpolation and cluster analysis methods, which give accurate data based on previously collected aerial data (Anderson et al. 2017). It was also found that UAV LiDAR and hyperspectral image infusion assist in evaluating the forest cover (burning and thinning) (Ravikumar et al. 2013). Eco-restoration success in abandoned mining terrain has been studied by comparing the leaf-off and leaf-on combined UAV imagery and airborne LiDAR. The results showed that leaf-off and leaf-on comparative analysis can potentially replace the airborne LiDAR survey system of vegetation monitoring (Moudrý et al. 2018). Use of cluster analysis with airborne LiDAR for regional biodiversity mapping has shown that there is a complex vegetation structure and persistent wildfires in the lower foothills' natural sub-regions, but most vegetation was disturbed by anthropogenic activities like timber harvesting. Hence, the method poses a greater potential for analyzing vegetation and forest cover and monitoring data. Most biomass cover analyses are done using lower-resolution data from LANDSAT, enhanced thematic mapper (ETM 6), and MODIS. LiDAR has also gained huge popularity in biomass cover study; for example, in biomass studies of the urban landscape, it has been found that vegetation growth has been increasing and decreasing because of the varied nature of the area. By varying the wavelength, one can obtain better object classification of land and vegetation; for example, at 1064 nm, vegetation gives a stronger reflection than soil and vice versa at 1550 nm.

A RADAR imaging sensor is the radio (microwave) wave detection of the objects on the surface of the Earth. These tools are very efficient in monitoring changes in vegetation and forest covers over spatio-temporal scales. The benefit of using RADAR is that it can provide images even in bad climate conditions like clouds, can easily identify the various minerals and types of vegetation cover, as well as penetrate water, soil, and walls. It is mostly used with complementary sensors such as infrared sensors to read the components of the Earth's surface. RADAR has a disadvantage in that the images produced have coarse textures as compared to optical sensors. This problem has been solved with the use of synthetic aperture RADAR SAR.

Szigarski et al. (2018) studied RADAR vegetation indices and their potential improvement. The study used a collaboration of active and passive microwave information to understand the dependencies of soil features such as surface roughness and moisture on vegetation. It was found that the improved RADAR indices showed less dependency on roughness and moisture, and correlation to moisture in plants and their structure.

Some available optical imaging sensors are multispectral, panchromatic, and hyperspectral systems. These sensors vary in the type of image detection, brightness, and colors of the image. Panchromatic works with black and white tones, multispectral uses few bandwidths, whereas hyperspectral uses dozens, even hundreds of bands to form images which are bright and make possible the identification of materials on Earth's surface. These data easily collaborate with other data obtained from other forms of spatial data resources to give high-quality resolution and images. For example, plant disease identification and detection such as grapevine leaf stripe have been detected using hyper spectral sensors by (Junges and Ducati 2018). The vine's leaf behavior was altered due to lack of photosynthesis and this was observed as an increase in green and red edge reflectance in the visible light range, whereas near-infrared reflectance decreased due to loss of cell structure and toxin concentration (Junges and Ducati 2018). This technique helps in the analysis of plant properties such as its compounds, nucleic acids, and metabolic processes in a non-destructive technique that uses photo pigmentation. Interferences from pathogens in plants are studied by using the spectral variation that occurs with physiology, moisture (1400–1930 nm near-infrared wave), and structure (700–1000 nm near-infrared wave) of plants. Change in pigmentation due to infestation varies between the developmental stage and advanced stages, and this helps in easy detection of the type and stage of disease in the plants. In the case of agriculture, this technique helps in finding the nutrient content, biomass, and disease infestation (Behrenfeld 2014; Bjorkman et al. 2018, Singhal et al. 2021). Optical imaging sensors have been used with UAVs and machine learning techniques. Horstrand et al. (2019) used a DJI Matrice 600 drone loaded with advanced processors, and a Specim FX10 hyperspectral camera to capture the hyperspectral images at a rate of 300 frames per second in the VNIR range in an agricultural field. They also calculated vegetation indices like NDVI for precision agriculture in Spain. This technique proved useful in assessing agriculture productivity with remote sensing (Yue et al. 2018). Some research uses optical imaging with deep learning; for example, Tan et al. (2019) studied the use of a random forest algorithm to analyze the heavy metal content in agricultural soils with the help of hyperspectral data. This technique helped in determining four heavy metals and proved to work better than traditional methods. Other uses of the random forest (RF) algorithm have been studied in classification (da Silva et al., 2018), wheat biomass (Yue et al. 2017), and vegetation mapping. Nagasubramanian et al. (2019) used 3D deep learning on hyperspectral images for plant disease identification and found that artificial neural network (ANN) models used the most sensitive wavelengths as a near-infrared region for vegetative health determination.

GIS has been used in policy making for environmental and forest management (Table 13.4) (Baral 2004; Meshesha et al. 2012) for the past few decades. It has proven to be a powerful tool to keep inventory records of resources, production, habitats and wildlife, harvest yield, ecosystem and landscape management, socio-economic parameters of foresters, analyzing of data, and decision making. Foresters implement the concept of sustainable forest management and must maintain and manage many large and small goals. From conservation to soil and water to managing wildlife and bio diversities, monitoring forest productions, and various Anthropocene factors that influence these objectives like deforestation, human lit fires, mining, etc. are all under the management of foresters. Addressing

Table 13.4 Advantages of using Geospatial tools in forest management.

- Cost-effective and user-friendly method
- Better application with little manpower training
- Cheap maintenance and time saving
- Ease of data management
- Easy surveying and monitoring of data through dense and undulated areas
- Broader visuals provide better controls and constant monitoring
- Future prediction of ecosystem changes
- Wide range of data processing such as location, navigation, tracking, mapping, timing, and thematic layering

Source: Adapted from Literature survey.

large areas of forest and keenly observing various forest processes, especially dense ones, and controlling all aspects of forest management is extremely hard. However, with the help of remote sensing data (Yang et al. 2018) forest areas can be analyzed. Identification of forest boundaries, location and areas, and level of degradation from natural and Anthropocene activities are all observed through RS.

Earlier the focus of forest management lay in assessing the products, especially timber production and management, but these views have changed with time and non-harvest related parameters are also included in the management practices like forest health, forest structure, hydrology of the area, bio-diversities. The inventory assessment of the whole forest is extremely difficult but with the coming of geospatial technology, this entire process has eased up. Comparing aerial photographs, finding routes within forests, annual production, or harvest of different types of trees can all be assessed. Some very generalized fields of application are forest health, resource and production inventory, natural hazard monitoring, resource conservation, strategic and participatory planning, wildlife inhabitant, wetlands, watersheds management, etc. (Baral 2004; Coops et al. 2008, 2020; Laamanen and Kangas 2011; Morgenroth and Visser 2013; Patenaude et al. 2005; Turcotte and Desrochers 2003; Wulder 1998).

The use of Geospatial applications in monitoring forest health is an evolving process (Table 13.5) (Dewar et al. 2009). Pest or insect attacks and other microbial diseases may be monitored on a plantation by visually analyzing geographical patterns, severity, location and direction of spread, frequency of occurrence, climatic parameters enhancing the disease, and insect attack (Baral 2004). Location mapping and monitoring of these patterns help in better dealing with pest and insect control, as well as warning concerned authorities about better management of the situation. Sharing this information can help to assess similar situations in other regions also. Use of ANN, hyperspectral and multispectral sensors, and geostatistical tools are commonly used for the analysis of pest infestation and climate change in vegetation and soil.

Table 13.5 Major applications of GIS in management of forests.

Application area in the forest	Use
Forest health	Insect, pest, biodiversity monitoring
Resource inventory and production	Quantification of forest yield and resource assessment
Natural hazards monitoring	Storms, wildfires, climatic change, droughts, etc.
Resource conservation	Conserving soil, water resources and forest products
Strategic and participatory planning	Developing strategies and inclusion of associated communities in planning
Wildlife	Conservation and monitoring of wildlife habitats
Wetlands and watersheds	Overall ecosystem, soil, and water conservation in the watershed
Recreation	Tourism

Source: Literature survey (Baral 2004; Laamanen and Kangas 2011; Morgenroth and Visser 2013; Patenaude et al. 2005; Turcotte and Desrochers 2003; Wulder 1998).

Inventory management of forest resources is very crucial to foresters. The basic application of Geospatial tools is in acquiring inventory data like distribution, quantity, and quality of forest resources (Sonti 2015). This helps in understanding the growth rate of resources, annual production, harvesting schedules, yield, and profit from resources. The equation of demand for ecological services and adequate supply is dependent on the proper management of inventories. Doing this manually was difficult, and quantifying loss from infected areas, illegal logging, and corruption were also difficult to monitor. However, RS and GIS applications help in providing a proper image of forest cover, estimation of areas of deforestation, and increased productivity from afforestation. Main forest products include wood (mainly timber and sal), fruits, bamboo, and honey. Anderson et al. (2017) found the vegetation biomass and cover across large plots in shrub and grass-dominated drylands using TLS and machine learning. The research showed that TLS could be highly effective in vegetation management, as it does not require classifying and separating individual plants. Many other techniques have been applied in combination to improve the monitoring of biomass in forest plots with the use of NDVI with MODIS. Linear and nonlinear RF models were applied, and it was found that post two years of planting the RF algorithm was improved using NDVI. In eucalyptus trees, use of RF techniques and radial basis function (RBF) have proved better tools in determining the diameter at breast height and total height (da Silva et al. 2021).

Natural hazards, mainly wildfires, have been a constant problem for foresters. Wildfires are dangerous for all forms of life both within the forest and outside the forest region. Although wildfires usually occur in dry seasons, global rise in temperature and frequent droughts have increased their occurrence. In the Pacific Northwest of the US, it was found that the wildfires are bound to increase as the air becomes dry and warm, and will have prolonged fire seasons in comparison to previous centuries. This may also result in insect and pest outbreaks, along with frequent droughts (Halofsky et al. 2020). Another example of this is the locust infestation on crops in countries like Pakistan and India in 2020 due to warmer climate that favored the breeding of locusts. Even so, no regular consistent data were easily available on causes of burning, climatic conditions for fire, and total loss of life and products. Through geospatial data, fire management has become better. Although forest fires are hard to prevent, a better study of influencing factors, climatic situation, and resource inventory would help in providing beforehand warning for better control measures and hazard management preparation, saving resources, cost, and lives. Fire mapping can help in figuring recurring fire incidences, cause, and prevention techniques, and by linking GIS to GPS, radio, and other portable equipment so other forest emergencies can be dealt with more easily.

To collect data, analyze characteristic information, and assess ecosystem services and biodiversity, and wildlife management data, use of ANN, CNN, and deep residual neural networks have proven to be an effective way of analysis. The use of drones has been useful in monitoring the movement of animals in the region and helping in minimizing possible threats like collisions with cars, to wildlife and humans. The use of drones and machine learning has been effective in the study of illegal poaching of animals. The study showed that the probability of detection decreased with enhanced biomass cover, but the use of automated software was found effective. Use of remote sensing, hyperspectral and multispectral data, LIDAR and RADAR, and data sensors also help in figuring out the healthy area for sensitive plant growth, delineating water bodies, and sustaining the wildlife zones. It also helps in designing corridors, buffers, and networks for better product harvesting and transportation without disturbing the forest and wildlife ecosystem. Overall, GIS helps in planning and developing strategies for harvesting and non-harvesting areas. Harvesting areas are mainly designed for commercial production as well as community recreation, whereas non-harvest zones include areas where wildlife and sensitive bio-diversities are prominent, or forest cover is extremely dense. Dangerous zones in the forest can be mapped out. The health of crop deterioration of harvest can also be monitored.

Geospatial tools are applied in watershed and wetland management, resource (water and soil) quality and quantity conservation, along with addressing landscape changes. Wetlands have unique ecosystems and biodiversity, which requires special protection and conservation. Conservation of the environment and other natural resources on the watershed level is crucial for ecological and socio-economic development. Use of indexing such as NDVI, NDWI, and SAVI may help in monitoring the biomass. The change in land use land cover (LULC) can be observed and sources like pollution urbanization can be easily identified. GIS can help stimulate a 3D model of various components of watersheds and wetlands. Changes in soil, water networks, biogeochemical cycles, forest, and the build-up of areas over time can be seen with the help of integrated use of GIS, RS, and their other advanced components.

Like forest cover, management Geospatial tools can also be used to assess another form of vegetation cover. Application of GIS and RS in monitoring LULC (Gashaw et al. 2017) and monitoring overall land degradation is very frequent and prominently used. Pollution mapping of all point and non-point sources, drainage networks, water quality, and climatic changes can all be mapped and modeled in different vegetation areas.

LULC is a global concern; alteration in the use of land from natural vegetative state to infrastructure development and selective agricultural practices has given rise to many environmental hazards, as discussed. Geospatial tools help us identify the changes in landscape and problems arising from these changes. Mapping of soil erosion susceptible areas, sources and reasons, hydrology, and climate changes integrated can help in proper conservation planning at watershed and community levels. Sediment yield in low land areas, reservoirs, and siltation in riverbeds can also be calculated using GIS-based software such as Universal Soil Loss Equation (USLE), Revised Universal Soil Loss Equation (RUSLE), and Modified Universal Soil Loss Equation (MUSLE), which help in monitoring the reservoir yield loss. New soil deposited areas may be used for revegetation or planned urbanization.

Water quality and conservation maps and models may be prepared using GIS, based on information attained by GPS and RS of drainage networks, built-up area, rainfall and evapotranspiration, temperature, frequency of droughts and floods, need and consumption of water, loss of vegetation cover, and potential runoff storage (Ahmad et al. 2014; Al-Adamat 2008; Sayl et al. 2016). This helps in the overall management of water resources in the rain-fed drainage system areas and drylands (Taffere et al. 2016). Sometimes interstate water distribution channels via reservoirs or dams are to be constructed to promote public welfare. Identifying the accurate elevation area–volume curve is the foremost challenge on which the optimal depth and minimum and maximum storage area of the reservoir could be based (Sayl et al. 2016; Weerasinghe et al. 2011), but understanding the ecological and socio-economic impact is particularly important in the selection of areas. Using Geospatial tools, maps, and models stimulating the present situation of LULC, water, ecology, and socio-economic conditions of inhabitants can be prepared. Future ecological and climatic predictions can be performed based on these (Weerasinghe et al. 2011).

RS and GIS also help in monitoring changes in the climate of different forms of vegetative areas. Warning of ecological hazards like flash floods, droughts, cloud bursts, heat waves, cold waves, and bush fires can be given beforehand (Choong and El-Shafie 2015; Van De Giesen et al. 2002; Jha et al. 2014; Paruch et al. 2015; Sayl et al. 2016). Through GIS based on DEM and GPS conservation measures and protection, plans can be made (Figure 13.3). Mapping Source, understanding reasons and the extent of influence of the climatic, ecological and land-use changes through Geospatial tools can help in identifying the potential revegetation zones and feasible solutions to make the ecosystem greener, as well as stable without compromising any component of the environment (Chowdary et al. 2009; El-Shafie et al. 2014; Gwenzi and Nyamadzawo 2014; Jha et al. 2014; Kahinda et al. 2008; Mwenge Kahinda et al. 2009; Pistrika 2010).

In summary, all the natural processes and Anthropocene activities have a combined effect on overall vegetation and forest cover. These processes threaten the sustainability of global greenery posing some serious challenges, but managing and monitoring the overall vegetation cover is not easy.

13.3.3 Types of Forest Data

13.3.3.1 Forest Extents

Every year a net change is observed in the forest cover as some forest areas are lost and some are gained. Although a global decline has been observed in the forest cover, a significant variation is observed in terms of regions across many countries. Additionally, not all

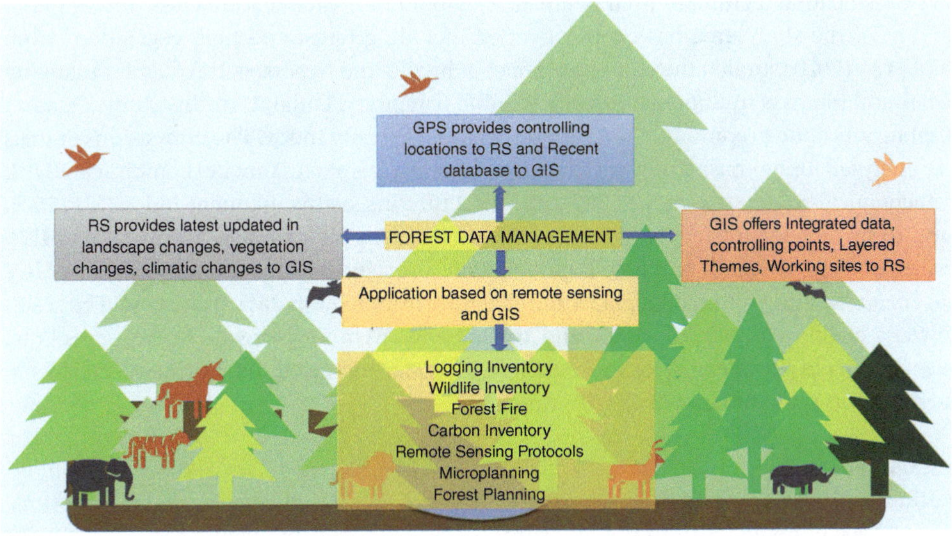

Figure 13.3 Application of remote sensing and GIS in forest data management.

types of forest reserves are equal in terms of value. Forest varies from primary forests to industrial or those commercially grown for production of timber, pulp, rubber, or fruit; they may also vary in terms of their carbon reserves, socio-economic benefits, and ecosystem services provided. Various indicators are studied to monitor the forest extent such as forest gain and loss area, loss of primary forest area, deforestation, and inventory of trees grown outside of original forest cover. There are certain limits to the forest extents such as most commercial plantations are not considered, as an orchard. Performed forest extension monitoring using airborne and satellite data from 1973 to 1976 in Ethiopia. The forest cover of the country decreased from 4.25 to 0.20% in a decade, with an annual loss of 163 600 ha. Hamzah et al. (2020) studied the mangrove forest changes around the Pahang coast, where the rate of erosion and accretion, along with carbon stock was estimated using SPOT 5 satellite images and vegetation indices like NDVI, and found that the coastline suffered from erosion processes. The Cherating and Penor Rivers suffered more with an estimate of 10.31 and 18.17 m/year.

13.3.3.2 Forest Cover

Forest cover is defined as the area covered in trees. Detection of forest cover change is crucial for continuous monitoring of environmental changes such as depletion of natural resources, loss of microbial diversities, deforestation, and carbon emission. Forest cover has been studied using satellite data obtained from LANDSAT imagery, but they have low coverage area and low repeat frequency compared to coarse spatial data such as moderate-resolution imaging spectrometer (MODIS) and high-resolution radiometer. Khan et al. (2017) studied the forest cover change detection in incomplete satellite images with deep neural networks and found that the use of combined tools leads to better analysis of change at finer resolution and automatically shows strong influencing features from raw data. Haque and Basak (2017) studied the change in land cover using GIS on Tanguar Haor, Sunamganj Bangladesh. The change was detected using NDVI, NDWI, and CVA based on the maximum-likelihood supervised classification technique used to create a signature for each feature class. It was found that 40% of the study area has been converted including forest and high vegetation cover. Malhi et al. (2020) studied the combined effect of in-situ and hyperspectral data for mapping species and biomass in Shoolpaneshwar Wildlife Sanctuary, Gujarat. In this study, Quadrat sampling was done to compute the Shannon-Weiner Diversity Index. The dimensions of trees were analyzed for biomass study, which lay in the sampling plots. A model combining NDVI, photochemical reflectance index (PRI), and structure insensitive pigment index (SIPI) was found to be best for retrieving Shannon-Weiner Diversity Index (Dempewolf et al. 2017). Here, the measurement of tree height growth within varying seasons was studied using UAV imagery along with ground data plates and structure from motion data processing. The result showed a medium growth of 68, 93, and 106 cm, and 26 cm for Norway spruce, Scots Pine, Silver Birch, and European Beech respectively. This study showed the efficiency of UAVs in studying short span tree growth monitoring.

13.3.3.3 Forest Fire

Another important hazard that is measured on both small and large scale is forest fires. Forest fires can be harmful to the environment as well as humans and other wildlife in the habitat. Forest fires are caused by climate change, season change, weather, human lit fires, lightning, etc. but the real problem is controlling the mass fire, which can grow out of proportion very quickly. Many such cases have been reported in Australia, India, China, and the US, and monitoring wildfires is important as they may cause heavy economic and ecological damage. Eruption of ecosystem services or loss of biodiversity are severe side effects of wildfires. To monitor and assess the wildfires many researchers have studied the indicators of wildfire creation using satellite data. The time-series studies help in analyzing the immediate changes in the atmosphere. Reiche et al. (2018) studied tropical forest cover loss using dense Sentinel-1 data and active fire alerts in Indonesia. The study revealed that the wildfires predating patterns were coinciding with forest cover loss.

13.3.3.4 Forest Health

Many natural and anthropogenic factors affect the resilience of forest biomes; and the relationship between the factors, causes, challenges, and effects are complex and multidimensional on a spatial–temporal scale. It is crucial to assess and monitor forest health, which can easily become affected by various stressors such as climate change, loss of microbes, etc. For better monitoring of forest health and diversities, better quality and quantity of data are required by forest managers and scientists, but the most conventional process of data collection is complex, costly, and inadequate. Lausch et al. (2018) studied the application of RS data on forest health and found that the use of combined RS data with network models will prove to be better for monitoring as compared to field study and remote sensing processed data (Figure 13.4). In their study on RS application in monitoring mangrove forests, Wang et al. (2019) found that the mangrove RS replicated those of forest RS with time lags. This

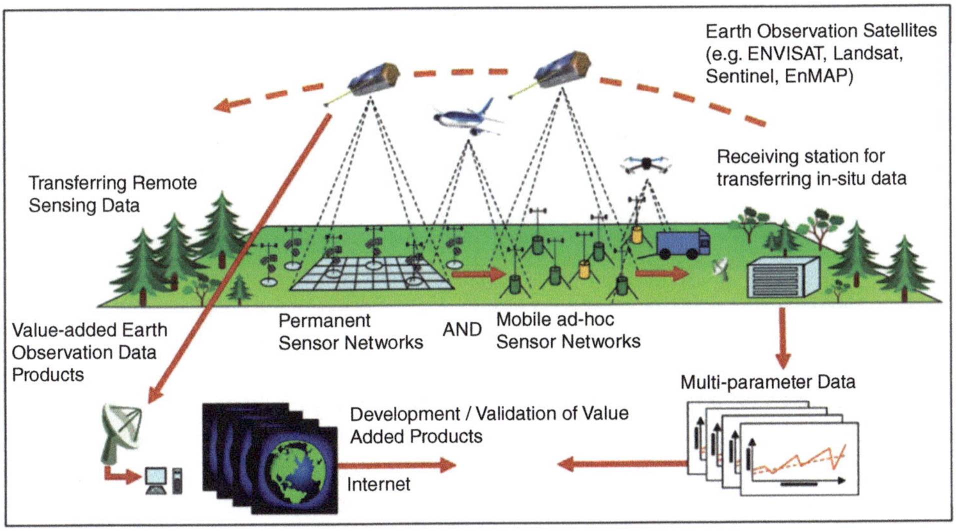

Figure 13.4 Linking different approaches (high frequency wireless sensor networks up to spaceborne satellites) with relative frequency monitoring, sensors, and different platforms of remote sensing to better describe, explain, predict, and understand forest health with remote sensing techniques, as well as improve the calibration and validation of remote-sensing data. Source: Adopted from Lausch et al. 2018.

was due to the mangrove forest appearing more elongated than the terrestrial forest and field study of mangroves is complex.

13.3.3.5 Forest Biophysical and Biochemical

Forests act as a carbon pool; and the distribution of nutrients, carbon, and biodiversity in the forest varies spatiotemporally. Moreover, the study of tree species, their biophysical properties, as well as biochemical features, have been complex and challenging. Varying biophysical properties such as forest composition, leaf area index, canopy gap distribution, and leaf angle distribution have been studied using remote sensing and GIS application with biochemical properties, which include nutrient cycle and their relationship with features and properties of forest stands and biomass. Much research has focused on the biophysical and biochemical properties using remote sensing data like LiDAR, hyperspectral and RADAR data. Many types of research have combined the remote sensing data with geostatistical and machine learning tools to provide patterns and predictions for future changes and study the deeper correlation among parameters. Dang et al. (2019) studied the forest aboveground biomass (ABG) estimation using a machine-learning regression algorithm (RF) in Vietnam. One hundred thirty-two spectral data were taken from Sentinel-2 imagery to predict ABG using RF algorithm, and around 11 variables were able to predict ABG with $R^2 = 0.81$ accuracies. The study showed increased efficiency of Sentinel-2 data with machine learning in analyzing ABG. Forkuor et al. (2020) studied the ABG in west African dryland forests using Sentinel-1 and -2 data and their derivatives like indices, biophysical and inventory data in random forest regression (RFR) and found S-2 data to be more efficient in analysis. Liu et al. (2019) estimated the forest stand mean height and AGB in China using SAR Sentinel-1B, multispectral Sentinel-2A, and DEM imagery. The ABG models were linked to field measurements of forest stand mean height and ABG using a random forest algorithm. The study showed the efficiency of using Sentinel data for developing the ABG models. Ota et al. (2017) studied the application of lightweight UAV photogrammetric point cloud in managed temperate coniferous forests. The study showed that the use of UAV photogrammetric point cloud was efficient in finding biophysical properties and could be good even without the estimation using spectral variable.

13.4 Summary

Vegetation and forest covers are the sources of survival for all forms of life on earth, as they provide many ecological products and services. Humans, livestock, and wild animals are all dependent on forest cover for food, fuel, and water security, hence managing the green cover is crucial. Forests affect the climate by channeling the biogeochemical cycles. It was found that canopy density and litter production help in managing the microclimate and enhancing the nutrient content of the soil. Forests are found to be the major source of carbon and nitrogen sink, especially the boreal and tropical forests. Unfortunately, increased Anthropocene activities have caused heavy destruction to vegetation cover and are responsible for overall land degradation. The increase in global warming is attributed to unmanaged exploitation of natural resources, especially urbanization and fuel extraction mostly from mining, wood harvesting, deforestation, and clearance of land for agriculture. These processes have led to a severe loss of microbial species, which help in balancing the nitrogen and other nutrient fixation in soil. The micro- and macro climates of any region are affected as the moisture retention and carbon sink deteriorates. Natural or human made hazards like wildfires and deforestation make the plot area into a carbon source instead of a sink. To fully identify, address, analyze, and understand the source and extent of damages that occur in the environment and their effect on the vegetation cover, geospatial tools are used. Integrated use of remote sensing, GIS, and its components have proven to be crucial. As per the recent research, the use of a combination of many modern tools such as sensors help in identifying the change in LULC, forest, agriculture, and even the climate change in the area. These sensors are based on the principle of reflection for which they may use the energy of their own (active) or sun (passive). These sensors use the platforms such as drones and satellites to create spectral images that could help in identifying the changes on the Earth's surface. These tools are effective enough to find the type of vegetation, the diseases in a crop's leaves, its area, and harvest cycles. They are also very helpful in tracking illegal mining, logging, or any other suspicious activities that could bring harm to the vegetation covers. These remote sensing tools have helped in preparing maps and models of the status of vegetation cover in many areas. This advanced technology has also helped in filling data gaps of environmental hazards and monitoring natural resources. Maps and models prepared based on Geospatial assessment, can help the policymakers to make revegetation plans to improve overall ecosystem health. Many international agencies like the UN, FAO, and IPCC have created a global platform for conferences and annual reports to monitor and discuss the latest environmental issues. These agencies help in identifying the crucial issues and in finding a better solution to tackle them. They also help in setting the agenda for countries to monitor misuse of resources and pollution generation.

References

Abdalla, M., Hastings, A., Chadwick, D.R. et al. (2018). Critical review of the impacts of grazing intensity on soil organic carbon storage and other soil quality indicators in extensively managed grasslands. *Agriculture, Ecosystems and Environment* 253(October 2017): 62–81. https://doi.org/10.1016/j.agee.2017.10.023.

Adio, A.F., Ojo, A.R. and Asinwa, I.O. (2010). Climate change: forestry, a step in the right direction. *ICEEA 2010–2010 International Conference on Environmental Engineering and Applications, Proceedings*, (ICEEA 2010), 68–71. https://doi.org/10.1109/ICEEA.2010.5596092

Agapiou, A. (2020). Estimating proportion of vegetation cover at the vicinity of archaeological sites using sentinel-1 and-2 data, supplemented by crowdsourced openstreetmap geodata. *Applied Sciences (Switzerland)* 10 (14): https://doi.org/10.3390/app10144764.

Ahiablame, L., Sinha, T., Paul, M. et al. (2017). Streamflow response to potential land use and climate changes in the James River watershed, upper Midwest United States. *Journal of*

Hydrology: Regional Studies 14 (August): 150–166. https://doi.org/10.1016/j.ejrh.2017.11.004.

Ahmad, A., El-Shafie, A., Mohd Razali, S.F., and Mohamad, Z.S. (2014). Reservoir optimization in water resources: a review. In: *Water Resources Management*, 28(11), 3391–28(11), 3405. Kluwer Academic Publishers https://doi.org/10.1007/s11269-014-0700-5.

Ahmad, A., Liu, Q.J., Nizami, S.M. et al. (2018). Carbon emission from deforestation, forest degradation and wood harvest in the temperate region of Hindukush Himalaya, Pakistan between 1994 and 2016. *Land Use Policy* 78 (July): 781–790. https://doi.org/10.1016/j.landusepol.2018.07.009.

Al Kafy, A., Dey, N.N., Al Rakib, A. et al. (2021). Modeling the relationship between land use/land cover and land surface temperature in Dhaka, Bangladesh using CA-ANN algorithm. *Environmental Challenges* 4 (May): 100190. https://doi.org/10.1016/j.envc.2021.100190.

Al-Adamat, R. (2008). GIS as a decision support system for siting water harvesting ponds in the basalt aquifer/NE Jordan. *Journal of Environmental Assessment Policy and Management* 10 (2): 189–206. https://doi.org/10.1142/S1464333208003020.

Ali, M. (2013). *Climate Change Impacts on Plant Biomass Growth*. Springer https://doi.org/10.1007/978-94-007-5370-9.

Allen, S.T., Brooks, J.R., Keim, R.F. et al. (2014). The role of pre-event canopy storage in throughfall and stemflow by using isotopic tracers. *Ecohydrology* 7 (2): 858–868. https://doi.org/10.1002/eco.1408.

Anand, A., Pandey, P.C., Petropoulos, G.P. et al. (2020). Use of hyperion for mangrove forest carbon stock assessment in bhitarkanika forest reserve: a contribution towards blue carbon initiative. *Remote Sensing* 12 (4): https://doi.org/10.3390/rs12040597.

Anderson, K.E., Glenn, N.F., Spaete, L.P. et al. (2017). Methodological considerations of terrestrial laser scanning for vegetation monitoring in the sagebrush steppe. *Environmental Monitoring and Assessment* 189 (11): 1–12. https://doi.org/10.1007/s10661-017-6300-0.

Apostol, B., Chivulescu, S., Ciceu, A. et al. (2018). Data collection methods for forest inventory: a comparison between an integrated conventional equipament and terrestrial laser scanning. *Annals of Forest Research* 61 (2): 189–202. https://doi.org/10.15287/afr.2018.1189.

Austin, A.T. and Ballaré, C.L. (2014). Plant interactions with other organisms: molecules, ecology and evolution. *New Phytologist* 204 (2): 257–260. https://doi.org/10.1111/nph.13062.

Austin, A.T. and Zanne, A.E. (2015). Whether in life or in death: fresh perspectives on how plants affect biogeochemical cycling. *Journal of Ecology* 103 (6): 1367–1371. https://doi.org/10.1111/1365-2745.12486.

Baral, H. (2004). *Applications of GIS in community-based forest management in Australia (and Nepal). Forestry*, (August), 40–60.

Bautista, S., Mayor, Á.G., Bourakhouadar, J., and Bellot, J. (2007). Plant spatial pattern predicts hillslope runoff and erosion in a semiarid mediterranean landscape. *Ecosystems* 10 (6): 987–998. https://doi.org/10.1007/s10021-007-9074-3.

Beckline, M. (2017). Perspectives of remote sensing and GIS applications in tropical forest management. *American Journal of Agriculture and Forestry* 5 (3): 33. https://doi.org/10.11648/j.ajaf.20170503.11.

Beckline, M., Yujun, S., Etongo, D. et al. (2018). Assessing the drivers of land use change in the Rumpi hills forest protected area, Cameroon. *Journal of Sustainable Forestry* 37 (6): 592–618. https://doi.org/10.1080/10549811.2018.1449121.

Behrenfeld, M. (2014). Climate-mediated dance of the plankton. *Nature Climate Change* 4: 880–887.

Behrenfeld, M. (2017). Annual boom-bust cycles of polar phytoplankton biomass revealed by space-based LiDAR. *Nature Geoscience* 10: 118–122.

van Bergen, T.J.H.M., Barros, N., Mendonça, R. et al. (2019). Seasonal and diel variation in greenhouse gas emissions from an urban pond and its major drivers. *Limnology and Oceanography* 64 (5): 2129–2139. https://doi.org/10.1002/lno.11173.

Bjorkman, A.D., Myers-Smith, I.H., Elmendorf, S.C. et al. (2018). Tundra trait team: a database of plant traits spanning the tundra biome. *Global Ecology and Biogeography* 27 (12): 1402–1411. https://doi.org/10.1111/geb.12821.

Bochet, E., Rubio, J.L., and Poesen, J. (1998). Relative efficiency of three representative matorral species in reducing water erosion at the microscale in a semi-arid climate (Valencia, Spain). *Geomorphology* 23 (2–4): 139–150. https://doi.org/10.1016/S0169-555X(97)00109-8.

Boillat, S., Scarpa, F.M., Robson, J.P. et al. (2017). Land system science in Latin America: challenges and perspectives. *Current Opinion in Environmental Sustainability* 26–27 (January): 37–46. https://doi.org/10.1016/j.cosust.2017.01.015.

Buchhorn, M., Smets, B., Bertels, L., De Roo, B. et al. (2020). *Copernicus Global Land Service.* Land Cover 100m: Collection 3 Epoch 2015, Globe. Version V3. 0.1 (Data set).

Buckland, C.E., Bailey, R.M., and Thomas, D.S.G. (2019). Using artificial neural networks to predict future dryland responses to human and climate disturbances. *Scientific Reports* 9 (1): 1–13. https://doi.org/10.1038/s41598-019-40429-5.

Calviño-Cancela, M. and Cañizo-Novelle, N. (2018). Human dimensions of wildfires in NW Spain: causes, value of the burned vegetation and administrative measures. *PeerJ* 2018 (9): https://doi.org/10.7717/peerj.5657.

Chave, J., Coomes, D., Jansen, S. et al. (2009). Towards a worldwide wood economics spectrum. *Ecology Letters* 12 (4): 351–366. https://doi.org/10.1111/j.1461-0248.2009.01285.x.

Chen, J., Luo, Y., Xia, J. et al. (2015). Stronger warming effects on microbial abundances in colder regions. *Scientific Reports* 5: 1–10. https://doi.org/10.1038/srep18032.

Chen, Z., Zhang, J., Xiong, Z. et al. (2016). Enhanced gross nitrogen transformation rates and nitrogen supply in paddy field under elevated atmospheric carbon dioxide and temperature. *Soil Biology and Biochemistry* 94: 80–87. https://doi.org/10.1016/j.soilbio.2015.11.025.

Chen, C., Park, T., Wang, X. et al. (2019). China and India lead in greening of the world through land-use management. *Nature Sustainability* 2 (2): 122–129. https://doi.org/10.1038/s41893-019-0220-7.

Choong, S.M. and El-Shafie, A. (2015). State-of-the-art for modelling reservoir inflows and management optimization. *Water Resources Management* 29 (4): 1267–1282. https://doi.org/10.1007/s11269-014-0872-z.

Chowdary, V.M., Ramakrishnan, D., Srivastava, Y.K. et al. (2009). Integrated water resource development plan for sustainable management of mayurakshi watershed, India using remote sensing and GIS. *Water Resources Management* 23 (8): 1581–1602. https://doi.org/10.1007/s11269-008-9342-9.

Clark, P.U., MacAyeal, D.R., Andrews, J.T., and Bartlein, P.J. (1995). Ice sheets playimportant role in climate change. *Eos, Transactions American Geophysical Union* 76 (27): 265–270. https://doi.org/10.1029/EO076i027p00265-02.

Coops, N.C., Wulder, M.A., Duro, D.C. et al. (2008). The development of a Canadian dynamic habitat index using multi-temporal satellite estimates of canopy light absorbance. *Ecological Indicators* 8 (5): 754–766. https://doi.org/10.1016/j.ecolind.2008.01.007.

Coops, N.C., Shang, C., Wulder, M.A. et al. (2020). Change in forest condition: characterizing non-stand replacing disturbances using time series satellite imagery. *Forest Ecology and Management* 474 (April): 118370. https://doi.org/10.1016/j.foreco.2020.118370.

Cornwell, W.K., Cornelissen, J.H.C., Amatangelo, K. et al. (2008). Plant species traits are the predominant control on litter decomposition rates within biomes worldwide. *Ecology Letters* 11 (10): 1065–1071. https://doi.org/10.1111/j.1461-0248.2008.01219.x.

Crockford, R.H. and Richardson, D.P. (2000). Partitioning of rainfall into throughfall, stemslow∠d interception effect of forest type, ground cover and climate. *Hydrological Processes* 14 (16–17): 2903–2920. https://doi.org/10.1002/1099-1085(200011/12)14:16/17<2903::AID-HYP126>3.0.CO;2-6.

Dang, A.T.N., Nandy, S., Srinet, R. et al. (2019). Forest aboveground biomass estimation using machine learning regression algorithm in Yok Don National Park, Vietnam. *Ecological Informatics* 50 (December 2018): 24–32. https://doi.org/10.1016/j.ecoinf.2018.12.010.

Daniel, C.J., Ter-Mikaelian, M.T., Wotton, B.M. et al. (2017). Incorporating uncertainty into forest management planning: timber harvest, wildfire and climate change in the boreal forest. *Forest Ecology and Management* 400: 542–554. https://doi.org/10.1016/j.foreco.2017.06.039.

De Baets, S. and Poesen, J. (2010). Empirical models for predicting the erosion-reducing effects of plant roots during concentrated flow erosion. *Geomorphology* 118 (3–4): 425–432. https://doi.org/10.1016/j.geomorph.2010.02.011.

Deka, J., Tripathi, O.P., Khan, M.L., and Srivastava, V.K. (2019). Study on land-use and land-cover change dynamics in eastern Arunachal Pradesh, N.E. India using remote sensing and GIS. *Tropical Ecology* 60 (2): 199–208. https://doi.org/10.1007/s42965-019-00022-3.

Dempewolf, J., Nagol, J., Hein, S. et al. (2017). Measurement of within-season tree height growth in a mixed forest stand using UAV imagery. *Forests* 8 (7): 1–15. https://doi.org/10.3390/f8070231.

Dewar, R.C., Franklin, O., Makela, A. et al. (2009). Optimal function explains forest responses to global change. *BioScience* 59 (2): 127–139. https://doi.org/10.1525/bio.2009.59.2.6.

Dijkstra, F.A., Hobbie, S.E., Reich, P.B., and Knops, J.M.H. (2005). Divergent effects of elevated CO_2, N fertilization, and plant diversity on soil C and N dynamics in a grassland field experiment. *Plant and Soil* 272 (1–2): 41–52. https://doi.org/10.1007/s11104-004-3848-6.

Dijkstra, F.A., Blumenthal, D., Morgan, J.A. et al. (2010). Contrasting effects of elevated CO_2 and warming on nitrogen cycling in a semiarid grassland. *New Phytologist* 187 (2): 426–437. https://doi.org/10.1111/j.1469-8137.2010.03293.x.

Djukem, W.D.L., Braun, A., Wouatong, A.S.L. et al. (2020). Effect of soil geomechanical properties and geo-environmental factors on landslide predisposition at Mount Oku,

Cameroon. *International Journal of Environmental Research and Public Health* 17 (18): 1–28. https://doi.org/10.3390/ijerph17186795.

Eid, A.N.M., Olatubara, C.O., Ewemoje, T.A. et al. (2020). Inland wetland time-series digital change detection based on SAVI and NDWI indecies: Wadi El-Rayan lakes, Egypt. *Remote Sensing Applications: Society and Environment* 19: 100347. https://doi.org/10.1016/j.rsase.2020.100347.

El-Shafie, A., El-Shafie, A.H., and Mukhlisin, M. (2014). New approach: integrated risk-stochastic dynamic model for dam and reservoir optimization. *Water Resources Management* 28 (8): 2093–2107. https://doi.org/10.1007/s11269-014-0596-0.

Etminan, M., Myhre, G., Highwood, E.J., and Shine, K.P. (2016). Radiative forcing of carbon dioxide, methane, and nitrous oxide: a significant revision of the methane radiative forcing. *Geophysical Research Letters* 43 (24): 12614–12623. https://doi.org/10.1002/2016GL071930.

FAO. (2002). *The state of food and agriculture 2002.*

FAO. (2012). *World Fisheries and Aquaculture*, Main Report. Food and Agriculture Organisation of the United Nations.

FAO. (2020). *Global Forest Resources Assessment 2020*, Main Report. Food and Agriculture Organisation of the United Nations.

FAO (2021). *Food Waste Index Report 2021*. UNEP.

FAO and UNEP. (2020). The state of the world's forests 2020. In *Forests, Biodiversity and People* (Issue 1). 10.2307/208372

Findeling, A., Ruy, S., and Scopel, E. (2003). Modeling the effects of a partial residue mulch on runoff using a physically based approach. *Journal of Hydrology* 275 (1–2): 49–66. https://doi.org/10.1016/S0022-1694(03)00021-0.

Foley, J.A., DeFries, R., Asner, G.P. et al. (2005). Global consequences of land use. *Science* 309 (5734): 570–574. https://doi.org/10.1126/science.1111772.

Forkuor, G., Benewinde Zoungrana, J.B., Dimobe, K. et al. (2020). Above-ground biomass mapping in west African dryland forest using Sentinel-1 and 2 datasets – a case study. *Remote Sensing of Environment* 236(October 2019): 111496. https://doi.org/10.1016/j.rse.2019.111496.

Fu, B.J., Wang, Y.F., Lu, Y.H. et al. (2009). The effects of land-use combinations on soil erosion: a case study in the loess plateau of China. *Progress in Physical Geography* 33 (6): 793–804. https://doi.org/10.1177/0309133309350264.

Garrett, R.D., Koh, I., Lambin, E.F. et al. (2018). Intensification in agriculture-forest frontiers: land use responses to development and conservation policies in Brazil. *Global Environmental Change* 53 (October 2017): 233–243. https://doi.org/10.1016/j.gloenvcha.2018.09.011.

Gascón, L.H. and Eva, H. (2014). *Field guide for forest mapping with high resolution satellite data. January* 2014: 79. https://doi.org/10.2788/657954.

Gashaw, T., Tulu, T., Argaw, M., and Worqlul, A.W. (2017). Evaluation and prediction of land use/land cover changes in the Andassa watershed, Blue Nile Basin, Ethiopia. *Environmental Systems Research* 6 (1): https://doi.org/10.1186/s40068-017-0094-5.

Gwenzi, W. and Nyamadzawo, G. (2014). Hydrological impacts of urbanization and Urban roof water harvesting in water-limited catchments: a review. *Environmental Processes* 1 (4): 573–593. https://doi.org/10.1007/s40710-014-0037-3.

Gyssels, G., Poesen, J., Liu, G. et al. (2006). Effects of cereal roots on detachment rates of single- and double-drilled topsoils during concentrated flow. *European Journal of Soil Science* 57 (3): 381–391. https://doi.org/10.1111/j.1365-2389.2005.00749.x.

Häger, A. (2012). The effects of management and plant diversity on carbon storage in coffee agroforestry systems in Costa Rica. *Agroforestry Systems* 86 (2): 159–174. https://doi.org/10.1007/s10457-012-9545-1.

Halofsky, J.E., Peterson, D.L., and Harvey, B.J. (2020). Changing wildfire, changing forests: the effects of climate change on fire regimes and vegetation in the Pacific northwest, USA. *Fire Ecology* 16 (1): https://doi.org/10.1186/s42408-019-0062-8.

Hamzah, M.L., Amir, A.A., Maulud, K.N.A. et al. (2020). Assessment of the mangrove forest changes along the Pahang coast using remote sensing and GIS technology. *Journal of Sustainability Science and Management* 15 (5): 43–58. https://doi.org/10.46754/JSSM.2020.07.006.

Haque, M.I. and Basak, R. (2017). Land cover change detection using GIS and remote sensing techniques: A spatio-temporal study on Tanguar Haor, Sunamganj, Bangladesh. *Egyptian Journal of Remote Sensing and Space Science* 20 (2): 251–263. https://doi.org/10.1016/j.ejrs.2016.12.003.

Hilmers, T., Friess, N., Bässler, C. et al. (2018). Biodiversity along temperate forest succession. *Journal of Applied Ecology* 55 (6): 2756–2766. https://doi.org/10.1111/1365-2664.13238.

Himanshu, S.K., Pandey, A., Yadav, B., and Gupta, A. (2019). Evaluation of best management practices for sediment and nutrient loss control using SWAT model. *Soil and Tillage Research* 192(August 2018): 42–58. https://doi.org/10.1016/j.still.2019.04.016.

Hishe, S., Lyimo, J., and Bewket, W. (2017). Soil and water conservation effects on soil properties in the middle Silluh Valley, northern Ethiopia. *International Soil and Water Conservation Research* 5 (3): 231–240. https://doi.org/10.1016/j.iswcr.2017.06.005.

Horstrand, P., Guerra, R., Rodriguez, A. et al. (2019). A UAV platform based on a hyperspectral sensor for image capturing and on-board processing. *IEEE Access* 7 (June): 66919–66938. https://doi.org/10.1109/ACCESS.2019.2913957.

Hungate, B.A., Jaeger, C.H., Gamara, G. et al. (2000). Soil microbiota in two annual grasslands: responses to elevated atmospheric CO_2. *Oecologia* 124 (4): 589–598. https://doi.org/10.1007/s004420000405.

IPCC. (2000). Nakicenovic, N. and Swart R. (eds.). Cambridge University Press, UK, (pp. 570). Available from Cambridge University Press, The Edinburgh Building, Shaftesbury Road, Cambridge CB2 2RU England.

IPCC (2013). Climate Change 2013. *IPCC Climate Change 2013.* Researchgate.Net. Available at: https://www.researchgate.net/profile/Abha_Chhabra2/publication/271702872_Carbon_and_Other_Biogeochemical_Cycles/links/54cf9ce80cf24601c094a45e/Carbon-and-Other-Biogeochemical-Cycles.pdf

IPCC (2017). Climate Change and Land Ice; IPCC Special Report on climate change, desertification, land degradation, sustainable land management, food security, and greenhouse gas fluxes in terrestrial ecosystems; Summary for Policymakers. *IPCC*, 1–15.

Izaurralde, R.C., Thomson, A.M., Morgan, J.A. et al. (2011). Climate impacts on agriculture: implications for forage and rangeland production. *Agronomy Journal* 103 (2): 371–381. https://doi.org/10.2134/agronj2010.0304.

Jenkins, M. and Schaap, B. (2018). Forest ecosystem services – Background Analytical Study. *United Nations Forum on Forests*, (April), 41. Available at: https://www.un.org/esa/forests/

wpcontent/uploads/2018/05/UNFF13_BkgdStudy_ForestsEcoServices.pdf%0Ahttp://www.unece.org/fileadmin/DAM/timber/meetings/20151102/English_new_FAO_logo/ECE_TIM_2015_6_FO_EFC_15_6.pdf

Jha, M.K., Chowdary, V.M., Kulkarni, Y., and Mal, B.C. (2014). Rainwater harvesting planning using geospatial techniques and multicriteria decision analysis. *Resources, Conservation and Recycling* 83: 96–111. https://doi.org/10.1016/j.resconrec.2013.12.003.

Johnson, S.L., Kuske, C.R., Carney, T.D. et al. (2012). Increased temperature and altered summer precipitation have differential effects on biological soil crusts in a dryland ecosystem. *Global Change Biology* 18 (8): 2583–2593. https://doi.org/10.1111/j.1365-2486.2012.02709.x.

Junges, A.H. and Ducati, J.R. (2018). Detection of grapevine leaf stripe disease symptoms by hyperspectral sensor. *Phytopathologia Mediterranea* 57 (3): 306–406. https://doi.org/10.14601/Phytopathol.

Kahinda, J.M., Lillie, E.S.B., Taigbenu, A.E. et al. (2008). Developing suitability maps for rainwater harvesting in South Africa. *Physics and Chemistry of the Earth* 33 (8–13): 788–799. https://doi.org/10.1016/j.pce.2008.06.047.

Katayama, N., Osada, Y., Mashiko, M. et al. (2019). Organic farming and associated management practices benefit multiple wildlife taxa: a large-scale field study in rice paddy landscapes. *Journal of Applied Ecology* 56 (8): 1970–1981. https://doi.org/10.1111/1365-2664.13446.

Keenan, R.J., Reams, G.A., Achard, F. et al. (2015). Dynamics of global forest area: results from the FAO global Forest resources assessment 2015. *Forest Ecology and Management* 352: 9–20. https://doi.org/10.1016/j.foreco.2015.06.014.

Khan, S.H., He, X., Porikli, F., and Bennamoun, M. (2017). Forest change detection in incomplete satellite images with deep neural networks. *IEEE Transactions on Geoscience and Remote Sensing* 55 (9): 5407–5423.

Kosmas, C., Karamesouti, M., Kounalaki, K. et al. (2016). Land degradation and long-term changes in agro-pastoral systems: an empirical analysis of ecological resilience in Asteroussia, Crete (Greece). *Catena* 147: 196–204. https://doi.org/10.1016/j.catena.2016.07.018.

Kumar, P., Rani, M., Pandey, P.C. et al. (2010). Monitoring of deforestation and forest degradation using remote sensing and GIS: a case study of Ranchi in Jharkhand (India). *Report and Opinion* 2 (4): 14–20. http://www.sciencepub.net/report/report0204/03_2578_report0204_14_20.pdf.

Laamanen, R. and Kangas, A. (2011). Large-scale forest owner's information needs in operational planning of timber harvesting - some practical views in Metsähallitus, Finnish state-owned enterprise. *Silva Fennica* 45 (4): 711–727. https://doi.org/10.14214/sf.101.

Ladd, B., Laffan, S.W., Amelung, W. et al. (2013). Estimates of soil carbon concentration in tropical and temperate forest and woodland from available GIS data on three continents. *Global Ecology and Biogeography* 22 (4): 461–469. https://doi.org/10.1111/j.1466-8238.2012.00799.x.

Lausch, A., Erasmi, S., King, D.J. et al. (2017). Understanding forest health with remote sensing, part II–a review of approaches and data models. *Remote Sensing* 9 (2): 1–33. https://doi.org/10.3390/rs9020129.

Lausch, A., Borg, E., Bumberger, J. et al. (2018). Understanding forest health with remote sensing, part III: requirements for a scalable multi-source forest health monitoring network based on data science approaches. *Remote Sensing* 10 (7): https://doi.org/10.3390/rs10071120.

Lenka, S., Lenka, N.K., Sejian, V., and Mohanty, M. (2015). Contribution of agriculture sector to climate change. In: *Climate Change Impact on Livestock: Adaptation and Mitigation* (ed. V. Sejian, J. Gaughan, L. Baumgard and C. Prasad), 37–48. Springer India https://doi.org/10.1007/978-81-322-2265-1_3.

Li, T., Tao, Q., Liang, C., and Yang, X. (2014). Elevated CO_2 concentration increase the mobility of Cd and Zn in the rhizosphere of hyperaccumulator *Sedum alfredii*. *Environmental Science and Pollution Research* 21 (9): 5899–5908. https://doi.org/10.1007/s11356-014-2560-1.

Lipinski, B., Hanson, C., Lomax, J., Kitinoja, L. et al, (2016). Toward a sustainable food system Reducing food loss and waste. *World Resource Institute*, (June), 1–40. Available at: http://unep.org/wed/docs/WRI-UNEP-Reducing-Food-Loss-and-Waste.pdf%5Cnhttp://ebrary.ifpri.org/cdm/ref/collection/p15738coll2/id/130211

Little, K.M., Willoughby, I., Wagner, R.G. et al. (2006). Towards reduced herbicide use in forest vegetation management. *Southern African Forestry Journal* 207 (1): 63–79. https://doi.org/10.2989/10295920609505254.

Liu, R., Wang, J., Shi, J. et al. (2014). Runoff characteristics and nutrient loss mechanism from plain farmland under simulated rainfall conditions. *Science of the Total Environment* 468–469: 1069–1077. https://doi.org/10.1016/j.scitotenv.2013.09.035.

Liu, H., Li, J., Li, X. et al. (2015). Mitigating greenhouse gas emissions through replacement of chemical fertilizer with organic manure in a temperate farmland. *Science Bulletin* 60 (6): 598–606. https://doi.org/10.1007/s11434-014-0679-6.

Liu, S.H., Zeng, G.M., Niu, Q.Y. et al. (2017). Bioremediation mechanisms of combined pollution of PAHs and heavy metals by bacteria and fungi: a mini review. *Bioresource Technology* 224: 25–33. https://doi.org/10.1016/j.biortech.2016.11.095.

Liu, Y., Gong, W., Xing, Y. et al. (2019). Estimation of the forest stand mean height and aboveground biomass in Northeast China using SAR Sentinel-1B, multispectral Sentinel-2A, and DEM imagery. *ISPRS Journal of Photogrammetry and Remote Sensing* 151 (September 2018): 277–289. https://doi.org/10.1016/j.isprsjprs.2019.03.016.

Luo, Y., Wan, S., Hui, D., and Wallace, L.L. (2001). Acclimatization of soil respiration to warming in a tall grass prairie. *Nature* 413 (6856): 622–625. https://doi.org/10.1038/35098065.

Lynch, J., Cain, M., Pierrehumbert, R., and Allen, M. (2020). Demonstrating GWP: a means of reporting warming-equivalent emissions that captures the contrasting impacts of short- and long-lived climate pollutants. *Environmental Research Letters* 15 (4): https://doi.org/10.1088/1748-9326/ab6d7e.

Ma, X., Li, Y., Li, B. et al. (2016). Nitrogen and phosphorus losses by runoff erosion: field data monitored under natural rainfall in Three Gorges Reservoir Area, China. *Catena* 147: 797–808. https://doi.org/10.1016/j.catena.2016.09.004.

Malhi, R.K.M., Anand, A., Mudaliar, A.N. et al. (2020). Synergetic use of in situ and hyperspectral data for mapping species diversity and above ground biomass in Shoolpaneshwar Wildlife Sanctuary. *Gujarat. Tropical Ecology* 61 (1): 106–115. https://doi.org/10.1007/s42965-020-00068-8.

Mamo, M. and Bubenzer, G.D. (2001). Detachment rate, soil erodibility, and soil strength as influenced by living plant roots. Part I: laboratory study. *Transactions of the American Society of Agricultural Engineering* 44 (5): 1167–1174.

Mannan, A., Liu, J., Zhongke, F. et al. (2019). Application of land-use/land cover changes in monitoring and projecting forest biomass carbon loss in Pakistan. *Global Ecology and Conservation* 17 (35): e00535. https://doi.org/10.1016/j.gecco.2019.e00535.

Martin, C., Pohl, M., Alewell, C. et al. (2010). Interrill erosion at disturbed alpine sites: effects of plant functional diversity and vegetation cover. *Basic and Applied Ecology* 11 (7): 619–626. https://doi.org/10.1016/j.baae.2010.04.006.

Mateo-Sagasta, J., Zadeh, S.M. and Turral, H. (2017). Water pollution from agriculture: A global review. *Executive Summary*, 35. Available at: http://www.fao.org/3/a-i7754e.pdf

McCarthy, N., Bentsen, N.S., Willoughby, I., and Balandier, P. (2011). The state of forest vegetation management in Europe in the 21st century. *European Journal of Forest Research* 130 (1): 7–16. https://doi.org/10.1007/s10342-010-0429-5.

Meraj, G., Romshoo, S.A., Ayoub, S., and Altaf, S. (2018). Geoinformatics based approach for estimating the sediment yield of the mountainous watersheds in Kashmir Himalaya, India. *Geocarto International* 33 (10): 1114–1138. https://doi.org/10.1080/10106049.2017.1333536.

Meshesha, D.T., Tsunekawa, A., and Tsubo, M. (2012). Continuing land degradation: cause-effect in Ethiopia's central Rift Valley. *Land Degradation and Development* 23 (2): 130–143. https://doi.org/10.1002/ldr.1061.

Mohammed, S., Al-Ebraheem, A., Holb, I.J. et al. (2020). Soil management effects on soil water erosion and runoff in Central Syria – a comparative evaluation of general linear model and random forest regression. *Water (Switzerland)* 12 (9): https://doi.org/10.3390/w12092529.

Morgenroth, J. and Visser, R. (2013). Uptake and barriers to the use of geospatial technologies in forest management. *New Zealand Journal of Forestry Science* 43 (16): 1–9.

Moudrý, V., Gdulová, K., Fogl, M. et al. (2018). Comparison of leaf-off and leaf-on combined UAV imagery and airborne LiDAR for assessment of a post-mining site terrain and vegetation structure: prospects for monitoring hazards and restoration success. *Applied Geography* 104(May 2018): 32–41. https://doi.org/10.1016/j.apgeog.2019.02.002.

Moussa, S., Omar, B.A. and El Hassan, E.B.(2020) Monitoring of long term land degradation trends based of LANDSAT observations – The case of land productivity trends in Souss-Massa Region. *2020 IEEE International Conference of Moroccan Geomatics, (Morgeo)*, pp. 1–5. https://doi.org/10.1109/Morgeo49228.2020.9121870

Mwenge Kahinda, J., Taigbenu, A.E., Sejamoholo, B.B.P. et al. (2009). A GIS-based decision support system for rainwater harvesting (RHADESS). *Physics and Chemistry of the Earth* 34 (13–16): 767–775. https://doi.org/10.1016/j.pce.2009.06.011.

Nagasubramanian, K., Jones, S., Singh, A.K. et al. (2019). Plant disease identification using explainable 3D deep learning on hyperspectral images. *Plant Methods* 15 (1): 1–10. https://doi.org/10.1186/s13007-019-0479-8.

Nowak, R.S., Ellsworth, D.S., and Smith, S.D. (2004). Functional responses of plants to elevated atmospheric CO_2 – do photosynthetic and productivity data from FACE experiments support early predictions? *New Phytologist* 253–280. https://doi.org/10.1111/j.1469-8137.2004.01033.x.

OECD-FAO (2012). Agricultural Outlook 2011–2020. *Organisation For Economic Co-Operation and Development*. https://doi.org/10.1787/agr_outlook-2011-en

Ota, T., Ogawa, M., Mizoue, N. et al. (2017). Forest structure estimation from a UAV-based photogrammetric point cloud in managed temperate coniferous forests. *Forests* 8 (9): 1–11. https://doi.org/10.3390/f8090343.

Ouyang, W., Shan, Y., Hao, F. et al. (2013). The effect on soil nutrients resulting from land use transformations in a freeze-thaw agricultural ecosystem. *Soil and Tillage Research* 132: 30–38. https://doi.org/10.1016/j.still.2013.04.007.

Pal, S.C. and Chakrabortty, R. (2019). Simulating the impact of climate change on soil erosion in sub-tropical monsoon dominated watershed based on RUSLE, SCS runoff and MIROC5 climatic model. *Advances in Space Research* 64 (2): 352–377. https://doi.org/10.1016/j.asr.2019.04.033.

Pan, Y., Birdsey, R.A., Fang, J. et al. (2011). A large and persistent carbon sink in the world's forests. *Science* 333 (6045): 988–993. https://doi.org/10.1126/science.1201609.

Pandey, R. and Tyagi, A.K. (2012). Particulate matter emissions from domestic biomass burning in a rural tribal location in the lower Himalayas in India: concern over climate change. *Small-scale Forestry* 11 (2): 185–192. https://doi.org/10.1007/s11842-011-9177-8.

Pandey, P.C., Srivastava, P.K., Chetri, T. et al. (2019). Forest biomass estimation using remote sensing and field inventory: a case study of Tripura, India. *Environmental Monitoring and Assessment* 191 (9): https://doi.org/10.1007/s10661-019-7730-7.

Pandey, S., Kumari, N., and Priya, S. (2021). Soil quality and pollution assessment around Jumar watershed of Jharkhand, India. *Arabian Journal of Geosciences* 14 (24): 2748. https://doi.org/10.1007/s12517-021-09091-y.

Pannkuk, C.D. and Robichaud, P.R. (2003). Effectiveness of needle cast at reducing erosion after forest fires. *Water Resources Research* 39 (12): 1–9. https://doi.org/10.1029/2003WR002318.

Paruch, A.M., Mæhlum, T., and Robertson, L. (2015). Changes in microbial quality of irrigation water under different weather conditions in Southeast Norway. *Environmental Processes* 2 (1): 115–124. https://doi.org/10.1007/s40710-014-0054-2.

Patenaude, G., Milne, R., and Dawson, T.P. (2005). Synthesis of remote sensing approaches for forest carbon estimation: reporting to the Kyoto protocol. *Environmental Science and Policy* 8 (2): 161–178. https://doi.org/10.1016/j.envsci.2004.12.010.

Peterson, T.C., Stott, P.A., and Herring, S. (2012). Explaining extreme events of 2011 from a climate perspective. *Bulletin of the American Meteorological Society* 93 (7): 1041–1067. https://doi.org/10.1175/BAMS-D-12-00021.1.

Pistrika, A. (2010). Flood damage estimation based on flood simulation scenarios and a GIS platform. *European Water* 30: 3–11.

Pohl, M., Alig, D., Körner, C., and Rixen, C. (2009). Higher plant diversity enhances soil stability in disturbed alpine ecosystems. *Plant and Soil* 324 (1): 91–102. https://doi.org/10.1007/s11104-009-9906-3.

Pokhriyal, P., Rehman, S., Areendran, G. et al. (2020). Assessing forest cover vulnerability in Uttarakhand, India using analytical hierarchy process. *Modeling Earth Systems and Environment* 6 (2): 821–831. https://doi.org/10.1007/s40808-019-00710-y.

Poudel, K.C. (2018). Silviculture for forest management in Nepal. *Banko Janakari* 4: 15–20. https://doi.org/10.3126/banko.v27i3.20537.

Rachid, G., Alameddine, I., Najm, M.A. et al. (2021). Dynamic Bayesian networks to assess anthropogenic and climatic drivers of saltwater intrusion: a decision support tool toward improved management. *Integrated Environmental Assessment and Management* 17 (1): 202–220. https://doi.org/10.1002/ieam.4355.

Raffel, S. (2016). *Climate communication and the exclusion of indigenous knowledge*. IEEE International Professional Communication Conference, 2016, November. https://doi.org/10.1109/IPCC.2016.7740496

Rahman, S.A., Sunderland, T., Roshetko, J.M., and Healey, J.R. (2017). Facilitating smallholder tree farming in fragmented tropical landscapes: challenges and potentials for sustainable land management. *Journal of Environmental Management* 198: 110–121. https://doi.org/10.1016/j.jenvman.2017.04.047.

Ravikumar, P., Aneesul Mehmood, M., and Somashekar, R.K. (2013). Water quality index to determine the surface water quality of Sankey tank and Mallathahalli lake, Bangalore urban district, Karnataka, India. *Applied Water Science* 3 (1): 247–261. https://doi.org/10.1007/s13201-013-0077-2.

Reiche, J., Verhoeven, R., Verbesselt, J., Hamunyela, E., Wielaard, N., and Herold, M. (2018). Characterizing tropical forest cover loss using dense Sentinel-1 data and active fire alerts. *Remote Sensing* 10 (5): 777.

Reubens, B., Poesen, J., Danjon, F. et al. (2007). The role of fine and coarse roots in shallow slope stability and soil erosion control with a focus on root system architecture: a review. *Trees – Structure and Function* 21 (4): 385–402. https://doi.org/10.1007/s00468-007-0132-4.

Ridder, R.M. (2007). Global Forest Resources Assessment 2010: Options and Recommendations for a Global Remote Sensing Survey of Forests. *Forest Resources Assessment Programme*, 68. Available at: http://www.fao.org/3/a-ai074e.pdf

Rustad, L.E., Campbell, J.L., Marion, G.M. et al. (2001). A meta-analysis of the response of soil respiration, net nitrogen mineralization, and aboveground plant growth to experimental ecosystem warming. *Oecologia* 126 (4): 543–562. https://doi.org/10.1007/s004420000544.

Saeed, S., Irfan Ashraf, M., Ahmad, A., and Rahman, Z. (2016). The bela forest ecosystem of district jhelum, a potential carbon sink. *Pakistan Journal of Botany* 48 (1): 121–129.

Sahani, U. and Behera, N. (2001). Impact of deforestation on soil physicochemical characteristics, microbial biomass and microbial activity of tropical soil. *Land Degradation and Development* 12 (2): 93–105. https://doi.org/10.1002/ldr.429.

Sayl, K.N., Muhammad, N.S., Yaseen, Z.M., and El-shafie, A. (2016). Estimation the physical variables of rainwater harvesting system using integrated GIS-based remote sensing approach. *Water Resources Management* 30 (9): 3299–3313. https://doi.org/10.1007/s11269-016-1350-6.

Sheik, C.S., Beasley, W.H., Elshahed, M.S. et al. (2011). Effect of warming and drought on grassland microbial communities. *ISME Journal* 5 (10): 1692–1700. https://doi.org/10.1038/ismej.2011.32.

Shuaibu, R.B. and Dagba, B.I. (2013). *Challenges of Forest Inventory and its Effects on Forest Management and Planning in Nigeria*. In: *Proceedings of the 36th Annual Conference of Forestry Association of Nigeria, held in Uyo*, Akwa Ibom State, Nigeria (04–09 November, 2013), vol. 1 (ed. L. Popoola, O.Y. Ogunsanwo, V.A.J. Adekunle, et al.), 200–206.

da Silva, C.A. Jr., Nanni, M.R., Shakir, M. et al. (2018). Soybean varieties discrimination using non-imaging hyperspectral sensor. *Infrared Physics and Technology* 89: 338–350. https://doi.org/10.1016/j.infrared.2018.01.027.

da Silva, A.K.V., Borges, M.V.V., Batista, T.S. et al. (2021). Predicting eucalyptus diameter at breast height and total height with UAV-based spectral indices and machine learning. *Forests* 12 (5): 1–13. https://doi.org/10.3390/f12050582.

Singhal, A., Pandey, S., Kumari, N. et al. (2021). Impact of climate change on soil microbes involved in biogeochemical cycling. In: *Climate Change and the Microbiome Sustenance of the Ecosphere, Soil Biology* (ed. D.K. Choudhary, A. Mishra and A. Varma), 63–94. Springer Cham https://doi.org/10.1007/978-3-030-76863-8_5.

Smets, T., Poesen, J., and Knapen, A. (2008). Spatial scale effects on the effectiveness of organic mulches in reducing soil erosion by water. *Earth-Science Reviews* 89 (1–2): 1–12. https://doi.org/10.1016/j.earscirev.2008.04.001.

Sonti, S.H. (2015). Application of geographic information system (GIS) in forest management. *Journal of Geography & Natural Disasters* 05 (03): https://doi.org/10.4172/2167-0587.1000145.

Srivastava, K., Pandey, P.C., and Sharma, J.K. (2020). An approach for route optimization in applications of precision agriculture using UAVS. *Drones* 4 (3): 1–24. https://doi.org/10.3390/drones4030058.

Steinauer, K., Tilman, D., Wragg, P.D. et al. (2015). Plant diversity effects on soil microbial functions and enzymes are stronger than warming in a grassland experiment. *Ecology* 96 (1): 99–112. https://doi.org/10.1890/14-0088.1.

Sujatha, E.R. and Sridhar, V. (2018). Spatial prediction of erosion risk of a small mountainous watershed using RUSLE: a case-study of the palar sub-watershed in Kodaikanal, South India. *Water (Switzerland)* 10 (11): 1–17. https://doi.org/10.3390/w10111608.

Szigarski, C., Jagdhuber, T., Baur, M. et al. (2018). Analysis of the RADAR vegetation index and potential improvements. *Remote Sensing* 10 (11): 1–15. https://doi.org/10.3390/rs10111776.

Taffere, G.R., Beyene, A., Vuai, S.A.H. et al. (2016). Reliability analysis of roof rainwater harvesting systems in a semi-arid region of sub-Saharan Africa: case study of Mekelle, Ethiopia. *Hydrological Sciences Journal* 61 (6): 1135–1140. https://doi.org/10.1080/02626667.2015.1061195.

Tan, K., Ma, W., Wu, F., and Du, Q. (2019). Random forest-based estimation of heavy metal concentration in agricultural soils with hyperspectral sensor data. *Environmental Monitoring and Assessment* 191 (7): https://doi.org/10.1007/s10661-019-7510-4.

Tewari, V.P. (2016). Forest inventory, assessment, and monitoring, and long-term forest observational studies, with special reference to India. *Forest Science and Technology* 12 (1): 24–32. https://doi.org/10.1080/21580103.2015.1018962.

Trisurat, Y., Eawpanich, P., and Kalliola, R. (2016). Integrating land use and climate change scenarios and models into assessment of forested watershed services in southern Thailan. *Environmental Research* 147: 611–620. https://doi.org/10.1016/j.envres.2016.02.019.

Turcotte, Y. and Desrochers, A. (2003). Landscape-dependent response to predation risk by forest birds. *Oikos* 100 (3): 614–618. https://doi.org/10.1034/j.1600-0706.2003.12234.x.

USGS (2007). Forest Affected by Acid Rain. *Water Science School*. Available at: https://www.usgs.gov/media/images/forest-affected-acid-rain (Accessed: 17 March 2021).

USGS (2020). *What is remote sensing and what is it used for?* Geological Survey (USGS): U.S Available at: https://www.usgs.gov/faqs/what-remote-sensing-and-what-it-used#publications.

Van De Giesen, N., Kunstmann, H., Jung, G., Liebe, J. et al. (2002). The GLOWA Volta project: integrated assessment of feedback mechanisms between climate, landuse, and hydrology, 151–170. https://doi.org/10.1007/0-306-47983-4_9

Wang, S. and Fu, B. (2013). Trade-offs between forest ecosystem services. *Forest Policy and Economics* 26: 145–146. https://doi.org/10.1016/j.forpol.2012.07.014.

Wang, Z., Hou, Y., Fang, H. et al. (2012). Effects of plant species diversity on soil conservation and stability in the secondary succession phases of a semihumid evergreen broadleaf forest in China. *Journal of Soil and Water Conservation* 67 (4): 311–320. https://doi.org/10.2489/jswc.67.4.311.

Wang, Y., Li, Y., Li, Y. et al. (2015). Intensive rice agriculture deteriorates the quality of shallow groundwater in a typical agricultural catchment in subtropical central China. *Environmental Science and Pollution Research* 22 (17): 13278–13290. https://doi.org/10.1007/s11356-015-4519-2.

Wang, Q. et al. (2019). Derivative hyperspectral vegetation indices in characterizing Forest biophysical and biochemical quantities. *Hyperspectral Indices and Image Classifications for Agriculture and Vegetation.* https://doi.org/10.1201/9781315159331-2.

Weerasinghe, H., Schneider, U.A., and Ow, A. (2011). Water harvest- and storage-location assessment model using GIS and remote sensing. *Hydrology and Earth System Sciences Discussions* 8 (2): 3353–3381. https://doi.org/10.5194/hessd-8-3353-2011.

Wulder, M. (1998). Optical remote-sensing techniques for the assessment of forest inventory and biophysical parameters. *Progress in Physical Geography* 22 (4): 449–476. https://doi.org/10.1191/030913398675385488.

Xue, J. and Su, B. (2017). Significant remote sensing vegetation indices: a review of developments and applications. *Journal of Sensors* 2017: https://doi.org/10.1155/2017/1353691.

Yang, Y., Erskine, P.D., Lechner, A.M. et al. (2018). Detecting the dynamics of vegetation disturbance and recovery in surface mining area via LANDSAT imagery and LandTrendr algorithm. *Journal of Cleaner Production* 178: 353–362. https://doi.org/10.1016/j.jclepro.2018.01.050.

Yergeau, E. and Kowalchuk, G.A. (2008). Responses of Antarctic soil microbial communities and associated functions to temperature and freeze-thaw cycle frequency. *Wiley Online Library* 10 (9): 2223–2235. https://doi.org/10.1111/j.1462-2920.2008.01644.x.

Yue, J., Yang, G., Li, C. et al. (2017). Estimation of winter wheat above-ground biomass using unmanned aerial vehicle-based snapshot hyperspectral sensor and crop height improved models. *Remote Sensing* 9 (7): 708. https://doi.org/10.3390/rs9070708.

Yue, J., Feng, H., Jin, X. et al. (2018). A comparison of crop parameters estimation using images from UAV-mounted snapshot hyperspectral sensor and high-definition digital camera. *Remote Sensing* 10 (7): https://doi.org/10.3390/rs10071138.

Zalidis, G., Stamatiadis, S., Takavakoglou, V. et al. (2002). Impacts of agricultural practices on soil and water quality in the Mediterranean region and proposed assessment methodology. *Agriculture, Ecosystems and Environment* 88 (2): 137–146. https://doi.org/10.1016/S0167-8809(01)00249-3.

Zhang, G.H., Liu, G.B., Wang, G.L., and Wang, Y.X. (2011). Effects of vegetation cover and rainfall intensity on sediment-bound nutrient loss, size composition and volume fractal dimension of sediment particles. *Pedosphere* 21 (5): 676–684. https://doi.org/10.1016/S1002-0160(11)60170-7.

Zhang, X., Fan, J., Liu, Q., and Xiong, D. (2018). The contribution of gully erosion to total sediment production in a small watershed in Southwest China. *Physical Geography* 39 (3): 246–263. https://doi.org/10.1080/02723646.2017.1356114.

Zhang, N., Zhang, Q., Li, Y. et al. (2020). Effect of groundcovers on reducing soil erosion and non-point source pollution in citrus orchards on red soil under frequent heavy rainfall. *Sustainability (Switzerland)* 12 (3): https://doi.org/10.3390/su12031146.

Zhu, H., Fu, B., Wang, S. et al. (2015). Reducing soil erosion by improving community functional diversity in semi-arid grasslands. *Journal of Applied Ecology* 52 (4): 1063–1072. https://doi.org/10.1111/1365-2664.12442.p.

14

Challenges and Future Possibilities Toward Himalayan Forest Monitoring

Ayushi Gupta[1], *Prashant K. Srivastava*[*,1], *K.V. Satish*[1], *Aashri Chauhan*[2], *and Prem C. Pandey*[2]

[1] *Institute of Environment and Sustainable Development, Banaras Hindu University, Varanasi, Uttar Pradesh, India*
[2] *Center for Environmental Sciences & Engineering, School of Natural Sciences, Shiv Nadar Institution of Eminence (Deemed to be University), Greater Noida, Uttar Pradesh, 201314, India*

14.1 Introduction

The Himalayan region is one of the most sensitive of the Earth's biodiversity hotspots. The glaciers in the Himalayas are the source of eight rivers, which are responsible for the welfare and sustenance of 1.4 billion people (Immerzeel et al. 2010). Indus, Brahmaputra, and Ganga are the three largest rivers originating from the Himalayan glaciers (covering 33,000 sq. km), which explains the why the Himalayas are regarded as the "Water Tower of Asia" that provides around 86,000,000 cubic metres of water annually). Climate change in the Himalayas, home of many sacred landscapes and monumental biodiversity, is likely to affect the well-being of ~20% of humanity (Shrestha et al. 2012). However, despite the remarkable environmental, cultural, and socio-economic importance of the Himalayas, and despite their rapidly increasing ecological deterioration, not much is known about actual variations in the two most critical climatic variables; temperature and rainfall. Lack of research and inventory known about the changes in these parameters might affect the ecosystems, including vegetation phenology. The Himalayas extend to about 3000 km in length and cover ~7.5 lakh km^2 of Pakistan, Nepal, Bhutan, and India (Mittermeier et al. 2004). The location of the Himalayan biodiversity hotspot with in the boundary of Hindu-Kush Himalayan region (HKH) is shown in Figure 14.1.

Unstable climate, uneven topography, continuous geomorphological changes, and altitudinal stratifications have generated unique landscapes, ecosystems, and biota in the Himalayas. There are 13 ecoregions in the Himalayas which have similar quality and quantity of environmental resources, which is 1.5% of the total ecoregions of the world (Olson and Dinerstein 2002). The height of the Himalayas is from 250 m in the foothills to 8,849 m

*Corresponding author's email: prashant.just@gmail.com

0000-0001-6048-8069 (Ayushi Gupta), 0000-0002-4155-630X (Prashant K. Srivastava), 0000-0001-6772-3914 (K.V. Satish), 0000-0002-0361-8549 (Aashri Chauhan), and 0000-0002-0049-1415 (Prem C. Pandey)

Advances in Remote Sensing for Forest Monitoring, First Edition. Edited by Prem C. Pandey and Paul Arellano.

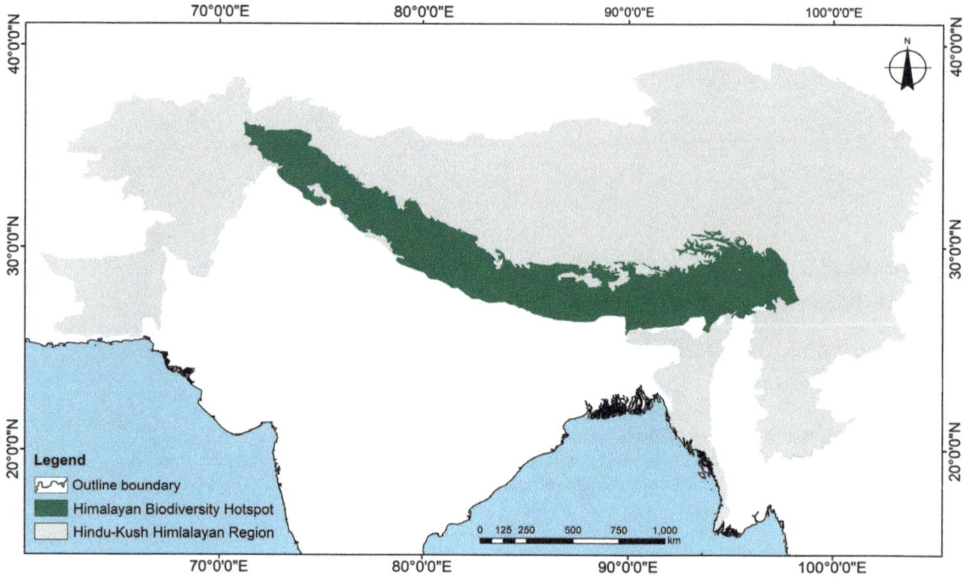

Figure 14.1 Map of the Himalayan biodiversity hotspot and Greater Hindu-Kush Himalayan Region.

Table 14.1 Distribution of Himalayan forest types along the elevation gradient.

S. No.	Forest type	Elevation
1	Tropical and subtropical broad-leaved forests	up to 1200 m
2	Subtropical Pine forests	1000–2000 m
3	Temperate evergreen broad-leaved forest, mixed coniferous, and sub-alpine forests	1500–3500 m

on the peak (Mount Everest), and they support a rich timberline of 3600 m. Himalayan forest types and their elevation limits are shown in Table 14.1.

Himalayan forests prevent soil erosion by holding the soil of the root zone and storing the water against frequent flooding in the plains. It is also a potential sink of carbon (Ghosh et al. 2016; Singh et al. 2021). If we look at the forest from root to apex, it includes living organisms, the stem and canopies extend vertically upward into the lower atmosphere, and roots penetrating down to the lowest soil profile (Waring and Running 2007), which is undergoing perpetual successional changes through growth and natural disturbances (Goodbody 2019). Due to rapid population growth, extensive logging, forest fires, and expansion of agriculture, the rate of net forest loss has risen by 40% in the past three decades (The State of The World's Forest 2020; Blackman 2015). This forest loss has weighty economic and environmental consequences, including biological and physiological effects on regional and global climate, rising of CO_2, diebacks caused by climate warming and drying, and increased dust pollution from bare soil and its effect on radiative forcing (Betts et al. 2008). Forests are widespread and long-lived ecosystems, and adapting forest management involves monitoring, planning at multiple spatial and temporal scales, and promotion of ecosystem-based adaptation to incorporate sustainable management (Bernier and Schoene 2009). Monitoring is a comprehensive process that includes estimation, presentation, and verification of the systematically collected data, and the derivation of information and knowledge at regular intervals to allow the constant access to changes over time (Morales-Hidalgo et al. 2017). The stages of forest data collection are depicted in Figure 14.2. The components of forest monitoring are given in Figure 14.3.

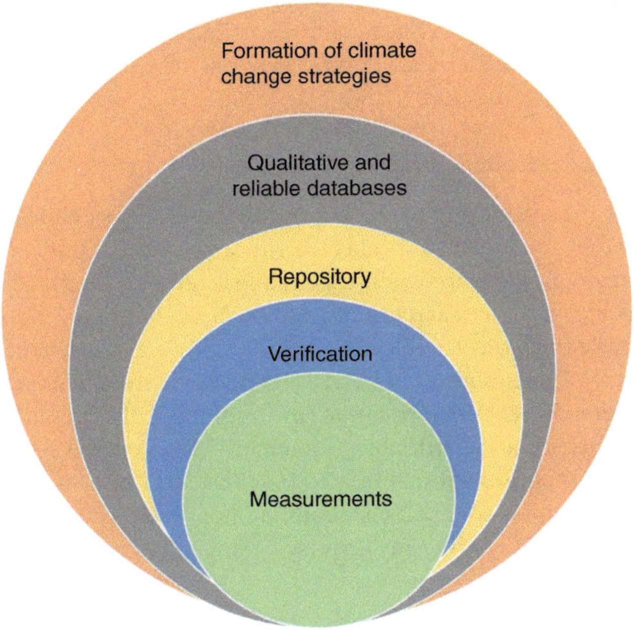

Figure 14.2 The stages of forest data collection and database generation. (Adopted from Food and Agriculture Organization (FAO)).

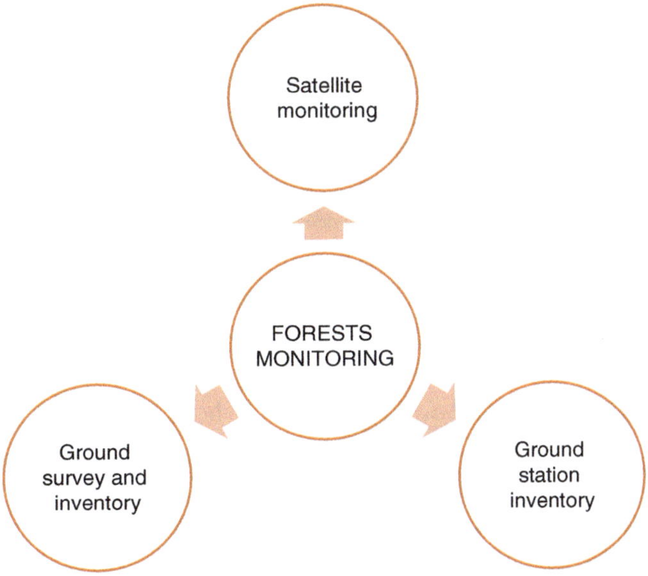

Figure 14.3 Components of forests montoring.

14.2 Component of Forest Monitoring

14.2.1 Satellite Monitoring

Satellite monitoring helps to identify and collect spatial data on land-use changes, emissions, and productions from forests as well other natural resources (Pandey and Sharma 2021). For example, monitoring of deforestation, afforestation, fragmentation and forest fires and growing stock is typically collected through the satellite monitoring system. Various countries have their satellite monitoring systems, and some organizations such as FAO request that various countries collect and monitor data based on forest management and changes. There are several missions which are in their way to serve remote sensing community in future (Pandey et al. 2020).

14.2.2 Ground Station Monitoring

For the monitoring of the large spatial areas by satellites, ground datasets are necessary. These ground data are collected either near the vicinity of the ground station or near the uncomplicated terrain since sampling is not possible at many sites in the Himalayas due to its harsh weather and terrain. Although satellites may measure a large or remote area at a single time, the variables such as soil moisture at different depths or under dense canopy forests, greenhouse gases (GHGs) emissions from degraded forest soil (which is exposed after a disaster), deforestation (by installing AI-based sensors for recording the sound of a chainsaw), and Earth's biogeochemical cycles can be accurately measured by ground stations. More ground stations need to be established near vulnerable zones, but must be carefully planned so that the stations are not destroyed at the time of a calamity, while giving valuable data during any extreme condition.

14.2.3 Ground Survey and Inventory

Ground surveying and preparing national inventory is the most important practice for countries and forest departments. Although satellite monitoring is most feasible, because it monitors large forest areas, ground surveying and inventory of the data collected are also very important. For example, we need species distribution monitoring, biodiversity monitoring, soil monitoring, and socio-economic monitoring of the forest.

In the Himalayan mountains, forest surveying and monitoring is more challenging and difficult than in plains. Due to undulating landforms, high terrain, dense forests, and unpredictable weather conditions it is a very challenging task for forest officials to cover uniform areas, which affects the necessary ground data required for validation of satellite monitoring outputs. Hence, we can say that ground surveying and monitoring are some of the key components of satellite monitoring.

14.3 Challenges in Satellite Monitoring

14.3.1 Forest Fire Monitoring

Over the last few decades, climate change and anthropogenic factors have had a significant impact on the environment. Forest fires are one of the factors which have an extreme impact on local and global ecosystems and cause huge damages to infrastructure, air quality and biodiversity; therefore, fire detection and monitoring of the disturbance type, size, and impact is becoming increasingly important (Tanase et al. 2018). Forest fires are caused by both natural and anthropogenic sources. Some natural sources such as lightning, the friction of rolling stones, rubbing of dry bamboo clumps, and volcanic explosions and are important for forest succession in some ways. Other than these natural causes, human activities like the collection of Non-Timber Forest Products (NTFPs), burning farm residues, campfires, and uncontrolled prescribed burning are the factors causing most forest fires (Satendra 2014). The coniferous forest in the Himalayan region is comprised of Firs (*Abies spp.*), Spruce (*Picea smithiana*), Pines (*Pinus roxburghii* and *Pinus wallichiana*) Cedar (*Cedrus deodara*), and are very prone to fire (Troup 1921). Western Himalayas are more prone to Forest fire than Eastern Himalayas as Eastern Himalayas receives more precipitation than Western Himalayas. Additionally, Western Himalayas are home to dominant species from Pinaceae family whose leaves are needle like which act as catalyzing agent in Forest Fire (Satendra 2014). Currently, the most common practices of monitoring forest fires in India are a participatory appraisal of local people and their training, which brings out changes to some extent in awareness and monitoring, but due to the rise in population and its increasing pressure on forests the problem of forest monitoring has become more complicated (Thapa et al. 2021). Making a network of fire lines is still the priority of forest departments, which is a very tough task in Himalayan regions and has become the most important cause of failure of forest fire monitoring (Martell et al. 1984). However, most of the methods and techniques currently used to achieve the goals of advanced forest fire monitoring (Arana-Pulido et al. 2018) use improved MODIS satellite systems; SMS alerts come from signals received by satellite and their transmission to Earth data-processing centers in the form of maps. MODIS satellites view the entire Earth's surface every few days in 36 spectral bands between 0.4 and 14.4 μm (Kumar et al. 2018). For forest fire studies, MODIS Terra 32-day composite time-series satellite images of India with a spatial resolution of 500 m acquired from the web-based satellite data archives of the Global Land Cover Facility (GLCF) and Earth Science Data Interface can be used (Maranganti 2009). However, some challenges still need to be focused on, such as the coarse resolution of the fire pixels, missing of smaller fire events by satellite's passing, and late SMS sent to the control room (Alkhatib, 2014). Other important challenges are the lack of proper fuel load management and fire risk zonation models, field testing, digital elevation model (DEM) resolution error, altitude determination error, atmosphere's transmissivity between the camera and fires.

14.3.2 Challenges in Land-Use Change Monitoring

Mountain ecosystems are among the most fragile ecosystems on Earth. Many mountain environments, including the Himalayan Mountains, which are among the most unstable mountain areas in the world, are firmly affected by drivers of global change such as land-use changes (Alkhatib 2014; Pandey et al. 2021). The accuracy of satellite images is extremely variable, due to geometric distortions introduced by topography, Earth curvature, sensor errors, and satellite viewing angle, and therefore requires corrections with the best possible accuracy. For example, the slopes that face the sun represent a higher radiance value as compared to the opposite slopes (Gao and Zhang 2009). The overall geolocation accuracy can vary significantly in the case of hilly/mountain areas or when images are acquired with large off-nadir angles. Therefore, getting precise terrain coordinates for large-scale mapping using very high-resolution images is a challenging task, even with today's technological advancements.

Topography and relief are the most critical components of any mountain system to help us understand surface processes and features associated with it. Relief in the form of DEMs is used for visual and mathematical analysis of topography and landforms, as well as modeling of surface processes. It provides easy extraction of commonly used terrain parameters such as slope, aspect, curvature, and slope length with finer details. Satellite images have proven their capabilities in the generation of DEM, and with the advent of high-resolution satellite data ranging from 10 m from SPOT-1 to 2.5 m in the case of Cartosat-1, high-resolution DEMs with submeter-level accuracy have been attempted. Other remote sensing techniques used to collect topographic information include interferometric synthetic aperture radar (InSAR), radar altimetry, and light detection and ranging (LiDAR), all of which can provide height accuracy in the range of submeter to centimeters.

However, in the Himalayan region, it is observed that there are challenges in obtaining a high-quality DEM mainly due to steep topography and shadows in case of optical remote sensing, and layover and foreshortening in case of microwave remote sensing. As a result, high-quality DEM with a spatial resolution less than 10 m and with a height accuracy of 34 m or higher is not available as a regular product (Agrawal et al. 2019).

14.3.3 Challenges in Species Distribution Monitoring

Species distribution models (SDMs) are of paramount importance in the conservation and management of biodiversity, and are considered one of the spatially quantifiable essential biodiversity variables (EBVs) (Reddy et al. 2021). Recent studies suggest that climate change already drives species extinctions by reducing species populations and their spatial distributions (Thuiller et al. 2008). The Himalayas are a biodiversity hotspot, and endemic and unique biota thrive here; however, it is considered one of the regions that is being impacted by twenty-first century climate change. In this situation, understanding current species distributions and their probable changes under projected climate change is necessary.

Monitoring and modeling species distributions in high-altitude regions provide numerous challenges, most importantly the sparse data availability due to poor accessibility of the terrain complexity (Maria and Udo 2017). Ecological modeling studies frequently use climate variables from numerical models to predict current/future species distributions (Hierro et al. 2005). However, they disregard important local biotic factors, which influence the actual species distribution (Thuiller 2004). Additionally, satellite imagery in hilly terrain is highly affected by variations in the topographic parameters such as altitude, slope, and aspect, which affect the brightness value of the satellite images having different spectral and spatial resolutions. Differential illumination results in significant variation in the spectral characteristics of other similar land cover features. Sun-facing illuminated slopes have higher spectral radiance or reflectance than the opposite slopes. Differential illumination effects in satellite imagery conceal the information on the north-facing leeward slopes in the Himalayan context. Unfortunately, SDM attempts in the Himalaya are in an immature state caused by huge ecological knowledge gaps and data limitations (Bobrowski et al. 2021). Collection of accurate species presence and absence geo-coordinates, and autecology of ecologically important species remains a challenge in the Himalayas. In general, species distribution modelers target such species for their research investigations, and precise gathering of such data is the most difficult job, and these investigations are further hardened by harsh environmental conditions. Data collection of alpine medicinal endemic species, temperate rare orchids, and birds are some examples of challenges. Modeling species' fundamental niches across the Himalayas is complicated by several data limitations (Bobrowski et al. 2018). The development of fine-grained climatic datasets that represent meso-climate conditions is considered one of the data deficits, and the development of such a data set needs the rigorous establishment of weather stations over the most complex terrains. On the other hand, development and validation of Earth observation-based SDM variables such as vegetation and other indices (i.e. soil moisture, leaf area, productivity, land surface temperature, evapotranspiration, water use efficiency, land surface phenology) are challenging research deficits. Variations in topographic parameters such as slope, aspect, and elevation have a large impact on image brightness values across different spectral and spatial resolutions, making topographical corrections for high spatio-temporal imagery difficult. Most of the conservation activities are frequently implemented at the local level, and fine-grained maps (i.e. species realized maps) are critical for such attempts. However, such maps and these attempts require AI-powered models and high-resolution datasets (Anand et al. 2022), which are both difficult. Investigation of species distributions under projected climate change scenarios needs robust climatic data. Developing such data for the Himalayas is difficult and challenging due to the large environmental heterogeneity within a small area. Finally, most of the SDM validations merely depend on evaluation statistics such as Area Under Curve (AUC) and True Skill Statistics (TSS), and need challenging ground reality verifications. There are several SDMs which are currently assessing the SDM situation across the globe with both merits and demerits.

14.3.4 Challenges in Climate Monitoring

Every growth, development, and harvesting of forest products, both timber and non-timber, are controlled by weather and climate. The establishment of a new forest stands either by artificial or natural processes, depends on weather conditions and their positive or negative impacts on soil conditions and seeds. As the plants grow, the growth and development of insects, viruses, pathogens, and diseases all depend on climate (De Melo-Abreu *et al.*, 2010). Monitoring humidity, precipitation, and temperature events are important when studying the relationships of evapotranspiration and gaseous exchange between forest and atmosphere (Tripathi et al. 2022).

Increase in air temperature, squeezing snow lines, shifts from solid to liquid precipitation, and upward shifts of vegetation demands monitoring in higher mountains, and lack of availability of ground stations and standard monitoring equipment are the most important challenges in the Himalayan climate monitoring (Haeberli et al. 2008; Beniston et al. 2018). Mountain hydropower, agriculture, and tourism are directly dependent on Himalayan headwaters, and they need to adapt and resist the water availability and its seasonal distribution because of the socio-economic development (Nelson et al. 2007; Beniston et al. 2018). Predicting the future weather and climate of the Himalayan region is challenging in any respect. These challenges arise from the incomplete knowledge of the physical parameters known for given changes and the lack of consistency in the datasets.

14.3.5 Challenges in Wildlife Monitoring

Geographic Information System (GIS) and Remote Sensing Technology (RST) play an important role in the conservation of wildlife. These techniques can help in understanding the interrelationships of biotic and abiotic environmental factors on population, growth, distribution, and extinction of the wildlife species. A combination of GIS, Global Positioning Systems (GPS), environmental sensors, Global Packet Radio Service (GPRS), and a range of analytical software tools may be used for wildlife tracking and studying the niche of wild animals in their habitats. In wildlife sciences and research there is a need for regular data collection of the movements of animals, which is a very challenging task for manual or ground studies (Acharyya and Bandyopadhyay 2017). However, the application of satellite remote sensing has the potential for monitoring wildlife habitats, but the successful applications are few (Skidmore 2017).

In ground-based sensors, the sensors can only observe phenomena in close vicinity. Energy sources such as solar or battery must be available in the desired size of the sensor. Other shadow effects and destruction by animals are also constraints (Handcock et al. 2009). In animal-based sensors, the distance of sensors to the static receiver, energy availability, and most importantly, the availability of different sizes of sensors for different morphological features of animals and birds, are the main constraints. For example, availability of small-sized sensors fitted with GPS, battery, and charger is very rare and costly. Animals that have larger neck sizes than mouth sizes such as boars, may not be able to hold collars for a long period (Ungar et al. 2005; Handcock et al. 2009).

A review article by Bilal Habib discussed about wildlife radio telemetry in India mentioned the development of radio telemetry in the past three decades. Radio telemetry technology was first introduced in the 1960s for wildlife tracking, and became popular after the availability of the Argos satellite in 1978. Previously it was used for large animals, due to the limited availability of large attachment devices. The ultra-high frequency (UHF) signals emitted by transmitters are received by polar-orbiting satellites. In India, the technology was brought in 1976 and it was used in Project Tiger for tracking sambar, nilgai, chital, and wild boar. After a long period of study, plain and tropical habitats show good results in wildlife monitoring using GPS technology. This technology was also used for tracking snow leopards, but after unsuccessful attempts due to difficult terrain, Himalayan regions have fewer studies using radio telemetry (Habib et al. 2014). Now the camera trapping method is showing better results in monitoring snow leopards. Additionally, GPS-telemetry discloses the regular crossing of black-eared Kites over the high-elevated Himalayas, which normally winters around Delhi, India (Kumar et al. 2020).

14.4 Challenges in Ground Survey and Inventory

14.4.1 Challenges in Forest Surveying

Forest surveys in mountainous areas are particularly difficult due to variable topography (Payne et al. 2017) and weather patterns. Seasonal taxonomic and floristic surveys are significantly more difficult and entirely dependent on nature's grace. Each year, the majority of field campaigns fail owing to natural disturbances such as cloud bursts, flash floods, thunderstorms, unexpected snowfall, avalanches, and landslides, which make access to field sites impossible. Carrying sophisticated field instrumentation and ration supplies to feed a field crew is indispensable, and doing so while traversing severe terrain is arduous and challenging. Additionally, poor connectivity to rural villages, difficulties obtaining visit and stay permission, gaining the backing of local residents, lack of protection from wildlife flash attacks, and a lack of health care emergencies are significant obstacles. Researchers still needs to map out forest burnt areas, demarcate or check forest boundaries, layout felling coups, prepare plantation maps or stock maps, detect and rectify encroachments and illicit possessions, and align extension roads and paths for forest protection, management, and administration.

It is very difficult to estimate the biomass growth rate per ha of natural forest with varied species composition in the case of the Hindu-Khush Himalayas (HKH) region due to lack of time-series data from the permanent plots on yield and growth for the given species (Upadhyay et al. 2005).

14.4.2 Challenges in Biodiversity Monitoring

Assessments of biodiversity, including phytosociology, species richness and exploration, wildlife monitoring, population studies, regular monitoring of seasonal vegetation dynamics, and phenological observations continue to be a significant scientific challenge in the Himalayas. These difficulties are exacerbated by a number of factors, including a lack of institutional support and collaboration, project budget constraints, ineffective monitoring and conservation measures, ineffective implementation of sustainable practices, and a scarcity of taxonomic expertise, which result in inaccurate species identification, and floristic and ecological studies. Climate and forest ecosystem interaction studies at alpine sites, such as the Global Observation Research Initiative in Alpine Environments (GLORIA) assessments (https://gloria.ac.at/home), frequently fail to precisely re-locate earlier plots, raising doubts about the data and interpretations. Indeed, such flaws eventually cause actual processes to be erroneous. Alternately, in recent years, detailed biodiversity monitoring operations at the local level using aircraft and unmanned aerial vehicles (UAVs) for fine-scale mapping of forest attributes have gained increased attention. Such operations include mapping species diversity, deforestation, degradation, fragmentation, fires, and invasive species because they provide detailed data with greater application value and are still an underexplored research field. Acquiring aerial photos using UAVs in mountainous regions is a challenging task by itself. Accurate above-ground biomass (AGB) estimation over various forest types is critical for understanding the carbon dynamics in the Himalayas, which may then be related to resource conservation and management. Estimates of biomass over the Himalayas are intermittent and based on small-sized plots, which may not minimize uncertainty in such estimations. Thus, while establishing 1-ha plots over forests in the Himalayas is a basic prerequisite for minimizing AGB estimation uncertainties, the core issue is that such inventories are extremely difficult and time-consuming to manage in view of accessibility concerns.

14.4.3 Challenges in Socio-Economic Survey

The present socio-economic stage of any community is indeed a result of socio-ecological succession, which was pioneered decades ago, and we understand this perspective by the socio-economic study of the different types of communities (McKean and Ostrom 2000). These communities, especially the tribal communities, play an important role in countries' environment and ecology. About 50% of the world's total tribal communities live in India. People who live in or near forest areas are dependent upon low cost and easily available resources from the forest for their welfare and sustenance. They have records of ancient knowledge and medicinal plants that have been supporting drug discovery for centuries. On the other hand, increasing population and development cause pressure on the carrying capacity of the forest. This anthropogenic pressure is harmful to forests and communities living near it (Rawat et al. 2009; Prakash 2015). The mountain cryosphere and its changes will affect the landscape, water resources, hydrological regimes, and the infrastructure. This impact is not only restricted to the mountain area but also affects the wide range of socio-economic repercussions (Prakash 2015). Forest structure and its monumental trees are the key elements to determine the capacity of forests to protect people and their assets against disasters and natural hazards (Prakash 2015). Mountain hydropower, agriculture, and tourism are directly dependent on Himalayan headwaters, and they need to adapt and resist the water availability and its seasonal distribution as a consequence of socio-economic development (Nelson et al. 2007; Vickers 2017; Beniston et al. 2018). To manage the mountain forests, a community-based approach is one of the best methods; but this approach also has a long list of constraints because this method is dependent upon manual survey, emotions and beliefs of the people. The major constraints in socio-economic based forest monitoring and management are reducing the gap between administration and communities, lack of documentation of trade and consumption of forest products, the bias of sampling in hilly regions (Dhyani and Dhyani 2020), lack of available resources in providing opportunities for income sources, and lack of awareness and coordination among people. Mountain forests, where ground activities are very challenging for its monitoring and management, reducing the gap between administration and community, the best tool to cope with these challenges is awareness (Roy et al. 2020).

14.4.4 Challenges in Forest Production Monitoring

Climatic variables such as temperature and precipitation may affect the forest structure and composition as well as production and development of flowering and biomass, which alter the quality and quantity of timber as well as NTFPs such as flowers, resins, gums, fruits, seeds, oils, and other valuable additions which may affect the growth of markets based of forest production. Forest management certification (SIDALC) is a relatively new type of formal, voluntary procedure. Under such a procedure, a certifier—who is a third-party inspector—gives a written assurance that the quality of forest management practiced by a defined manager or group conforms to specified standards. Certification is often followed by verification of the chain of custody of products from certified forests, and labeling of the products, so that they can be proven not to have been mixed with, or substituted by, products from other forests. In this way, certification attempts to link market demands for forest products produced to high environmental and social standards, with producers who can meet such demands. Forest Certification is the new approach from the markets to consumers of the forest goods in which the monitoring and certification of the forests have to be done for sustainable management, linking the quality of products, biodiversity conservation, and standards (Demske et al. 2016). To apply all these issues, monitoring the ability of forests for sustainable production is a very challenging task. It needs a survey and inventory of increments and yield of forest trees species. Forest certification started in 2002 in India on private plantations of Uttar Pradesh and West Bengal, but it dissolved after one year. Forest certification needs precise data about forest production at spatial and temporal scales; this requires the role of satellite monitoring and remote sensing in sustainable forest management and forest certification. The recent application of LiDAR technology in forest certification of Nepal is a breakthrough approach in forest production management and monitoring. LiDAR provides the 3-D structure of forest stands and it can penetrate the vegetation to calculate the commercial volume of biomass produced in the forest, and most importantly, survey and data collection for forest certification. Due to high costs, LiDAR is still in the research and development phase in India (Asmare 2013). Section 5.3 explains LiDAR as a future opportunity.

14.5 Future Possibilities in Forest Monitoring

14.5.1 Application of High Spatial and Spectral Resolution Satellites and Cameras

Shankar Sharma, in his study of comparison of Tropical Rainfall Measurement Mission (TMPA) and Integrated Multi-Satellite Retrievals for Global Precipitation Measurement (IMERG) on precipitation across Nepalese Himalaya, concluded that even after the various updated sensors and improved rainfall retrieval algorithms, IMERG product only shows slight improvement to older generation TMPA products (Sharma et al. 2020). The fuel classification scheme followed a site-specific fuel model within the study area, which is suitable for fire behavior prediction and spatial simulation. The Mallinis et al. (2014) study suggests that the high spatial resolution might be more decisive than the high spectral resolution in Mediterranean fuel type mapping. Fuel type classification as shown in Figure 14.4 is crucial for estimating ignition and fire propagation for sustainable forest management of these wildfire-prone environments. The newly developed Hyperspectral sensors are used to estimate content metabolite undestrucutively (Gupta et al., 2021a,b).

14.5.2 Application of Drones and Aircraft

Forests, which span over 30% of the Earth's surface, are one of the planet's most diversified natural ecosystems. Tropical forests cover over half of these areas and are located between 23.5 degrees north and south of the equator. They play an important role in climate control as they act as carbon storage and sink areas (Murrins Misiukas et al. 2021). Despite their immense advantages, drones still face enormous challenges and issues; drones are constrained by battery life and hence by limited geographical coverage, as well as payload

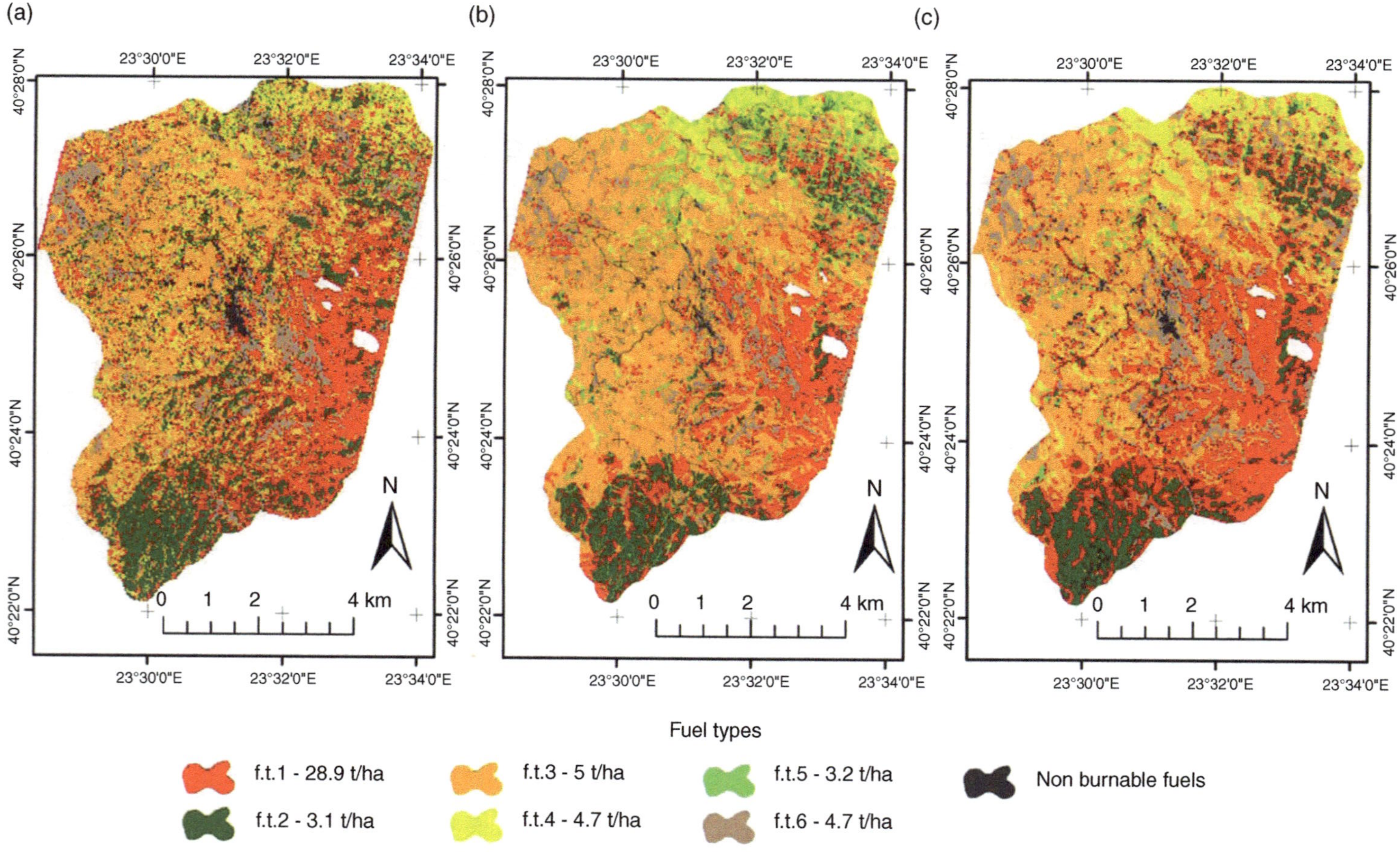

Figure 14.4 Fuel type classification obtained from the EO-1 Hyperion; (a) Quickbird, (b) Landsat TM, and (c) data using the SVMs classification algorithm (Mallinis et al. 2014).

weight (Manfreda et al. 2018) and sensitivity to some adverse weather circumstances (e.g. wind, precipitation, and rapid and severe changes in light conditions) (Hakala et al. 2013; Pádua et al. 2017). The radiometric and geometric quality of the data are the key technological obstacles. Users must be aware of data slathering issues caused by platform motion, as well as the importance of capturing accurate information on platform attitude (i.e. three-dimensional [3D] information describing the aircraft's roll, pitch, and yaw) for each data point in order to understand angular illumination and measurement characteristics. Because picture brightness values fluctuate based on the angle of lighting and view, this information is also required to assure long-term radiometric data quality (Anderson and Gaston 2013). Due to their small size and low weight, drones are severely limited in the quantity of equipment electronics they can carry onboard. This reduces the quality of imaging sensors that can be fitted into a small drone, which, when combined with the high cost of professional small imaging sensors, makes it difficult to obtain certain types of data and, as a result, conduct certain types of analysis that organizations and end-users may want to conduct. The high-resolution image is shown in Figure 14.5 (Paneque-Gálvez et al. 2014). Terrain and alpine winds might cause system disruption or product noise. Unless protected by special laws, drones can collide with towering structures and trees, causing issues (Berie and Burud 2018). Due to policies and regulations, drone operations cannot be carried out easily. For this reason, researchers were not able to explore the application possibilities of drones (Xiang et al. 2019). Most common image pre-processing and processing applications still need to be automated so that companies and end-users can perform sophisticated analysis faster. This involves stitching pictures across densely wooded regions, as well as geometric and radiometric adjustments if a high degree of precision is required, such as in time-series analysis. However, makers of remote sensing software are quickly improving this problem (Paneque-Gálvez et al. 2014). It is difficult to reliably identify species and individuals from a drone, especially in subtropical and tropical forests with deep canopies and great species diversity (Zhang et al. 2016).

The operation of drones in areas where illicit logging and farming, poaching, illegal drug manufacturing, land infringement, or military operations may occur could pose major hazards to the safety of drone operators and other community members. Drones are the future of technology and they will occupy the market to generate a billion-dollar economy. Almost all business areas, including precision agriculture, logistics, and infrastructures, are upgrading and adopting drones. Future technologies will focus on extending endurance, payload capacity, improving human–drone interactions, and establishing clear laws and regulations for the safe and secure operation of drones. Aside from that, combining artificial intelligence with drone technology will allow the drone to make judgments without relying on human operators (Singhal et al. 2018).

14.5.3 Application of LiDAR

LiDAR remote sensing is a game-changing technique, and it has a lot of potential for preserving and managing India's precious forest wealth. Because of its exact data gathering and precision, LiDAR technology has permeated most sectors, and it is now one of the most widely used remote sensing technologies in the world. Researchers have worked for decades to discover viable ways for measuring forest structure and tree heights (Behera and Roy 2002).

Figure 14.5 Examples of imagery gathered by small drones that show the extremely high spatial resolution that can be achieved. (a) Danau Girang (Sabah, Malaysia); (b) Chitwan National Park (Nepal); (c) Palm oil plantation by river (Indonesia); (d) Recently logged forest (Indonesia). Source: Imagery provided by Conservation Drones Paneque-Gálvez et al. 2014, MDPI, CC BY 4.0.

The developing technology known as LiDAR allows for the collection of this data from aircraft. LiDAR data gives detailed information not just about vegetation structure but also about Earth-surface characteristics underneath the vegetation cover. LiDAR provides researchers with geo-referenced information on these plant and Earth-surface elements across a wide area. While LiDAR has several advantages, it also has significant drawbacks that make it challenging to employ.

LiDAR data is still more costly per acre than aerial photography when utilized for forest inventory. LiDAR data sets need specialized software and abilities for processing and analysis. Although the potential for employing LiDAR is growing, not all of the characteristics needed to solve contemporary forestry concerns, particularly those pertaining to forest inventory, can be extracted from LiDAR data, and the models are not necessarily understood enough to generalize findings from local research.

LiDAR will probably not reach the ground if the vegetation is so dense that light cannot penetrate it, such as in tropical forests, and dense undergrowth may be mistaken for bare ground. The height of vegetation derived using LiDAR data is frequently lower than the height acquired by photogrammetry and fieldwork. Area-based LiDAR estimations require high-quality forest inventory data, specialist software for processing LiDAR point clouds is frequently necessary, and the related sequence of processes for creating acceptable forest inventory features can be statistically difficult and time-consuming. A LiDAR-based inventory, like any other forest inventory, will depart from reality over time, especially if harvesting or other types of disturbance occur (Thatcher and Swindell 2013).

Because of the effects of refraction, LiDAR pulses may be influenced by heavy rains or low-hanging clouds; however, the data produced can still be used for analysis. It may be inferred that LiDAR is an accurate instrument for assessing terrain, vegetation height, and cover, as well as complicated aspects of canopy structure and function. It is believed that the discoveries mentioned will be replicated in a range of biomes, with comparable outcomes. It might be effective in recognizing habitat traits linked to certain species, such as rare and endangered species. Different structural complex indicators might be beneficial in identifying places with high biodiversity potential, therefore offering inputs to Gap Analysis Projects (GAP). Identifying different types of forest structures and canopy gaps that are linked to different fire behavior might aid in fire prevention. It would help to further the research of canopy science and the physical characteristics of plant canopy structure in the future. The existence of individual creatures and the diversity of animal groups may be strongly reliant on the three-dimensional spatial structure of vegetation; particularly in systems with large biomass build-up. In forests, distinct three-dimensional characteristics are inevitably connected with certain animal and bird species. Forest canopy structure, which can be precisely measured using LiDAR data, may be linked to other functional aspects such as forest productivity (Kelly and Tommaso 2015).

The future of LiDAR for forest applications is dependent on a variety of factors, including decreasing prices, new advancements to overcome discrete LiDAR constraints, and new approaches to deal with growing data sizes.

14.5.4 Carbon Credits and Attracting Funds for Nations

Forest ecosystem services such as carbon sequestration, and the associated Green House Gas (GHG) market are emerging as a very advanced and encouraging option to facilitate new financial resources to sustainable forestry management (Wunder and Wertz-Kanounnikoff, 2009). Markets for forest ecosystem services have been endorsed in international policy-making practices as an approach to environmental conservation and human well-being (Corbera and Brown 2008). In conventional markets, there were not any incentives provided for the conservation of the benefits provided by ecosystem services and the cost of degradation (Costanza et al. 1997); and this leads to the rise of discussions on forest carbon markets.

Forest carbon markets are those markets commercializing emission rights generated through carbon dioxide fixation by forest ecosystems, which emerged primarily due to international institutions and their efforts to regulate and mitigate carbon emissions such as carbon dioxide in the UNFCC and Kyoto Protocol. However, these markets are as problematic as they are needed, emphasizing competing views regarding social acceptance to climate change scale and certainty, and plantation of trees far away from the actual source of emission (Corbera and Brown 2008).

Evolution in the institutions and markets for forest ecosystem services are attempting to change the behavior of forest managers through incentives (Landell-Mills and Porras 2002).

There are three types of institutional efforts through which forest carbon and its offsets are commercialized in both developing and developed countries. These are first, forest carbon trading through Clean Development Mechanism (CDM), in which traders can gain Certified Emission Reductions (CER's) against the GHG reduction. These are generated through Kyoto Markets-Joint Implementation and CDM under UNFCCC-Kyoto Protocol. Second is the Non-Kyoto Markets, Chicago Climate Exchange (CCE), by which investors may purchase emission reduction offsets from verified emission reduction projects. Third, is Non-Kyoto Retail Markets, in which project managers may sell carbon forestry offsets to retailers who insure its credibility from third-party verification (Corbera and Brown 2008).

In forest carbon trading, carbon credits are just like the value of any company in share market stocks. The carbon credits are like stock and the price of this stock is dependent upon the forest cover, which is valued as carbon stock of forest biomass. The investors who have these stocks are at a loss when their forest stocks are burned due to fire and reduced due to any catastrophic event or by deforestation, and have to pay for that loss as per REDD. While any investor may also gain or receive payments by increasing its forest cover or stocks through afforestation or conservation (Chakraborty 2021).

Due to India's Joint Forest Management (JFM), India has the highest number of registrations for Afforestation–Reforestation Clean Development Mechanism (AR–CDM). The forest cover of India captures 16% of India's carbon dioxide emissions, and India has an ambitious target to increase this to 2.5–3.3 billion tons CO_2 (equivalent) by a 2030 target of 33% forest cover of the country and 66% in Himalayan regions. Hence, some critical steps need to be implemented regarding the domestic carbon market (Dube and Chatterjee 2020). Carbon markets might be the key to unlocking much-needed cash for these efforts, but several obstacles must be overcome before donors can expect the private sector to play a substantial role in forestry finance such as REDD credits. Deforestation is considered as the second important contributor toward climatic change, as it accounts for the largest emissions of GHGs, mainly CO_2; up to 20% of total global emission in the atmosphere in developing nations (Ebeling and Yasué 2008). This is largely owing to the significant risks associated with monetizing forest carbon credits at the moment, most notably the risk that the credits will not be delivered due to tree loss, or that cap and trade or REDD policies will change, causing the price of carbon credits to fall. These steps are necessary to build forests as assets with high returns creating money flow from governments and private agencies. These steps will enhance investments in research and technology developments for better forest monitoring and management.

14.6 Conclusion

In this chapter, we have explored the environmental, cultural, and socio-economic importance of the Himalayan forest ecosystem, which is prone to rapidly increasing ecological deteriorations due to deforestation, forest fire, soil erosion, climate change, and various anthropogenic disturbances and challenges to monitoring these consequences for better management of the Himalayan forest. Forest loss causes economic and environmental consequences, including biological and physiological effects on regional and global climate, rising of CO_2, diebacks caused by climate warming and drying, and increased dust pollution from bare soil and its effect on radiative forcing. These losses promote the need for forest management, which involves monitoring and planning at multiple spatial, temporal scales, and promotion of ecosystem-based adaptation to incorporate sustainable management. Forest monitoring is not a linear approach, but a holistic approach that incorporates satellite monitoring, ground station monitoring, and ground survey and inventory.

This chapter includes challenges that come under all three types of monitoring, as stated. Satellite monitoring, which is the most developed monitoring technology, having the ability to cover a large area of untouched and difficult terrains in a single period of time, also have some challenges such as course resolution of fire pixels, missing of smaller fire events, lack of field of testing, and altitude and atmosphere transmissivity errors in forest fire monitoring. In land-use change, species distribution and biodiversity monitoring, shadow, topography, and relief are the most critical components of any mountain system to understand surface processes and features. Ground stations also play a major role in the monitoring of forest ecosystems, especially in the monitoring of climate variability, wildlife habitats, and socio-economic-based monitoring, and emphasized on forest production and utilization, forest survey, and preparation of the inventory.

Future possibilities are based on case studies and ongoing projects in which updated technologies such as high-spatial and spectral resolution satellites, laser, and radar-based sensors, and drones are used. Some policies interventions and proposals made by international organizations are needed to focus on raising funds by facilitating research and development for conservation and management of most volatile forest resource of the mountain ecosystem. This can be achieved by upgrading technology such as high resolution sensors, hyperspectral remote sensing, which needs to be included in Himalayan forest monitoring, and conservation to overcome the current challenges in Himalayan monitoring. Development of databases should also be given priority based on ground survey, which will help scale up data at a national level and help maximize conservation efforts.

References

Acharyya, N. and Bandyopadhyay, J. (2017). Remote sensing and GIS for wildlife management. In: *National Level Seminar on Defaunation and Conservation*, 65–71. Tuscon, AZ, USA: Tucson Herpetological Society.

Agrawal, S., Raghavendra, S., Kumar, S., and Pande, H. (2019). Geospatial Data for the Himalayan Region: Requirements, Availability, and Challenges. In: *Remote Sensing of Northwest Himalayan Ecosystems*, 471–500). Springer.

Alkhatib, A. (2014). A review on forest fire detection techniques. *International Journal of Distributed Sensor Networks* 10: 1–12.

Anand, A., Srivastava, P.K., Pandey, P.C. et al. (2022). Assessing the niche of *Rhododendron arboreum* using entropy and machine learning algorithms: role of atmospheric, ecological and hydrological variables. *Journal of Applied Remote Sensing* 16 (4): 042402.

Anderson, K. and Gaston, K.J. (2013). Lightweight unmanned aerial vehicles will revolutionize spatial ecology. *Frontiers in Ecology and the Environment* 11 (3): 138–146.

Arana-Pulido, V., Cabrera-Almeida, F., Perez-Mato, J. et al. (2018). Challenges of an autonomous wildfire geolocation system based on synthetic vision technology. *Sensors* 18 (12): 3631.

Asmare, M.F. (2013). *Airborne LiDAR Data and VHR WORLDVIEW Satellite Imagery to Support Community Based Forest Certification in Chitwan*. Nepal: University of Twente.

Behera, M. and Roy, P.J.C.S. (2002). LiDAR remote sensing for forestry applications: the Indian context. *Current Science* 83 (11): 1320–1328.

Beniston, M., Farinotti, D., Stoffel, M. et al. (2018). The European mountain cryosphere: a review of its current state, trends, and future challenges. *The Cryosphere* 12: 759–794.

Berie, H.T. and Burud, I. (2018). Application of unmanned aerial vehicles in earth resources monitoring: focus on evaluating potentials for forest monitoring in Ethiopia. *European Journal of Remote Sensing* 51: 326–335.

Bernier, P. and Schoene, D. (2009). Adapting Forests and their Management to Climate Change: An Overiew. *Information Service of Unasylva (FAO) Rome, Italy*, ISSN: 0041-6436.

Betts, R., Sanderson, M., and Woodward, S. (2008). Effects of large-scale Amazon forest degradation on climate and air quality through fluxes of carbon dioxide, water, energy, mineral dust and isoprene. *Philosphical Transactions of the Royal B Society* 363: 1873–1880.

Blackman, A. (2015). Strict versus mixed-use protected areas: Guatemala's Maya biosphere reserve. *Ecological Economics* 112: 14–24.

Bobrowski, M., Bechtel, B., Böhner, J. et al. (2018). Application of thermal and phenological land surface parameters for improving ecological niche models of *Betula utilis* in the Himalayan region. *Remote Sensing of Environment* 10: 814.

Bobrowski, M., Weidinger, J., Schwab, N., and Schickhoff, U.J.E.M. (2021). Searching for ecology in species distribution models in the Himalayas. *Ecological Modelling* 458: 109693.

Chakraborty, L. (2021). Mainstreaming Climate Change Commitments through Finance Commission's Recommendations. *NIPFP Working Paper Series* 341: 1–23.

Corbera, E. and Brown, K.J.W.D. (2008). *Building institutions to trade ecosystem services: marketing forest carbon in Mexico. World Development*, 36: 1956–1979.

Costanza, R., d'Arge, R., Groot, R.D. et al. (1997). The value of the world's ecosystem services and natural capital. *Nature* 387: 253–260.

De Melo-Abreu, J., Daldoum, M., Andrews, P. et al. (2010). *Applications of Meteorology to Forestry and Non-forest Trees*, 1–107. Springer.

Demske, D., Tarasov, P.E., Leipe, C. et al. (2016). Record of vegetation, climate change, human impact and retting of hemp in Garhwal Himalaya (India) during the past 4600 years. *World Development* (26): 1661–1675.

Dhyani, S. and Dhyani, D. (2020). Local Socio-economic Dynamics Shaping Forest Ecosystems in Central Himalayas. In: *Socio-Economic and Eco-Biological Dimensions in Resource Use and Conservation*, 31–60). Springer.

Dube, L.C. and Chatterjee, S. (2020). Forest carbon in climate change supermarket: is India prepared to sail? *Environmental Sciences Proceedings* 3: 41.

Ebeling, J. and Yasué, M. (2008). Generating carbon finance through avoided deforestation and its potential to create climatic, conservation and human development benefits. *The Royal Society Publishing* 363: 1917–1924.

Gao, Y. and Zhang, W. (2009). A simple empirical topographic correction method for ETM+ imagery. *International Journal of Remote Sensing* 30: 2259–2275.

Ghosh, B., Meena, V., Alam, N. et al. (2016). Impact of conservation practices on soil aggregation and the carbon management index after seven years of maize–wheat cropping system in the Indian Himalayas. *Agriculture, Ecosystems & Environment* 216: 247–257.

Goodbody, T.R.H. (2019). *Assessing the Role of Digital Aerial Photogrammetry for Characterizing Forest Structure and Enhancing Forest Inventories*, 12–38. University of British Columbia.

Gupta, A., Singh, P., Srivastava, P.K. et al. (2021a). Development of hyperspectral indices for anti-cancerous Taxol content estimation in the Himalayan region. *Geocarto International* 8: 1–14.

Gupta, A., Srivastava, P.K., Petropoulos, G.P., and P. Singh. (2021b). Statistical Unfolding Approach to Understand Influencing Factors for Taxol Content Variation in High Altitude Himalayan Region. *Forests* 12 (12): 1726.

Habib, B., Shrotriya, S., Sivakumar, K. et al. (2014). Three decades of wildlife radio telemetry in India: a review. *Animal Biotelemetry* 2: 1–10.

Haeberli, W., Clague, J.J., Huggel, C., and Kääb, A. (2008). *Hazards from lakes in high-mountain glacier and permafrost regions: climate change effects and process interactions*. 2010: 439–446.

Hakala, T., Honkavaara, E., Saari, H. et al. (2013). Spectral imaging from UAVs under varying illumination conditions. International Archives of the Photogrammetry, Remote Sensing and Spatial Information Sciences. In: *International Society for Photogrammetry and Remote Sensing (ISPRS)*. Vol. XL-1/W2, 189–194.

Handcock, R.N., Swain, D.L., Bishop-Hurley, G.J. et al. (2009). Monitoring animal behaviour and environmental interactions using wireless sensor networks, GPS collars and satellite remote sensing. *Sensors* 9: 3586–3603.

Hierro, J.L., Maron, J.L., and Callaway, R. (2005). A biogeographical approach to plant invasions: the importance of studying exotics in their introduced and native range. *Journal of Ecology* 93: 5–15.

Immerzeel, W.W., Van Beek, L.P., and Bierkens, M. (2010). Climate change will affect the Asian water towers. *Science* 328: 1382–1385.

Kelly, M. and Tommaso, S.D.J.C.A. (2015). Mapping forests with Lidar provides flexible, accurate data with many uses. *California Agriculture* 69: -14, 20.

Kumar, A., Singh, N., and Solanki, R. (2018). Evaluation and utilization of MODIS and CALIPSO aerosol retrievals over a complex terrain in Himalaya. *Remote Sensing of Environment* 206: 139–155.

Kumar, N., Gupta, U., Jhala, Y.V. et al. (2020). GPS-telemetry unveils the regular high-elevation crossing of the Himalayas by a migratory raptor: implications for definition of a "Central Asian Flyway.". *Scientific Reports* 10: 1–9.

Landell-Mills, N. and Porras, I.T. (2002). Selling Forest Environmental Service. In: *Market-Based Mechanisms for Conservation and Development*, 261–290. Routledge.

Mallinis, G., Galidaki, G., and Gitas, I.J.R.S. (2014). A comparative analysis of EO-1 Hyperion, Quickbird and Landsat TM imagery for fuel type mapping of a typical Mediterranean landscape. *Remote Sensing* 6: 1684–1704.

Manfreda, S., McCabe, M.F., Miller, P.E. et al. (2018). On the use of unmanned aerial systems for environmental monitoring. *Remote Sensing* 10: 641.

Maranganti, S. (2009). *Vegetation Change Detection in India Using MODIS Satellite Images.* Department of Computer and Information Science, Linköping Universitet SE-581 83 Linköping, Sweden September. Independent thesis - Advanced level, (pp. 1–103).

Maria, B. and Udo, S. (2017). Why input matters: Selection of climate data sets for modelling the potential distribution of a treeline species in the Himalayan region. *Ecological Modelling* 359: 92–102.

Martell, D.L., Drysdale, R.J., Doan, G.E., and Boychuk, D.J.I. (1984). An evaluation of forest fire initial attack resources. *Interfaces* 14: 20–32.

McKean, C.C.A.G.M. and Ostrom, E. (2000). *People and Forests: Communities, Institutions, and Governance. MIT Press* ISBN 0-262-07201-7.

Mittermeier, R.A., Gil, P., Hoffman, M. et al. (2004). Hotspots revisited: earth's biologically richest and most endangered terrestrial ecoregions cemex. *Regional Deevelopment, Spatial Planning And Strategic Governance* 392: 808–822.

Morales-Hidalgo, D., Kleinn, C., and Scott, C. (2017). Voluntary guidelines on national forest monitoring. AGRIS. Available at: https://agris.fao.org/agris-search/search.do?recordID=XF2018000947

Murrins Misiukas, J., Carter, S., and Herold, M.J.R.S. (2021). Tropical forest monitoring: challenges and recent progress in research. *Remote Sensing* 13: 2252.

Nelson, D.R., Adger, W.N., and Brown, K.J.A.R.E.R. (2007). Adaptation to environmental change: contributions of a resilience framework. *The Annual Review of Environment and Resources* 32: 395–419.

Olson, D.M. and Dinerstein, E. (2002). The Global 200: priority ecoregions for global conservation. *Annals of the Missouri Botanical Garden* 89: 199–224.

Pádua, L., Vanko, J., Hruška, J. et al. (2017). UAS, sensors, and data processing in agroforestry: a review towards practical applications. *International Journal of Remote Sensing* 38: 2349–2391.

Pandey, P.C. and Sharma, L.K. (ed.) (2021). *Advances in Remote Sensing for Natural Resource Monitoring*. Wiley.

Pandey, P.C., Koutsias, N., Petropoulos, G.P. et al. (2021). Land use/land cover in view of earth observation: data sources, input dimensions, and classifiers—a review of the state of the art. *Geocarto International* 36 (9): 957–988.

Pandey, P.C., Balzter, H., Srivastava, P.K. et al. (2020). Future perspectives and challenges in hyperspectral remote sensing. In: *Hyperspectral Remote Sensing*, 429–439. Elsevier.

Paneque-Gálvez, J., McCall, M.K., Napoletano, B.M. et al. (2014). Small drones for community-based forest monitoring: an assessment of their feasibility and potential in tropical areas. *Forests* 5: 1481–1507.

Payne, D., Spehn, E.M., Snethlage, M., and Fischer, M. (2017). Opportunities for research on mountain biodiversity under global change. *Current Opinion in Environmental Sustainability* 29: 40–47.

Prakash, R. (2015). Medicinal plants used by tribal communities: a study of Uttarakhand Himalayan region. *International Journal of Humanities and Social Science Invention* 4 (2): 55–61.

Rawat, Y.S., Vishvakarma, S.C., and Todaria, N.J.B. (2009). Fuel wood consumption pattern of tribal communities in cold desert of the Lahaul Valley, northwestern Himalaya, India. *Bioenergy* 33: 1547–1557.

Reddy, C.S., Kurian, A., Srivastava, G. et al. (2021). Remote sensing enabled essential biodiversity variables for biodiversity assessment and monitoring: technological advancement and potentials. *Conservation* 30: 1–14.

Roy, N., Roychoudhury, S., Nautiyal, S. et al. (2020). Socio-economic and eco-biological dimensions in resource use and conservation: epilogue. In: *Socio-Economic and Eco-Biological Dimensions in Resource use and Conservation*, 547–553. Springer.

Satendra, K. (2014). *Forest Fire Disaster Management*. New Delhi: *Ministry of Home Affairs Reports*.

Sharma, S., Khadka, N., Hamal, K. et al. (2020). How accurately can satellite products (TMPA and IMERG) detect precipitation patterns, extremities, and drought across the Nepalese Himalaya? *Earth and Space Sciences* 7: e2020EA001315.

Shrestha, U.B., Gautam, S., and Bawa, K. (2012). Widespread climate change in the Himalayas and associated changes in local ecosystems. *PlosOne* 7: e36741.

Singh, C., Chauhan, N., Upadhyay, S.K. et al. (2021). The Himalayan natural resources: challenges and conservation for sustainable development. *Phytochemistry* 10: 1643–1648.

Singhal, G., Bansod, B., and Mathew, L. (2018). Unmanned aerial vehicle classification, applications and challenges: a review. *Creative Commons* https://doi.org/10.20944/preprints201811.0601.v1.

Skidmore, A. (2017). *Environmental Modelling with GIS and Remote Sensing*. CRC Press eBook ISBN 9780429219405.

Tanase, M.A., Aponte, C., Mermoz, S. et al. (2018). Detection of windthrows and insect outbreaks by L-band SAR: a case study in the Bavarian forest national park. *Remote Sensing of Environment* 209: 700–711.

Thapa, S., Chitale, V.S., Pradhan, S. et al. (2021). Forest Fire Detection and Monitoring. In: *Earth Observation Science and Applications for Risk Reduction and Enhanced Resilience in Hindu Kush Himalaya Region*, 147–167. Cham: Springer.

Thatcher, T. and Swindell, B. (2013). *Yellowstone River Land Use Mapping and Analysis*, 1–110. DTM Consulting Reports, Inc.

Thuiller, W. (2004). Patterns and uncertainties of species' range shifts under climate change. *Global Change Biology* 10: 2020–2027.

Thuiller, W., Albert, C., Araujo, M.B. et al. (2008). Predicting global change impacts on plant species' distributions: future challenges. *Perspectives in Plant Ecology, Evolution and Systematics* 9: 137–152.

Tripathi, A.K., Pandey, P.C., Sharma, J.K. et al. (2022). Climate change and its impact on Forest of Indian Himalayan region: a review. *Climate Change* 207–222.

Troup, R.S. (1921). *The Silviculture of Indian Trees*. Clarendon Press. *Current Science* 80: 887–889.

Ungar, E.D., Henkin, Z., Gutman, M. et al. (2005). Inference of animal activity from GPS collar data on free-ranging cattle. *Management* 58: 256–266.

Upadhyay, T., Sankhayan, P.L., and Solberg, B.J.A. (2005). A review of carbon sequestration dynamics in the Himalayan region as a function of land-use change and forest/soil degradation with special reference to Nepal. *Agriculture, Ecosystems & Environment* 105: 449–465.

Vickers, N. (2017). Animal communication: when I'm calling you, will you answer too? *Current Biology* 27: R713–R715.

Waring, R. and Running, S.J.F.E. (2007). Forest ecosystem analysis at multiple time and space scales. *Forest Ecosystems* 1–16.

Wunder, S. and Wertz-Kanounnikoff, S. (2009). Payments for ecosystem services: a new way of conserving biodiversity in forests. *Journal of Sustainable Forestry* 28: 576–596.
Xiang, T.-Z., Xia, G.-S., and Zhang, L. (2019). Mini-unmanned aerial vehicle-based remote sensing: techniques, applications, and prospects. *IEEE Geoscience Remote Sensing Magazine* 7: 29–63.
Zhang, J., Hu, J., Lian, J. et al. (2016). Seeing the forest from drones: testing the potential of lightweight drones as a tool for long-term forest monitoring. *Biological Conservation* 198: 60–69.

Web Sources

https://www.fao.org/redd/areas-of-work/national-forest-monitoring-system/en.

Index

Advances in Remote Sensing for Forest Monitoring, First Edition. Edited by Prem C. Pandey and Paul Arellano.

t

u

v

Printed in the USA/Agawam, MA
September 16, 2026

910661.186